William Stirling

Outlines of Practical Histology

William Stirling

Outlines of Practical Histology

ISBN/EAN: 9783337396282

Printed in Europe, USA, Canada, Australia, Japan

Cover: Foto ©berggeist007 / pixelio.de

More available books at **www.hansebooks.com**

OUTLINES

OF

PRACTICAL HISTOLOGY:

A Manual for Students.

BY

WILLIAM STIRLING, M.D., Sc.D.,

BRACKENBURY PROFESSOR OF PHYSIOLOGY AND HISTOLOGY IN THE OWENS COLLEGE,
AND PROFESSOR IN THE VICTORIA UNIVERSITY, MANCHESTER; EXAMINER IN
PHYSIOLOGY IN THE UNIVERSITIES OF OXFORD AND EDINBURGH, AND
FOR THE FELLOWSHIP OF THE ROYAL COLLEGE OF SURGEONS, ENGLAND.

With 368 Illustrations.

SECOND EDITION, REVISED AND ENLARGED

PHILADELPHIA:
P. BLAKISTON, SON & CO.,
1012 WALNUT STREET.
1894.

PREFATORY NOTE TO THE SECOND EDITION.

In the Second Edition of this Handbook I have endeavoured to introduce a succinct account of the more recent Histological Methods. Most of these I have tested, a matter of no inconsiderable labour. The fruitful methods of Golgi, Ramón y Cajal, and Ehrlich are fully set forth in the text.

I have availed myself freely of the Journals and Text-books published at home and abroad, and more especially useful to me have been the works of Garbini and Ramón y Cajal. For most of the additional illustrations I am indebted to the *Gewebelehre* of Schiefferdecker and Kossel.

In conclusion, I have to thank my Senior Demonstrator, James A. Menzies, M.B., C.M., for assistance in reading the proof-sheets.

WILLIAM STIRLING.

PHYSIOLOGICAL LABORATORY, THE OWENS COLLEGE,
MANCHESTER,

PREFACE.

THE present work was written primarily to meet the wants of the students attending the course of instruction in "Practical Histology" in The Owens College, but it is hoped that it will be found useful also to students of medicine and science in other colleges and universities.

The Exercises printed in small type are intended for Senior Students, and for those attending the course of "Advanced Histology" given in The Owens College.

Although a drawing accompanies almost every exercise, still this is not intended to relieve the student from what is a most important part of the training in Practical Histology, viz., that the student should make sketches of his preparations. It serves very little useful purpose to give students sections ready prepared, and ask them merely to mount them. Hence, considerable stress has been laid on *methods*, as a knowledge of these is, after all, one of the most important parts of a practical training in Histology. Many methods have been tried and found wanting, and accordingly those only are introduced which the author, after experiment, has found to be reliable.

The author is indebted to the various Manuals of Histology published in this country, and also to those of Friedländer and Eberth, Stöhr, Fol, Kölliker, Martinotti, Francotte, Bolles Lee and Henneguy, Orth, and Edinger, as well as the various Microscopical Journals, British and Foreign.

Most of the illustrations have been drawn from microscopical preparations made in the class of Practical Histology, but a few of them are taken from preparations kindly presented to me by my friend Professor Swaen of Liège. I am greatly indebted to my demonstrator, Arthur Clarkson, M.B., C.M., for a considerable number of the drawings, and also to my pupil, Mr. C. E. M. Lowe, for similar services. A few of the drawings were made by myself. To F. W. Stansfield, M.B., Ch.B. (Vict.), my thanks are due for assistance in reading the proof-sheets.

I have also to express my obligations to several scientific instrument makers and others, including Messrs. Zeiss, Hartnack, Verick, Leitz, Reichert, Jung, Hicks, Eternod, Zimmermann, Hume, Beck, Swift & Son, Gardner, and A. Fraser.

Finally, I have to thank my publishers for the liberal manner in which the work is illustrated.

WILLIAM STIRLING.

PHYSIOLOGICAL LABORATORY, THE OWENS COLLEGE,
    MANCHESTER,

CONTENTS.

PART I.

CHAP.		PAGE
I.	APPARATUS REQUIRED	1
II.	THE MICROSCOPE AND ITS ACCESSORIES	5
III.	NORMAL OR INDIFFERENT FLUIDS	24
IV.	DISSOCIATING FLUIDS	24
V.	HOW TO TEASE A TISSUE	26
VI.	FIXING AND HARDENING FLUIDS	27
VII.	DECALCIFYING FLUIDS	36
VIII.	PREPARING TISSUES FOR MICROSCOPICAL EXAMINATION	38
IX.	EMBEDDING	40
X.	SECTION CUTTING	48
XI.	FIXATIVES	60
XII.	STAINING REAGENTS	63
XIII.	CLARIFYING REAGENTS	82
XIV.	MOUNTING FLUIDS AND METHODS	85
XV.	INJECTING BLOOD-VESSELS AND GLAND-TUBES	89
XVI.	EXAMINATION OF FRESH TISSUES AND FLUIDS	92

PART II.

LESSON		
I.	MILK, GRANULES, FIBRES, AND VEGETABLE ORGANISMS	97
II.	THE BLOOD	106
III.	HUMAN BLOOD, CRYSTALS FROM BLOOD, AND BLOOD PLATELETS	115
IV.	EPITHELIUM (STRATIFIED) AND ENDOTHELIUM	124
V.	COLUMNAR, SECRETORY, AND TRANSITIONAL EPITHELIUM	131
VI.	CILIATED EPITHELIUM	135
VII.	KARYOKINESIS OR MITOSIS	141
VIII.	CELLULAR AND HYALINE CARTILAGE	146
IX.	THE FIBRO-CARTILAGES, WHITE AND YELLOW	151
X.	CONNECTIVE TISSUE	156
XI.	TENDON	163

CONTENTS.

LESSON		PAGE
XII.	ADIPOSE, MUCOUS, AND ADENOID TISSUES—PIGMENT-CELLS	168
XIII.	BONE, OSSEOUS TISSUE	174
XIV.	BONE AND ITS DEVELOPMENT	182
XV.	MUSCULAR TISSUE	189
XVI.	STRIPED OR STRIATED MUSCLE	193
XVII.	NERVE-FIBRES	202
XVIII.	NERVE-CELLS, NERVE-GANGLIA, AND PERIPHERAL TERMINATIONS OF MOTOR NERVES	213
XIX.	THE HEART AND BLOOD-VESSELS	223
XX.	THE LYMPHATIC SYSTEM, SPLEEN, AND THYMUS GLAND	234
XXI.	TONGUE, TASTE BUDS, SOFT PALATE	246
XXII.	TOOTH, ŒSOPHAGUS	251
XXIII.	SALIVARY GLANDS AND PANCREAS	256
XXIV.	THE STOMACH	266
XXV.	THE SMALL AND LARGE INTESTINE	272
XXVI.	LIVER	285
XXVII.	TRACHEA, LUNGS, THYROID GLAND	294
XXVIII.	KIDNEY, URETER, BLADDER, SUPRARENAL CAPSULE	303
XXIX.	SKIN AND EPIDERMAL APPENDAGES	315
XXX.	SPINAL CORD	328
XXXI.	MEDULLA OBLONGATA, CEREBELLUM, CEREBRUM	347
XXXII.	THE EYE	358
XXXIII.	EAR AND NOSE	371
XXXIV.	TERMINATION OF NERVES IN SKIN AND MUCOUS MEMBRANES	376
XXXV.	THE TESTIS	381
XXXVI.	OVARY, FALLOPIAN TUBE, UTERUS	388
XXXVII.	MAMMARY GLAND, UMBILICAL CORD, AND PLACENTA	393
XXXVIII.	TO MAKE PREPARATIONS RAPIDLY FROM FRESH TISSUES	396

APPENDIX.

A.—SOME GENERAL WORKS OF REFERENCE	406
B.—TABLES OF MAGNIFYING POWER OF OBJECTIVES AND OCULARS	408
C.—LIST OF MAKERS, BRITISH AND FOREIGN, OF MICROSCOPES, MICROTOMES, CHEMICALS, AND HISTOLOGICAL REAGENTS	409
D.—WEIGHTS AND MEASURES, TABLE OF EQUIVALENTS	410
INDEX	411

LIST OF ILLUSTRATIONS.

FIG.		PAGE
1.	English form of slide. (*Zeiss.*)	1
2.	Square and round cover-glasses. (*Zeiss.*)	1
3.	Pinewood cabinet. (*Swift.*)	2
4.	Paper tray for slides. (*Jung.*)	2
5.	Scissors. (*Beck.*)	3
6.	Cover-glass lifter	3
7.	Cornet's cover-glass forceps	3
8.	Section-lifter. (*Beck.*)	4
9.	Leitz's microscope, No. 5. (*Leitz.*)	6
10.	Zeiss's large stand. (*Zeiss.*)	8
11.	Iris, diaphragm. (*Zeiss.*)	9
12.	Diaphragms. (*Reichert.*)	9
13.	Abbe's condenser. (*Swift.*)	11
14.	Leitz's IA stand. (*Leitz.*)	12
15.	Swift's college microscope. (*Swift.*)	14
16.	Hartnack's No. III.	15
17.	Zeiss's camera lucida. (*Zeiss.*)	17
18.	Abbe's camera lucida. (*Zeiss.*)	17
19.	Chevalier's camera lucida. (*Verick.*)	18
20.	Malassez's camera. (*Verick.*)	19
21.	Eye-piece micrometer scale. (*Zeiss.*)	20
22.	Eye-piece micrometer. (*Zeiss.*)	21
23.	Microscope lamp. (*Swift.*)	21
24.	Dissecting microscope. (*Verick.*)	22
25.	Reichert's dissecting microscope	22
26.	Cover-glass tester. (*Zeiss.*)	23
27.	Glass-thimble. (*Beck.*)	26
28.	Photophore. (*Stirling.*)	27
29.	Mayer's embedding bath. (*Jung.*)	42
30.	Embedding box. (*Stirling.*)	45
31.	Celloidin embedding box. (*Stirling.*)	46
32.	Valentine's knife. (*Beck.*)	49
33.	Rutherford's freezing microtome. (*Gardner.*)	50
34.	Cathcart's freezing microtome. (*Fraser.*)	52
35.	Planing-iron. (*Reichert.*)	53
36.	Malassez's microtome. (*Verick.*)	54
37.	Cambridge rocking microtome	54
38.	Minot's microtome. (*Zimmermann.*)	55
39.	Silk band for serial sections. (*Zimmermann.*)	56
40.	Thoma's sledge microtome. (*Jung.*)	56
41.	Malassez' microtome. (*Verick.*)	57
42.	Williams' microtome. (*Swift.*)	57
43.	Swift's ether microtome. (*Swift.*)	58
44.	Ranvier's hand-microtome. (*Reichert.*)	59
45.	Vulcanite rings. (*Eternod.*)	77
46.	Ring for stretching membranes. (*Eternod.*)	77
47.	Ranvier's support. (*Ranvier.*)	82
48.	Capped balsam bottle. (*Beck.*)	86
49.	Turntable. (*Beck.*)	88
50.	Turntable with slide.	88
51.	Hand-centrifuge. (*Muencke.*)	94
52.	Milk globules.	99
53.	Potato starch. (*Stirling.*)	100
54.	Rice starch. (*Blyth.*)	101
55.	Air, fat granules in water. (*Ranvier.*)	101

LIST OF ILLUSTRATIONS.

FIG.		PAGE
56.	Silk, wool, cotton, linen. (*Stirling*.)	102
57.	Cells from onion-bulb. (*Schiefferdecker*.)	103
58.	Mould. (*Ainsworth Davis*.)	104
59.	Yeast-cells. (*Landois and Stirling*.)	104
60.	Micrococci. (*V. Jaksch*.)	104
61.	Blood of frog. (*Ranvier*.)	107
62.	Amphibian blood-corpuscles. (*Landois and Stirling*.)	107
63.	Action of acetic acid on blood. (*Stirling*.)	108
64.	Action of water on blood-corpuscles. (*Stirling*.)	108
65.	Action of syrup on frog's blood. (*Stirling*.)	108
66.	Tannic acid on blood. (*Stirling*.)	109
67.	Boracic acid on blood. (*Stirling*.)	109
68.	Blood-corpuscles of fish and bird. (*Stirling*.)	110
69.	Amœboid movements of leucocytes. (*Landois*.)	111
70.	Moist chamber. (*Ranvier*.)	111
71.	Acetic acid on colourless corpuscles. (*Stirling*.)	112
72.	Human blood. (*Ranvier*.)	116
73.	Human blood. (*Landois*.)	116
74.	Human red corpuscles. (*Ranvier*.)	116
75.	Copper hot stage. (*Stirling*.)	118
76.	Reichert's warm stage. (*Reichert*.)	118
77.	Crenation of corpuscles. (*Stirling*.)	119
78.	Fibrils of fibrin. (*Landois*.)	120
79.	Rat's hæmoglobin. (*Stirling*.)	120
80.	Guinea-pig's hæmoglobin. (*Landois and Stirling*.)	120
81.	Hæmin crystals. (*V. Jaksch*.)	121
82.	Leukæmic blood. (*V. Jaksch*.)	121
83.	Blood platelets. (*Landois and Stirling*.)	123
84.	Isolated epithelial squames	125
85.	Squames of newt's epidermis. (*Stirling*.)	126
86.	Hard palate of cat. (*Stirling*.)	127
87.	Prickle-cells of epidermis. (*Stirling*.)	128
88.	Isolated prickle-cells. (*Ranvier*.)	128
89.	Endothelium of central tendon. (*Ranvier*.)	129
90.	Omentum of young rabbit, silver nitrate. (*Stirling*.)	129
91.	Omentum of cat, silvered. (*Stirling*.)	130
92.	Columnar and goblet cells. (*Stirling*.)	132
93.	Isolated columnar cells. (*Stirling*.)	133
94.	Liver-cells. (*Cadiat*.)	133
95.	Transitional cells. (*Stirling*.)	134
96.	Gas chamber. (*Gscheidlen*.)	136
97.	Gas chamber. (*Gscheidlen*.)	137
98.	Ranvier's moist chamber. (*Verick*.)	137
99.	Ciliated epithelium. (*Stirling*.)	137
100.	Frog's ciliated cell. (*Ranvier*.)	138
101.	Ciliated epithelium, ox. (*Stirling*.)	138
102.	V.S. ciliated epithelium	139
103.	Goblet-cells. (*Stirling*.)	139
104.	V.S. frog's tongue. (*Stirling*.)	140
105.	Connective-tissue corpuscle. (*Stöhr*.)	142
106.	Mitosis	143
107.	V.S. epidermis of salamander. (*Stirling*.)	144
108.	Mitosis. (*Stirling*.)	144
109.	Cellular cartilage. (*Stirling*.)	146
110.	Hyaline cartilage. (*Stöhr*.)	147
111.	Hyaline cartilage of thyroid. (*Schiefferdecker*.)	148
112.	Costal cartilage. (*Stöhr*.)	149
113.	V.S. articular cartilage. (*Stirling*.)	150
114.	Branched cartilage-cells. (*Stirling*.)	151
115.	V.S. intervertebral disc. (*Stirling*.)	152
116.	White fibro-cartilage. (*Stöhr*.)	153
117.	T.S. epiglottis. (*Ranvier*.)	154
118.	Elastic cartilage ear of horse. (*Schiefferdecker*.)	155
119.	Transition from hyaline to elastic cartilage. (*Schiefferdecker*.)	155

LIST OF ILLUSTRATIONS.

FIG.		PAGE
120.	Elastic fibres. (*Stöhr.*)	157
121.	T.S. ligamentum nuchæ. (*Stirling.*)	158
122.	Elastic fibres. (*Stirling.*)	158
123.	Fenestrated membrane. (*Stöhr.*)	159
124.	Areolar tissue. (*Schiefferdecker.*)	159
125.	Omentum of dog. (*Schiefferdecker.*)	160
126.	Hypodermic syringe. (*Hicks.*)	161
127.	Coarsely granular cells. (*Stirling.*)	162
128.	Cell-spaces in areolar tissue. (*Stirling.*)	162
129.	T.S. tendon. (*Stirling.*)	163
130.	L.S. tendon. (*Schiefferdecker.*)	164
131.	Fibrils of tendon. (*Stirling.*)	165
132.	Tendon of rat's tail. (*Stirling.*)	165
133.	Tendon cells. (*Ranvier.*)	165
134.	T.S. tendon of rat's tail. (*Stirling.*)	166
135.	Endothelial sheath of tendon. (*Stirling.*)	166
136.	Cell-spaces of diaphragm. (*Stirling.*)	167
137.	Fat-cells. (*Stirling.*)	169
138.	Fat-cells. (*Schiefferdecker.*)	169
139.	Empty fat-cells. (*Stirling.*)	170
140.	Fat-cells with margarine crystals.	170
141.	Development of fat-cells. (*Stirling.*)	171
142.	Mucous tissue. (*Stirling.*)	172
143.	Adenoid tissue. (*Stöhr.*)	173
144.	Pigment and guanin cells. (*Stirling.*)	173
145.	T.S. metacarpal bone. (*Ranvier.*)	175
146.	T.S. femur. (*Ranvier.*)	176
147.	T.S. dense bone. (*Ranvier.*)	177
148.	L.S. dense bone. (*Stirling.*)	177
149.	T.S. metacarpal bone. (*Stöhr.*)	178
150.	T.S. Haversian canal. (*Schiefferdecker.*)	179
151.	Sharpey's fibres. (*Stirling.*)	179
152.	L.S. injected bone. (*Stirling.*)	180
153.	Cancellated bone. (*Stirling.*)	180
154.	Cancellated bone. (*Stirling.*)	181
155.	T.S. embryonic bone. (*Ranvier.*)	182
156.	T.S. fœtal bone. (*Stöhr.*)	183
157.	L.S. developing bone. (*Stöhr.*)	184
158.	V.S. tibia. (*Stirling.*)	185
159.	Developing bone. (*Ranvier.*)	186
160.	Membranous bone. (*Stöhr.*)	186
161.	Marrow cell. (*Stirling.*)	187
162.	Isolated smooth muscle. (*Landois and Stirling.*)	190
163.	Bladder of frog. (*Stirling.*)	190
164.	T.S. smooth muscle. (*Stirling.*)	191
165.	Cement of smooth muscle. (*Stirling.*)	191
166.	T.S. smooth muscle. (*Schiefferdecker.*)	192
167.	Striped muscle of frog. (*Stirling.*)	194
168.	Muscle of fibrils. (*Schiefferdecker.*)	195
169.	Tendon and muscle. (*Landois and Stirling.*)	196
170.	Muscular fibre. (*Ranvier.*)	196
171.	Fibril of hydrophilus. (*Ranvier.*)	196
172.	Crab's muscle. (*Stirling.*)	197
173.	T.S. muscle. (*Krause.*)	197
174.	T.S. muscle. (*Stirling.*)	197
175.	Injected muscle. (*Landois and Stirling.*)	197
176.	T.S. and L.S. injected muscle. (*Owsjannikow.*)	198
177.	T.S. injected muscle. (*Stirling.*)	199
178.	Heart muscle.	199
179.	T.S. frozen muscle.	199
180.	Polariser. (*Zeiss.*)	201
181.	Medullated nerve-fibre. (*Stirling.*)	203
182.	Non-medullated nerve-fibre. (*Ranvier.*)	204
183.	Fresh nerve-fibre. (*Obersteiner.*)	205

LIST OF ILLUSTRATIONS.

FIG.		PAGE
184.	Nerve-fibre after osmic acid. (*Schwalbe.*)	205
185.	Frog's nerve-fibre. (*Obersteiner.*)	206
186.	Ranvier's crosses. (*Obersteiner.*)	207
187.	Ranvier's cross and Frommann's lines. (*Obersteiner.*)	207
188.	Peripheral nerve-fibre with axis-cylinder	208
189.	Intercostal nerve with Ranvier's crosses	208
190.	T.S. nerve. (*Eichhorst.*)	209
191.	Non-medullated nerve. (*Obersteiner.*)	209
192.	Sympathetic nerve. (*Schiefferdecker.*)	210
193.	Neuro-keratin network.	210
194.	Frommann's lines. (*Obersteiner.*)	211
195.	Nerve-fibre of frog. (*Obersteiner.*)	211
196.	T.S. nerve-fibres, osmic acid. (*Stirling.*)	212
197.	Spinal ganglion. (*Cadiat.*)	214
198.	Cells of spinal ganglion. (*Obersteiner.*)	214
199.	Bipolar nerve-cell of skate.	215
200.	Spinal ganglion nerve-cell. (*Ranvier.*)	216
201.	Cervical sympathetic ganglionic cell. (*Stirling.*)	217
202.	Multipolar nerve-cell of spinal cord. (*Obersteiner.*)	218
203.	Nerves of frog's sartorius. (*Mays.*)	220
204.	End-plates of lizard. (*Kühne.*)	221
205.	End-plate of lizard. (*Stirling.*)	221
206.	Sympathetic nerve-cell. (*Ranvier.*)	222
207.	Heart muscle. (*Stirling.*)	224
208.	Purkinje's fibres. (*Ranvier.*)	225
209.	Endocardium. (*Ranvier.*)	226
210.	T.S. tricuspid valve. (*Ranvier.*)	226
211.	L.S. human aorta. (*Ranvier.*)	227
212.	T.S. artery. (*Stirling.*)	227
213.	Endothelium of artery and vein. (*Stirling.*)	228
214.	Capillaries. (*Obersteiner.*)	229
215.	Small artery. (*Stirling.*)	229
216.	Arteriole. (*Obersteiner.*)	229
217.	T.S. small artery and vein. (*Stirling.*)	230
218.	Artery of brain. (*Obersteiner.*)	231
219.	Arteriole, silvered. (*Ranvier.*)	231
220.	Capillary, silvered. (*Landois and Stirling.*)	232
221.	Developing blood-vessels. (*Stirling.*)	233
222.	L.S. lymph gland. (*Ranvier.*)	235
223.	Lymph sinuses. (*Stirling.*)	237
224.	Central tendon. (*Ranvier.*)	238
225.	Pleural surface of diaphragm. (*Stirling.*)	238
226.	Stomata. (*Stirling.*)	239
227.	Tonsil. (*Stöhr.*)	240
228.	Thymus. (*Stirling.*)	241
229.	Injected thymus. (*Cadiat.*)	242
230.	Elements of thymus. (*Cadiat.*)	242
231.	Human spleen. (*Stöhr.*)	243
232.	Elements of splenic pulp. (*Stöhr.*)	244
233.	Reticulum of spleen. (*Cadiat.*)	244
234.	T.S. tongue of cat. (*Stirling.*)	246
235.	Filiform papilla. (*Stöhr.*)	247
236.	Fungiform papilla. (*Stöhr.*)	247
237.	Crypt of tongue. (*Schenk.*)	248
238.	Injected tongue of cat. (*Stirling.*)	248
239.	Papillæ foliatæ. (*Stirling.*)	249
240.	V.S. Papillæ foliatæ. (*Stöhr.*)	250
241.	V.S. tooth.	252
242, 243.	Enamel prisms. (*Landois and Stirling*)	253
244.	T.S. fang of tooth. (*Landois and Stirling.*)	253
245.	Development of tooth. (*Landois and Stirling.*)	254
246.	Later stage of 245. (*Stöhr.*)	254
247.	Later stage of 246. (*Stöhr.*)	254
248.	T.S. œsophagus. (*Stirling.*)	255

LIST OF ILLUSTRATIONS.

FIG.		PAGE
249.	Lobules of submaxillary gland. (*Stirling.*)	260
250.	Acini of submaxillary gland. (*Stirling.*)	260
251.	T.S. duct of salivary gland. (*Landois and Stirling.*)	261
252.	Resting serous gland. (*Heidenhain.*)	261
253.	Human submaxillary gland. (*Heidenhain.*)	262
254.	T.S. pancreas. (*Stirling.*)	263
255.	T.S. fresh pancreas. (*Heidenhain.*)	264
256.	V.S. stomach. (*Stöhr.*)	268
257.	V.S. mucous membrane of stomach. (*Stirling.*)	269
258.	T.S. gastric glands. (*Landois and Stirling.*)	270
259.	V.S. pyloric glands. (*Heidenhain.*)	270
260.	T.S. small intestine. (*Stirling.*)	274
261.	L.S. Peyer's patch. (*Landois and Stirling.*)	275
262.	Injected small intestine. (*Stöhr.*)	276
263.	Scheme of 262. (*Mall.*)	277
264.	Injected villi. (*Stirling.*)	277
265.	Auerbach's plexus. (*Stirling.*)	278
266.	Auerbach's plexus. (*Cadiat.*)	278
267.	Section of 266. (*Cadiat.*)	279
268.	Meissner's plexus. (*Stirling.*)	279
269.	V.S. duodenum. (*Stöhr.*)	280
270.	L.S. large intestine. (*Schenk.*)	281
271.	Lieberkühn's gland. (*Heidenhain.*)	281
272.	T.S. villus. (*Heidenhain.*)	282
273.	Intestinal villus. (*Kultschitzky.*)	283
274.	Villus absorbing fat. (*Stirling.*)	283
275.	T.S. liver of pig. (*Stirling.*)	286
276.	Liver-cells. (*Heidenhain.*)	287
277.	T.S. portal canal. (*Stirling.*)	287
278.	Human liver. (*Stöhr.*)	288
279.	Liver of frog. (*Stirling.*)	289
280.	Injected liver of rabbit. (*Stirling.*)	290
281.	Interlobular bile-duct. (*Landois and Stirling.*)	290
282.	Bile-ducts injected. (*Cadiat.*)	291
283.	T.S. trachea of cat. (*Stirling.*)	295
284.	L.S. trachea of cat. (*Stirling.*)	296
285.	T.S. bronchus. (*Stirling.*)	296
286.	T.S. human bronchus. (*Hamilton.*)	297
287.	V.S. lung and pleura. (*Hamilton.*)	298
288.	Silvered lung of kitten. (*Hamilton.*)	299
289.	Injected lung. (*Stirling.*)	299
290.	T.S. dried lung. (*Stirling.*)	300
291.	T.S. fœtal lung. (*Stirling.*)	300
292.	T.S. thyroid gland. (*Cadiat.*)	301
293.	L.S. Malpighian pyramid.	303
294.	Scheme of renal tubules. (*Klein.*)	304
295.	Glomerulus. (*Stirling.*)	308
296.	Rodded epithelium. (*Landois and Stirling.*)	308
297.	Irregular renal tubule. (*Landois and Stirling.*)	308
298.	T.S. apex of Malpighian pyramid. (*Stirling.*)	309
299.	Blood-vessels of kidney. (*Landois and Stirling.*)	310
300.	Henle's tubule. (*Landois and Stirling.*)	311
301.	Cells from renal tubule. (*Landois and Stirling.*)	311
302.	T.S. ureter. (*Stöhr.*)	312
303.	V.S. bladder. (*Stöhr.*)	313
304.	T.S. penis. (*Stirling.*)	313
305.	V.S. suprarenal capsule. (*Stöhr.*)	314
306.	V.S. skin of palm. (*Stirling.*)	317
307.	Human epidermis. (*Ranvier.*)	318
308.	T.S. sweat gland. (*Stirling.*)	319
309.	V.S. hair-follicle. (*Landois and Stirling.*)	322
310.	T.S. hair-follicle. (*Landois and Stirling.*)	324
311.	V.S. injected skin. (*Taguschi.*)	324
312.	T.S. nail. (*Landois and Stirling.*)	326

LIST OF ILLUSTRATIONS.

FIG.		PAGE
313.	Axillary gland. (*Landois and Stirling.*)	327
314.	T.S. spinal cord. (*Cadiat.*)	334
315.	T.S. anterior cornu of cord (*Obersteiner.*)	335
316.	T.S. white matter of cord. *Obersteiner.*)	335
317-20.	T.S. spinal cord at various levels. (*Obersteiner.*)	336
321.	Neuroglia cell. (*Ranvier.*)	343
322.	L.S. spinal cord to show collateral fibres. (*Kölliker.*)	347
323.	T.S. medulla oblongata. (*Henle.*)	348
324.	T.S. medulla oblongata. (*Edinger.*)	348
325.	Leaflet of cerebellum. (*Stirling.*)	350
326.	V.S. cerebellum. (*Obersteiner.*)	350
327.	V.S. cerebrum. (*Obersteiner.*)	352
328.	V.S. frontal convolution. (*Obersteiner.*)	352
329.	V.S. injected cerebrum. (*Obersteiner.*)	354
330.	Cell of Purkinje. (*Obersteiner.*)	356
331.	T.S. cerebrum of rat. (*Cayal.*)	356
332.	V.S. cerebrum. (*Edinger.*)	357
333.	V.S. cornea. (*Landois and Stirling.*)	359
334.	Cornea corpuscles. (*Ranvier.*)	359
335.	Nerves of cornea. (*Ranvier.*)	360
336.	V.S. cornea. (*Ranvier.*)	360
337.	Cornea cell-spaces. (*Ranvier.*)	361
338.	V.S. sclerotic and choroid. (*Stöhr.*)	361
339.	T.S. sclerotic and cornea. (*Landois and Stirling.*)	362
340.	Lens fibres. (*Landois and Stirling.*)	364
341.	V.S. retina. (*Ranvier.*)	366
342.	V.S. cochlea. (*Ranvier.*)	372
343.	V.S. cochlear duct. (*Landois and Stirling.*)	373
344.	V.S. olfactory mucous membrane. (*Stöhr.*)	375
345.	Olfactory cells. (*Landois and Stirling.*)	375
346.	Tactile cells. (*Ranvier.*)	377
347.	Tactile discs. (*Ranvier.*)	378
348.	End-bulb	378
349.	Pacini's corpuscle. (*Ranvier.*)	379
350.	Endothelium of 349. (*Stirling.*)	379
351.	T.S. of 349. (*Stirling.*)	379
352.	Wagner's corpuscle. (*Landois and Stirling.*)	380
353.	Wagner's corpuscle. (*Ranvier.*)	381
354.	Organ of Eimer. (*Stirling.*)	381
355.	T.S. testis	382
356.	Tubule of testis. (*Landois and Stirling.*)	382
357.	Spermatogenesis of rat. (*Stirling.*)	386
358.	Spermatozoa. (*Landois and Stirling.*)	387
359.	Epididymis. (*Schenk.*)	387
360.	V.S. ovary. (*Turner.*)	389
361.	Ovum	390
362.	T.S. Fallopian tube. (*Schenk.*)	391
363.	T.S. fimbriated end of 362. (*Stirling.*)	391
364.	V.S. uterus. (*Stirling.*)	392
365.	T.S. mamma. (*Stirling.*)	394
366.	T.S. active mamma. (*Stirling.*)	394
367.	Colostrum. (*V. Jaksch.*)	394
368.	Placental villus. (*Cadiat.*)	395

[The illustrations indicated by the word "*Stöhr*" are from Stöhr's *Lehrbuch der Histologie;* by "*Cadiat,*" from Cadiat's *Traité d'Anatomie Générale;* by "*Ranvier,*" from Ranvier's *Traité Technique d'Histologie;* by "*Schenk,*" from Schenk's *Grundriss der normalen Histologie;* by "*Obersteiner,*" from Obersteiner's *Anleitung beim Studium des Baues der nervösen Centralorgane;* by "*v. Jaksch,*" from v. Jaksch's *Klinische Diagnostik;* by "*Edinger,*" from Edinger's *Zwölf Vorlesungen über den Bau der nervösen Centralorgane;* by "*Landois and Stirling,*" from their *Text-Book of Physiology;* by "*Schiefferdecker,*" from *Das Mikroskop,* by "*Behrens, Kossel, and Schiefferdecker.*"]

PRACTICAL HISTOLOGY.

PART I.

I.—APPARATUS REQUIRED.

THE student of Practical Histology must be provided with the following **apparatus:**—

1. **A Compound Achromatic Microscope** capable of magnifying from about 50 to 300 or 450 diameters linear.

2. **Glass Slides.**—The most convenient size is 3 inches by 1 inch (or 76 × 26 mm.), with ground edges made of the best flatted crown-glass. About two gross will be required (fig. 1).

FIG. 1.—English form of Slide.

It is convenient to have two dozen or so of a larger size, 3 inches by 1½ inches.

3. **Cover-Glasses.**—Some are square and others circular (fig. 2).

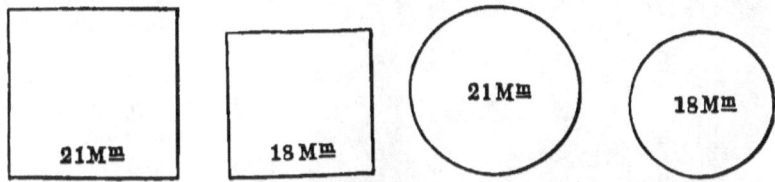

FIG. 2.—Square and Round Cover-Glasses, showing the most convenient sizes.

The student should be provided with both sorts. Keep the square ones for balsam preparations, and the circles for those that require ringing, *i.e.*, those mounted in glycerine or Farrant's solution. Only extra-thin covers (or those sold as No. 1) should be used. It is well to measure the thickness of the covers, and to use for mounting only those that are less than .006 inch in thickness.

FIG 3.—Pinewood Cabinet to hold Sixty Slides.

Get half-an-ounce of $\tfrac{3}{4}$-inch circles, and the same weight of $\tfrac{5}{8}$-inch squares. It is convenient to have a few circles and squares somewhat larger, viz., 1 inch and $1\tfrac{1}{4}$ inch in diameter, for mounting particularly large sections. When a large number of sections are mounted under one cover-glass, as in serial preparations, it is well to have oblong cover-glasses of a still larger size.

FIG. 4.—Paper Tray to hold Slides.

4. **A Wooden Cabinet**, fitted with trays for holding the mounted specimens. It should be capable of holding at least 60 slides, and the slides should lie on the flat. A convenient form is shown in fig. 3. Some prefer the flat compressed paper trays shown in fig. 4.

5. **Two Mounted Needles** in handles. The student can easily make these himself. Fix a sewing-needle into the end of a pen-holder, allowing only about $\tfrac{1}{4}$ inch of the needle to project. The needles should always be kept bright and polished. A convenient form is to fix a sewing-needle into a strong wooden "crochet" needle.

6. **A Dissecting Case**, but failing that, a pair of strong **forceps**,

APPARATUS REQUIRED.

and also a finer pair, the latter with long narrow points; a stout and a fine pair of **scissors**, and a **scalpel**. It is well to have a straight pair of scissors, and also a pair curved on the flat (fig. 5.) It is convenient to have a pair of forceps like those shown in fig. 6 for lifting and applying a cover-glass to a preparation, or like those in fig. 7 for holding a cover-glass on which a thin film with bacteria is spread.

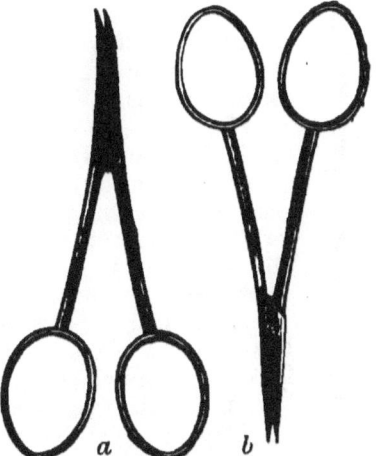

FIG. 5.—(*a*.) Scissors curved on the flat; (*b*.) fine straight pair.

7. Camel-Hair Brushes, at least two, the smaller crow-size, and one somewhat larger.

8. A Razor, which is not to be hollow-ground. It must be kept very sharp, and stropped frequently. It is better to have one ground flat on one side.

9. Watch-Glasses.—Instead of the ordinary-sized glasses, the student should provide himself with at least four 3 inches in diameter.

10. A Section-Lifter.—This may be made by beating out the end of a piece of copper wire ($\frac{1}{10}$ inch thick) until a thin plate is formed.

FIG. 6.—Cover-Glass Lifter.

The plate is then bent at an angle to the stem. It is better, however, to purchase one made of German silver (fig. 8).

11. Drawing Materials.—As great importance is attached in

FIG. 7.—Cornet's Cover-Glass Forceps.

this laboratory and in this course to making drawings of the microscopic objects, each student must provide himself with a *drawing-book*—a quarto, with unruled paper, and containing 150 pages or thereby, is sufficient. Suitable *drawing-pencils*, including an H.B., and a harder one, *e.g.,* H.H.H.; both must be of a good quality of lead. After a sketch has been made in pencil, the sketches should be coloured. This may be done either with coloured pencils or water-colours. The latter are greatly to be preferred.

12. Slips of white **bibulous paper**, 3 inches by ½ inch, to soak up any superfluous fluid, and to be used for irrigation. For irrigation purposes use small *triangular* slips.

13. Small **glass pipettes**, which the student should make for himself by heating in a gas-flame and drawing out a piece of narrow glass-tubing at two places, close to each other, leaving a small part of the tube of the original width, which acts as the bulb of the pipette. Several may be made at a time, and their capillary ends sealed in the flame, and kept until they are required.

14. A pair of narrow glass rods drawn to a point to tease tissues in such metallic solutions as gold chloride or silver nitrate, which act on metallic instruments.

15. **Labels** for the slides.—It is well to have a large number of pieces of paper cut, 3 inches by 1 inch, as temporary labels, on which is written the name of the preparation. Each label is placed under its appropriate slide in a tray. These labels are merely temporary. This is specially desirable where the slides have to be "ringed," as in this process a permanent label is apt to be displaced or destroyed. In the case of balsam preparations, they may be labelled at once with the small square permanent labels. In every case the preparation should be labelled, and the label should bear not only the name of the tissue or organ, but the direction of the section and the medium in which it is mounted, and, if desired, the date of mounting. Labels are now printed so cheaply, that for half-a-crown a student can have a thousand labels printed with his own name.

FIG. 8.—Section-Lifter.

16. **Reagents.**—The student should also be provided with the following reagents, placed in a small wooden framework on the work-table. Only those reagents that are most frequently used need be provided for in the framework; the others can be supplied as required.

Small bottles—two ounces or thereby—not too tall, and provided with a glass rod, are necessary. The glass rod has a bulge at the junction of its upper and middle thirds, and this bulge prevents it from falling into the bottle, and, at the same time, acts as stopper for the bottle. Failing this, a piece of glass rod passed through the cork will answer the purpose.

(1.) **Normal Saline**, or ·6 per cent. salt solution. Dissolve 6 grms. of pure common salt in 1000 cc. of water. As this fluid is apt to undergo change, it should not be kept too long.

(2.) **Glycerine** (either pure or equal parts of glycerine and water).
(3.) **Balsam**, either Canada balsam or dammar (p. 85).
(4.) **Farrant's Solution** (p. 85).
(5.) **Dilute Acetic Acid** (2 per cent.).
(6.) **Hæmatoxylin Solution** (p. 68).
(7.) **Picro-Carmine** (p. 66).
(8.) **Clove-Oil** or **Xylol**.—This should be provided with a small brush fixed on the end of a wooden rod perforating the cork.

17. Other Apparatus is required, but in a well-equipped laboratory special articles are supplied as required; they are referred to in the context. They include a dissecting microscope, photophore, mounting block, warm stage, eye-piece micrometer, lamp, turntable, polarising apparatus, camera lucida, &c., &c.

II.—THE MICROSCOPE AND ITS ACCESSORIES.

1. An account of the optical principles on which the microscope is constructed is purposely omitted. The compound microscope consists of a **stand** fixed to a heavy, usually horse-shoe shaped, foot. The stand (fig. 9) carries a **stage** to support the microscopic preparation, the **mirror** or arrangement for illuminating the object, together with the body **tube**; the latter consists of a long brass tube, or one tube telescoped into another. To the lower end of this tube is fixed a combination of lenses, constituting the lens or **objective**, while at its upper end is the **eye-piece**.

2. The tube is blackened inside, and to its lower end is screwed the **objective**, consisting usually of several lenses screwed together. By means of it a magnified inverted aerial image is produced in the body of the tube. The lenses on the objective should not be unscrewed. At least two objectives are required—a low power and a high power.

At the upper end of the tube is the **ocular** or **eye-piece**, composed of two plano-convex lenses, the one next the eye of the observer being called the *eye-glass*, the lower one the *field-glass*. The two lenses, with their convex surfaces downwards, are fixed in a brass tube which slips into the upper end of the tube of the microscope.

3. The tube itself is supported in a vertical position on the stand, so that it can be moved upwards and downwards vertically, to bring the objective near to the object, and thus bring the latter clearly into focus. This arrangement is termed the **adjustment**. The mode in which this is accomplished varies. In the cheaper microscopes the tube of the microscope is moved inside another tube fixed to the stand (fig. 9). This is done by means of the hand, while in the more expensive microscopes there is a rack-and-pinion movement for raising or depressing the tube (fig. 10).

Usually there are two adjustments—one the **coarse adjustment**, whether it be by rotating one tube inside the other or by a rack-and-pinion movement; it is used to bring the outlines of the object dimly into focus. The other, the **fine adjustment**, is a fine screw, usually placed at the upper and back part of the pillar of the stand of the microscope. By it the object is brought accurately into focus.

4. The stand is provided with a horizontal solid table or **stage**, placed at a convenient height, and on which the object to be examined is placed. The stage consists of a flat plate of brass blackened on its under surface, or of a glass plate fixed on a black ground. There are two **clips** which are used for fixing a preparation in a definite position on the stage. It is perforated by a circular aperture in its centre, into

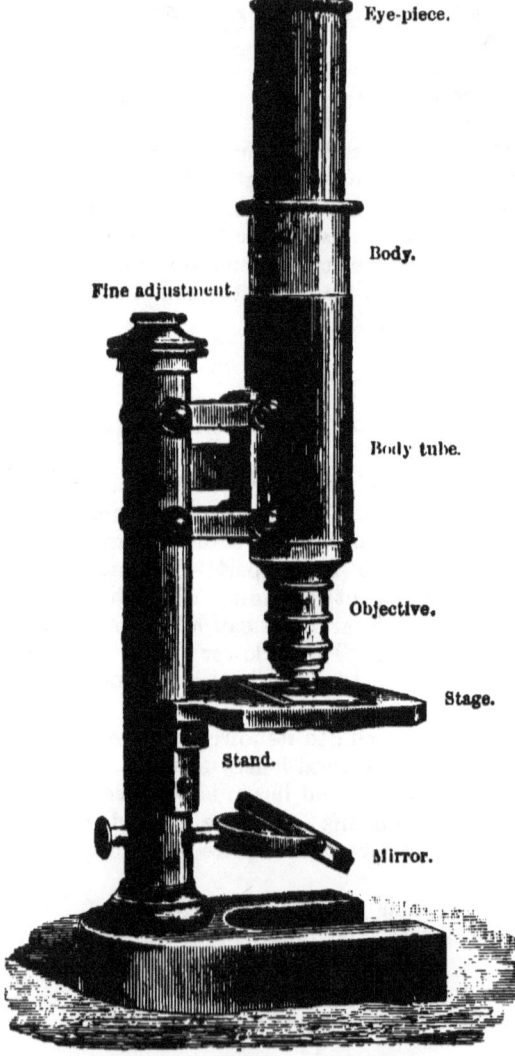

FIG. 9.—Leitz's Microscope, No. V.

which can be fitted diaphragms of different sizes, or a blackened circular brass plate with holes of different sizes—from a pinhole to half an inch—rotates under the stage, so that the desired size of hole —or **diaphragm**—can be brought under the central aperture in the stage. The aperture in the stage must not be too small; it should be sufficiently large to enable a section of the spinal cord to be seen as a whole. A small aperture is used with high powers and a large aperture with low powers. In the more expensive, and in some of the cheaper microscopes also, the stand is provided with a joint, so that the microscope can be inclined as shown in fig. 10.

5. **Illumination of the Object.**—Under the stage is placed a mirror, movable in all directions, and which is usually provided with a flat and a concave surface. When it is available, diffuse light—never direct sunlight—reflected from a white cloud, and a northerly exposure are to be preferred. For ordinary illumination the concave side of the mirror is used. The light is reflected from it, and is transmitted through the hole in the stage, the object on the stage, and the tube of the microscope, to the eye of the observer. The flat mirror is used along with a sub-stage condenser (§ 10).

6. **Direct and Oblique Illumination.**—When the light is reflected from the concave mirror, it strikes the object nearly vertically; this is called *direct* or central illumination. But sometimes it is of importance to detect very fine variations on the surface of the object; then for this purpose *oblique* illumination is practised. This may be done by tilting the mirror slightly, so that the rays of light fall somewhat obliquely on the object. In this case there must be no small diaphragm in or under the stage. This may also be done in the more expensive microscopes by introducing a diaphragm which permits light to pass only at its sides, its centre being blocked. This is known as a central stop-diaphragm, which shuts off all the axial, and transmits only the marginal rays, causing what is called *dark-ground* illumination (fig. 12).

7. **The Diaphragm**, of which there are two forms in common use. The most common form is a blackened metallic plate—*disc diaphragm*—perforated with holes of different sizes, placed under the stage, and so arranged as to rotate on a pivot. The edge of the plate usually projects a little beyond the stage, so that it can be readily rotated by the finger, so as to bring the appropriate aperture under the hole in the stage. The diaphragm is usually provided with a slightly projecting pin, which gives a click when the hole in the diaphragm is exactly centred. Another form—*cylindrical diaphragm*—consists of a small brass cylinder, into which can be

fitted small diaphragms perforated by apertures of different sizes (figs. 10, 12). The cylinder, with its diaphragm, is fitted into a slot under the stage. These cylinder diaphragms should be so arranged as to be easily changed.

Perhaps the most convenient form of all is the *iris diaphragm*, which can be adapted to any of the larger microscopes (fig. 14), and in which any size of aperture desired is obtained by turning a small milled head. The new form (fig. 11), as made by English makers, Zeiss, and others, and also by Leitz, is an admirable substitute for interchangeable diaphragms, as by it the aperture can be readily increased or diminished. The smallest aperture is about 0.5 mm., and the largest equal to the full aperture of the condensing system.

FIG. 10.—Zeiss's Large Jointed Stand, fitted with drawtube, rack and pinion for the coarse adjustment, a double nose-piece, and cylindrical diaphragm.

The management of the diaphragm is most important in order to obtain distinct definition of an object. The one general rule—but one which is very frequently neglected in practice by the student—is when employing a low power to use a large aperture, and when employing a high power, a small or medium aperture of the diaphragm.

When one wishes to observe a brightly coloured object with a homogeneous immersion lens—the object lying in the tissues, such as stained fungi or mitotic figures—then remove the diaphragm and allow a flood of light to reach the sub-stage condenser, which is absolutely necessary in this case.

If, however, it be desired to see certain peculiarities of structure —as in bacteria—on a black background, then use a central stop-diaphragm, *i.e.*, one with a central stop (fig. 12), always with an

Abbe's condenser. The object is then seen brilliantly illuminated on a black background.

8. Objectives.—For ordinary work every microscope requires to have two objectives of different magnifying powers; one of these, when used with an ordinary ocular, should magnify about 60–75

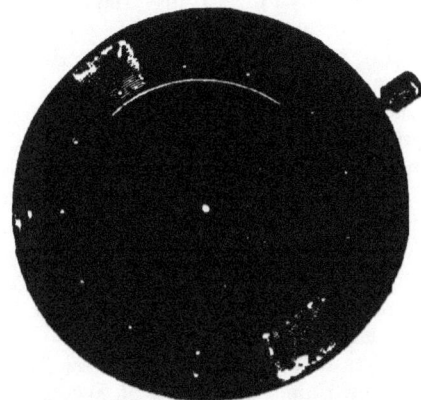

FIG. 11.—Iris Diaphragm, showing a small aperture.

FIG. 12.—Cylindrical Diaphragms (1, 2, 3), a central stop-diaphragm (5).

diameters linear. This is spoken of as the **low power**, and in these pages is indicated by the letter (**L**). The other should magnify from 350–400 diameters linear, and is called the **high power**, indicated by the letter (**H**).

If an English make of lens is preferred, let them be an "inch" and a "quarter inch." The term "one-inch objective" has no direct relation to the distance between the object and the lens, but indicates that such a lens possesses the same magnifying power as a single lens of one-inch focus. Swift's new high-angled 1-inch lens is a very good low-power lens. If Continental lenses be preferred, and if Zeiss's be selected, let them be A and D, equal to $\frac{2}{3}$-inch and $\frac{1}{6}$-inch respectively. Zeiss makes lenses AA and DD of slightly better quality, which cost a few shillings more. If Leitz's lenses be preferred, use Nos. 3 and 7; and if Hartnack's, Nos. 3 and 7—3 being the weaker lens. Crouch's lenses are excellent.

Leitz numbers his objectives 1–9, and Zeiss A–F. 1 and A are the weakest objectives, and the magnifying power increases up to 9 and F.

The microscopes of Reichert of Vienna are also excellent. The stands are all provided with a universal thread or screw, so that the lenses of Zeiss, of English makers, or of others may be adapted to them.

For certain special purposes much higher powers are required, but for ordinary work these lenses are sufficient.

The following patterns of microscopes are to be commended :—

Leitz, Stands Nos. V., III. 17, IV. 19, V. 23, with objectives 3 and 7.

Zeiss, Stands Nos. VI. and VII., with objectives C and E.

9. Eye-Pieces or Oculars.—Two eye-pieces are required, one medium length and the other shorter, the latter being the more powerful. The two most useful eye-pieces are Nos II. and IV. of foreign makers (of Leitz, however, I. and III.), or A and C of English makers. English makers speak of a deep and a shallow eye-piece; the former (IV. or C) is shorter, and is the more powerful; the latter (II. or A) is longer, and is a weaker eye-piece.

It is to be remembered that the eye-piece only magnifies the image of the preparation formed by the objective, and does not magnify the preparation itself. Hence any fault in the lens, and consequently in the image in the tube, is magnified by the eye-piece. Moreover, the field is not so bright as with a weaker eye-piece. Hence it is expedient to use rather a weak eye-piece. There is an exception to this rule in the case of apochromatic lenses (p. 13), which yield a magnification of the object by means of special eye-pieces, without any loss in the brightness or sharpness of the image.

These comprise the essential parts of the microscope, but if expense be no objection, the stand may be provided with a hinge-joint, which enables the microscope to be inclined at any angle (figs. 10, 14). In some microscopes the tube can be elongated by a *draw-tube* (fig. 14), which, when it is elongated, increases the magnifying power of the instrument. It is well to have the draw-tube with a scale engraved on it, as is done in the more expensive instruments.

10. Sub-Stage Condenser.—When working with high powers, this is essential, more especially in connection with bacteriological work. **Abbe's condenser** is by far the most convenient form. Fig. 14 shows one of the more expensive stands fitted with a sub-stage condenser.

The essential feature of Abbe's illuminating apparatus is a condenser system of very short focus, which collects the light reflected by the mirror into a cone of rays of very large aperture and projects it on the object (fig. 13).

The cone of light is usually reduced by diaphragms of suitable size, or by means of an iris diaphragm. By means of the rack-work the diaphragm can be placed excentrically and oblique illumination obtained. This apparatus can be used for ordinary work, but it is specially useful for the investigation of bacteria. The rays of light from the condenser are brought to a focus in the object, so that an enormous amount of light is concentrated on the object. The angle of aperture of Abbe's condenser is 120°. The full aperture of the illuminating cone of rays is only used when observing deeply-stained bacteria with objectives of large aperture. In every other case a diaphragm of suitable size is introduced so as to diminish the cone of light. When the diaphragm is placed excentrically, oblique illumination is obtained, while, with a central-stop diaphragm all the axial rays are cut off, and dark-ground illumination is obtained.

11. Dry and Immersion Lenses.—By the term *dry lens* is meant one in which air is the medium between the lens and the object, or at least the cover-glass on the object. In immersion lenses some fluid intervenes between the lowest lens of the objective and the cover-glass of the object, and the liquid chosen is water, or a medium of higher refractive index, such as cedar-wood oil, glycerine, or a mixture of fennel and ricinus oils. These oils have nearly the same refractive index as the cover-glass.

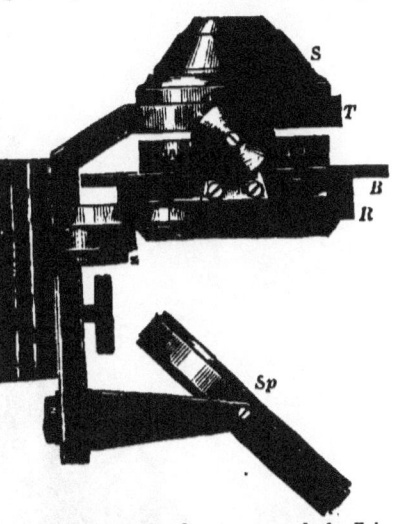

FIG. 13.—Abbe's Condenser, as made by Zeiss and Swift. (*S.*) condenser system. Milled head for throwing the diaphragm out of centre; (*Sp.*) mirror.

In virtue of their greater refractive power, these liquids, especially the oils, refract more of the rays passing through the object, and cause these rays of light to enter the lens, so that they increase the amount of light transmitted to the eye of the observer. In contrast with dry lenses, they have a larger angle of aperture and a greater resolving power, and are employed only for the highest magnifying powers. Powers above $\frac{1}{8}$ of an inch should be oil-immersion lenses. *Oil-immersion* or *homogeneous immersion* lenses are to be preferred to water ones. The oil-immersion are fast displacing water-immersion lenses.

In using an **oil-immersion lens**, by means of a glass rod, place a small drop of thick **cedar-oil** on the cover-glass and a small drop on the lowest lens of the objective, and slowly depress the objective until it touches the cupola of the drop, and focus. Cedar-oil has a refractive index nearly the same as that of crown-glass, so that almost all the rays of light passing through the object reach the lens and pass up the tube of the microscope. Care must be taken that the drop of oil does not run on to the cement of the preparation, else it will dissolve it. It is better to use marine glue to seal up the preparation, as it is not dissolved by cedar-oil. A very fine

FIG. 14.—Leitz's Stand Ia., with a triple nose-piece and Abbe's condenser.

linen cloth moistened with benzene may be used to remove the oil from the lens. After use, the cover-glass must be wiped dry, as well as the lens, by means of a clean piece of wash-leather. These lenses are very good, but very dear, and are not required for ordinary work. Latterly, instead of oil, a mixture of glycerine and chloral hydrate has been recommended. The cedar-oil is not so easily removed from the cover-glass. It may be best removed by a well-washed linen rag moistened with benzene. The lens itself must be most carefully cleaned. A lens moistened with glycerine is best cleaned with alcohol.

12. **Angle of Aperture.**—Lenses may be of narrow or wide angle of aperture. The angle of aperture is the angle formed by the outermost rays coming from a luminous point placed in the focus of an object, and which enter not only the lowest lens of the objective, but pass throughout the entire system of the lenses of the objective. Lenses with a large angle of aperture (130° and upwards), therefore, will admit more light, *i.e.*, more of the oblique rays will enter the system of lenses. These lenses are well suited for resolving fine lines on the surface of an object, such as the striæ on the scales of insects' wings and the markings on diatoms, but those parts of the object superficial to or deeper than the focus are not sharply defined. Such a lens is said to have greater power of *resolution*. For ordinary histological work, a high-power (350–400) lens of medium (80°–100°) angular aperture is to be preferred, for such a lens has greater *penetrating power*, *i.e.*, the focal plane is deeper, so that with it one can see with tolerable distinctness parts of the object lying immediately above and below the true focus of the lens. It is to be remembered that the angle of aperture has nothing to do with the magnifying power of the lens.

13. **Abbe's Apochromatic Lenses.**—These are dry or immersion, and are used as high-power lenses when very exact definition is required. They are very expensive, and are constructed of a peculiar kind of glass. These objectives secure the union of *three* different colours of the spectrum in one point of the axis, *i.e.*, they remove the so-called "secondary spectrum," and they correct the spherical aberration for two different colours. The images projected by them are nearly equally sharp with all the colours of the spectrum. As there is a very great concentration of light by these objectives, they permit of the use of very high eye-pieces, thus giving high magnifying power with relatively long focal length. The natural colours of objects are reproduced unaltered by these objectives.

Zeiss has constructed a series of **compensating oculars** to be used with these lenses. They are classified as 1, 2, 4, 8, 12, 18, and 27, according to their magnifying power. The eye-pieces of extremely low power are designated—

(i.) *Searchers*, which reduce to its lowest limits the available magnification with each objective, thus facilitating the preliminary examination of objects, diminishing the labour of searching for particular parts of the specimen. Thus, No. 1 enables an objective to be employed with its own initial magnifying power, *i.e.*, as if it were used as a simple lens without an ocular (*Zeiss*).

FIG. 15.—New College Microscope, by Swift & Sons—Tripod Stand.

(ii.) The *working eye-pieces* begin with a magnifying power of 4. The most useful are 4 and 6. Other eye-pieces are—

(iii.) *Projection eye-pieces*; but they do not concern us here.

14. In selecting a microscope, there are many points to be attended to.

(A.) **Mechanical Parts.**—It should be quite stable, so that it cannot be readily upset. To this end the stand should be solid, and either of the tripod (fig. 15) or horse-shoe pattern (figs. 9, 10). Very good stands of the tripod pattern are made by Messrs. James Swift & Sons (fig. 15) and Messrs. Crouch. Fig. 15 has a glass stage, the body-tube is cloth-lined, which gives a smooth and steady action. The stage should be at a convenient height, so that when the ulnar edge of the left hand is resting on the table, the thumb and forefinger of the same hand can conveniently grasp and move the slide on the stage. The stage itself should be a little broader than the length of the slide. The slide can be fixed on the stage by means of two brass clips, which are fitted into holes at the two posterior angles of the stage. The pillar of the microscope may be fitted with a joint to enable the instrument to be inclined, if desired; but of course this cannot be used when fluids are being examined; still in many instances it is convenient (figs. 10, 14). Fig. 16 shows a convenient form made by Hartnack of Potsdam, and one very extensively used by students.

It is highly inconvenient to have to screw and unscrew a lens every time a change of lens is required. This is obviated by using a **nose-piece** or **revolver** (figs. 10, 14), which is screwed to the lower end of the tube of the microscope. The high and low powers are fitted to this framework, and can be rotated under the tube as they are required. Nose-pieces are made for attaching two or three (figs. 10, 14) or more objectives, and they can be adapted to any microscope.

FIG. 16.—Stand III. of Hartnack, with joint and condenser.

One must be cautious in using a nose-piece, and take care to raise the tube high enough to allow the objective to revolve without touching the cover-glass, which is especially apt to happen on using a high power after a low power, the lenses not being of the same length.

Test the mechanical parts that they are all solid and work well. Raise and lower the tube by means of the fine adjustment, and do this to the full extent of the threads on the screw, noting particularly if there is any lateral movement of the tube while this is being done.

(B.) **Optical Parts.**—See that all the parts are properly centred by looking through the tube after removal of the eye-piece. Put on the high power; use a medium eye-piece, and focus a microscopic preparation; any thin section of a tissue will do. The field should be large, well illuminated, and flat. If there are specks in it, clean all the lenses to see that these specks are not due to dust on the lenses. If the centre of a flat object in the field does not come into focus at the same time as the periphery of the object, then the lens should be rejected on account of its spherical aberration. If coloured rings appear in an object, then the lens must also be rejected, as it is not perfectly achromatic. The definition ought to be sharp and distinct.

The **qualities of a good lens** are the following:—

The **definition** should be good, *i.e.*, the correction for spherical and chromatic aberration should be perfect. The outlines of objects should be sharply defined and not blurred, and there should be no coloured halos or fringes round the object. If so, the chromatic aberration of the lens is not perfectly corrected.

Flatness of field, *i.e.*, all parts of the object in the field at the same time should be seen with equal distinctness, and, of course, this can only be so when the field is flat. If the central parts are sharply focussed, the peripheral parts not, then the field is not flat.

Resolving power, which depends on the angle of aperture. This is the power to render visible surface markings, superficial lines, or structural details. It is better to have a lens of large angle of aperture for this purpose.

Penetrating power, or the power to see objects in several planes in the same preparation at the same time. It represents the focal depth of the lens and its power of focussing images from different planes in the vertical range. Narrow angled lenses are more suitable for this purpose.

Working Distance, *i.e.*, the distance between the lens and the object. It has no direct relation to the focal length of a lens. Wide angle lenses have a shorter working distance than narrow angle lenses. A good lens should combine all the foregoing qualities, but of these qualities definition is all important.

15. Drawing of Microscopic Objects.—In this laboratory every student is required to make sketches of his preparations. This is of the utmost importance, not only from the point of view of the student, but also of the teacher.

(A.) **Freehand Sketching.**—The student must provide himself with a *drawing-book* (p. 3) and with suitable pencils (H.B. and H.H.H.), and also with either coloured crayons or water-colours. It is far more important that an outline sketch in pencil be made accurately portraying the shape of the object, than that an indifferent sketch should be covered with a smear of colour. It is astonishing how some students, after protesting that they cannot "draw," succeed in delineating their preparations when they have given the matter a fair trial.

(B.) Various Forms of Camera Lucida.—Sometimes, however, a drawing has to be done to scale, or its outlines and details accurately portrayed. There are many devices for this purpose. Some of these instruments are by no means easy to work with, but they are excellent for tracing the outlines of objects and showing the exact relation of one part to another, and the relative size of the parts of an object.

(i.) **Zeiss's Camera Lucida** (fig. 17).—A collar (c) with a vertical rod attached (a), carrying a bar (b) which bears the prism or camera (K). The collar is slipped over the tube of the microscope and the eye-piece inserted. Focus the object, and rotate the prism until it is above the eye-piece; the prism covers one-half of the latter.

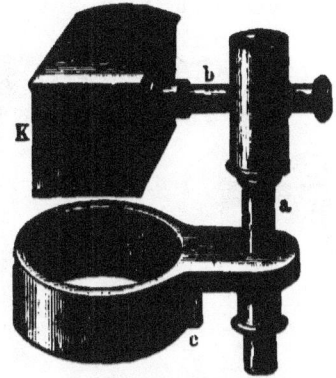

FIG. 17.—Zeiss's Camera Lucida. (c.) collar; (a.) and (b.) supporting bars; (K.) camera.

The drawing-board is inclined at an angle of 20°, and in looking down the microscope the paper and the image of the object are seen simultaneously.

(ii.) **Camera Lucida** (*Abbe*).—This apparatus is made by Zeiss (fig. 18). The apparatus is screwed to the tube of the microscope.

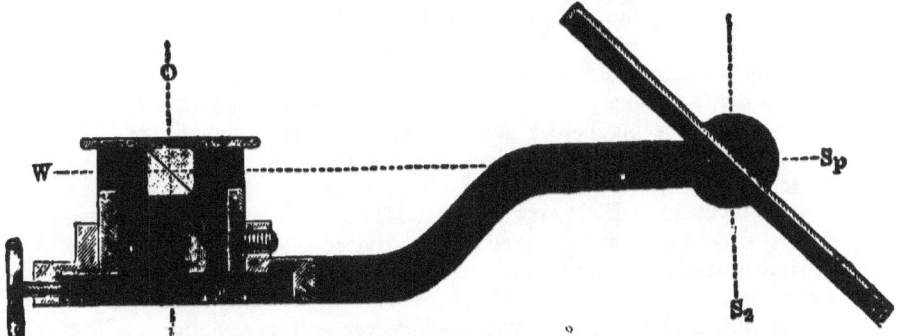

FIG. 18.—Abbe's Camera Lucida. (Sp.) mirror; (W.) line of reflected light; (i.) collar for screwing over eye-piece; (O.) light to eye of observer; (S_2.) light reflected on paper.

The direction and reflection of the rays from the object and those from the paper are shown in the figure. In all camera-lucida drawings it is important to regulate the brightness of the illumination

B

of the paper; this is effected in this instrument by smoke-tinted glasses, which fit into the prism mounting. In other cases this is done by means of a blackened cardboard shade.

(iii.) **Chevalier's Camera Lucida** (fig. 19).—The microscope remains erect, the eye-piece is removed, and in its place is inserted the tube of the camera. Place a sheet of white paper at the side of the microscope, and directly under the horizontal part of the camera. The instrument itself consists of two tubes at right angles to each other, and at the angle is a prism, which reflects at a right angle in a horizontal direction the light coming vertically from an object on the stage. These horizontal rays have to pass through what is practically an eye-piece of a microscope, and reach a small prism at the free end of the horizontal part of the instrument. This prism is so arranged that the rays coming from the object are reflected into the eye of an observer, who, on looking through the prism, sees an image of the object upon the sheet of paper on the table. The observer looks through this prism, but he must so adjust his position as to bring one-half of the pupil over the prism, and thus with one-half of the pupil view the object on the paper, while the other half of the pupil receives the rays of light coming from the pencil and the paper. The distance between the eye of the observer and the paper is about 10 inches. It is by no means so easy to sketch the outline of an object with this camera. The eye should be protected from other rays of light by means of a blackened shade. The difficulty encountered by the student is to see the object and the point of the pencil at the same time.

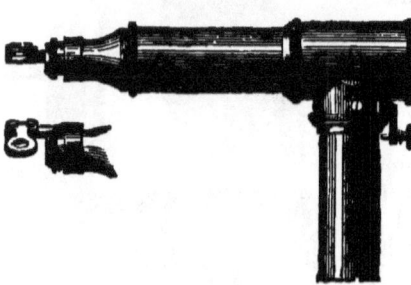

FIG. 19.—Camera Lucida of Chevalier or Oberhauser.

(iv.) **Camera of Malassez** (fig. 20).—This consists of two prisms; one, the ocular prism, is fixed, while the second and larger can be moved round a horizontal axis. This enables the instrument to be used in two different positions of the microscope. If one is examining a fluid, the microscope is erect and not inclined. The instrument is fixed to the upper part of the tube of the microscope. Slip the collar over the tube and insert the eye-piece. The image is projected on the table to the right of the microscope; but the paper must be placed at an angle, *i.e.*, on an inclined plane, horizontal to the direction of the rays from the camera. This is necessary to avoid one side of the figure being larger than the other. If it be desired to incline the microscope at an angle of 45° (fig. 20), the camera is so placed that an image is thrown on paper placed behind the foot of the microscope, the prism being turned to an angle of 45°. This camera is particularly easy to work with. It is sometimes important not to have too much light on the paper; this is avoided by using the plane side of the mirror.

16. Magnifying Power of a Microscope.—This will vary with the objective and ocular used, and also with the length of the draw-tube. The magnification is usually determined when an image is seen at the range of normal distinct vision, *i.e.*, 10 inches or 25 centimetres. As this is about the height of a Hartnack's stand or Zeiss's stand, supposing these forms of instrument to be used, place a sheet of white paper on the table.

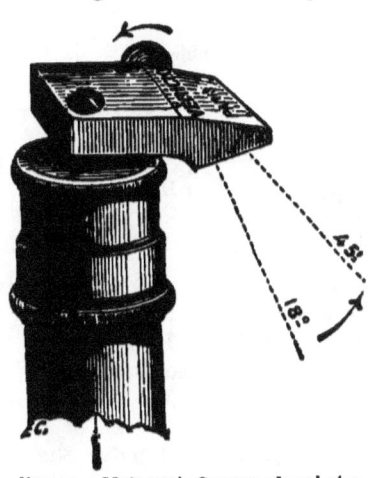

FIG. 20.—Malassez's Camera placed at a variable angle.

Let the microscope be upright, with ocular and lens in place. Supposing we use a No. 7 objective and No. 3 ocular. Begin with the draw-tube in, place a stage-micrometer on the stage of the microscope, and focus the scale upon it. The stage-micrometer is like a glass slide with a fine scale engraved upon it. The English ones are generally subdivided into thousandths of an inch, and the Continental ones into hundredths of a millimetre. Look through the microscope, but keep both eyes open, and part of the scale will be seen on the white paper. With a pencil mark off say ten interspaces, or use a pair of compasses to measure this distance. Measure off this distance on a millimetre scale, and suppose the distance thus measured be 28 millimetres. This means that .1 millimetre has been magnified to appear equal to 28 millimetres, *i.e.*, 280 times. The magnifying power for other combinations of lenses should be determined, and a table made for future reference and use.

Magnifying Power.—This may be increased—

(*a*) By using a higher objective.
(*b*) By using a higher eye-piece.
(*c*) By pulling out the draw-tube and increasing the distance between the objective and the eye-piece.

Construct a table—according to the following scheme—of the

magnifying powers of the different combinations of the objective and ocular supplied with your microscope :—

Objective.		Ocular.	
		Shallow, or A or 2.	Deep, or C or 4.
Draw-tube in	{ Half-inch or A { One-sixth inch or D	... ...	... ...
Draw-tube out	{ Half-inch or A { One-sixth inch or D	... ...	... ...

17. Measurement of the Size of a Microscopic Object.—Suppose a red blood-corpuscle is to be measured. Keep both eyes open; with one look down the tube of the microscope, and the other eye will see an image of the corpuscles on a sheet of white paper placed beside the microscope. With a pencil mark off the outlines of a corpuscle, or a sketch may be made of one with a camera lucida. Remove the preparation, and for it substitute a stage-micrometer, leaving all the other parts, microscope and paper, as they were. On looking down the tube of the microscope, the lines on the micrometer scale are seen. Make a drawing of these lines. This may be kept for future use, but on the scale should be noted the combination of ocular and objective, and the extent to which the draw-tube has been elongated. The distance between the lines on the micrometer scale being known, say $\frac{1}{100}$th of a millimetre, it is easy to calculate what part of the distance between any two lines corresponds to the size of the corpuscle.

The most expeditious plan is to use an **eye-piece micrometer**. A circular flat piece of glass, with a scale ruled on it (fig. 21), is inserted in the ocular between the field-glass (FG) and the eye-glass (EG) (fig. 22). With this eye-piece focus the scale on a stage-micrometer—its markings must be parallel to those in the eye-piece—and count the number of divisions of the latter that correspond to one of the former. This must be determined for each combination of lenses with a known length of tube. Suppose the stage-micrometer to be divided into $\frac{1}{100}$ths of a millimetre, and that one of these divisions corresponded to three of those in the ocular, then each of the spaces in the ocular micrometer is equal to $\frac{1}{300}$th of a millimetre, or .0033 millimetre.

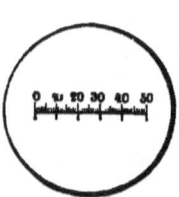

FIG. 21.—Scale introduced into Eye-piece Micrometer (*Zeiss*).

The histological unit of measurement is 1000th part of a millimetre or **micro-millimetre**, expressed shortly as a **micron**, or by the Greek letter μ. Thus, each space of the ocular micrometer is equal to 3.3 μ.

With these data it is easy to estimate the size of any object. The object is placed on the stage and focussed, always, of course, with the same combination of lenses and ocular. If other lenses be used, the value of the ocular divisions must be determined for this particular combination. Once determined, they can be noted for future use.

18. Artificial Illumination.—White daylight is, of course, to be preferred, but it is not always available. When artificial light has to be used, a gas flame, *e.g.*, an Argand burner, or other artificial light, may be used. An ordinary paraffin lamp with a flat wick and arranged to burn steadily does very well. An unsteady flame is

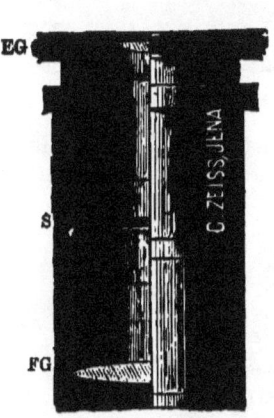

FIG. 22.—Micrometer Eye-piece. (*EG.*) eye-glass; (*FG.*) field-glass; (*S.*) scale.

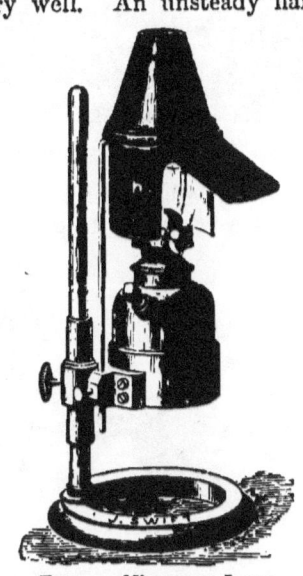

FIG. 23.—Microscope Lamp.

very injurious to the eyes. It requires some care to find the exact distance at which the lamp should be placed from the mirror. If very intense light be required to examine a small part of a preparation, turn the wick edge on to the mirror. Between it and the mirror place a screen of white paper. If the direct rays from a lamp are used, to correct the yellow rays place a sheet of pale blue glass, or a glass globe containing a weak solution of ammonio-sulphate of copper, between the light and the mirror, or a thin blue glass may be placed on the stage under the preparation. A paraffin-oil lamp may be used. Fig. 23 shows a cheap paraffin lamp sold by Swift & Sons. A remarkably good light is obtained from the

Welsbach gas-burner, the rays being transmitted through a large globe containing a dilute blue-coloured solution of ammonio-sulphate

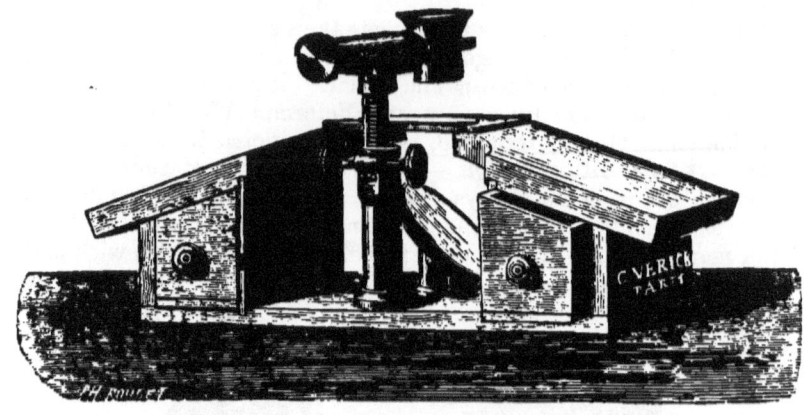

FIG. 24.—Dissecting Microscope by Verick.

of copper. The lamp recently constructed by **Kochs-Wolz** of Bonn consists of a petroleum or gas flame, covered by an opaque chimney

FIG. 25.—Dissecting Microscope by Reichert.

provided with a reflector, which directs the rays of light through a hole in the chimney. Into the hole in the chimney is fitted a glass rod, which, bent in a gentle curve, conducts the light from the flame to the under surface of the stage, the mirror in this case not being used.

Perhaps the most useful lamp is that known as the **albo-carbon light**. It is the one I am in the habit of using when—especially in winter—good daylight is not available. Ordinary gas is used, but the gas has to traverse a chamber containing naphthaline before it reaches the burner. A very white light is thus obtained, and can be used without the intervention of blue glass or a blue copper solution.

If expense be no bar, then a small incandescent electric light is most useful.

19. Dissecting Microscope (figs. 24, 25).—This is very useful. The lenses usually employed magnify from 5 to 20 times linear, and are fitted into a framework which can be raised or depressed, so as to bring the object distinctly into focus.

20. Method of Measuring the Thickness of Cover-Glasses.— Thick cover-glasses are of no use, and thin ones—extra thin, so called—are to be preferred. A convenient thickness is 0.16 mm. For high powers this is essential, and it is well to measure the thickness of the cover-glasses, and to reject all those above a certain thickness. This is conveniently and rapidly done by means of the instrument shown in fig. 26. A clip projecting from the box fixes the cover-glasses, and the thickness is given by an indicator moving over a divided circle on the lid of the box. The divisions show hundredths of a millimetre.

FIG. 26.—Cover-glass Tester.

21. Camera Obscura Shade.—When one has to continue observing a microscopic object for a long time, it is convenient to shade the eyes from all light reaching them, except that transmitted through the eye-piece. This is best done by a blackened screen either fixed to the microscope or arranged in front of the microscope.[1] Some observers place the microscope in a dark chamber, allowing light to fall upon the mirror through an aperture in its front wall Flögel[2] has designed a camera obscura for this purpose.

[1] Schiefferdecker, *Archiv f. wiss. Mikr.*, p. 180, 1892.
[2] *Zoolog. Anzeiger*, by V. Carus, p. 566, 1883.

III.—NORMAL OR INDIFFERENT FLUIDS.

Not unfrequently **living** tissues or **fresh** tissue elements have to be examined in as natural a condition as possible, and obviously, if a fluid require to be added to it, the fluid must be of such a nature that it will not injuriously affect the tissue or its elements. These fluids are spoken of as **indifferent** or **normal fluids**. When fresh, these fluids cause very slight changes in the tissues. Amongst those used are—

1. **Normal Saline.**—Dissolve 6 grams of pure sodic chloride in 1000 cc. water.

2. **Kronecker's Fluid.**—It consists of

Distilled water	100 cc.
Sodic chloride	.6 gram.
Soda	.06 ,,

3. **Blood Serum.**—The blood is allowed to clot, and the serum, after a day or so, is poured away from the clot. In a laboratory provided with a "centrifugal apparatus" any red blood-corpuscles can be got rid of by "centrifugalising" the serum. This fluid does not keep long, and must be fresh. It has been suggested to add a piece of camphor to it, but this only helps to preserve it for a short time. Iodine is sometimes added to serum to form *iodised serum*, but this is by no means an indifferent fluid (p. 25).

4. **Aqueous Humour.**—With a narrow triangular knife puncture the cornea of a freshly-excised ox's eyeball and collect the aqueous humour which exudes. If only a small quantity be required, it may be obtained by puncturing the excised eye of a frog with a fine capillary glass pipette (p. 4).

5. **Fluid of Ripart and Petit :—**

Camphorated water	75 cc.
Distilled water	75 ,,
Glacial acetic acid	1 ,,
Acetate of copper	0.30 gram.
Chloride of copper	0.30 ,,

It is very useful for examining animal cells.

IV.—DISSOCIATING FLUIDS.

These fluids dissolve or soften the cement substance or interstitial material, *e.g.*, of epithelium, connective tissue, and thus facilitate the separation of the histological elements from each other. In some cases, in isolating tissue elements, it is well not to have too much fluid in proportion to the morsel of tissue, else hardening is more apt to take place.

1. Strong Iodised Serum.—A strong solution should be kept, which may be diluted as required. To the amniotic fluid of a cow or to blood-serum add crystals of iodine, and keep it in a stoppered bottle. Shake it frequently. At first very little iodine is dissolved, but after a time (15-20 days) the solution becomes much stronger, *i.e.*, it becomes of a deep-brown tint.

For dissociating a tissue, a **weak iodised serum** is used. A little of the strong fluid is added to fresh serum until the latter has a light-brown colour. If an object be placed in the dilute iodised serum and the brown colour fades, more of the strong solution must be added. In using this fluid, take a very small piece of tissue, the size of half a pea or less, and place it in 5 cc. of the fluid in a glass-stoppered bottle. After a day or two it may be dissociated with needles, but the brown tint must be maintained; more strong fluid must be added if putrefaction is to be prevented.

2. Dilute Alcohol (Ranvier's *Alcool au tiers*).—This fluid, devised by Ranvier, and sometimes called "one-third alcohol," is of the greatest possible service, and is one of the best dissociating fluids we possess. Mix 1 part of 96 per cent. alcohol with 2 parts of distilled water. It dissociates epithelial and other tissues in 24-36 hours. Use a small quantity of the fluid in proportion to the tissue.

3. Chromic Acid.—One gram in 1000 of water. This requires two days to a week, according to the tissue placed in it.

4. Potassic Bichromate.—Two parts in 1000 of water, *i.e.*, .2 per cent. It is very useful for dissociating epithelium and the nerve-cells of the spinal cord. It does so in 2-3 days.

5. Ammonium Chromate (5 per cent.).—It is used for dissociating the "rodded" cells of the renal tubules, cells of salivary glands, Purkinje's fibres of the heart, &c. It acts in 24-36 hours, and the chromate must be well washed out of the tissues if they are to be preserved.

6. Caustic Potash.—Thirty to 35 parts in 100 of water. It acts in 20-30 minutes, rapidly destroying connective tissue. Water must not be added to the dissociated tissue, else the tissues are rapidly dissolved. It is used for isolating the fibres of smooth muscle, or heart-fibres. Examine the dissociated tissues in the dissociating fluid. As a rule, tissues so dissociated cannot be preserved, but there are certain exceptions.

7. Dilute Osmic Acid (.1 per cent.).—It acts in 24-48 hours according to circumstances, and is well suited for cells containing fat globules.

8. Landois' Fluid:—

Neutral ammonium chromate	.	.	.	5 grms.
Potassium phosphate	.	.	.	,,
Sodium sulphate	.	.	.	,,
Distilled water .	.	.	.	100 cc.

This is specially useful for the central nervous system. Small pieces must lie in it for 1–5 days. (See Lesson on Nervous System.)

9. Methyl Mixture or Schiefferdecker's Fluid:—

Methylic alcohol .	.	.	.	5 cc.
Glycerine	.	.	.	50 ,,
Distilled water .	.	.	.	100 ,,

It is better to prepare it fresh, but it can be preserved in a glass-stoppered bottle. The methylic alcohol rapidly evaporates. The tissues remain in it for several days, and it is specially useful for the retina and central nervous system. I find that it isolates epithelial cells in one to two days. It acts very much like Ranvier's alcohol.

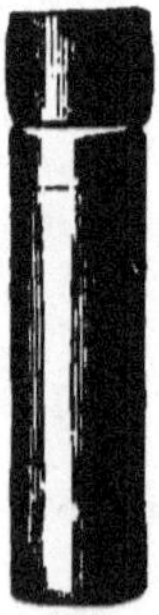

10. Other fluids are referred to in the text, *e.g.*, **baryta water**, and **10 per cent. sodic chloride** for tendon. This dissolves the cement substance of epithelial cells and the mucinoid cement of connective tissue.

11. **Digestion Methods**, both gastric and tryptic, are used (see text).

FIG. 27.—Glass Tube or Thimble for dissociating small pieces of a tissue.

The special uses of the above-named fluids are referred to in text under the tissues or organs, for which each is specially adapted.

General Directions for Dissociating Tissues.—Always use a very small piece of the tissue or organ, and place it in a not too large quantity of the dissociating fluid. Small glass thimbles (fig. 27) are very useful for this purpose.

V.—HOW TO TEASE A TISSUE.

To separate by means of needles the elementary parts of a tissue is by no means an easy task. The tissue must be seen distinctly, and the needles must be so used as not to break up the parts, but only to separate them. The process may be done by the unaided eye or with the aid of a lens or dissecting microscope (figs. 24, 25). The light must be good, and an appropriate background for the object should be selected. If the tissue is colourless, use a black background, *e.g.*, black paper; if coloured, a white one. In the latter case, however, the shadows interfere with exact vision, and it is better to support the slide upon an object raised slightly above the white background. This is readily accomplished by placing it over a white porcelain capsule, or on a photophore (fig. 28), into which is slipped a piece of white paper.

In a laboratory, one of the most convenient ways is to have the tables painted of a black or very dark green colour, but the painting must be flat, with no shining varnish. At the edge of the table is painted a white strip 1½ inches broad. Some prefer to burn into the surface of the table solid paraffin blackened by means of lamp-black. Others use porcelain slabs one-half black and the other white.

Photophore (fig. 28).—This is a small wooden box, 5 cm. high, 9 cm. long, and 9 cm. broad. The upper part is formed of glass. The front wall of the box is wanting. Placed obliquely within the box is a mirror, which reflects the light upwards through the glass cover and the slide to the eye of the observer.

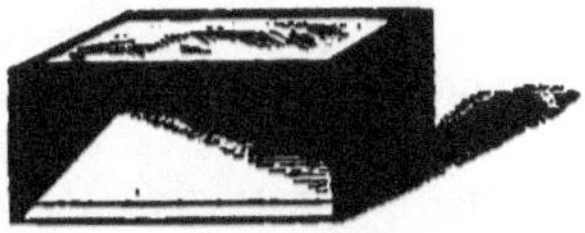

FIG. 28.—Photophore.

In dissociating with needles, we must have some knowledge of the arrangement of the parts of the object to be teased, such as the direction of the fibres, &c. Take a *small* piece only. Always tease one end of the tissue, and fix the latter with one needle while the parts are separated with the other needle.

One of the most convenient combinations is that of Eternod, which combines a photophore with a turntable. The wheel of the latter can be removed, and the block forms not only a photophore, but also a surface on which tissues of different colours can be teased and mounted.

VI.—FIXING AND HARDENING REAGENTS.

Most of the tissues and organs must be hardened in suitable fluids before they can be cut into sections. A large number of fluids of various kinds are used, each organ or tissue requiring its own appropriate fluid. Some organs, *e.g.*, bone, are too hard to be cut in their natural condition; they must be decalcified by appropriate fluids. Amongst others, the following fluids are required, but others are referred to in the text.

A. Alcohol.

Alcohol is one of the most important hardening fluids used

either by itself, or to complete the hardening processes begun with other fluids. Moreover, it is almost universally used for the preservation of tissues already hardened. There are three kinds of alcohol used for histological purposes, viz., absolute alcohol, rectified spirit, and methylated spirit.

(*a.*) **Absolute Alcohol** is alcohol without water, but that sold usually contains 96 per cent. of pure alcohol, and is sufficiently strong for microscopical purposes. If an absolutely water-free alcohol is desired, place well-dried potassic carbonate in the alcohol. This rapidly absorbs the moisture. Or powdered and heated cupric sulphate—a white powder—is added. If water is present, it absorbs it, and becomes blue again. Lowne uses slips of gelatine.

(*b.*) **Rectified Spirit** contains 84 per cent. of spirit and 16 per cent. of water.

(*c.*) **Methylated Spirit** contains a little wood naphtha, and is nearly as strong as the ordinary absolute alcohol, and may be made stronger by placing some well-dried carbonate of potash in it. The carbonate of potash absorbs any water present in the alcohol. Others use cupric sulphate well heated in a metallic capsule, until it becomes a white powder. When it becomes blue in the spirit it must be replaced by new $CuSO_4$. It can be heated and used again.

Alcohol is used of **various strengths for hardening purposes.**

Seventy-five per Cent. Alcohol.—To every 75 cc. of absolute alcohol add 25 cc. of distilled water.

Seventy per Cent. Alcohol.—To every 70 cc. of absolute alcohol add 30 cc. of distilled water.

Fifty per Cent. Alcohol.—Take equal volumes of absolute alcohol and distilled water.

To Harden in Alcohol Alone.—The tissues are placed for 12–24 hours in the weaker (50 per cent.) alcohol, and passed through the stronger alcohols (in each a day or thereby), and finally kept in 95 per cent. alcohol until they are required.

For certain special purposes the tissues, *e.g.*, glands and structures for the preservation of mitotic figures, are "fixed," and at the same time hardened by being placed at once, and while as fresh as possible, in absolute or 96 per cent. alcohol.

B. Chromium and its Compounds.

1. **Chromic Acid (Stock Solution).**—It is well to prepare a strong solution—10 per cent.—and to keep this as a stock to be diluted when required. Dissolve 10 grams of fresh chromic acid in 90 cc. of distilled water.

2. **Half per Cent. Solution of Chromic Acid.**—To 50 cc. of the 10 per cent. solution add 950 cc. of distilled water, or dissolve 1 gram of chromic acid in 200 cc. of water. Similar solutions

containing .3 and .2 per cent. chromic acid are frequently used, and can readily be made from the stock-bottle.

3. **Chromic Acid and Spirit, or Klein's Fluid.**—Mix 2 parts of chromic acid (.6 per cent. *i.e.*, 6 grams in 1000 cc.) with 1 part of methylated spirit. This should be made fresh, and kept from the light. If the fluid be changed often it hardens tissues in 8–10 days.

4. **Chromic Acid and Bichromate Solution.**—Dissolve 1 gram of chromic acid and 2 grams of potassic bichromate in 1500 cc. of water.

5. **Müller's Fluid.**—Dissolve 25 grams of potassium bichromate and 10 grams of sodium sulphate in 1000 cc. of water. Solution takes place slowly at the ordinary temperature. Pound the ingredients in a mortar, add the water, and warm until they are dissolved. It takes five to seven weeks to harden a tissue, according to the size of the tissue placed in it. This fluid is very extensively used, as it penetrates into the tissues and hardens them equally throughout. To prevent the formation of fungi, place a piece of camphor in the solution. Müller's fluid preserves blood-corpuscles in their original form, and they retain their yellow colour. It also shrivels very slightly the tissue elements.

6. **Müller's Fluid and Spirit.**—Müller's fluid 3 parts, and methylated spirit 1 part. Mix, and allow the mixture to cool before using it. When tissues are hardened in it they should be kept in the dark.

7. **Erlicki's Fluid.**—Dissolve 2.5 grams of potassium bichromate and .5 gram cupric sulphate in 100 cc. water. It should be prepared fresh. It hardens more quickly than Müller's fluid, and after the first day or two its action is greatly facilitated by keeping the tissues in it at 40° C. Ten days or so will suffice for hardening under these conditions. Experience, however, has shown that the process of rapidly hardening tissues at a comparatively high temperature is not so satisfactory as that conducted at a lower temperature.

8. **Potassium Bichromate.**—Make a 2 per cent. solution by dissolving 10 grams of the salt in 500 cc. of water. It takes from three to seven weeks to harden tissues, and is one of the best hardening fluids for the central nervous system.

9. **Ammonium Bichromate** is used in the same way and of the same strength. It takes much longer to harden than **8**. (See Central Nervous System.)

10. **Ammonium Chromate** (5 per cent. solution).—This is used for hardening the kidney and other secretory glands, the mesentery of the newt, &c. (see text).

Precautions in Connection with Chromium Compounds.—Solutions of pure *chromic acid* do not penetrate well into tissues, therefore the pieces of tissue must be small. In the case of tissues placed in *Müller's fluid* or *potassic bichromate*, they are hardened very slowly. In the case of most organs 3-4 weeks suffice, but in the case of the brain and spinal cord it takes several months to harden tissues and organs in these fluids. The fluid must be frequently changed, and it must be large in amount. The formation of fungi on its surface may be prevented by adding a piece of camphor, thymol, or naphthaline to the fluid. The hardening process is accelerated when the fluid and tissue are kept at 30°-40° C., but the result is not so satisfactory as by the slower cooler process.

Except for special purposes, *e.g.*, Weigert's method for the central nervous system, the chromic acid salts must be thoroughly washed out of the hardened tissue or organ. This is done by leaving them for many hours or days in a stream of running water. In all cases it is well to keep the fluids and tissues in a *cool* place, and in some cases in the *dark*, *e.g.*, **3** and **6**.

In connection with the hardening of tissues, especially those of the central nervous system, it is, with right, insisted upon that the hardening fluid should be frequently changed. I find, however, and my observations are borne out by other observers, that after the first change of fluid a very satisfactory result is obtained by placing the organ to be hardened in a very large volume of the hardening fluid, *e.g.*, the spinal cord of a cat or dog in a litre of the hardening reagent, and leaving it under proper conditions until the cord is hardened. Stir the fluid from time to time.

Chromic acid seems to form a compound with the tissues, so that it is not easily removed from them. Tissues hardened in chromic acid are not readily stained by carmine, so that for this purpose it is better to use hæmatoxylin or safranin.

C. Acids and Acid Mixtures.

1. Picric Acid.—Make a cold saturated solution of picric acid in water. There should always be a large excess of crystals on the bottom of the bottle. Place a small piece of tissue in the solution for 6-24 hours. The tissues are to be afterwards washed in 70 per cent. alcohol—not water—and transferred to 95 per cent. alcohol.

2. Kleinenberg's Picro-Sulphuric Acid.—To 100 cc. of a cold saturated watery solution of picric acid add 2 cc. of concentrated sulphuric acid. This causes a copious precipitate. After twenty hours filter, and to the filtrate add 300 cc. of distilled water.

Tissues are placed in this fluid for a comparatively short space of time—from a few minutes to two or six hours. The time should never exceed six hours. The liquid must be changed if it becomes turbid. The hardening process is completed in alcohol. It is well adapted for embryological work.

3. Picro-Nitric Acid (*P. Mayer*).—Some substitute nitric acid for the sulphuric acid, but for our purposes there is no advantage in this. Small pieces of tissue not over $\frac{1}{2}$ cm. in diameter—preferably those that contain little connective tissue—are hardened for 1-3 hours. They are washed with alcohol until most of the yellow colour disappears. Sections can be stained with hæmatoxylin.

4. Nitric Acid (*Altmann*).—Use a 3 per cent. watery solution of pure nitric acid. This has a sp. gr. of 1.02. Use as small pieces as possible, and leave them in the mixture just until they are fixed, i.e., from a quarter to half an hour. Strong solutions, if they act too long, dissolve the chromatin. The tissues are then hardened successively in 70, 80, and 90 per cent. alcohol. It is particularly useful for preserving the nuclei of cells, mitotic figures, embryological tissues, and the retina.

5. Perènyi's Fluid :—

Nitric acid (10 per cent.)	40 cc.
Chromic acid (0.5 per cent.)	30 ,,
Alcohol	30 ,,

It is a light greenish-blue liquid, and is specially useful for hardening ova and embryos. Time, 4-6 hours for a small embryo and 4-12 hours for the tissues of vertebrates. The tissue is transferred direct to 75 per cent. alcohol without previous washing in water. Borax carmine may be added to it, and then this mixture hardens and stains at the same time (*Garbini*).

6. Chromo-Formic Acid (Rabl's Fluid).—To 200 cc. of $\frac{1}{3}$ per cent. chromic acid add four to five drops formic acid. It must be freshly prepared, and fresh tissues—small pieces—are placed in it for 12-24 hours. The tissues are thoroughly washed in water, and hardened in alcohol of gradually increasing strength. Sections can be stained in hæmatoxylin and safranin. It is specially useful for the study of mitosis and nuclei generally. It has this advantage, that tissues hardened in it do not afterwards darken.

7. Chromo-Acetic Acid (*Flemming*).—As chromic acid by itself is apt to cause shrinking of some of the more delicate textures, it has been proposed to mix it with a substance which has an opposite effect, such as acetic acid. The following combination is specially recommended by Flemming for fixing the achromatic spindle in cells :—

Chromic acid	$\frac{1}{4}-\frac{3}{4}$ gram.
Glacial acetic acid	$\frac{1}{10}$ cc.
Water	100 ,,

The tissues must be small in size, not above 3-5 mm. in diameter.

They are hardened for twenty-four hours in this mixture, and afterwards washed for twenty-four hours in water, and the hardening completed successively in 70, 80, and 90 per cent. alcohol—each for 12-24 hours. Sections can be stained with hæmatoxylin.

8. Osmic Acid.—Make a 1 per cent. solution by dissolving 1 gram in 100 cc. of distilled water. This substance is rather expensive, and is sold in sealed glass tubes. Carefully clean the surface of the tube, snip off one end of it, and place it with the requisite quantity of water in a glass-stoppered bottle, which has been carefully cleaned, so that it does not contain a trace of organic matter. Organic matter decomposes it very rapidly. It takes several hours to dissolve. Some prefer to use normal saline solution instead of water to dissolve it. Its vapour is very irritating to the eyes and mucous membranes generally. It should be preserved in yellow-coloured bottles and kept in the dark. If a yellow bottle is not available, cover a bottle with brown or black paper.

9. Osmic Acid Vapour.—This is most useful for thin membranes and glands. Tissues stain readily after its use.

10. Chromo-Aceto-Osmic Acid (Flemming's Fluid).—To 45 cc. of 1 per cent. chromic acid add 12 cc. of 2 per cent. osmic acid, and then 3 cc. of glacial acetic acid. This can be kept for a long time, and need not be kept in the dark. It is specially useful for fixing the figures in cell-division or mitosis, and for many other purposes. It "fixes" tissues in from a few hours to twenty-four hours or longer; but the pieces must be small, 2-3 mm. thick, as it does not penetrate deeply. This is Flemming's "strong formula."

A weaker fluid is sometimes used, and is prepared as follows:—

Osmic acid (1 per cent.)	10 c.c.
Glacial acetic acid (1 per cent.)	10 ,,
Chromic acid (1 per cent.)	25 ,,
Water	55 ,,

They must be thoroughly washed by being kept in running water for twelve hours, and then hardened in the various strengths of alcohol, 70, 80, and 90 per cent., each for twenty-four hours. Sections should be stained as soon as possible after they are made, as on keeping they do not stain so well. They may be stained with safranin, hæmatoxylin, or methyl-violet, but safranin is the best (p. 75), it stains the chromatin of the cells a bright red. Tissues hardened in it are not well adapted for teasing.

11. Fol's Solution.—This is a modification of 10. It contains less osmic acid, and is used more generally. To 2 cc. of 1 per cent. osmic acid add 25 cc. of 1 per cent. chromic acid, five parts of 2 per cent. glacial acetic acid, and 68 cc. water.

Precautions with Osmic Acid or Solutions containing it.—If a tissue is to be hardened in a certain fluid, and to be treated with osmic acid afterwards, that tissue had better not be put into alcohol if it contains fat or fatty particles, as the alcohol dissolves the fat. Thus, a tissue hardened in Müller's fluid may be put into osmic acid, and the fat-cells will still be blackened.

It has recently been shown that the prolonged action of turpentine, toluol, xylol, ether, and creasote, but not clove-oil or chloroform, will decolorise particles of oil (fat) blackened by osmic acid. This is most important in connection with the osmic method for studying the absorption of fat in the small intestine (Lesson XXV.).

Tissues containing fat may therefore be embedded in paraffin after being passed through chloroform (p. 43).

Osmic acid fixes fresh tissues very rapidly, but it does not penetrate deeply; therefore the tissues must be cut into very small pieces to get them fixed throughout by the acid. It has the power of differentiating tissues, the nuclei become yellow, fat, and the nervous system black. After twenty-four hours or so, the tissues are thoroughly washed in water and hardened in 90 per cent. alcohol. To avoid the blackening which is apt to occur in tissues still containing a trace of osmic acid, it has been proposed to treat the tissues with a weak solution of cyanide of potassium, and then to harden in alcohol. It enters into the composition of several hardening fluids (p. 32), and **Hermann's fluid** (Lesson XXXV.).

D. Other Hardening Fluids.

1. Bichloride of Mercury or Corrosive Sublimate.—A saturated watery solution contains about 5 per cent. of the salt; but it is much more soluble in alcohol, especially alcohol of 50 to 60 per cent. Make a saturated watery solution, and also a saturated alcoholic solution. A saturated solution in 0.5 NaCl solution is also used.

A cold saturated solution is best made as follows:—Place 60 grams of it in 1000 cc. of water, and dissolve with the aid of heat. Filter the warm solution and allow it to cool. On cooling, long white needle-shaped crystals of the sublimate separate. The supernatant fluid is used, and the tissues, which must not be more than $\frac{1}{2}$ cm. in diameter, remain in the fluid from 1–3 hours, according to their size. After fixation they are hardened in 70, 80, and 90 per cent. alcohol. No metallic instruments are to be used. Use glass or wooden instruments.

This is a most excellent hardening reagent, and it hardens tissues with great rapidity, so that tissues must not be left in it for too long a time. For small pieces, a quarter of an hour or thereby is sufficient; for large pieces, one to two hours. The pieces when fixed become whitish throughout.

For *glands* and glandular structures generally, a half-saturated alcoholic solution is most useful, *i.e.*, to 50 cc. of a saturated alcoholic solution add 50 cc. of 70 per cent. alcohol. Vignal recommends that to 100 cc. of this mixture there be added 5 to 6 drops of nitric acid. The pieces of glands, 4 mm. cubes, are hardened in one hour or so. The hardening is completed in alcohol.

N.B.—All the corrosive sublimate must be washed out of the tissue—by alcohol, not water—otherwise the sections will be dotted with small black specks or star-like or needle-shaped crystals of the salt, or the salt may be removed by adding a few drops of an alcoholic solution of iodine to the alcohol used to complete the hardening, or before complete hardening in alcohol it may be steeped for 2–3 days in 70 per cent. alcohol, to which a few drops of tincture of iodine is added.

The action of the salt may be accelerated by placing the tissues in the fluid heated to 38° C. Sections stain readily with all the usual staining reagents.

N.B.—Do not use metallic instruments to transfer the tissues from one vessel to another. Use glass or wood or horn.

2. Other Fluids for special purposes are mentioned in the text.

General Directions on Hardening.

The tissues should be taken from the body as soon as possible after death, and transferred as soon as possible to the hardening fluid.

Any blood adhering to the parts may be removed by washing them in normal saline solution.

With a sharp razor cut the tissues in the same plane in which they are afterwards to be cut for sections. The tissue must be cut into *small* pieces, *i.e.*, $\frac{1}{4}$ to $\frac{3}{4}$ inch cubes, except in the case of tissues to be hardened in Müller's fluid. They will harden better if they are small, say 1 cm. square.

The best way is to suspend the tissues in the upper half of the fluid, which should always be many times the volume—15–20—of the tissue. If it is inconvenient to suspend them, cover the bottom of the jar or wide-mouthed bottle in which they are placed with an old washed rag or blotting-paper to prevent the tissue from resting directly on the glass.

The liquid must be changed within the first twenty-four hours, and again on the second day, then on the fourth, eighth, and twelfth

FIXING AND HARDENING REAGENTS.

day, and once or twice afterwards, and on each occasion alter the position of the tissues. If the fluid becomes turbid, change it at once (p. 30).

Label each bottle carefully, and place it in a cool place. Keep all chromic acid and solutions containing it or its salts in the dark.

Tissues hardened in alcohol and picric acid must not be placed in water, but directly into the various strengths of alcohol, beginning with 50 per cent. and rising to 95 per cent. Wash out by means of alcohol as much of the picric acid as possible.

For other tissues, hardened in chromium salts (p. 30), the excess of these salts may be removed from them by washing in water (for certain special purposes this is omitted), and they are then transferred first to weak spirit, in which they may remain a few days, and then to the stronger alcohols (p. 28). To avoid the deposits which occur in chromic acid preparations when they are placed in alcohol, *keep them in the dark.* If kept in the dark, as Hans Virchow has shown, there is no deposit formed when a tissue hardened in chromic acid or Müller's fluid is placed in spirit. Kept in this way, I find that the alcohol remains quite clear and no deposit forms.

Scheme for the Hardening of Tissues (Garbini).

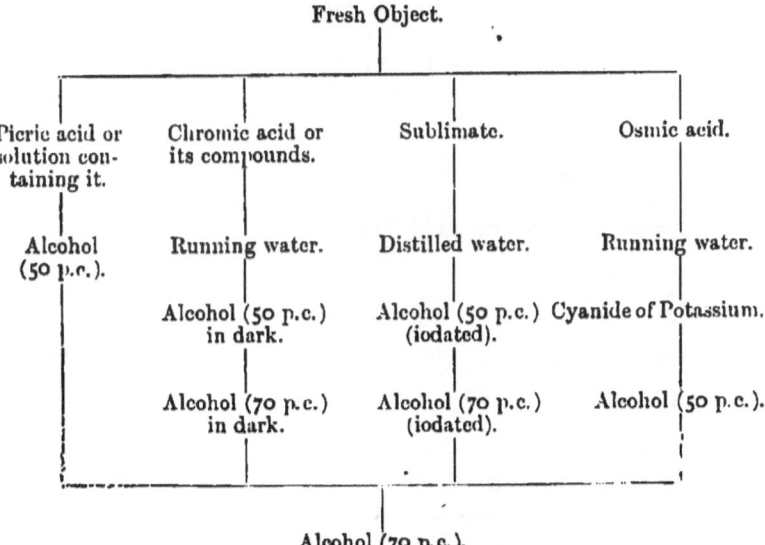

Garbini divides hardening reagents into the two following classes, according as the tissues—after hardening—stain readily or do not stain readily with carmine :—

Tissues stain readily after hardening in
- Vapour of osmic acid.
- Corrosive sublimate.
- Picric acid and solutions containing it.
- Nitric acid.
- Silver nitrate (weak solution).
- Alcohol.

Tissues stain with difficulty after hardening in
- Chromic acid and its compounds.
- Osmic acid in solution.
- Silver nitrate (strong solution).
- Chloride of gold.
- Perchloride of iron.

Alcohol, picric acid and its compounds, and nitric acid, coagulate the albumen of the tissues. Chromic acid, or solutions containing it, or its salts form chemical compounds with substances in the tissues. Osmic acid, mercuric chloride, and gold are decomposed when they come in contact with the tissues, and are deposited in the tissues as an inorganic precipitate.

VII.—DECALCIFYING FLUIDS.

General Directions.—Use small parts of the tissue or organ, and a large quantity of the decalcifying solution. Renew the latter often. In all cases harden the tissue thoroughly, *e.g.*, in alcohol or Müller's fluid, before it is decalcified. After decalcification every trace of the decalcifying fluid must be thoroughly removed by prolonged washing (for a day or two) in water or other fluid. The tissue is then finally hardened in alcohol.

1. **Chromic Acid.**—.1 to .5 per cent.
2. **Chromic and Nitric Acid Fluid.**

Chromic acid	1 gram.
Water	200 cc.
Nitric acid	2 ,,

3. **Chromic Acid and Hydrochloric Acid Fluid** (*Bayerl's Fluid*).

Chromic acid	1 part.
HCl	1 ,,
Water	100 parts.

This is specially good for young bones and for ossifying bone, but it is solely for decalcifying.

4. **Saturated Solution of Picric Acid.**—The solution must be

saturated and *kept saturated.* This is done by keeping some crystals of picric acid in the bottom of the bottle. There must be a large volume of fluid, and the bone should be suspended in the fluid. It usually requires a fortnight to decalcify a small bone. Picric acid acts as a fixing, hardening, and staining reagent.

After the bone is decalcified it should be washed and kept in spirit—not water—until no more yellow stain is given up to the alcohol. This rule, however, is not rigidly followed.

5. **v. Ebner's Fluid.**—This fluid prevents the ground substance of bone from swelling up. Sections should be examined in 10 per cent. solution of sodic chloride, if it be desired to see the fibrillar structure of bone, but sections of bone softened in this way may be mounted in other media, if it be desired to see the other details in the structure of bone. (See Bone.) It has the following composition:—

Alcohol	500 cc.
Water	100 ,,
Sodic chloride	2.5 grams.
Hydrochloric acid	2.5 cc.

6. **Picro-Sulphuric Acid or Picro-Nitric Acid** (p. 31).

7. **Arsenic Acid.**—I find that a 4 per cent. solution of this acid rapidly decalcifies a bone at 30° to 40° C. The tissues, after hardening in alcohol, are well preserved and stain readily.

8. **Chromo-Osmic Acid** (*Haug*).

Osmic acid (1 per cent.)	10 cc.
Chromic acid (1 per cent.)	25 ,,
Water	65 ,,

Very useful for delicate tissues, *e.g.*, very young developing teeth. Wash in water and harden in 70 per cent. alcohol.

9. **Phloroglucin Method.**—Dissolve with the aid of gentle heat 1 gram of phloroglucin in 10 cc. pure non-fuming nitric acid. To the red fluid add 100 cc. of 10 per cent. watery solution of nitric acid. The following mixture acts more slowly:—

Phloroglucin	1 gram.
Nitric acid	5 cc.
Alcohol	70 ,,
Water	30 ,,

Phloroglucin itself does not decalcify, it only protects the tissues from the action of the nitric acid. By this method decalcification can be done very rapidly, even in a few hours. I find that a portion of a human clavicle about an inch in length is softened in 12–16 hours, the outer surface being stained slightly red.

Methods.—The bone should be cut into short pieces and placed in a *large* volume of the fluid. If chromic acid be used, the bone must be first hardened in this fluid. Place it in .1 per cent. chromic acid for twenty-four hours; renew the fluid, but use .2 per cent., and after a week use .5 per cent.; shake the vessel from time to time, to bring new fluid into contact with the tissue.

If a more rapid process is desired, after the bone has been two or three days in dilute chromic acid (.2 per cent.), use the chromic and nitric acid fluid. Decalcification requires about fourteen days.

To test for the removal of all the salts, push a needle into the bone, or make a section with a blunt razor. Obstruction in either case denotes that the bone has not been sufficiently decalcified. In most cases, the bony tissues should be hardened before they are decalcified. This is specially the case in connection with bone softened in chromic acid. It is better to harden them first in Müller's fluid, and then to decalcify them in chromic and nitric acid fluid (p. 36).

Bone decalcified in chromic acid must be thoroughly washed in running water for many hours to remove all the chromic salts, and is then transferred to 70 per cent. spirit and *kept in the dark*, otherwise there will be a copious deposit. Renew the spirit, and transfer the tissue to strong alcohol, still keeping it *in the dark*.

If a bone is to be softened in picric acid, it may be placed at once in this fluid, with the precaution indicated at p. 37. It need not be kept in the dark, but it is better to remove as much as possible of the yellow stain by means of alcohol. It decalcifies somewhat more slowly than chromic acid.

VIII.—METHOD OF PREPARING TISSUES AND ORGANS FOR MICROSCOPICAL EXAMINATION.

As most of the tissues require to be hardened, and it is frequently impossible to obtain human tissues sufficiently fresh, recourse must be had to the fresh tissues of animals. As frequently as possible, however, human tissues should be secured. Most of the tissues may be obtained from a cat, rabbit, or guinea-pig, and for certain special purposes the dog, frog, newt, and salamander are used.

The cat, rabbit, or guinea-pig—or, better, all three—are killed by chloroform. The animals are placed in an air-tight box—a large saucepan does very well—along with a sponge saturated with chloroform. Small animals may be chloroformed under a bell-jar. As soon as the animal is dead, open the thorax by a longitudinal incision through the costal cartilages—right and left—raise the sternum, expose the pericardium, open it, and with a pair of scissors make a snip into the right auricle of the heart, and allow the animal to bleed freely.

It is best to begin by removing the brain and spinal cord. They are hardened in Müller's fluid or potassic bichromate (2 per cent.), and must be placed in a large volume of fluid. A few spinal ganglia should also be found and hardened in the same way.

Remove the trachea and lungs, and fill the lungs and trachea of the rabbit with a $\frac{1}{4}$ per cent. solution of chromic acid.

This is readily effected by tying a funnel into the trachea and pouring in the fluid. By squeezing the lungs gently much air is forced out, and the fluid gradually runs into and distends the lungs, which, when distended, are placed in a large volume of the same fluid. The chromic acid and spirit mixture may be used instead of pure chromic acid.

Fill the windpipe and lungs of the guinea-pig with $\frac{1}{4}$ per cent. silver nitrate. (See Lungs.)

Remove the heart, and harden it in alcohol, after washing away any blood with normal saline.

The central tendon of the diaphragm may be preserved for silvering. (Lesson IV.)

The omentum and mesentery, if desired, are silvered. (Lesson IV.)

Open the abdomen, remove the liver, cut it into small pieces; harden some pieces in Müller's fluid, and others in spirit.

Take out the tongue and œsophagus; harden them in Müller's fluid.

Open the stomach and intestine, wash away any food residues by means of normal saline. Harden part of the stomach—cardiac and pyloric—in absolute alcohol, other pieces in Müller's fluid, and others in corrosive sublimate, and small pieces in osmic acid. (See Lesson on Stomach and Intestine.)

The duodenum and small and large intestine are hardened in the same way, although the bichromate and chromic acid mixture (p. 29) is particularly good for the small intestine.

The salivary glands and pancreas are removed and hardened by the methods given under these headings, *i.e.*, small pieces are placed in each of the following solutions:—Absolute alcohol, osmic acid, corrosive sublimate, &c.

Remove the lower jaw, cut it into short pieces, place it in .2 per cent. chromic acid for a few days, and then decalcify it in chromic and nitric fluid. This will yield sections of softened tooth.

Remove the kidneys, cut one longitudinally and the other transversely. Using the kidneys of different animals, harden pieces of each in the following fluids:—Müller's fluid, chromic acid and spirit, ammonium chromate, and corrosive sublimate. Other methods of preparing the kidney are referred to. (Lesson on Kidney.)

The bladder is best hardened in chromic acid and spirit mixture, or in Müller's fluid.

Harden the spleen, without cutting into it, in Müller's fluid.

The suprarenals may be hardened in picro-sulphuric acid. (Lesson on Suprarenal Capsules.)

Small lymphatic glands from the region of the neck or submaxillary region are hardened in alcohol, while others are injected with silver nitrate and osmic acid. (Lesson on Lymphatics.)

If desired, the large nerve-trunks may be removed and hardened as indicated in Lesson on Nerves, or the smaller branches of nerves may be used for showing the effects of the action of certain reagents on nerve-fibres.

Remove some of the long bones, leaving in each case the periosteum attached to the bone. Cut the bones into pieces about $\frac{1}{2}$ inch long, and place them for a week in $\frac{1}{3}$ per cent. chromic acid, and then decalcify them with picric acid, or chromic and nitric acid fluid, or Ebner's fluid. (Lesson XIII.)

In every case decalcify the ends of the bones, so as to have a section which will demonstrate the relation between the articular cartilage and the osseous tissue.

Place small pieces of striped muscle in $\frac{1}{6}$ per cent. chromic acid, and other pieces in alcohol.

Nerves, with the precautions given in Lesson on Nerves, are hardened in osmic acid, potassic bichromate (2 per cent.), alcohol, or picric acid.

For the methods of hardening the eye, ear, nose, see the Lessons on these subjects.

The testis—very small pieces—is best hardened in Flemming's mixture, and larger pieces in Müller's fluid.

For the methods of hardening the ovary, Fallopian tube, and uterus, see the Lessons on these subjects.

N.B.—Label every bottle, and write on the bottle the name of the hardening fluid used, and the dates on which it was changed.

IX.—EMBEDDING.

This is necessary for many tissues; the piece of tissue may be either too small to be conveniently held in the hand, or its parts may tend to fall asunder before or after they are cut.

There are two methods, one **simple embedding**, where the tissue is simply fixed or placed in another medium to hold it while it is being cut, and the other **interstitial embedding**, where the substance used for the embedding process is made to penetrate into the interstices of the tissue.

A. **Simple Embedding.**—1. The tissue may be clamped between two pieces of **carrot**, scooped out to receive it, or in **elder pith**, or (what is very convenient) between two pieces of amyloid or **waxy liver** hardened in alcohol.

2. **Paraffin.**—It is sometimes desirable to surround the tissue with paraffin or some such medium. The embedding medium should be about the same degree of hardness as the tissue.

Two **paraffins** are required, a hard paraffin melting at 60° and a soft one at 45° C. For use they may be mixed as follows:—

I.	Hard paraffin	. . . .	1 part
	Soft paraffin	. . . .	1 ,,
II.	Hard paraffin	. . . .	1 ,,
	Soft paraffin	. . . .	2 parts.

Two parts of the hard paraffin and one of the soft yield a mixture which cuts well when the temperature of the room is 21° C. (70° F.), but a softer paraffin is easily made by mixing two parts of the hard paraffin with one part of chrisma or vaseline. The mixture can be made softer by the addition of a little more vaseline, and harder by adding more paraffin. The paraffin mixture is heated on a water-bath or sand-bath until it melts, but its temperature is raised as little as possible above its melting-point. It is convenient to melt it in a porcelain dish with a wooden handle. The tissue is removed from alcohol, the surplus alcohol removed by wiping it with blotting-paper, until the surface is dry. It is then placed in melted paraffin, and retained in it until the paraffin solidifies. The melted paraffin can be run into embedding boxes of paper (fig. 30), or the embedding L's may be used (p. 44), but this simple method is now but rarely used. It has been almost entirely displaced by the following method.

B. **Infiltration Method or Interstitial Embedding.**

1. **Embedding in Gum.**—The tissues after being hardened must have all their alcohol removed by prolonged soaking in water. They are then transferred to gum mucilage, or a mixture of gum and syrup, in which they can be preserved until they are required for freezing, if freezing is to be the process used for cutting the sections.

Tissues saturated with and embedded in gum mucilage may be hardened in alcohol and then cut. The sections are placed in water, which dissolves out the gum.

2. **Saturation with, or Infiltration with, and Embedding in Paraffin**—**Interstitial Embedding.**—In this case the embedding

medium is made to penetrate into the tissue, and when it sets, it thus supports all its component parts. This method is extremely valuable, especially for brittle and friable tissues, and is largely used. Moreover, the tissues once embedded can be kept in a box, each duly labelled, for any length of time.

Make a mixture of two parts of hard paraffin and one part of soft; place the mixture in a small copper pan or capsule in a hot-air oven, kept at a constant temperature by means of a gas regulator. The gas supply must be so arranged that the thermometer

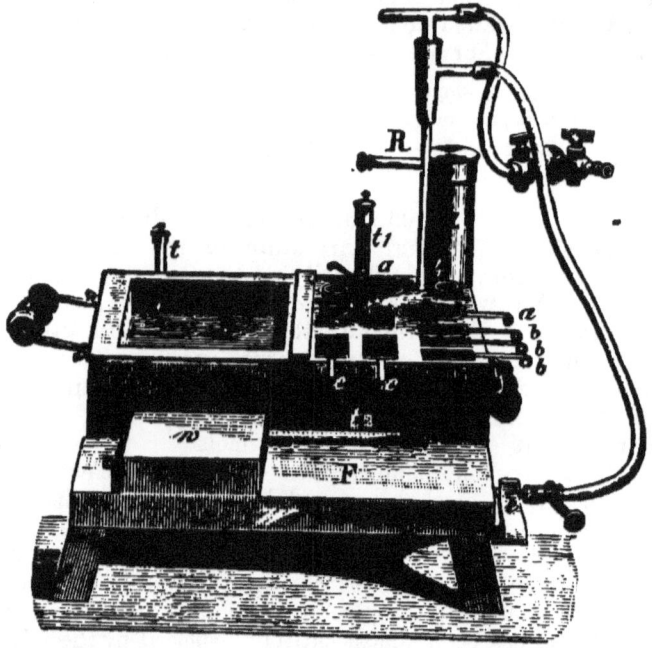

FIG. 29.—Mayer's Paraffin Embedding Bath, as made by Jung of Heidelberg.

steadily registers at most 1° C. above the melting-point of the paraffin. Or the paraffin may be melted and kept melted in a little copper vessel, placed in a hot-water bath, kept at a constant temperature, as shown in fig. 29. The temperature is kept constant by means of a gas regulator, R; Z is for filling the instrument with water; a, b, c, are embedding vessels and pots.

The tissues to be saturated and embedded should not be large, and they must be thoroughly **dehydrated**; keep them, therefore, several hours in absolute alcohol. Place them direct into turpentine, creosote, benzol, toluol, or xylol—some use chloroform, but turpentine, toluol, or xylol is to be preferred. The turpentine or xylol penetrates into the tissue, displaces the alcohol, and makes the tissue itself transparent. If the tissue be very small, this will be done in an hour or so; if it be larger, of course a longer time will be required. Thus the time may vary, according to the size and nature of the tissue, from one to six hours.

Transfer the specimens from the clarifying medium, and place them in the melted paraffin, where they may remain 2–10 hours, according to the size and nature of the tissue. By the end of that time they will be thoroughly impregnated or saturated with the paraffin.

For delicate tissues, however, it is better not to transfer them direct from the clarifying medium to pure paraffin, but to place them first of all in a mixture of toluol and paraffin, or turpentine and paraffin, for an hour or two. In the thermostat at 50°–55° C. the toluol gradually evaporates, so that nearly pure paraffin remains, and the saturation with paraffin has been accomplished more gradually. The object is then finally placed in pure paraffin for several hours.

Some prefer chloroform as a clarifying agent and as a solvent for paraffin instead of toluol or turpentine. Objects transferred from absolute alcohol at first float on the surface of the chloroform, but as the latter penetrates them and displaces the alcohol they sink. They are then embedded in a chloroform paraffin mixture and finally in pure paraffin.

The time (in hours) required for immersion in the several fluids for pieces of tissue of various sizes is approximately as follows (*Böhm and Oppel*):—

	Small Objects less than 1 mm. diam.	Objects about 5 mm. diam.	Larger Objects about 5 mm. diam.	Very large Objects.
Absolute alcohol	2	6	24	A longer time, but at the expense of their quality.
Toluol or turpentine	½	2–3	3–4	
Objects are now placed in the thermostat or bath.				
Toluol or turpentine-paraffin	1	4	6	
Paraffin	½	2	3–4	

Objects may remain longer in the paraffin without damage, provided they have been completely dehydrated.

The method of interstitial embedding is particularly useful for tissues stained "in bulk" or *en masse*. The tissues or organ may be stained before the process is begun, or sections may be cut and stained afterwards.

To Stain "in bulk" before embedding.—Pieces of the tissue a few millimetres square are placed in borax-carmine or Kleinenberg's logwood for 10–24 or 48 hours, according to the size and nature of the tissue. If they be placed in borax-carmine, the pieces are transferred to acid alcohol for 24 hours, and then transferred to various strengths of alcohol, and finally to absolute alcohol, by which they are completely dehydrated. The tissues stained with Kleinenberg's logwood are well washed in spirit, and transferred to absolute alcohol to be dehydrated. These dehydrated stained masses are then placed in turpentine or xylol, and then in melted paraffin, as described above.

Process of Embedding in Paraffin.—This is the same both for the simple and the saturation methods. Use embedding L's, which consist of two L-shaped pieces of lead about $\frac{1}{2}$ inch high, the long arm about 2 inches long and the short one $\frac{3}{4}$ inch. Their inner surfaces are moistened with glycerine, and the L's themselves are placed on a piece of glass, coated with glycerine, to enable the paraffin to separate easily, with the long limb of the one in contact with the short limb of the other, thus making a rectangular box, the size of which can be increased as required. Fill the trough with melted paraffin.

Take the tissue—if for **simple embedding**—direct from alcohol; dry its surfaces with bibulous paper, to remove any alcohol which would prevent the paraffin from adhering to it; pour the melted paraffin into the trough, transfix the tissue with a fine pin, plunge it into the paraffin just when the latter begins to set at the edges, move the tissue in the still fluid paraffin to one end of the trough, and hold it there until the mass sets around it.

If the tissue has been previously **saturated** with paraffin, the trough is filled as before, and the tissue, saturated with paraffin, is taken, by means of a hot needle, from the fluid warm paraffin, and fixed in the trough in the same way as described above. It is not always necessary to transfix the tissue with a pin or needle, but it is sometimes convenient to do so. Insert the needle in the direction of the cutting plane, thus indicating afterwards (when the mass is set) the direction in which the section is to be cut. A little paraffin may be poured into the trough, and when it just begins to set, the tissue is laid on it, and another layer of melted paraffin is poured over it as soon as its surfaces are set. Place the whole under the tap, and allow cold water to run over it to accelerate the cooling, as paraffin cooled rapidly is more homogeneous and cuts better than when it is cooled slowly. In about half-an-hour the paraffin has set, and can be removed from the mould.

Embedding Boxes (fig. 30) may be used. These are readily made from a rectangular piece of writing-paper folded to the size required. The paper is first folded along the lines aa' and bb', then along cc' and dd', always folding the paper towards the same side. The diagonals AA'-DD' are indented by means of the point of a lead pencil, or the paper is folded along these lines. These corners are then bent up between the fore-finger and thumb, and then bent round, so as to be applied to the sides AB and CD of the oblong, and are fixed there by turning down the flaps ff'.

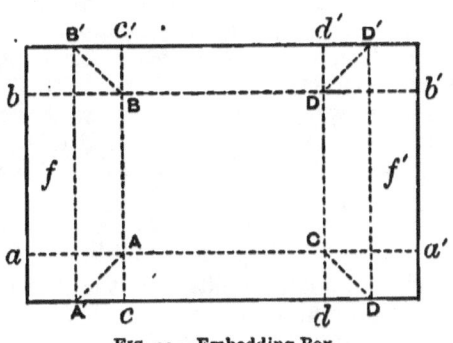

FIG. 30.—Embedding Box.

Embedding in Paraffin.—The following scheme shows the stages:—

(1.) Harden, either from the first or subsequent to other agents in absolute alcohol.
(2.) Xylol or turpentine (24 hours).
(3.) In warm paraffin fluid at 50° C. (1–12).
(4.) Embed and allow paraffin to cool.
(5.) Cut sections, and in some cases fix them on slide with a fixative (p. 60).
(6.) Remove paraffin by turpentine or xylol.
(7.) Alcohol to remove xylol.
(8.) Add water and then stain, &c.

3. Embedding in Celloidin.—This method is specially valuable where the parts of an organ when cut into sections are apt to fall asunder. It is specially valuable in such as those of the ovary, central nervous system, retina, &c.

Celloidin is a form of nitro-cellulose or pyroxylin, or solidified collodion, and is sold in two forms, one in tablets and the other in cuttings. E. Schering's is the best, and the form sold in "cuttings" is to be preferred. Do not let its solution dry, as it is then difficult to redissolve it.

This method was invented by Duval, who used collodion, and improved by Schiefferdecker. Prepare the solutions of celloidin by dissolving the latter in equal parts of absolute alcohol and ether. The first, or weaker solution, is made of a thin consistence like collodion duplex, the other is made stronger, until it has a thick syrupy consistence.

The hardened tissue is placed for some time in absolute alcohol, and then for several days, or until it is completely saturated, in a mixture of equal parts of absolute alcohol and ether. After this it is placed in a glass-stoppered bottle in the thin solution of celloidin until it is completely saturated with it (2–3 days). Transfer the tissue to syrupy celloidin, and let it remain there for several days until the tissue is thoroughly infiltrated with celloidin.

Some use three strengths of celloidin solution :—

(1.) A thick syrup consistence.
(2.) One part of (1) diluted with 2 vols. of ether.
(3.) One part of (2) diluted with 2 vols. of ether.

After thorough infiltration the tissue has to be embedded. For this purpose make a paper box (fig. 30); or use a pill-box, or embed it on a cork, thus. Take a cork corresponding to the size of the object, roughen one end of it, and surround it with a piece of paper fastened by a pin (fig. 31). Moisten the roughened surface with absolute alcohol, and on it place the tissue infiltrated with celloidin, and surround the latter with the thick solution of celloidin. Allow it to stand until the celloidin begins to harden on the surface. This takes place in less than an hour. Place the box or cork, as the case may be, in 80 per cent. alcohol for 24–48 hours, which hardens the celloidin to such a consistence that it can be cut like a stiff cheese, but the sections must be cut with a knife wetted with 70 per cent. alcohol. It is immaterial which microtome is used, as long as the knife is moistened with not too strong alcohol. The sections may be transferred to alcohol or water, and stained with any suitable dye. Some dyes stain the celloidin, especially the aniline dyes, and others do not.

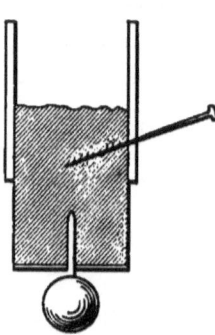

FIG. 31.—Embedding Paper Box with Weight for Celloidin with a lead "Sinker."

A section after being stained may be mounted in glycerine or balsam, but in the latter case absolute alcohol cannot be used to dehydrate them, as celloidin is soluble in this fluid. Alcohol (95 per cent.) must therefore be used for this purpose. Moreover, the sections must be clarified by origanum or bergamot oil—not by oil of cloves, which has a solvent action on the celloidin—and mounted in balsam. (See also Clarifying Reagents.)

Objects embedded in celloidin may be kept ready for cutting for an indefinite time in 75–80 per cent. alcohol.

In cutting sections embedded in celloidin, the knife must be so placed as to cut with as much of the blade as possible. The hardened celloidin is fixed to a piece of cork, which is clamped in the microtome.

If, however, it be desired to cut sections embedded in celloidin by the freezing method, the following procedure must be adopted. After the tissue embedded in celloidin has been hardened in alcohol, whereby it becomes not only "hardened" but somewhat milky in appearance, the alcohol must be got rid of, which is done by keeping it in running water for twenty-four hours, when it is transferred to a freezing mixture of gum and syrup (p. 49). This freezing fluid gradually penetrates the cheesy-like mass, and displaces the water. Such a preparation can be frozen in an ordinary microtome.

For **embedding and cutting in celloidin**, the following are the stages :—

(1.) Organ hardened, either at first or subsequently, in absolute alcohol.
(2.) Then in a mixture of equal parts of absolute alcohol and ether (1–2 days).
(3.) In dilute celloidin mixture (1–5 days).
(4.) In thick celloidin mixture (1–5 days).
(5.) Object placed on cork and exposed to air to dry.
(6.) Then in 80 per cent. spirit (1 day).
(7.) Cut sections.
(8.) Stain and wash them.
(9.) Dehydrate them in 96 per cent. alcohol (and perhaps 1–2 minutes in absolute alcohol).
(10.) Clarify with origanum oil, xylol, cedar or bergamot oil.
(11.) Mount in balsam.

Suppose a piece of the human spinal cord is to be embedded in celloidin. After being hardened properly it is transferred from absolute alcohol to ether, for not more than twenty-four hours. It is then placed for 6–8 days in solution of celloidin (3), in solution (2) four or five days and in solution (1) two or three days. It is then embedded in celloidin as described above, and when it begins to harden by evaporation of the ether and alcohol it is transferred to 80 per cent. alcohol for three or four days, which finally hardens it, but it must be completely covered with spirit. If it is to be preserved for some time before it is cut, keep it in 70 per cent. alcohol.

Sections of an organ embedded in celloidin may also be clarified by means of creosote, or a mixture of 1 part of creosote and 3 of xylol, which, however, must be quite water-free.

Scheme for Infiltration and Embedding (Böhm and Oppel).

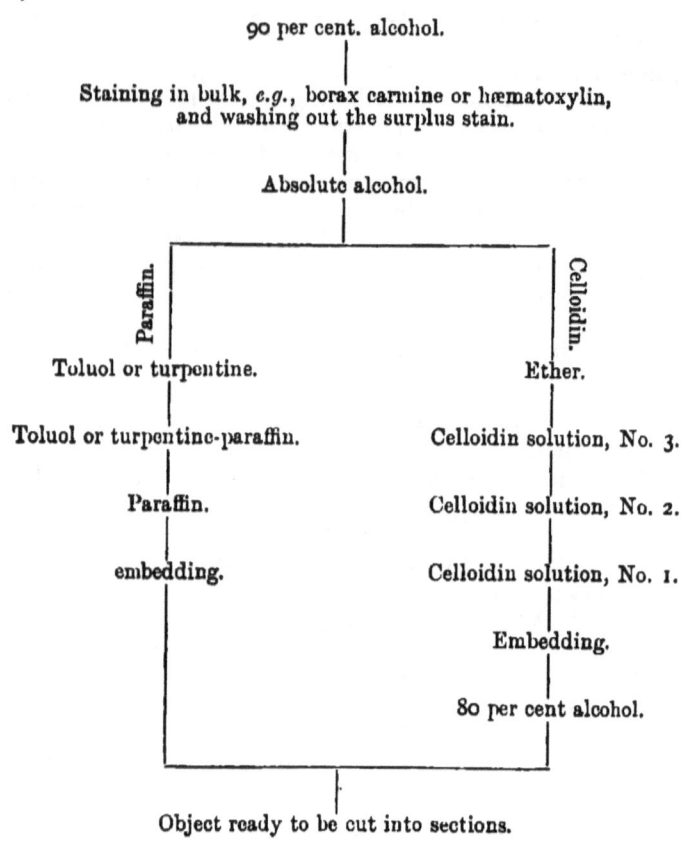

X.—SECTION CUTTING.

1. With a Razor.—It is absolutely essential to have a sharp razor if sections are to be cut by hand, and it is well that the student should practise this method.

Moisten the blade of the razor with 90 per cent. alcohol, and place some alcohol in an oblong glass vessel. Dip the cutting blade or razor into the alcohol after every third or fourth section. Grasp the razor tightly with the right hand, so that its blade is horizontal, its edge directed to the operator, whose fingers are pressed against the

SECTION CUTTING.

back of the blade, the back of the hand being directed upwards. The razor is made to glide through the tissue, cutting as thin sections as possible, which are placed in alcohol. The tissue to be cut may be embedded in any of the ways already described, and it is held in the left hand while being cut.

A razor ought to be sharp and free from notches in its cutting edge. A razor with notches in its edge causes striated bands in the section. Notches can readily be detected by drawing the razor across one's nail or over a piece of cardboard. In sharpening the razor use a soft hone moistened with water, then lay the razor flat on the hone and draw it diagonally from heel to point, with the edge forwards.

The razor must be frequently stropped to keep it sharp, and it should not be hollow-ground; it is better to be flat on one side.

2. **Valentin's Knife** (fig. 32) consists of two parallel blades, which can be placed at a greater or less distance from each other by

FIG. 32.—Valentin's Knife.

means of the screw (*a*). The blades are first set apart at the required distance; the thickness of the section depends on the distance between them. The knife was formerly much used by pathologists for making sections of fresh organs. It is now rarely used.

3. **Microtomes.**—For many purposes some form of freezing microtome will be found most convenient for teaching purposes; where a large number of sections is required it is indispensable.

Preparation of Hardened Tissues for Cutting by Freezing.—In order to secure the full advantages of cutting sections by freezing, the tissues must be previously soaked in and saturated with proper "freezing fluids." If the tissues be kept in alcohol, first remove all the alcohol by soaking them for twenty-four hours in running water. After this the tissue is soaked in gum mucilage, or, what is preferable, a mixture of gum and syrup. The best receipt is that of Hamilton.

Make a syrup of 28.5 grams of pure cane-sugar in 30 cc. of water; boil and saturate it with boracic acid. Allow it to cool and filter. Place 45.6 grams of gum acacia in 2400 cc. of cold water, allow it to dissolve, saturate it with boracic acid by boiling, and filter.

Freezing Fluid.

Take of the syrup	.	.	.	4 parts.
,, ,, mucilage	.	.	.	5 ,,
,, ,, water	.	.	.	9 ,,

Boil and saturate while hot with boracic acid. Filter through muslin.

The tissues are soaked for twenty-four hours or longer in this fluid, *i.e.*, after removal of all alcohol from them. The longer they are kept in it the better; in fact, tissues may be kept permanently ready for freezing in this fluid.

Rutherford's Ice-Freezing Microtome (fig. 33).—By means of a finely-graduated screw a brass plug can be raised or lowered inside

FIG. 33.—Rutherford's Freezing Microtome adapted for Freezing with Ether.

a brass cylinder. At the top of this cylinder is a stage or plate (B). The plug has a small flattened brass knob screwed into it, so as to catch the frozen mass, and prevent it from being detached (K). Practically this is the arrangement of the microtome devised by the late A. B. Stirling, curator of the Anatomical Museum of the University of Edinburgh. To this arrangement Professor Rutherford added an ice-box (C), which surrounds the upper part of the

SECTION CUTTING.

brass cylinder or well. This box is provided with an exit-tube (H), to allow the water resulting from the melting of the ice to escape. The size of the cylinder varies from 1-2 inches in diameter, but for most purposes one with a diameter of 1 inch will be found sufficient. The ice-box is covered on the outside with a thick layer of gutta-percha. Professor Hamilton, of Aberdeen, first suggested the addition of a glass top to be screwed upon the plate of the instrument.

By far the most convenient cutting tool for use with this microtome is an ordinary planing-iron fitted with a handle, as recommended by Delépine (fig. 35).

Above all, the tissue must have been properly hardened, and previously steeped in a freezing fluid, either gum mucilage or gum and syrup, after removal of all alcohol from it (p. 50).

In using this instrument, screw the plug down to the necessary depth, thus making a well of the required depth—at least half the depth of the cylinder—and into the well drop a few drops of glycerine, or put a little lard round the line of contact of the plug and the cylinder. This is to prevent any of the fluid passing down between the plug and the cylinder.

Fill the ice-box with a mixture of pounded ice and salt, and pack it well around the central brass cylinder. Keep a cork in the exit tube H, and only allow the fluid to flow away when it accumulates in large amount. In a short time the temperature of the plug is greatly reduced. Pour into the well a little mucilage (BP), sufficient to form a layer about $\frac{1}{8}$ inch thick, and allow this to freeze. The piece of tissue taken from the freezing mixture is lifted with a pair of forceps, and put into the well, so that it touches and adheres to that part of the well farthest away from the operator.

When the tissue is fixed, fill up the well with mucilage and cover it with a piece of sheet india-rubber, and keep the latter in position by a weight. This is to prevent the entrance of the freezing mixture into the well.

Supposing the tissue to be frozen, the operator seizes the elevating screw P with his left hand, and in his right holds the planing iron, which is fixed in a wooden handle. With the left hand the operator turns the screw, *i.e.*, elevates the tissue, while as rapidly as he chooses with his right hand the planing iron, firmly pressed on the glass plate at (about) an angle of 45°, is pushed rapidly forwards and drawn backwards, and in a few seconds twenty or thirty sections accumulate on the upper surface of the planing iron. By means of a large camel's-hair brush they are transferred to a large quantity of water, whereby the gum contained within them is dissolved and the sections themselves uncurl. One cannot cut all tissues with equal rapidity. In the case of many tissues and organs (as elastic tissue, lung, kidney, &c.), after they are completely frozen, a quarter of an hour will suffice, if a planing iron be employed, to cut more than a thousand sections. This, however, cannot be done with sections of the cerebrum, cerebellum, or spinal cord. With these and with some other organs it is better to cut each section singly.

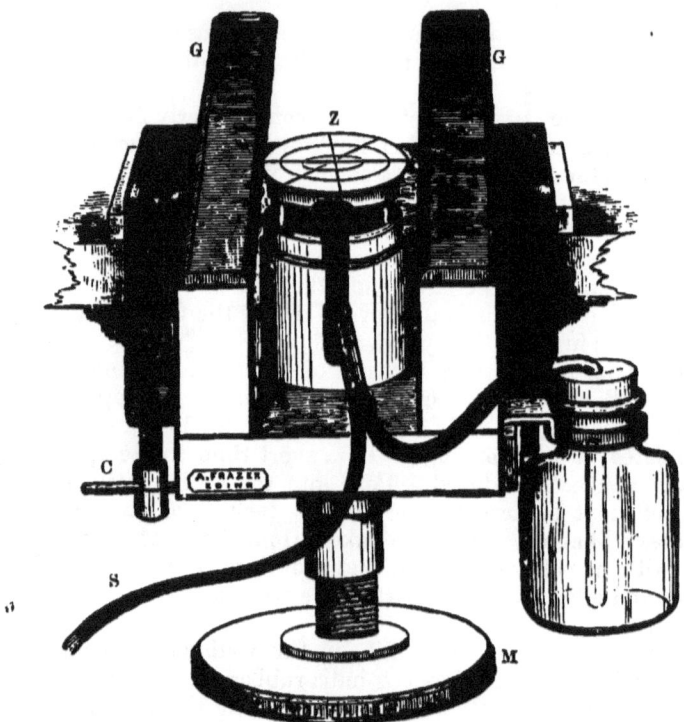

FIG. 34.—Cathcart's Freezing Microtome.

If one has several tissues to cut, the one tissue can be embedded above the other in the instrument at the same time.

This instrument, however, has other advantages, as it can be used also as an ether-freezing microtome. Place the tissue, previously saturated with the freezing fluid, upon the zinc plate Z, and cover it with mucilage. By means of an ether-spray producer N, direct a spray of anhydrous ether from the bottle O against the under surface of the zinc plate Z.

In order to economise ether, any non-volatilised is collected and returned by the tube T to the bottle O.

SECTION CUTTING.

In using the instrument in this way, however, the piece of tissue must be thin, not more than 5-7 mm. in thickness, or thereby.

Cathcart's Freezing Microtome (fig. 34).—The instrument is screwed to a table by means of a clamp (C). Fix by means of gum the tissue to be frozen, not more than 1 cm. thick, upon the zinc plate. Fill the bottle with ether and put the spray apparatus under the zinc plate. Work the spray apparatus until the mucilage and tissue are frozen.

Sections are cut by means of a flat knife pushed along on the glass supports (GG), the tissue being raised by the large milled head placed underneath (M).

The sections are removed from the knife by means of a camel's-hair pencil, and placed in water.

Fig. 35 shows Cathcart's microtome clamped to a table, and the method of cutting sections by means of a planing iron. The planing iron is used in exactly the same way as for Rutherford's microtome.

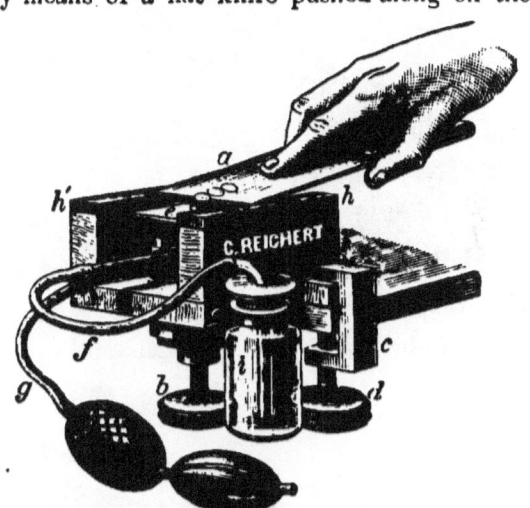

FIG. 35.—Showing how to use a Planing Iron for Cutting Sections.

Roy's Microtome (Freezing), Modified by Malassez (fig. 36).— This instrument is extremely convenient, the cutting instrument being an ordinary razor. Instead of ether, Malassez uses chloride of methyl, which is preserved in a stout metallic flask. A stream of the methyl chloride is directed against the under surface of the plate on which the tissue to be frozen is placed. It freezes much more rapidly than with ether.

Cutting a Continuous Series of Sections in Paraffin.—For this purpose the "rocking microtome" (fig. 37) of the Cambridge Scientific Instrument Company, or Minot's microtome, is most useful.

Embed the tissue in paraffin in the usual way; place the razor in position, and fix the embedded tissue to the end of the brass cap on the horizontal bar. Move the brass cap, with its adherent embedded tissue, forwards, until it touches the knife-edge. The horizontal bar has an axle which moves in a V-shaped pivot. At

Fig. 36.—Malassez's Modification of Roy's Microtome for Freezing.

the end of the instrument is a knobbed bar, which depresses and

Fig. 37.—Cambridge Rocking Microtome.

raises the horizontal bar, and at the same time moves a toothed

SECTION CUTTING. 55

wheel, which is pushed round by a small catch, whereby the bar bearing the pivot is raised. At the same time the embedded tissue in paraffin on the horizontal bar is brought down into contact with the edge of the razor.

The piece of paraffin should have its sides squared, and the two faces looking upwards and downwards should be coated with soft paraffin, *i.e.*, with a low melting-point (48° C.). This is to enable the one section to adhere to the other. On working the instrument, the sections come off in "chains" or ribbons, and can be caught upon a plate of glass. If the sections tend to curl up, they may be "flattened" by being placed in not too warm water.

Minot's Microtome.—In this microtome, which is one of the best microtomes yet invented for embedded tissues, as shown in fig. 38, the knife is fixed while the embedded tissue—in paraffin—fixed

FIG. 38.—Minot's Microtome.

to a circular disc, is moved vertically upwards and downwards by means of a wheel. The embedded tissue is fixed to a disc which can be moved around three axes, and thus the tissue can be cut in any desired plane. The thickness of the sections is regulated by a special toothed-wheel mechanism, which is so arranged that sections can be cut varying from $\frac{1}{300}$, $\frac{1}{150}$, $\frac{1}{100}$, $\frac{1}{75}$, $\frac{1}{60}$, to $\frac{1}{50}$ mm. Instead of the knife supplied with the apparatus, a razor may be used.

The sections can be received on a silk ribbon, as shown in fig. 39. This is clamped to the apparatus. The sections are received on the ribbon, which is rotated by means of a milled head.

Jung's or Thoma's Microtome.—In this instrument the tissue to be cut is fixed in a clamp, the knife is fixed in another heavy clamp which moves on planed surfaces. After each section the tissue is pushed up an inclined plane by means of the milled head on the extreme right of the figure (fig. 40).

Malassez's Modification of Roy's Microtome (fig. 41).—Sometimes it is desired to cut sections of a tissue while it is under fluid, e.g., alcohol. This can be done as shown in fig. 41.

FIG. 39.—Silk Band for Catching the Chain of Serial Sections made by Minot's Microtome.

The microtome is made to move on its base, and can be placed

FIG. 40.—Thoma's Sledge Microtome, as made by Jung.

vertically in such a way that the razor and the piece of tissue to be cut come to lie in a vessel filled with alcohol.

SECTION CUTTING.

FIG. 41.—Malassez' Modification of Roy's Microtome for Cutting Sections under a Fluid.

FIG. 42.—Williams' Ice-Freezing Microtome.

58   PRACTICAL HISTOLOGY.

Williams' Freezing Microtome (fig. 42).—This consists of a wooden, non-conducting tub for holding the freezing mixture. Vertically in the centre of this rises a brass cylinder, into whose upper end the roughened brass plate on which the tissue is frozen can be screwed. The lid of the box is formed by a glass plate fitted

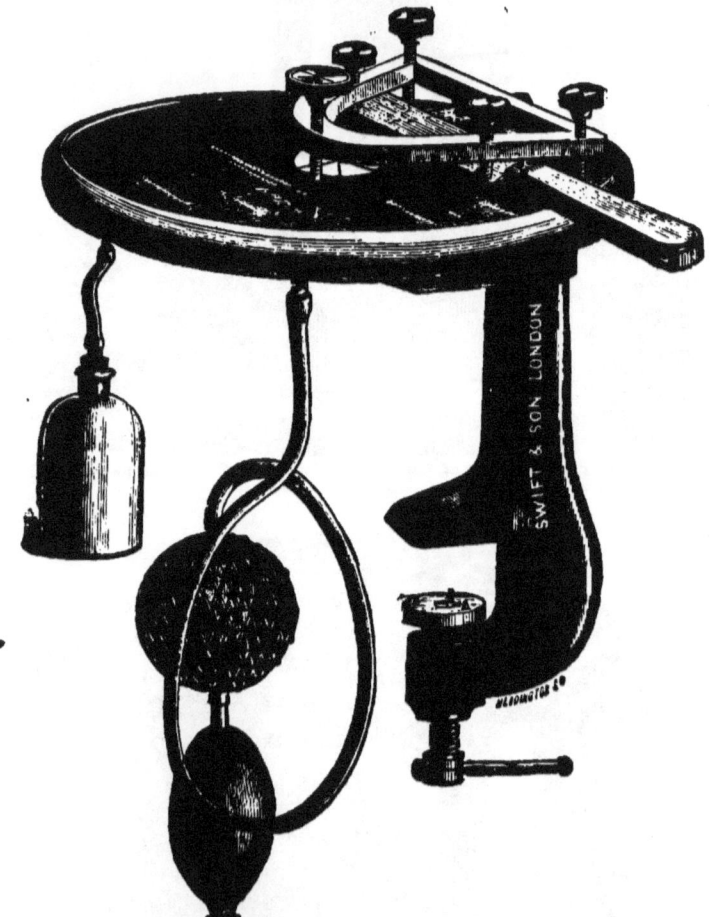

FIG. 43.—Swift's Ether-Freezing Microtome.

into a framework or kind of cap for the tub. In the centre of the glass plate is a circular hole into which the freezing-plate projects. In this instrument the tissue remains fixed, while the knife or blade is depressed by the movement of a screw. The knife or razor is fixed in a brass tripod frame-work or knife-carrier.

CUTTING SECTIONS.

In using the instrument, fill the tube with the freezing mixture of ice and salt, and when the central brass pillar becomes sufficiently cooled, pour on it a little mucilage, and when this is frozen place on it the tissue, which must not be more than 1 cm. in thickness, and pencil some mucilage on it. When it is solid, the knife in the tripod is used to cut the sections; and as the front leg of the tripod consists of a screw, this is turned, and thus the cutting edge is brought to touch the tissue.

Swift's Freezing Microtome (fig. 43).—A modification of the previous instrument is shown in fig. 43, which is adapted for freezing with ether.

Hand-Microtome (fig. 44).—When only a few sections are required, this instrument, invented by Ranvier, is extremely convenient. The

FIG. 44.—Ranvier's Hand-Microtome.

tissue is embedded in the well of the instrument in paraffin or elder pith, and sections made by means of a razor, as shown in the figure. The tissue is gradually raised by means of the milled head.

Section-Flatteners.—Sometimes the sections saturated with paraffin when cut exhibit a great tendency to curl up. This can partly be avoided by pressing the section as it is cut gently against the knife, by means of a camel's-hair brush. Several section-flatteners attached to the cutting-knife have been used for this purpose. Take a wire 1 mm. in diameter, heat it in a flame, and bend it twice at right angles, the distance between the angles being about an inch. The free ends are then bent round in the form of a hook. These hooks serve to fix the frame on the back of the razor, forming, as it were, a spring-clip. The part of the wire between the right angles is so arranged that it lies parallel to, and about one-hundredth of an inch from, the edge of the knife. In cutting, the section has to pass between the spring and the knife, and is thus largely prevented from curling up.

Curled-up paraffin sections may be made to uncurl by being placed in water at about 40° C. (Gaskell).

Cutting a Continuous Series of Sections in Celloidin.—The tissue, embedded in celloidin, is clamped in a microtome, *e.g.*, that of Jung, and section after section is made. The knife must pass at an acute angle through the celloidin, and must be moistened with 80 per cent. alcohol. This is easily effected from a wash-bottle. Schanze of Leipzig supplies such a bottle provided with a valve, which facilitates the outflow of a gentle stream of alcohol upon the cutting blade. Weigert's method of arranging and fixing the sections on a slide is the best. Each section in celloidin as it is made is laid upon a narrow strip of curl-paper by means of a camel's-hair pencil. The curl-paper is kept moist by being placed on a plate covered with blotting-paper well moistened with 80 per cent. alcohol. The sections are laid upon the curl-paper in the order desired.

A slide is coated with a layer of *thin* collodion, and when it is dry, the celloidin sections on the curl-paper are transferred to it. This is done by lifting up the curl-paper, and placing it, sections lowermost, upon the coating of collodion on the slide. Press on the whole with a piece of dry blotting-paper. The sections adhere to the slide, and the curl-paper is removed. Dry the sections with blotting-paper, and pour over them a layer of thin collodion. They are now permanently fixed, and can be stained on the slide in any way that may be desired. This is an extremely convenient method for serial sections of the central nervous system.

XI.—FIXATIVES AND SUBSEQUENT TREATMENT OF SECTIONS.

Further Treatment of Sections.

This depends on how the sections have been made, and whether they have or have not been previously stained. Paraffin sections must be freed from paraffin. If they are unstained, they must be stained. In most cases it is found advantageous to fix paraffin sections to the slide by means of a "**fixative.**" Many sections can thus be fixed on one slide, and treated simultaneously. The series of events will then be for **unstained paraffin sections** :—

    (1.) Fixation on a slide.
    (2.) Removal of paraffin.
    (3.) Staining the section.
    (4.) Mounting the specimen.

FIXATIVES AND SUBSEQUENT TREATMENT OF SECTIONS. 61

To Fix Paraffin Sections on a Slide. There are several "**fixatives**" for serial sections, but the following will be found the most useful.

(1.) **Collodion and Clove-Oil.**—Mix one part of collodion with three parts of clove-oil. By means of a brush paint a thin layer on a slide, and on it place the sections. Heat gently over the flame of a lamp, to fix them firmly and drive off the clove-oil.

(2.) **Albumen and Glycerine** (*P. Mayer*).—Mix filtered fresh white of egg with an equal volume of glycerine, add a little carbolic acid or morsel of thymol to prevent putrefaction. White of egg filters very slowly. A very thin layer is painted on the slide, and made smooth by means of a clean glass rod, which is thus prepared to receive the sections. The sections are flattened on the albuminised surface by means of a fine brush, care being taken that no air-bubbles remain under the sections. Warm the slide to a temperature just sufficient to coagulate the albumen (70° C.). This may also be done by holding the slide for a few seconds over a jet of steam.

Such substances as acids and alkalies which dissolve the albumen must not be applied to the sections, nor must the sections be stained with such substances as picrocarmine, which also dissolve the albumen.

(3.) **Method of Gaule.**—This method depends on capillary attraction. The slide is moistened with water or weak spirit, and on this the paraffin sections are carefully spread out. The surplus spirit or water is removed by blotting-paper, and the slide placed in a thermostat at 50° C. for twenty-four hours. Sections so dried are heated for a moment above the melting-point of the paraffin, and are then firmly fixed on the slide.

To Remove Paraffin from the Sections.—Sections of tissues soaked and embedded in paraffin and fixed on a slide are placed in turpentine, toluol, or xylol. The extraction of the paraffin requires some time, and takes place more rapidly when the temperature is raised. The slides may be fitted into a zinc framework and lowered into a bath of turpentine or toluol. Clove-oil must not be used if collodion and clove-oil have been used as a fixative. In that case clarify with creosote and turpentine. The turpentine dissolves out the paraffin.

After this, if the tissue has been previously stained in bulk, before it was embedded, drive away the turpentine with xylol or clove-oil, and mount the section in balsam.

If, however, the sections are from an unstained tissue, after dissolving out the paraffin with turpentine, the latter must be displaced by absolute alcohol, and the slides are passed through alcohols of various strengths and then into water. *i.e.*, provided the sections are to be stained in a watery solution of a dye. The sections are then coloured *in situ* on the slide. If the sections are to be mounted in balsam, they must go through the same process in the reverse order, viz., increasing strengths of alcohol—a clarifying agent, clove-oil, or xylol—and finally they are mounted in balsam.

Scheme for the Further Treatment of Paraffin Sections.

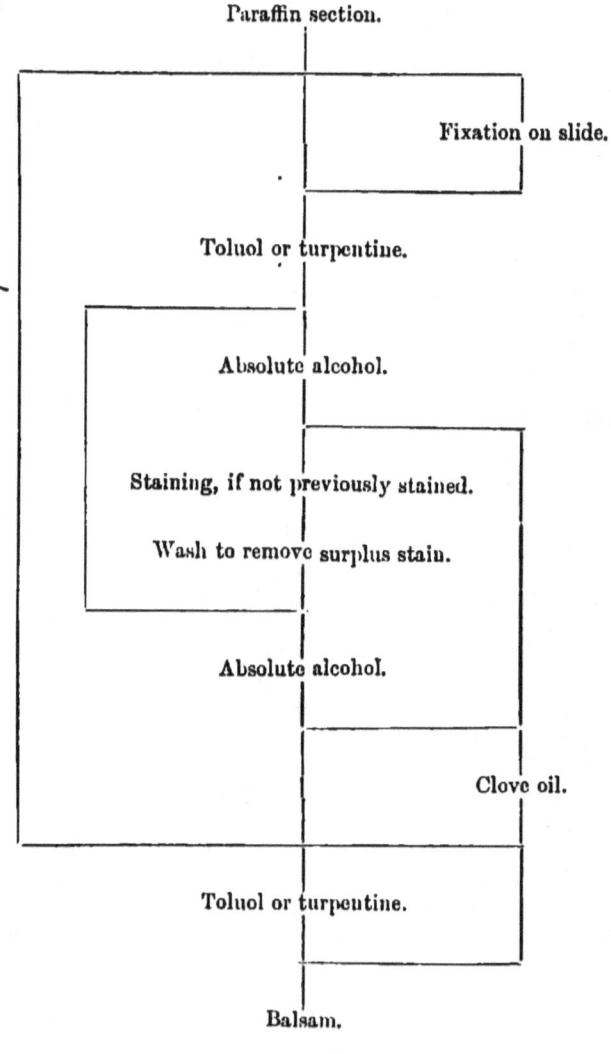

XII.—STAINING REAGENTS.

Staining.—This process depends on the fact that different tissues, or different parts of the same tissue, have an affinity for certain dyes, and not for others. Thus some dyes stain only the nuclei, others however may cause a uniform stain, all the tissues being of the same colour. By using some decolorising reagent, it is possible to remove the stain from certain parts of the preparation, leaving other parts stained.

A thin section of a tissue or an organ, as a rule, when examined shows but little differentiation of its several parts. Only in cases where pigment is naturally present is this difference very marked. Some substances when applied to the section stain one part and leave other parts unaffected, thus enabling one to differentiate more easily the several parts of a section.

Those substances which stain the nuclei chiefly have been called **nuclear stains**. The section is placed in a weak solution of the dye, *e.g.*, hæmatoxylin; and after it seems to be sufficiently stained, the surplus dye is removed by thoroughly washing the section in water or alcohol, a part of the dye remaining united with the chromatin of the nucleus and colouring the latter. Such stains may also colour to a less degree some other parts of the section. Amongst nuclear stains are carmine, hæmatoxylin, and some of the aniline colours.

When a section is stained, it is called **Section Staining**, but the tissue may be stained in bulk before the sections are made (p. 44) **—staining in bulk.**

A. Carmine and its Compounds.

Carmine.—In order to obtain a strong solution of this dye, certain solvents require to be employed. It is readily soluble in ammonia, yielding an ammoniacal solution, which may be made strong or weak. The ammoniacal solution may be diluted to any extent required with water, and practically the best results are obtained by allowing sections to remain for a long time (24–48 hours) in a weak solution.

1. Strong Ammoniacal Carmine Solution.—Rub up in a mortar 2 grams of *pure* carmine with a few drops of water, add 5 cc. of strong liquor ammoniæ, mix thoroughly, and add 100 cc. of water. Place the whole in a bottle, and after a day or so any undissolved carmine is filtered off and the clear fluid kept as a stock solution. This solution may be diluted to any required extent. If it smell strongly of ammonia, the excess of ammonia must be allowed to evaporate. When the solution becomes neutral it is very liable to undergo putrefaction, but this may be avoided by placing a small piece of thymol in it to preserve it.

2. Frey's Carmine.—Ordinary carmine has two drawbacks: it is apt to undergo putrefaction, and as the ammonia escapes the carmine is precipitated.

Carmine	0.3 gram.
Distilled water	30 cc.

Dissolve the carmine in the water, adding ammonia drop by drop until solution is complete. Then add—

Glycerine	30 cc.
Alcohol	4 ,,

and shake the mixture. Keep it in a stoppered bottle. It has no advantage, as far as coloration is concerned, over ordinary carmine, but it can be kept for a long time unchanged.

3. Alcoholic Borax Carmine (*Grenacher*).

Carmine	3 grams.
Borax	4 ,,
Water	100 cc.

Dissolve the borax in the water and add the carmine, which is quickly dissolved, especially with the aid of gentle heat. Add 100 cc. of 75 per cent. alcohol, and filter.

4. Watery Borax-Carmine.—Rub up in a mortar 8 grams borax with 2 grams carmine, and add 150 cc. water. After twenty-four hours decant and filter. Tissues to be stained in bulk—*e.g.*, after hardening with corrosive sublimate—are placed for twenty-four hours or longer in this fluid. They are then transferred to acid alcohol (1 per cent. HCl in 70 per cent. spirit for twenty-four hours), and then into alcohol.

If it be desired to stain in bulk without bringing the tissue into contact with water, then use :—

5. Carmine-Solution (*P. Mayer*).—For staining in bulk, and also for sections, Mayer recommends the following :—Suspend 4 grams carmine in 15 cc. water, and then add 30 drops hydrochloric acid, gently heating the mixture. Add 95 cc. alcohol (85 per cent.) and boil. Neutralise with ammonia and on cooling filter.

6. Borax Carmine (*Grenacher*).

Carmine	1 gram.
Borax	2 grams.
Distilled water	200 cc.

The borax dissolves the carmine. The whole is placed in a porcelain capsule and heated to boiling, when the fluid becomes of a dark-purplish or bluish-red. Add a few drops of 5 per cent. acetic acid, until the colour becomes more like that of carmine dissolved in ammonia. Let it stand for twenty-four hours, and filter. Add a drop or two of carbolic acid to preserve it. This gives a diffuse stain, so that the sections have to be treated with acid alcohol (p. 65).[1]

The original receipt is

Carmine	.5 to .75 gram.
Borax	2 ,,
Water	100 cc.
Alcohol (70 per cent.)	100 ,,

Borax-carmine is chiefly used for staining tissues "in bulk." Small pieces of tissue, to ½-inch cubes or larger, may be left in it for days, and they do not become over-stained. It gives by itself a diffuse stain; hence to get its effect concentrated upon the nuclei, for which it has a special affinity, the pieces of tissue must be placed for twenty-four hours or thereby in 70 per cent. alcohol containing 1 per cent. of hydrochloric acid.

Acid Alcohol.—This is called **acid alcohol**, and is prepared thus—

Hydrochloric acid	1 cc.
Alcohol	70 ,,
Water	30 ,,

When tissues are placed in the acid alcohol, they change in colour to a bright scarlet. A certain amount of the surplus carmine is extracted, but the nuclei become intensely stained.

This method is particularly valuable for a large number of organs, and especially where nuclear staining is desired.

7. Alum Carmine.—Dissolve 5 grams of potash-alum in 100 cc. water. Add 1 gram carmine, and boil for a quarter of an hour. Make up the bulk with water and filter. Add a drop of carbolic acid to preserve it, as fungi rapidly form in it. It has the advantage of not over-staining tissues left in it for a long time.

8. Lithium Carmine (*Orth*).

Carmine	2.5 grams.
Saturated solution of lithium carbonate	100 cc.

Dissolve the carmine in the cold saturated solution of lithium carbonate; solution occurs very quickly. It gives a diffuse stain, to nearly all tissues very rapidly, and the sections must, therefore, be transferred, without previous washing in water, to acid alcohol (p. 65). They can then be mounted in glycerine or balsam as desired. The nuclei are stained a brilliant red. It cannot be used for sections fixed on a slide by means of white of egg.

Application.

(1.) Stain (2–3 minutes).
(2.) Wash out surplus dye in acid alcohol (½–1 minute). *i.e.*, in 100 cc. of 70 per cent. spirit + 1 cc. HCl.
(3.) Remove all acid by prolonged washing in water.
(4.) Alcohol, oil, balsam.

[1] *Archiv f. Mik. Anat.*, vol. xvi. p. 363.

9. Picro-Lithium Carmine.—This is even preferable to the foregoing, because, in addition to staining nuclei red, it stains certain other parts yellow.

Lithium carmine	50 cc.
Saturated solution of picric acid . . .	100 ,,

Mix the two slowly. If, after trying it, one or other colour is too pronounced, add a little more of the other.

The sections are to be treated with acid alcohol like the foregoing. The acid alcohol, however, ultimately extracts the picric acid. This is avoided by not leaving them too long in acid alcohol. If the picric stain be removed, it may be restored at once by dipping the section in absolute alcohol to which a little picric acid has been added.

10. Picro-Carmine.—This most valuable reagent was introduced by Ranvier, and has the great advantage of giving a double stain without the use of acid or alkali.

(*a.*) *Ranvier's method* of preparing it is as follows:—To a saturated watery solution of picric acid add a saturated ammoniacal solution of carmine until precipitation just appears, *i.e.*, until saturation. The fluids must be well mixed. Leave it exposed in shallow vessels to crystallise, but protect it from the dust. Crystals are deposited, and also some amorphous carmine. After several weeks, when its bulk is reduced to one-third, decant the liquid, filter, and evaporate it to dryness on a water-bath. Redissolve it and the crystalline deposit in water, filter, and evaporate to dryness. The brown powder so obtained is dissolved in the proportion of 1 per cent. in water. This fluid, prepared in this way, gives very satisfactory results.

(*b.*) *Stöhr's Method.*—A very good solution is obtained by this method. To 50 cc. water add 5 cc. liquor ammoniæ and 1 gram carmine, which is rapidly dissolved. After complete solution, add 50 cc. of a saturated solution of picric acid. Set the mixture aside for two or three days in a large open flat evaporating dish, and after this time filter. To the filtrate add a drop of chloroform to prevent the formation of fungi.

Precautions.—Preparations stained with picro-carmine are not liable to be over-stained, and may be mounted in Farrant's solution or glycerine acidulated with formic acid (1 per cent.) [formic acid sp. gr. 1.16].

In staining sections with picro-carmine, cover the section with picro-carmine, and after a few minutes remove the surplus pigment. *On no account should the section be placed in water.* Water rapidly extracts the picric acid and leaves the preparation stained with the carmine only. Add a drop of Farrant's solution or formic glycerine, and cover. Such sections improve with keeping, and the surplus picro-carmine is really an advantage, for, after a time, the nuclei become more differentiated—red, and other parts yellow.

If it be desired to mount sections stained with picro-carmine in balsam, the alcohol through which they are passed, or the clove-oil, must contain some picric acid to restore the yellow colour.

11. Carmine and Dahlia Fluid (*Westphal*).—Dissolve 1 gram of carmine in 2.5 per cent. of alum. Take of this—

Alum carmine	100 cc.
Glycerine	100 ,,
Sat. sol. of dahlia in absolute alcohol	100 ,,
Acetic acid	20 ,,

This fluid is specially useful for staining the granular cells ("Mastzellen") of the liver.

12. Cochineal (*Csokor*).

Powdered cochineal	50 grams.
Alum	5 ,,
Water	500 cc.

Dissolve the alum in the water, add the cochineal (*Coccus cacti*), and boil. Evaporate down to two-thirds of its original volume. Filter. Add a few drops of carbolic acid to prevent the formation of fungi. This is an excellent nuclear stain, especially for the central nervous system.

It stains nuclei of a violet-red and does not overstain, so that sections may be left in it for many hours. It, however, does not stain well objects that naturally stain with difficulty.

Application.
    (1.) Stain for an hour or more.
    (2.) Wash in water.
    (3.) Alcohol, oil, balsam.

13. Indigo Carmine (*Merkel*).

Solution A.—Carmine	2 grams.
Borax	8 ,,
Water	130 cc.
Solution B.—Indigo carmine	8 grams.
Borax	8 ,,
Water	130 cc.

Keep the solutions separate (B is apt to develop a precipitate). When required, mix equal volumes of A and B. The mixture undergoes a change within a week, and hence it is better to make it fresh when required. Sections must remain in it at least twenty-four hours—with advantage longer—but the results are certainly satisfactory. The stained sections are placed for half-an-hour or thereby in a saturated solution of oxalic acid, which extracts the superfluous pigment. Mount sections in Farrant's solution or balsam. (See Stomach.)

B. Hæmatoxylin and Logwood.

Hæmatoxylin.—This substance was introduced to the notice of histologists by Böhmer. It is one of the most valuable nuclear staining reagents we possess, and this is specially the case when its violet-blue stain is set off by contrast with a ground stain of eosin, picric acid, or other appropriate dye.

1. Hæmatoxylin (*Böhmer*).—Make a solution containing—

Hæmatoxylin	1 gram.
Absolute alcohol	100 cc.

Make a second solution of—

Alum	5 grams.
Distilled water	100 cc.

Add drop by drop the first solution to a little of the second until a deep-violet colour is obtained. The fluid is placed for fourteen days in an open vessel, protected from dust, and exposed to the light and air, when it becomes of a bluish tint. This fluid is said to "ripen." Oxidation processes take place whereby the hæmatoxylin is converted into hämatëin. Filter. Add a fragment of thymol to preserve it. It stains tissues in 5-15 minutes.

It is sufficient to make a saturated solution of the crystals of hæmatoxylin in a small quantity of absolute alcohol. Add a few drops to a 1 per cent. solution of alum, which yields a light violet-coloured fluid. Expose the fluid to light and air, when it becomes blue. It is well to prepare this fluid several weeks beforehand to enable it to "ripen."

The ordinary manipulative procedure for hæmatoxylin staining is as follows:—

(1.) Stain the section (2-10 minutes).
(2.) Wash in distilled water.
(3.) Allow section to remain 12-24 hours in distilled water.
(4.) Remove water with alcohol,—add ethereal oil, balsam.

Böhmer's hæmatoxylin is well adapted for staining sections. It is best adapted for tissues hardened in sublimate, alcohol, picric or nitric acid, and not so good for those from chromium salts or osmic acid. Sections to be stained with it before being placed in the dye are better to be placed first in water or in 1 per cent. alum solution. If taken direct from alcohol they may contain a deposit after being stained. After staining, the sections are thoroughly washed in ordinary water. Hæmatoxylin stains the chromatin of the nuclei a deep blue, and other parts of some tissues light blue, e.g., the matrix of hyaline cartilage.

2. Strong Nucleus-Staining Hæmatoxylin (*Hamilton*).

Hæmatoxylin	12 grams.
Alum	50 ,,
Glycerine	65 cc.
Distilled water	130 ,,

Boil, and while hot add 5 cc. liquid carbolic acid. Allow the mixture to stand in the sunlight for at least a month.

3. Delafield's Hæmatoxylin.—To 100 cc. of a saturated solution of ammonia alum add drop by drop a solution of 1 gram hæmatoxylin dissolved in 6 cc. absolute alcohol. Expose to the air and light for a week. Filter. Add 25 cc. glycerine and 25 cc. of methylic alcohol. Allow it to stand exposed to the light for a long time. Filter.

This solution stains extremely rapidly, and may be greatly diluted when it is used. It keeps for a very long time, and stains well even tissues which have been hardened in chromic or osmic acid.

4. Kleinenberg's Hæmatoxylin.—(1.) Make a saturated solution of calcium chloride in 70 per cent. alcohol. Shake it well, and allow it to stand. Decant the saturated solution, add alum to excess, shake it well, allow it to stand for a day or so, and filter.

(2.) Make a saturated solution of alum in 70 per cent. alcohol. Filter.

(3.) To one volume of the filtrate from (1) add eight volumes of (2).

(4.) To (3) add drop by drop a saturated solution of hæmatoxylin in absolute alcohol, until it becomes of a decided purple colour, but not too dark, as the solution becomes darker by keeping and exposure to light. It should be prepared at least a few months before it is wanted.

It may be diluted to any extent by adding the mixture (1) or (2). As it contains much spirit, sections placed in it must be covered, *i.e.*, protected from evaporation, else the spirit will rapidly evaporate.

This logwood stain is particularly valuable when it is required to stain a tissue or an organ "in bulk." This a diluted solution will do in 24–48 hours, provided the pieces be not too large.

5. Acid Hæmatoxylin (*Ehrlich*).

Hæmatoxylin	1 gram.
Absolute alcohol	30 cc.

To the solution add—

Glycerine	60 cc. ⎫ Saturated
Distilled water	60 ,, ⎬ with alum.
Glacial acetic acid . . .	3 ,,

Dissolve the hæmatoxylin in the alcohol, add the glycerine and water, and then the acid. At first the solution is light-red, but when it has been exposed to the air for 2-3 weeks it gets bluish. It does not over-stain, and the tissues stained with it, when exposed to the light, become violet or bluish tinted. It may be used either for sections or for staining in bulk, and in the latter case it does not tend to over-stain.

Application.—Sections from alcohol—

(1.) Stain (3-5 minutes).
(2.) Wash in alcohol (90 per cent.).
(3.) Alcohol, oil, balsam.

Garbini finds it better to place the sections after staining in—

(2.) Distilled water.
(3.) Solution of carbonate of lithia (.25 per cent.).
(4.) Alcohol.

6. Glycerine Hæmatoxylin (*Renaut*).—Saturate perfectly neutral glycerine with potash-alum, and to it add drop by drop a saturated solution of hæmatoxylin in 90 per cent. alcohol, until a deep-violet tint is obtained. About one-fourth part of the hæmatoxylin solution has to be added. Let it stand exposed to the light for weeks, and filter.

7. Eosin Hæmatoxylin (*Renaut*).—Add drop by drop a concentrated watery solution of eosin to 200 cc. of glycerine saturated with potash-alum. Filter. Add drop by drop an alcoholic solution of hæmatoxylin. Expose to light for weeks, and filter.

8. Heidenhain's Hæmatoxylin.

(1.) This is used for staining in bulk. Prepare ½ per cent. watery solution of hæmatoxylin, which must not be kept too long. Boil the hæmatoxylin in water and allow it to cool. Place the hardened tissue in this dye for 24-48 hours.

(2.) Transfer it for 24 hours to .5 per cent. watery solution of yellow chromate of potash (24-48 hours). This causes dark clouds in the fluid, so that the chromate must be frequently changed.

(3.) Wash carefully in water. The tissues are then hardened in alcohol, and may be embedded in paraffin. It is best adapted for objects hardened in absolute alcohol, *e.g.*, salivary glands, pancreas (Lesson XXIII.), or in picric acid. Besides tinting the nuclei a greyish-blue, the protoplasm of the cells has a fine steel-gray tint, but the sections must not be too thick.

STAINING REAGENTS.

9. Hämatëin. It is well known that a solution of hæmatoxylin ($C_{16}H_{14}O_6$) after being prepared must stand some time to "ripen" before it is ready for use. The substance ultimately formed chiefly by the action of the air is hämatëin ($C_{16}H_{12}O_6$), which occurs in commerce in the form of a brown powder which is soluble in alcohol or water. P. Mayer recommends the following solution of this body in alum or **Häm-alum**: [1]—

(a.) Hämatëin . . . . . 1 gram.
   Alcohol (90 per cent.) . . . 50 cc.

   Dissolve by heating.

(b.) Alum . . . . . . 50 grams.
   Distilled water . . . . 1000 cc.

Mix the fluids (a) and (b). Allow the mixture to settle, and use the clear supernatant fluid as a stain. If it be too strong, dilute with distilled water, or, better still, with alum solution.

I have used this extensively during the last year for staining in bulk, and find it to be an excellent dye.

Logwood.—Staining solutions were formerly, and sometimes are, made from logwood chips.

10. Logwood (*Mitchell's*).—The tannic acid is removed.

Place finely-ground logwood (2 oz.) in a funnel; pack it well, and allow water to percolate through it until it flows away with but little colour. Allow the water to drain away; spread the logwood on a board to dry.

Dissolve alum (9 oz.) in 8 oz. of water. Moisten the dry logwood, pack it again tightly into a funnel, and pour on the alum solution. Close the lower end of the funnel, and allow the alum solution to extract the dye from the logwood for forty-eight hours.

Allow the coloured fluid to flow off, and percolate 4 oz. of water through the logwood in the funnel. Add a few drachms of glycerine and rectified spirit. Dilute largely when using, so that staining will take place slowly.

11. Other solutions of hæmatoxylin, including *Weigert's*, are referred to in the text. (Lessons on Central Nervous System and Salivary Glands.)

General Statement regarding Hæmatoxylin.—Hæmatoxylin is specially useful for staining tissues hardened in chromic acid. Ehrlich's hæmatoxylin is much to be commended. In all cases it is better to clarify sections stained with hæmatoxylin by means of xylol—not clove-oil—before mounting in balsam.

Hæmatoxylin and logwood are amongst the best nuclear stains we possess, and the tissues which stain best are those hardened in alcohol; but those also from Müller's fluid stain well. With those

---

[1] *Zeitsch. f. wiss. Mik.*, viii. p. 341.

hardened in watery solutions of chromic acid it is otherwise. It is sometimes very difficult to get them to stain. This may sometimes be effected by soaking the sections previously in a dilute solution of sodic carbonate.

They are also very valuable in double and treble staining.

In the case of preparations stained by hæmatoxylin or logwood, over-staining may be got rid of by placing the sections in dilute acetic acid. This will rapidly extract the surplus stain, but at the same time, if allowed to act too long, it will make the remaining part red. Great care should be taken afterwards by thorough washing of the sections in water to remove every trace of the acid.

C. Eosin.

1. Eosin.—This substance is readily soluble in alcohol and water. Make a 5 per cent. watery solution. It gives a beautiful diffuse rosy hue, and stains very quickly, in a minute or two. The stronger solution can be diluted as required. It forms one of the best ground-stains in contrast to logwood or one of the numerous aniline dyes. When using it as a double stain, e.g., logwood and eosin, stain the section first of all in logwood, and if it is to be mounted in balsam, clarify with clove-oil in which a little eosin has been dissolved. Sections stained with it can be mounted in balsam, Farrant's solution, or glycerine. It is a specific stain for the hæmoglobin of red blood-corpuscles, as it stains it, even after hardening in chromic salts, a copper-red colour, while it also stains the granules of certain leucocytes of the blood of a reddish tint. It is used very extensively, and in a very dilute solution is a good stain for cartilage and striped muscle.

D. The Aniline Dyes.

Watery or alcoholic solutions of the aniline dyes stain sections with great rapidity. The word "stain" is perhaps not quite the right word to use. It is rather a process of imbibition than staining proper. One of the difficulties in using aniline dyes is the rapidity with which sections have to be transferred from one liquid to another. They are not used for staining in bulk. After staining and dehydrating, it is best to clarify the sections (except in special cases) with cedar or bergamot oil or xylol. The sections are mounted in balsam, not in glycerine, as the latter dissolves the dyes.

They are amongst the most valuable so-called staining reagents we possess, and although many of them do not yield permanent preparations—the colour fading after a time—still the results obtained by their use are so important that it behoves the student to use them frequently. It is most important that they should be obtained from reliable sources.

The aniline colours are divided by Ehrlich into *acid*, *basic*, and *neutral* compounds. Of the three, the basic colours are most used, as they are excellent nuclear stains. Some of them have special affinities for certain tissues, and, as is well known, they are of the utmost value in bacteriological investigations. They may be kept in drop bottles. All of them are soluble in alcohol, and most of them in water; and so powerful are they, that usually a 1 per cent., or even a much weaker solution, suffices to stain tissues in a few minutes. Some of them, according to Ehrlich's researches, stain better when they are mixed with a mordant.

Aniline-Oil and Aniline-Water (*Ehrlich's Method*).—Shake up excess of pure aniline-oil with excess of water, and allow it to stand. The most of the oil sinks to the bottom. This solution should not be kept too long; in fact, it is better to make it fresh. Filter a little of the aniline-water into a watch-glass. To the fluid in the watch-glass add ten or twelve drops of a concentrated alcoholic solution of any of the aniline dyes it is desired to use in this way.

Very few of the preparations stained by aniline dyes can be preserved in glycerine or Farrant's solution. A 50 per cent. solution of acetate of potash keeps their colours well, but it is a medium which it is difficult to keep tightly under the cover-glass. Many of them can be preserved in balsam, but most of them must be kept away from the action of acids. In a few cases the colour is partly extracted and fixed with dilute acids, but in such cases the free acid must be removed from the section before it is finally mounted.

For convenience these colours may be grouped as follows :—

(A.) **Violet Aniline Colours.**

Methyl Violet.—Dissolve 2 grams in 100 cc. and filter. This gives a 2 per cent. solution, which may be diluted as required. A 0.1 per cent. solution is, in many cases, sufficient. It does best for porous textures. Sections are left to stain in it for several hours, washed with water, and then with alcohol, until no more colour comes away, and mounted in balsam. It stains intercellular substances but slightly, while cells, and especially their nuclei, are stained by it.

Gentian Violet always answers very well for staining cell-nuclei, but it seems to be better for hardened preparations than the previous dye.

Dahlia is used in the same way as the preceding.

(B.) **Blue (and Purple) Aniline Colours.**

Aniline Blue.—Make a 1 per cent. watery solution, adding a few drops of absolute alcohol. This is useful for the glands of the stomach and for a double stain with safranin.

Methylene Blue.—Make a saturated watery solution. Rectified spirit may be added to make it keep. It is not very largely used in bacteriology, *e.g.*, for the tubercle bacillus, but largely for double-staining of tissues, in contrast to red. It stains axis cylinders of nerve fibres. (See Nervous System and Epithelial Cement.)

Sections to be mounted in balsam are best clarified by cedar-oil.

Better results are obtained by using a very dilute solution made as follows (Garbini):—Add 10 drops of a saturated alcoholic solution to 100 cc. of a solution of caustic potash (1 in 10,000). Leave the sections to stain (12-24 hours). Leave them in absolute alcohol (6-8 hours), and then in oil of cloves (2-4 hours)—xylol—xylol-balsam.

Spiller's Purple.—Use Spiller's purple No. 1. Rub up 2 grams in a glass mortar with 10 cc. of alcohol, and add 100 cc. of distilled water. It is used as a double stain, and for staining the fibrin in coagulated blood. The stain requires to be pretty deep, as it is washed out by alcohol. If a section stained by it is to be mounted in balsam, use cedar-oil to clarify it.

(C.) Green Aniline Colours.—Amongst these are *iodine green*, *methyl green* (1 per cent.), *aniline green*, and *aldehyde green*.

Iodine-Green is used as a 5 per cent. filtered watery solution. It stains nuclei and developing cartilage green, and makes a good contrast stain. It is not very readily extracted by spirit, and does not soon fade.

Methyl-Green.—This is a nuclear stain. Make a 1 per cent. solution in distilled water, and add 25 cc. of absolute alcohol. Belgian observers, more particularly Carnoy, use this for fresh tissues in the following manner:—

Methyl-green	1 gram.
Glacial acetic acid	1 cc.
Distilled water	100 ,,

Add a few drops of this liquid to a watch-glassful of an indifferent fluid, *e.g.*, normal saline.

The others will be referred to in the text.

(D.) Red Aniline Colours.—They are very numerous.

Rosaniline Acetate, Sulphate, and Hydrochlorate (Magenta).— The term *fuchsin* is sometimes applied to the one, sometimes to the other, but the acetate is more soluble in water. They give a rather diffuse stain, but are useful for nuclear staining, elastic fibres, and blood-corpuscles.

Rub up a little (1 gram) in a glass mortar with rectified spirit (20 cc.). After solution add 20 cc. of distilled water.

Magenta for Blood-Corpuscles.—Dissolve .1 gram of magenta in 5 cc. of rectified spirit and 15 cc. water, and add 20 cc. glycerine.

Acid Fuchsin is a specific colouring-matter for the nervous system. It was introduced by Weigert, but it has been largely displaced by Weigert's hæmatoxylin copper stain.

Safranin is specially used as a nuclear stain, and very largely for the study of mitosis. It is specially useful for tissues hardened in Flemming's mixture.

Safranin	1 gram.
Absolute alcohol	100 cc.
Water	200 ,,

It may also be used as a much stronger alcoholic solution, 5 per cent. in 70 per cent. alcohol. There are several varieties of this dye, and some of them are of little value as dyes. Therefore it is important to obtain a good sample. That sold as safranin-O can usually be relied upon, and is to be obtained from Dr. George Grübler, Leipzig.

In using this dye, the sections are, as a rule, left for several hours in the solution—even twenty-four hours or longer—and are then placed in ordinary alcohol or acid alcohol (p. 65)—containing .5 per cent. hydrochloric acid—to remove the surplus stain. If this be properly done, the nuclei—the chromatin of the nuclei only—are stained. Sections may be mounted in balsam.

Application.

(1.) Stain sections in 1 per cent. watery solution of safranin (1-24 hours).
(2.) Wash rapidly in water.
(3.) Wash in absolute alcohol or acid alcohol.
(4.) Absolute alcohol, oil, balsam.

Sometimes it is useful to use it in aniline-water, after Ehrlich's method (p. 73).

As shown by Martinotti, it colours black or dark-purple elastic fibres hardened in 0.2 per cent. chromic acid (Lesson X.).

(E.) Brown Aniline Colours.

Bismarck Brown or Phenylene Brown.—This is but slightly soluble in water, but it forms a good ground-stain in contrast to hæmatoxylin, and is useful for staining plasma cells and the cells of the cerebrum. It preserves its colour when mounted in Farrant's solution or balsam.

(*a*.) Boil **Bismarck brown** with 100 of water ($= 3-4$ per cent.). Filter and add one-third its volume of absolute alcohol.

(*b*.) Or use a concentrated alcoholic solution in 40 per cent. alcohol ($= 2-2\frac{1}{2}$ per cent.).

Application.

(1.) Stain (5 minutes).
(2.) Wash out in strong spirit.
(3.) Alcohol, oil, balsam.

It does not tend to over-stain. It is a nuclear stain, and the nuclei are brown, and the protoplasm light brown. Such preparations are well adapted for photographic reproductions. Some use a solution in 75 per cent. alcohol.

Vesuvin is much more soluble, and is used in the same way as Bismarck brown.

(F.) **Other Aniline Dyes.**

Aniline Blue-Black.—This has a remarkable power of staining nerve-cells, as shown by Sankey. It is best adapted for staining fresh nerve-tissues, although it is also used for staining in bulk. It has little power of diffusing through them. (Lessons on Central Nervous System.)

TABLE *showing some of the Aniline Dyes in most Common Use, arranged according to their Colour.*

Red.	Green.	Blue.	Yellow.	Violet.	Brown and Black.
Safranin. Eosin. Fuchsin. Ponceau RR.	Methyl-green. Iodine green. Aldehyde green.	Methylene blue. Aniline blue. Quinoline.	Orange.	Methyl-violet. Violet BBBBB. Dahlia.	Bismarck brown. Nigrosin. Aniline blue-black.

E. Metallic Substances.

1. **Nitrate of Silver.**—This substance possesses the property of forming a compound with intercellular substance, which darkens, and becomes brown or black, on exposure to light. It is unequalled for the study of the cement substance of epithelium and endothelium, and for the cell-spaces of the cornea, and connective tissue generally. The tissue, however, must be fresh—the fresher the better. Use a glass or horn rod to manipulate the tissue in the fluid, not metallic instruments.

Make a 1 per cent. solution, *i.e.*, 1 gram of silver nitrate is dissolved in 100 cc. of distilled water. This is kept as a stock. A $\frac{1}{4}$ or $\frac{1}{2}$ per cent. solution is the strength usually employed.

If it be desired to "silver" a part of the omentum or mesentery, this membrane should be pinned out with hedgehog spines, without being stretched, on a piece of flat cork with a large hole in it, so that the solution can get to both sides of the membrane. A very convenient plan is to pass the membrane over a porcelain or ebonite ring, and fix it with another ring in the manner in which a skin is fitted on a drum (figs. 45 and 46).

Lave the membrane gently in distilled water to remove any

chlorides, and place it in a dilute solution of the reagent ($\frac{1}{4}$-$\frac{1}{3}$-$\frac{1}{2}$ per cent.). The tissue soon—5-10 minutes—becomes white, and as soon as it looks grayish remove it and wash it in distilled water. Place it in ordinary water, and expose it to good daylight, when it rapidly becomes brown. It can then be preserved in spirit until it is required.

A membrane may be stained *in situ*, *e.g.*, the central tendon of the diaphragm of a rabbit, or the mesentery, *e.g.*, of a frog. Open the abdomen, irrigate the membrane by allowing distilled water to fall on it from a pipette. This removes all substances that might combine with the silver and give rise to illusive appearances. Drop on the silver solution by means of a pipette. Treat the membrane as in the previous case.

FIG. 45.—A. Vulcanite rings used for stretching a membrane which is to be silvered; B. Conical rings, the one inside the other; C. A single ring.

2. Negative Method (*Recklinghausen*).—This is the usual method by which the intercellular cement substance is stained black—a *weak* solution being used—as for the study of endothelium. The membrane ought to be kept stretched, *e.g.*, over the mouth of a porcelain capsule.

The following modification is recommended by Thanhoffer, and it works well. After the membrane is exposed to the action of silver nitrate, it is washed with a 2 per cent. solution of acetic acid.

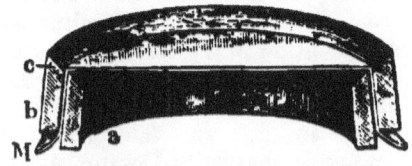

FIG. 46.—Section of A. *a*. Inner, and *b*. Outer ring; M. Stretched membrane.

3. Positive Method (*His*) is specially designed for showing lacunæ and lymphatic cavities, *e.g.*, the lymphatics of the skin and cornea. Place the tissue for several hours in 1 per cent. $AgNO_3$ in the dark, wash in water, transfer in dark to 3·5 solution of sodic chloride. Wash in water, expose to light. The spaces are found filled with black granules. Examine in glycerine.

Sometimes silver nitrate is used in a solid form, *e.g.*, for the cornea. This will be referred to afterwards in treating of the cell-spaces of the cornea itself. (Lesson on Eye.)

Silver preparations show the cement substance of epithelium

stained black as "silver lines," and they may be mounted in glycerine or balsam, either unstained or after staining with logwood, picro-carmine, or other dye as desired.

In order to stain the lining endothelium of the vascular system, a solution of silver nitrate is used. The special precautions required are referred to in the text. (Lesson on Blood-Vessels.)

4. Golgi's Method.—In this method parts of the central nervous system hardened in potassic bichromate are treated for many days with silver nitrate or mercuric chloride to demonstrate the processes of nerve-cells. (Lesson on Nervous System.)

5. Nitrate of Silver and Osmic Acid (*Golgi*).—This is specially useful for the nervous system. Place a fresh nerve of a rabbit just killed in

Potassic bichromate (2 per cent.) . . . 10 parts
Osmic acid (1 per cent.) . . . 2 ,,

for an hour; tease the nerve, and let it remain in the mixture for another hour. Transfer for 8 hours to .5 per cent. silver nitrate and then to alcohol.

B. Gold Chloride.

Gold Chloride.—This substance has rendered particular service, especially in connection with the terminations of nerves. It is used as $\frac{1}{2}$–2 per cent. watery solution. Various methods are employed, according to the end desired.

1. Acetic Acid Method.—Place a small piece of perfectly fresh tissue, 2 mm. cubes, *e.g.*, hyaline cartilage or a small cornea, in $\frac{1}{2}$ per cent. solution in a glass thimble for half-an-hour, keeping it in the dark all the time. These small glass thimbles are particularly useful, and are better to be somewhat broader relatively than those shown in fig. 26. The tissue will become yellow; wash it thoroughly in distilled water, and expose it to bright daylight in distilled water slightly acidulated with acetic acid. In a day or two it will become of a purplish or violet-brown colour. Sections can then be made and mounted in glycerine.

2. Loewit's Method.—To one part of formic acid (sp. gr. 1.16) add two parts of distilled water. Place small pieces of the fresh tissue (1–2 mm. in thickness), *e.g.*, tendon from a rat's tail, in this mixture for $\frac{1}{2}$–1 minute, until they become somewhat transparent; transfer them to a glass thimble or watch-glass containing 1 per cent. gold chloride for 15–20 minutes, *i.e.*, until they have become yellow throughout. During this process, the tissue should be exposed to light as little as possible. Place the tissue in formic acid (1 : 3), and keep it in the dark for twenty-four hours. Afterwards place it for twenty-four hours in pure formic acid, and keep it also in the dark. Wash it thoroughly with water, and mount in glycerine or balsam.

3. **Ranvier's Lemon-Juice Method.**—The fresh tissue is placed for 5-10 minutes in the freshly expressed and filtered juice of a lemon, until it becomes transparent. Rapidly wash it in distilled water, transfer it to 1 per cent. gold chloride solution for from ten minutes to one hour; the time depends on the tissue under investigation. Wash with water and place the tissues in 50 cc. of water containing two drops of acetic acid, and expose them to light, when reduction takes place. Or the tissue may be placed in formic acid (1 : 3) after being treated with lemon-juice and gold chloride, and kept in the dark for twenty-four hours. The latter plan is in many cases to be preferred, especially where the retention of the superficial epithelium is not desired.

4. **Boiled Gold Chloride.**—For some purposes, especially for studying the terminations of the nerves in sensory surfaces, this method of Ranvier has yielded me the best results.

Make as required—fresh—a mixture of four parts of gold chloride (1 per cent.) and one part of formic acid. Boil the mixture and let it cool. Place the fresh tissues (small pieces) in it for ten minutes to one hour. Wash in water, and place in formic acid (1 : 4 water), and keep in darkness, where the reduction takes place.

5. **Rapid Reduction of Gold Chloride.**—A tissue may be left in gold chloride (1 per cent.) for half-an-hour or more, and then transferred to a strong solution of tartaric acid and heated to $45°$ or $50°$ C., when it rapidly becomes of a purplish-brown colour, usually in the course of a quarter of an hour.

Although it has been stated that for gold chloride preparations the tissues should be fresh, Drasch, in his researches on the nerves of the intestine and those of the circumvallate papillæ, points out that he obtained the best results with tissues twenty-four hours after death; the tissues, however, must have been kept cool. Any one who has had the privilege of studying the preparations of Drasch cannot but have been impressed with the beauty of specimens prepared by his method.

6. **Gold Chloride and Chromic Acid** (*Kolossow*).—Place the tissues for 2-3 hours, according to their size, in 1 per cent. gold chloride acidulated with hydrochloric acid (1-100). Wash the tissues with water, and keep them in the dark in chromic acid ($\frac{1}{50}$ to $\frac{1}{100}$ per cent.) for 2-3 days. Wash out the chromic acid thoroughly.

7. **Method of Ciaccio** is good for the termination of nerves in cornea and muscles. Place a small piece of tissue, not more than 2 cubic mm. in size, in fresh juice of lemon (5 minutes): wash; place in

1 per cent. solution of *chloride of gold and cadmium* (30–60 minutes) in the dark. Wash; then in 1 per cent. formic acid (24 hours) in the dark, and then for 12 hours in sunlight. Finally for 24 hours in pure formic acid. Wash. Tease and mount in glycerine.

F. Double, Treble, or Multiple Stainings.

It is possible to stain a section so that the several parts of it may be differently stained. This may be done either by staining successively with different stains, or by mixing the dyes in one fluid, and staining the section with the mixture, whereby one part takes up one of the dyes and another part one of the other dyes. Thus one gets an elective and differential stain. It is possible thus to combine a nuclear stain with one which stains only the protoplasm of the cells, such as eosin or orange.

Amongst double stains are the following :—

1. Picro-carmine (p. 66), one of the most valuable we possess, and which we owe to Ranvier; and **Picro-litho-carmine**. If the section is to be mounted in glycerine or Farrant's solution, do *not wash it in water*. If it is to be mounted in balsam, the alcohol in which it is washed should contain picric acid—the same result is obtained by using clove-oil with picric acid dissolved in it—otherwise only a carmine stain is obtained.

2. Carmine and Aniline Blue.—Stain a section in borax-carmine, and then in very dilute aniline blue, specially useful for the glands of the fundus of the stomach. In other tissues the nuclei are red and the perinuclear parts blue.

3. Hæmatoxylin and Eosin.—Hæmatoxylin is a nuclear stain,— the sections are first stained in it, and then in eosin, which stains the general protoplasm of the cell. (See also p. 72.)

4. Aniline-blue and Safranin (*Garbini*).—The section is transferred from water.

Manipulation.

Aniline-blue sol. (.5 per cent.), 2–4 minutes.
Wash in water.
Lithium carbonate (.5 per cent.), a few minutes.
Hydrochloric acid (.5 per cent.), a transparent blue colour is produced.
Wash in water.
Safranin (1 per cent.), 10 minutes.
Dehydrate in *methylic* alcohol.
Clarify in oil of cloves (2 parts) and cedar-oil (1 part).
Xylol-balsam.

This is especially useful for some of the salivary glands—thus in the sub-maxillary one set of cells is red, the other blue; in the stomach the parietal cells are red and the inner blue; the epithelial cells of the villi are blue, the goblet cells reddish; in a hair-follicle the sheath of Henle is an intense red, and the sheath of Huxley blue.

5. Ehrlich-Biondi-Heidenhain Stain.

Saturated watery solution of orange . . 100 cc.
,, ,, ,, acid fuchsin . 20 ,,
,, ,, ,, methyl-green . 50 ,,

To get complete saturation it is necessary to have an excess of the crystals for several days. Each fluid is saturated separately. Before use, the solution is diluted in the proportion of 1 in 100 with water, and then on the addition of acetic acid must be bright red. It is better to obtain the mixture from Dr Grübler.

Application.

(1.) Harden the organ in corrosive sublimate.
(2.) Stain sections in the dilute solution (12-24 hours).
(3.) Wash quickly in 90 per cent. alcohol.
(4.) Dehydrate in absolute alcohol.
(5.) Xylol-balsam.

It is especially useful for sections containing many leucocytes, and is best used for paraffin sections fixed on a slide. Red blood-corpuscles are stained red, resting nuclei blue, mitotic figures and nuclei of leucocytes green-violet.

General Remarks on Staining.—Filter the staining fluid. When possible, use a *weak solution* of the dye, and thus let the sections *stain slowly* in a fairly large amount of the fluid. Place a piece of blotting-paper on the inside of a large watch-glass, pour some of the diluted stain into the watch-glass, and place the sections in it. The sections should lie as flat as possible, and not overlap each other. The sections may be moved gently in the fluid by means of a needle. Cover the watch-glass with another glass of the same size, or set it aside in a moist chamber, e.g., on a plate covered by a bell-jar, with a piece of moistened blotting-paper attached to the inside of the jar (fig. 47). After staining, the sections are to be carefully washed in distilled water to remove any trace of surplus dye (except in the case of picro-carmine). Sections stained with hæmatoxylin should be washed for a long time in water before they are mounted in glycerine or Farrant's solution.

Be careful not to over-stain the tissue, except in those cases where the excess can be again removed, e.g., with the aniline dyes by means of alcohol or acid alcohol.

All acids should be removed from the sections before they are placed in the dye.

It is convenient on many occasions that the student should rapidly stain his sections on a slide, but he should also be taught to practise the slower method of staining sections in very dilute solutions of a dye.

There is one method of staining sections which may be profitably impressed upon the student, viz., that so strongly insisted upon by Ranvier. Suppose any delicate object—isolated epithelial or other cells—to be mounted in a watery medium; a drop of a solution of picro-carmine is placed at one side of the cover-glass. As the fluid evaporates at one side of the cover-glass the picro-carmine slowly diffuses under the cover-glass and stains the preparation. Exposed to the air, the preparation would soon become dry. This must be corrected. This is best done by placing the slides to be stained in this way on a stage with several shelves

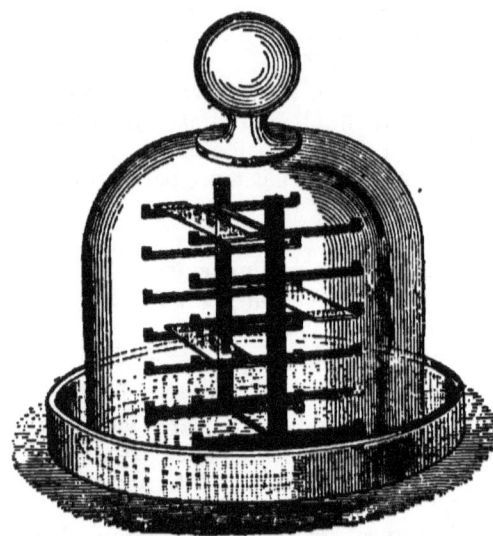

FIG. 47.—Support of Ranvier for Holding Slides Placed under a Bell-Jar.

(fig. 47), the whole being placed on a plate moistened with a few drops of water and covered by a bell-jar. This forms a moist chamber.

After the cells are stained, glycerine may then be applied at the side of the cover-glass, with the same protective precautions, so that the preparation can be finally mounted and preserved in glycerine.

XIII.—CLEARING OR CLARIFYING REAGENTS.

Glycerine, Farrant's Solution, and Glycerine Jelly.—When any one of these reagents is used for mounting preparations, no other clarifying substance is used.

Balsam Preparations.—When a preparation is to be mounted in balsam, be it Canada balsam or dammar, some clarifying reagent has to be added to the preparation before the balsam is applied. Before applying the balsam the tissues must have been rendered transparent.

The following substances are most commonly used:—Oil of cloves, a mixture of creosote and turpentine, turpentine, creosote, xylol, cedar-oil, bergamot oil, lavender oil, origanum oil, &c.

Oil of Cloves has this advantage, that it clarifies rapidly and does not evaporate, so that sections may be left exposed to the air in it for some time. It renders the sections very hard. It, however, is not so satisfactory for aniline dye specimens, as it is apt to abstract their colour. Moreover, it becomes yellow with age.

Creosote is specially useful for preparations which one does not desire to harden in alcohol; do not use metallic instruments.

Creosote and Turpentine.—When the fluids are mixed a cloudiness appears, but this disappears on keeping. It is much cheaper than clove-oil, but it rapidly evaporates (one part creosote to four of turpentine).

Xylol is perhaps the best, and is specially useful with aniline dyes. In these cases the balsam—Canada or dammar—should also be dissolved in xylol.

Cedar-Wood Oil clarifies very slowly. It does not, however, abstract the aniline dyes, and is used for special purposes, as indicated in the context.

Origanum Oil is used for clarifying sections embedded in celloidin.

Xylol-Aniline Oil.—Equal parts of xylol and aniline are used for clarifying sections under certain conditions without the previous use of alcohol. (Weigert's method, Lesson III.)

Carbolic Acid and Xylol.—A mixture of 1 part of carbolic acid and 3 of xylol is used to clarify celloidin sections (p. 47). The section can be taken from 70 per cent. alcohol, and does not require to be further dehydrated. To remove the water from the mixture, keep in the bottom of the bottle containing it a thick layer of previously-heated copper sulphate.

General Remarks.—Although several essential oils are used for clarifying purposes, it is not immaterial which one is used. Thus clove-oil may be used for clarifying sections stained with animal or vegetable dyes (carmine), while it is inapplicable for aniline staining, as it dissolves aniline dyes.

In many cases the result may be obtained more gradually by using a mixture of half alcohol and half essential oil.

Moreover, clove-oil dissolves celloidin, so that it cannot be used when the section fixed on the slide contains either celloidin or collodion. Oil of bergamot does not dissolve celloidin.

84   PRACTICAL HISTOLOGY.

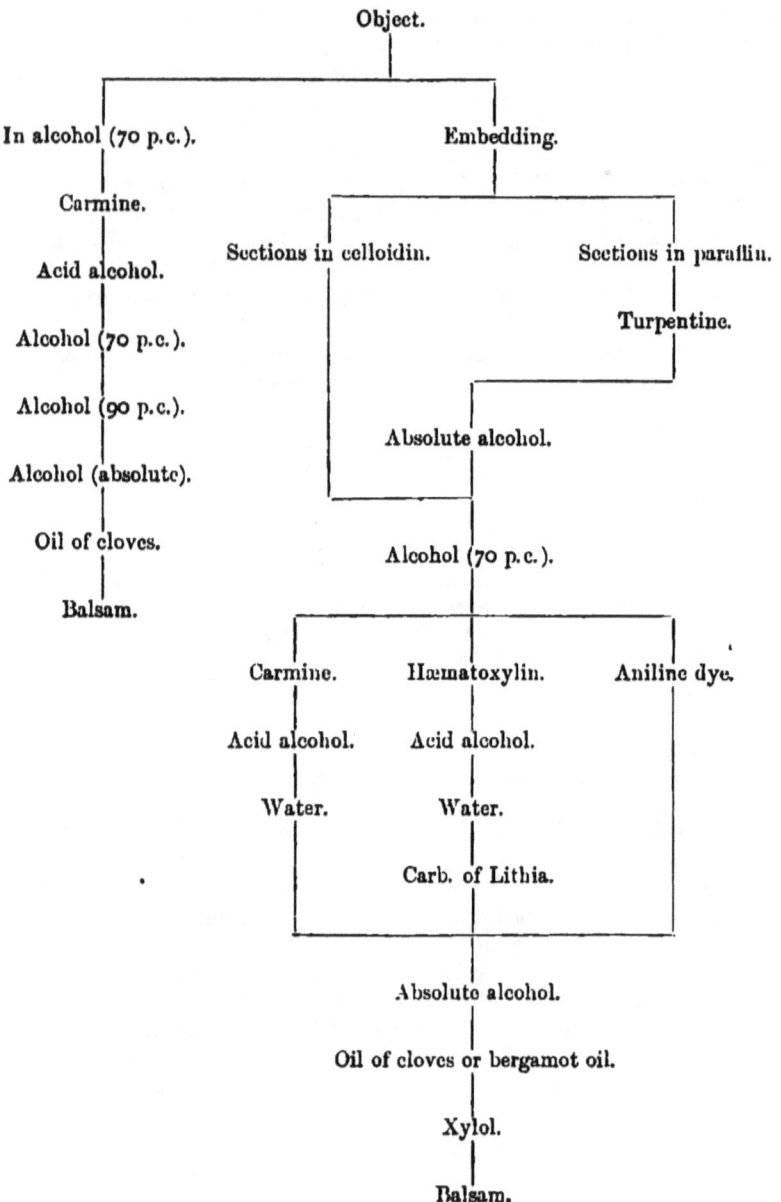

XIV.—MOUNTING FLUIDS, AND METHODS.

The fluid chosen will depend on the nature of the tissue and other circumstances.

1. If a section is to be mounted direct from water, glycerine, Farrant's solution, or glycerine jelly may be used.

2. If a section is to be mounted in balsam, it must have *every trace of water removed* by alcohol, and the alcohol must be displaced by one of the clarifying reagents—xylol, clove-oil, &c.—already mentioned.

Glycerine.—Pure glycerine is only used for such tissues as have been previously hardened. In the case of tissues—delicate tissues which have not been previously hardened or fixed—the direct application of pure glycerine would injure them. In this case, the best way is to mount the object in normal saline, and at one edge of the cover-glass to place a drop of a mixture of equal parts of glycerine and water. Put the preparation in a plate covered by a bell-jar—an extempore moist chamber. The glycerine slowly penetrates as the water evaporates.

Some tissues are rendered too transparent by glycerine, and, moreover, it is very difficult to seal up and keep tight glycerine preparations.

Glycerine and Formic Acid.—This is sometimes used, especially for picro-carmine preparations. It is made by adding formic acid to dilute glycerine (1 per cent.).

Farrant's Solution.—This is for many preparations far more serviceable than glycerine, as it does not render some tissues so transparent as glycerine, and the preparations can be easily sealed up or " ringed."

Preparation (*Hamilton's receipt*).—Make a saturated solution of arsenious acid in water by boiling. After standing for twenty-four hours filter. Take equal quantities of water, glycerine, and arsenious water, and to the mixture add picked gum-arabic. Let the latter dissolve until a thick syrupy fluid is obtained, which takes about a week at an ordinary temperature, but it must be stirred frequently. Filter slowly through filter-paper, which must be frequently changed.

Glycerine Jelly.—Melt it in hot water, place a drop on the section, apply a cover-glass, and gently press it down. It gelatinises in a few minutes.

Canada Balsam.—Place some Canada balsam in a capsule or wide-mouthed bottle near a fire or in a warm chamber (65° C.) until it becomes hard. Let it cool. This dry balsam is to be dissolved in some medium. Some use chloroform, others benzol, others a mixture of both, or turpentine as a solvent. In any case, the solvent is added until a fairly thin fluid is obtained. Perhaps the best solvent of all is xylol. It requires nearly twice its volume of xylol. Filter through paper. The balsam should be kept in a "capped" bottle (fig. 48) instead of a stoppered one. If it gets too thick, add a little xylol.

Dammar Lac (*Klein*).

Gum dammar	$1\frac{1}{2}$ oz.
Gum mastic	$\frac{1}{2}$,,
Turpentine	2 ,,
Chloroform	2 ,,

Dissolve the dammar in the turpentine, and filter; the mastic in the chloroform, and filter. Mix the two solutions and filter again.

There must be no moisture in the bottles, and the mixture must be kept in 'capped" bottles, else the chloroform will evaporate.

Xylol-Balsam.—Dry ordinary Canada balsam in a sand-bath, to drive off all the moisture, and until it becomes vitreous.

FIG. 48.—Capped Bottle for Balsam.

If it be spread out in a thin layer in a tin vessel, this is usually accomplished in two hours or so, but the balsam must not be overheated or change its colour and become brown. Dissolve the dried balsam in an equal volume of xylol. Perhaps this is the best form of balsam to use.

Balsam, when prepared, should be kept in a glass bottle with a ground-glass cap.

To Place a Section on a Slide.—By far the most convenient method is to place the section in a basin of water. Hold the slide perpendicularly by the edges in the left hand, plunge the slide into the water until it is about three-fourths immersed, and with a mounted needle pull the section on to the slide, and at the same moment raise the latter out of the water. The section adheres to the glass, and if it be folded at one end, dip this end in the water, when it floats out quite flat. Do not attempt to spread out the folds on the slide by means of a needle.

Hold the slide vertically to allow the water to drain off, and remove with a rag or well-washed cloth the remainder of the water close up to the section.

It may be stained on the slide. After the staining is complete, remove the surplus dye by means of bibulous paper, taking care, however, that the section itself does not adhere to the absorbent paper.

If the section is to be mounted in glycerine or Farrant's solution, add a drop of either of these reagents and apply a cover-glass.

If the section is to be mounted in balsam, remove as much as possible of the surplus water or dye, as the case may be, and pour methylated spirit upon the section. Allow it to remain on the section for a minute or so, and drain it off at one end of the slide. Apply fresh methylated spirit again, and finally absolute alcohol. This is done to secure complete dehydration. The frequent and prolonged application of strong spirit removes all the water.

Remove as much of the spirit as possible, but do not allow the section to dry. It is now ready to be cleared up.

With a brush insinuate a drop of the clarifying reagent—clove-oil or xylol—*under* one corner of the section, and allow the xylol to flow *under* the whole of the section. It will gradually diffuse into the tissue; and if the process be watched under the microscope with a low power, the section will be seen to become gradually more transparent, while the spirit will be seen as fine globules driven out into the essential oil. The success of the process depends on complete removal of the water by spirit, and the complete removal of the latter by the essential oil used as the clarifying reagent. If any opacity remains, and it looks milky or like an emulsion, there has been either water or spirit, or both, left in the section.

More of the essential oil is placed on the section, so that it is completely bathed in it and rendered quite clear by it. Pour off the superfluous oil, remove the surplus close up to the edge of the section, add a drop of balsam, apply a cover-glass, and the process is complete.

In some cases it is convenient to put the drop of balsam on the cover-glass, and then to invert this on the clarified preparation.

In all cases where it is directed to **mount in balsam**, this process must be gone through, viz.—

(1.) Stain the section.
(2.) Wash it in water.
(3.) Treat the section with strong alcohol (96 per cent.) to remove water (3–5 minutes).
(4.) Absolute alcohol (3–5 minutes).
(5.) Clarify with an essential oil to remove all the alcohol.
(6.) If the section be not on a slide already, place it on a slide by means of a lifter. Remove surplus oil with blotting-paper.
(7.) Add balsam, cover the section with a cover-glass.
(8.) If desired, the hardening of the balsam may be hastened by gently warming the preparation on a water-bath.

Sometimes it is not convenient to stain, dehydrate, and clarify a section on a slide. In this case the sections are stained, dehydrated, and clarified in watch-glasses, the sections being transferred from one fluid to the other, and finally to the slide by means of a "lifter" (p. 3).

Sometimes the one method is adopted, sometimes the other.

To Clean a Microscopic Preparation.—Any excess of *balsam* round the edge of a preparation may be moved with a cloth dipped in benzol.

In the case of a preparation mounted in *glycerine*, any excess of the latter must be removed with great care, otherwise the cement will not adhere to the glass.

With preparations mounted in *Farrant's solution*, leave them in an airy dry place for ten days or longer; this gives the medium time to harden at the edges, and fixes the cover-glass pretty firmly to the slide. Place the slide in a basin of water, and with a camel's-hair brush brush away from the edge of the cover-glass every trace of the medium. There is no fear of disturbing the cover-glass. Lave the slide in fresh water, and then wipe it thoroughly dry. It is better to wash a number of slides at a time.

To cement or "Ring" the Specimens.

Balsam Preparations need not be touched. They keep perfectly without being covered in by coating the edge of the cover-glass with an adhesive and resistant cement. If it be desired to cement them, a thin coating of Hollis's glue must first be applied, and after it is dry the cement is laid on as directed for preparations mounted in Farrant's solution.

To Ring a Slide.—The slide should be fixed on a turntable, the centre of the circular cover-glass corresponding to the centre of the brass disc of the table. The slide is fixed in position by means of two brass clips (figs. 49, *a*, *b*, 50).

For *Farrant's preparations* or *glycerine preparations*, lay on a ring of white zinc cement with a goat's-hair brush. The disc

FIG. 50.—Showing how slide is to be centred on the Turntable.

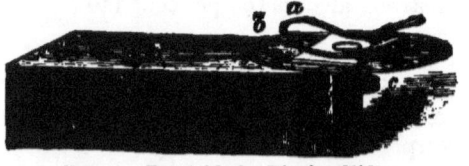

FIG. 49.—Turntable for Ringing Slides.

is made to revolve with the fore-finger of the left hand, but not too quickly, and a coating of the cement is laid on evenly. The fore-finger is applied to the smaller disc (*c*).

The turntable should be heavy and mounted on a pin-point centre-piece. The brushes must not be too large, and should be washed immediately after use in the same fluid as is used to dissolve the cement. Thus, for zinc-white the brush is to be washed in benzol or xylol, and for gold-size in turpentine, and for Farrant's solution in water.

White Zinc Cement.—Dissolve 3 oz. of dammar in 3 oz. of benzol, and add 200 grains of finely-ground oxide of zinc. Mix the whole thoroughly, and strain through several folds of muslin. It is perhaps more convenient to purchase the cement.

Mounting Block.—It is important that the section be placed in the centre of the slide. As a guide for this purpose, cut a piece of paper the size of the slide, and draw diagonal lines from corner to corner of it; they will intersect in the centre. Or the piece of paper may be gummed by means of Hollis's glue between two slides.

XV.—INJECTING BLOOD-VESSELS AND GLAND-TUBES.

Transparent Injection Masses.—At the present time, histologists use transparent injections, consisting of a vehicle—which may be water, glycerine, or gelatine—and a colouring matter. Most commonly gelatine is used as a vehicle. The colouring matter of most red injections is carmine. In this case, the secret is to have the mass as neutral as possible.

1. Carter's Carmine Injection.

Carmine	1 dr.
Strong solution of ammonia	2 fl. drs.
Glacial acetic acid	86 mins.
Solution of gelatine (1 to 6 water)	2 oz.
Distilled water	1½ ,,

Rub up the carmine with a little water in a mortar, add the remainder of the water, and then add the ammonia, and stir until the carmine is dissolved. Add the glacial acetic drop by drop, stirring thoroughly. Add the gelatine solution, and stir briskly.

2. Ranvier's Method.—The following method yields excellent results. Mix 2–5 grms. of pure carmine with a little distilled water in a stoppered bottle, and add ammonia solution, drop by drop, until the carmine is dissolved, which occurs when the liquid becomes transparent. Shake up the liquid to get it homogeneous.

Weigh 5 grms. of dry Paris gelatine (Coignet's), and place it in distilled water for one hour. At the end of this time it is swollen up and soft. Remove it from the water, wash it in water, and place it in a beaker in a water-bath. When the gelatine is dissolved by the water which it has absorbed, add to it—stirring vigorously—the solution of carmine, which yields an ammoniacal solution of carmine in gelatine.

When the carmine mixture is on the water-bath make a solution of—

Distilled water	2 parts.
Glacial acetic acid	1 part.

Pour the acid drop by drop into the mass, stirring thoroughly all the time with a glass rod. The acid is to neutralise the excess of ammonia. This requires great attention. It is by the odour that one recognises when the fluid is neutralised. As the acid is added the ammoniacal odour diminishes, and there is at last a faint acid odour. This is the moment to stop adding the acid. Towards the end of the operation it is best to dilute the acid somewhat.

Filter the mass through *new* flannel.

3. Carmine Gelatine Mass (*Carter's*) (Fearnley's method).

Carmine	3 grams.
Strong ammonia	6 cc.
Glacial acetic acid	6 ,,
Coignet's French gelatine . . .	7 grms.
Water	80 cc.

Cut up the gelatine into small pieces and place it in 50 cc. of the water to swell up, *i.e.*, for four or five hours. Rub up the carmine in a mortar with a little water and add the ammonia. Let it stand for two hours and then pour it into a bottle, rinsing the mortar with the remainder of the water. Place the swollen-up gelatine, and any remaining water unabsorbed by it, on a water-bath until it melts. To the dark purple carmine fluid add the acid (a few drops at a time), mixing the two thoroughly, and as soon as the fluid changes to a crimson stop adding the acid. To the melted gelatine add the crimson carmine little by little and keep stirring all the time.

This mass may be kept in a cool place for a long time if its surface be covered with methylated spirit. Before using it, dissolve it on a water-bath, and filter it through fine flannel wrung out of hot water. The best gelatine to use is French gelatine—Coignet's.

4. Blue Mass.—
The mass is made with gelatine coloured with soluble Prussian blue or Brücke's blue. It is very difficult to obtain a pure sample of Brücke's blue, but this can now be had from Dr. Grübler of Leipzig. Use a saturated watery solution of Brücke's blue.

Weigh 5 grms. of gelatine, and treat it exactly as described for the carmine mass of Ranvier. Take 125 cc. of the blue solution and heat it on a water-bath, and when the gelatine is fluid and still on the water-bath, add the warm blue solution in a small quantity at a time and stir briskly. The glass rod should show no granules on it when it is withdrawn from the mass. Filter the mass through new flannel. Even the best gelatine gives a precipitate at first, but it disappears with heat.

Other injection masses are used, *e.g.*, a watery solution of Brücke's blue, or gelatine and silver nitrate. These are referred to in the text.

Brass Syringe.—Many good injections have been made with a brass syringe. The syringe should have a long barrel, and be warmed by repeatedly sucking up hot water before the injecting fluid is drawn into it. When the injection mass is forced into the blood-vessel, the pressure should be applied steadily, and should not be so great as to rupture the small blood-vessels.

When the blood-vessels of an animal are to be injected, deeply narcotise an animal with chloroform, *e.g.*, a rabbit or a rat; make a vertical incision through the skin from the lower part of the neck to the ensiform cartilage, cut through the sternal cartilages, turn up the breast-bone, and pull the sides of the thorax apart to reveal the contents of the chest. Open the pericardium and make a snip into the right ventricle. Tie a ligature round the upper part of the sternum to prevent escape of the injection through divided vessels. Wash the blood out of the chest. Snip off the apex of the heart, whereby the cavity of the left ventricle is opened into.

Insert a cannula into the left ventricle and push its nozzle into the aorta. Tie it firmly into the aorta with a stout thread. Place the animal in a bath in warm water at 40° C. If a syringe is to be used, by means of a pipette fill the cannula with the injecting mass, and attach the syringe and force the mass onwards into the blood-vessels. This is done by slow, steady pressure. It takes fifteen or twenty minutes to make a good injection. Any sudden increase of pressure is apt to cause rupture of blood-vessels and consequent extravasation of the injection mass. We can judge when a part is well injected by the colour of semi-transparent parts, such as the gums or the skin. They must be deeply coloured by the injection mass if the injection is successful.

Continuous Air Pressure.—Most frequently injections are now made by continuous air pressure. The apparatus used should consist of a tin trough sufficiently large to contain the animal to be injected, and contain sufficient water to cover it. The water is kept at 40° C. by means of a gas-burner or spirit-lamp. In the same trough are placed the injection masses in Wolff's bottles. Each Wolff's bottle is connected to a large air-chamber into which water can flow from the water-tap, and thus compress the air. The pressure within this cylinder can be registered by means of a manometer. The compressed air acts on the surface of the injection mass in the Wolff's bottle, and forces it through a tube which is attached to the cannula fixed in the aorta. The large cylinder for the compressed air may be made of tin, or one of the large stone jars used by spirit merchants, or a carboy may be used.

After the tissues are injected, they should be cooled rapidly by being placed in running water. After the mass is completely set the injected organs are cut into small pieces and hardened in alcohol.

The methods of **interstitial** injection of fluids and the **puncture** methods are referred to in the text.

XVI.—EXAMINATION OF FRESH TISSUES AND FLUIDS.

In examining a fresh tissue or organ snip off a small part with scissors and tease it in normal saline, or one of the indifferent fluids mentioned in Chapter III. A convenient plan with some organs, *e.g.*, lymphatic gland or liver, is to make a fresh cut and scrape the surface with the blade of knife, and then examine the scrapings.

If it be desired to study the tissue elements, it may be placed in one of the macerating media mentioned in Chapter IV., the particular fluid selected depending, of course, on what object is sought to be obtained.

If it be desired to render certain parts of the tissue more transparent, it may be examined in glycerine.

Again acetic acid (1–2 per cent.) may be added to a fresh tissue. It has the double action of making connective tissue swell up and become transparent, thus making nuclei more evident, while it also shrivels the latter somewhat. Albuminous granules are dissolved by it, while oil globules are not, so that it may be useful occasionally in determining the nature of the granules in protoplasm. Elastic fibres are not affected by it, and thus can readily be distinguished from the white fibres of connective tissue. Groups of micro-cocci are also not affected by it.

The tissue may be stained by **acetic fuchsin**. This is made as follows:—To a 2 per cent. solution of acetic acid add sufficient fuchsin to give a saturated red colour (*Kahlden*). This reagent not only makes the nuclei visible, but it stains them as well.

Sometimes a weak watery solution of iodine makes the outlines of

EXAMINATION OF FRESH TISSUES AND FLUIDS.

tissue elements more distinct. It is used in the form known as **Lugol's Solution** diluted with water.

Iodine	1 part.
Potassic iodide	2 ,,
Water	100 ,,

Weak-Alkalies (1–3 per cent.) dissolve most tissues with the exception of elastic fibres, pigment, fat, and bacteria.

The vapour of osmic acid or the fluid itself (1 per cent.) may be used. It blackens fatty particles.

Finally, the tissue may be stained by means of a watery solution of methyl-green, methyl-violet 5B, acetic fuchsin, or methylene-blue in the form of **Löffler's Methylene-blue**.

Concentrated alcoholic solution of methylene blue	30 cc.
Caustic potash (0.10 p.c.)	100 ,,

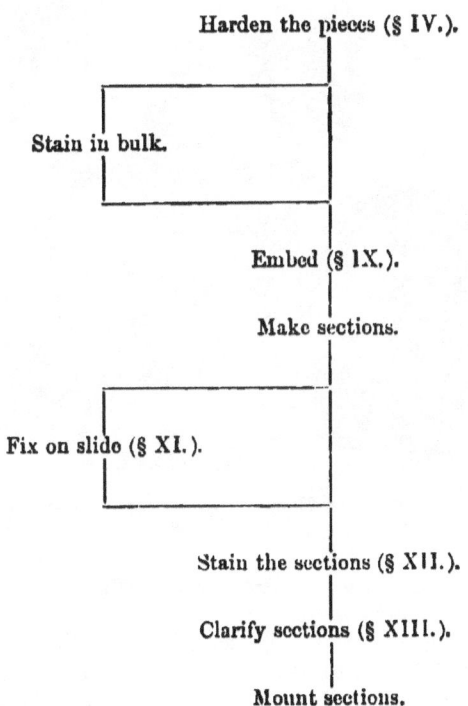

Scheme for Living or Fresh Objects.

Harden the pieces (§ IV.).

Stain in bulk.

Embed (§ IX.).

Make sections.

Fix on slide (§ XI.).

Stain the sections (§ XII.).

Clarify sections (§ XIII.).

Mount sections.

94 PRACTICAL HISTOLOGY.

Fresh Fluids with Suspended Particles.—In the examination of a fluid for suspended particles, *e.g.*, bacteria, such as cells, membranes, &c., especially if these be few in number, it is well to place it in a

FIG. 51.—Hand Centrifuge made by Muencke, Luisen-strasse, 58, Berlin, N.W.
It costs £3, 10s.

conical glass and allow the deposit to subside. It can then be removed with a glass pipette.

Centrifugal Apparatus.—To collect the sediment or suspended particles, a **centrifugal apparatus** is most useful. By means of it the deposit can readily be collected at the bottom of a test-tube.

EXAMINATION OF FRESH TISSUES AND FLUIDS.

Fig. 51 shows a form of **hand-centrifuge** devised by Litten[1] and Muencke of Berlin. The figure is reduced to ⅓ the natural size. One revolution of the wheel R—its teeth fit into the thread of the vertical axis S—causes 50 revolutions of the disc M with the 4 glass tubes G. The wheel can readily be turned 100 times per minute, which gives 5000 revolutions per minute for the disc. Fig. II. shows the disc in full rotation, and fig. III. the form of glass vessel used.

By means of this instrument, corpuscles of light specific gravity, such as blood-corpuscles, albumen, micro-cocci, as well as crystals, e.g. oxalates, can readily be obtained in the form of a sediment, and it is therefore especially useful for the investigation of the deposits in urine and exudations. It is also very useful for obtaining tubercle bacilli from sputum.

Synoptical Statement.

Processes required for Preparing a Specimen for Microscopic Examination, e.g., the Spinal Cord of a Dog or Cat (Garbini).

Cut the cord into pieces about 2 centimetres in length, wash them in normal saline to remove all blood. Place them in

1. Bichromate of potash (2 p.c.) . . . 10 days ⎫
2. Wash in running water . . . . 12 hours ⎪ Hardening.
3. Alcohol (50 p.c.) . . . . . 1 day ⎬
4.   ,,   (70 p.c.) . . . . . 4 days ⎪
5.   ,,   (90 p.c.) . . . . . 1 day ⎪
6.   ,,   (absolute) . . . . . 1 day ⎭
7. In xylol or chloroform . . . . 12 hours ⎫ Embedding.
8. In paraffin on warm bath . . . 6–8 hours ⎬
9. Embed in paraffin. ⎭
10. Cut sections with microtome, and fix them on a slide ⎫ Cutting.
    with fixative. ⎬
11. Remove paraffin by washing in toluol or turpentine, and ⎪
    then in absolute alcohol to remove essential oil. ⎭

12. Place in 70 p.c. alcohol. ⎫   12. Place in distilled water.
13. Stain in strong carmine ⎪ Staining with animal or vegetable dyes.   13. Pour on a few drops of
    or hæmatoxylin (5–15 ⎬   the aniline dye (3–5
    minutes). ⎪   minutes.)
14. Wash in water, then in ⎪   14. Wash in water and then
    70 p.c. alcohol, and de- ⎪   in absolute alcohol until
    hydrate in absolute alco- ⎪   the section has the de-
    hol (5–10 minutes). ⎭   sired tint.

[1] *Deutsch. med. Wochensch.*, No. 23, 1891.

15. After removing surplus alcohol, pour on a few ⎫ Dehydra-
    drops of essential oil (origanum, xylol). ⎬ tion and
16. Remove oil and mount in xylol-balsam. ⎭ mounting.

N.B.—In making a balsam preparation, the sections must always pass through four groups of fluids.

$$1 \longrightarrow 2 \longrightarrow 3 \longrightarrow 4$$
Watery liquid—Alcohol—Essential oil—Balsam.
$$1 \longleftarrow 2 \longleftarrow 3 \longleftarrow 4$$

Never take an object from one group to another of the series without passing it through the intermediate group; it must be passed from the 1st to the 4th, and on the return from the 4th to the 1st.

PART II.

LESSON I.

MILK, GRANULES, FIBRES, AND VEGETABLE ORGANISMS.

1. Examine the Microscope, the objectives, and the eye-pieces.

(*a.*) Select the objective and ocular required. For a high power (**H**), if a Zeiss' microscope be used, select the objective D and the ocular 2 ; if Hartnack's, the objective No. 7 and the eye-piece III. See that the lenses are clean. Place the ocular in the tube, and screw the **H** lens to the lower end of the tube, and leave it half an inch above the level of the stage. For a low power (**L**) use No. 2 ocular of Zeiss or III. of Hartnack, and objective A or No. 3 respectively. In using a low power, the lens must be $1\frac{1}{2}$ inches above the stage to begin with.

(*b.*) With the microscope in front of you, with high-power lens on it, arrange the concave side of the mirror under the stage so as to reflect a beam of light up the tube of the microscope into the eye, looking in at the ocular. Turn the sub-stage diaphragm until a small aperture in it is under the aperture in the centre of the stage. If any specks are visible on looking through the microscope, rotate the ocular ; if they move, of course they are on the ocular itself. Clean the outer surfaces of the lenses of the ocular with a piece of clean wash-leather, which should be kept tied to the microscope and used for no other purpose than cleaning the lenses. Replace the ocular, and if specks are still present and move when the ocular is moved, they must be on the inner surface of the eye-glass or field-glass of the latter. This is easily determined by rotating the eye-glass of the ocular alone, while looking through the microscope, and observing if the specks do or do not move with it. Clean the inner surfaces of these lenses. A general dimness indicates that the objective itself is dirty. The light used should not be direct sunlight, but preferably light reflected from a white cloud.

The rule with regard to the use of the diaphragm must never be neglected, viz., to use a small aperture with a high power, and a large aperture with a low power.

2. Clean a Slide and Cover-Glass.

(*a.*) *The Slide.*—Seize the slide by its edges with the thumb and forefinger of the left hand, dip one half of it into water, withdraw it, and with a clean old handkerchief rub both wetted surfaces at once until they are clean and dry. Reverse the slide, still holding it by its edges, and dip the other end in water, and clean its surfaces as before. Lay the slide upon some clean, suitable background, white or black paper, or on the photophore.

(*b.*) *The Cover-Glass.*—Sometimes the covers have a thin film on them; this may be got rid of by placing them in strong sulphuric acid, and subsequently removing every trace of acid by water. Dip the cover-glass in water, take it between two folds of a handkerchief held between the thumb and forefinger of the right hand, and rub both surfaces at once. After it is cleaned, do not lay it flat, but tilt it up against some convenient object.

The first lesson is devoted to the examination of a few simple objects—some of which are occasionally found as foreign bodies in microscopical preparations—with a view to familiarise the student with the use of the microscope.

3. Milk.

—By means of a glass rod, place on the *centre* of the slide a small drop of milk diluted with three or four volumes of water. To find the centre of the slide, use the mounting block (p. 89).

Apply a Cover-Glass.—Seize the cover-glass by the edge by means of a pair of forceps with broad points. The pattern shown in fig. 6 is convenient. The edge of the cover-glass opposite to the forceps is allowed to touch the slide close to the drop of fluid, the edge opposite being gradually and evenly lowered by depressing the forceps until the fluid touches the under surface of the cover-glass. By lowering the cover-glass thus gently and obliquely the entrance of air-bubbles is avoided. Place the object on the stage right under the lens.

Focus the Object (**H**).—The objective is still half an inch above the stage. While looking into the eye-piece of the microscope, seize the tube of the latter between the thumb and adjoining fingers of the right hand, and with a screwing or twisting movement of the tube from left to right, gradually depress the tube until the outlines of the object are indistinctly seen. This is the *coarse adjustment*. The focussing process is facilitated by keeping the slide and object moving slightly. This can readily be done by moving the slide with the thumb and forefinger of the left hand, the ulnar margin of the palm conveniently resting on the table.

Now use the *fine adjustment*, and bring the outlines of the object in the field sharply into view.

It is of the greatest importance that the student should be taught to describe the objects which he sees, and also to make sketches of them. To facilitate the description of isolated objects, the following heads may be adopted:—

    a. Shape.
    b. Border.
    c. Surfaces (upper and lower).
    d. Size.
    e. Colour.
    f. Transparency and relation to light.
    g. Contents.
    h. Effects of reagents.

In the object under examination there is a large number of minute bodies floating in a fluid. Describe the appearance of the floating particles under the following heads:—

(*a.*) *Shape.*—The milk globules (fig. 52) are spherical, as can be shown by touching the edge of the cover-glass with a needle, and then observing them as they rotate in the field of the microscope. Moreover, if one be focussed, its outline comes gradually into focus, and disappears gradually, while optically with regard to light these bodies behave as globules, and not as discs.

(*b.*) *Border.*— Smooth and regular.

(*c.*) *Surfaces.*— Elevate the objective by means of the fine adjustment until the upper surface of the globules comes into view; depress it again slowly and then examine the globule throughout its entire thickness, until its under surface is brought into view. Both surfaces are smooth.

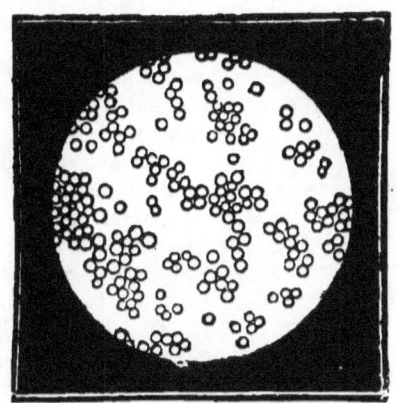

FIG. 52.—Milk Globules, × 400.

(*d.*) *Size.*—The globules are not all of the same size. If desired, measure their actual size (p. 20).

(*e.*) *Colour.*—The smaller ones appear colourless, but some of the larger may have the slightest tinge of a faint yellow.

(*f.*) *Transparency and relation to light.*—They are transparent, because the outline of a subjacent one can be seen through a globule lying over it. Notice also the highly refractile character of each globule, characteristic of an oil droplet (fig. 55, 3).

(*g.*) *Contents.*—They appear homogeneous and uniform, and no included body is to be seen. Each globule is, in fact, a globule of oil.

(*h.*) *Effects of reagents.*—To one side of the cover-glass apply a drop of *acetic acid*. To the opposite edge of the cover-glass apply the apex of a triangular piece of blotting-paper. The blotting-paper sucks up some of the milk, and the acid runs in at the opposite side to supply its place. This is the process of *irrigation*. Move the slide to bring into focus a part of the field which has been acted on by the acid, and note that the corpuscles, instead of floating about singly as before, are now aggregated into small groups. The acid seems to have altered the surfaces of the globules, so that they adhere to each other. The acid is said to act on the casein envelopes of the globules, and to soften or dissolve them. This preparation is not to be preserved.

Make sketches of these objects before and after the action of reagents.

4. Potato-Starch Granules.—With the blade of a knife gently scrape the surface of a freshly-cut raw potato; place the matter so obtained in a drop of water on a slide. Remove any coarse fragments, and apply a cover-glass. Focus the object (H).

(*a.*) Observe that the granules (fig. 53) are ovoid bodies of unequal size, not equal at the two ends, clear, and with a sharp outline. Near the smaller end of each granule notice a small spot, the "nucleus" or hilum, round which are concentric layers, giving rise to the appearance of fine concentric lines arranged with relation to the nucleus. The lines are more numerous on one side of the hilum than the other. Sketch two or three of the granules.

FIG. 53.—Granules of Potato-Starch.

(*b.*) Irrigate the corpuscles with a diluted solution of iodine in iodide of potassium. Each granule becomes blue as the iodine reaches it. This is due to the formation of iodide of starch. If the iodine be too strong the granules appear black.

5. Rice-Starch (H).—Examine a little rice-starch in the same way. Notice the much smaller irregular granules. Each granule is polygonal, mostly five or six sided. The granules are stained blue by iodine (fig. 54). Make sketches of the starch corpuscles.

6. Gamboge and Brownian Movement (H).—Rub up a small piece of solid gamboge in water, until the latter has a faint yellow appearance. Place a drop on a slide, cover, and examine.

(*a.*) Observe granules of various sizes and shapes floating in the

MILK, FIBRES, VEGETABLE ORGANISMS.

field. The larger ones appear coloured yellow, but the finer ones cannot be seen distinctly as yellow bodies.

(*b.*) Note the finer granules, which exhibit a slow dancing movement. They are never at rest. This is called Brownian movement, and appears to be due to inequalities of temperature in different strata of the fluid. This movement is exhibited by all finely-divided particles suspended in a fluid, *e.g.*, China ink, Berlin blue, provided the particles be small enough, and the fluid in which they are suspended be not too viscid. The fine granules in salivary corpuscles exhibit this movement. (Lesson IV. **1**, *d*).

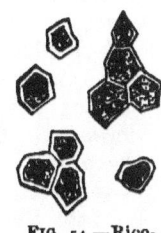

FIG. 54.—Rice-Starch.

7. Air-Bubbles in Gum or Water (H).—Make a solution of gum

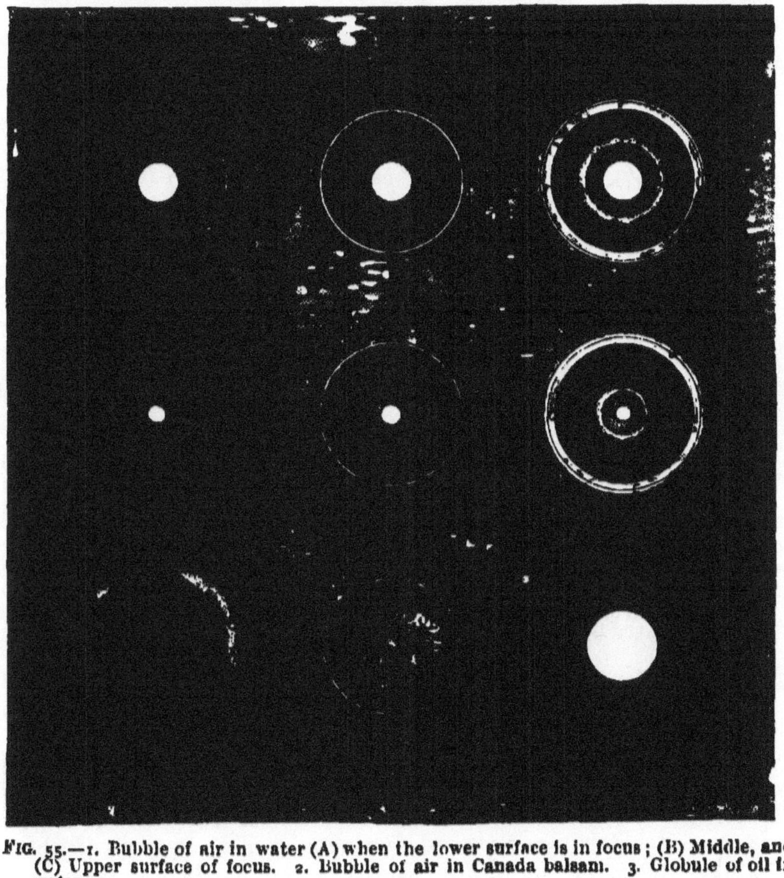

FIG. 55.—1. Bubble of air in water (A) when the lower surface is in focus; (B) Middle, and (C) Upper surface of focus. 2. Bubble of air in Canada balsam. 3. Globule of oil in water.

mucilage and shake it up in a test-tube with air until it forms fine bubbles. Place a drop of it with its included bubbles on a slide, cover, and examine.

(*a.*) Observing larger and smaller bubbles, especially note how the appearance of the **bubble varies with the elevation or depression of the lens.** Sketch these appearances. When the tube is depressed, the bubble has a small, clear centre, and a wide, black, sharp, refractile margin (fig. 55, A), because so much of the light is refracted by the air, and does not pass through the bubbles of air into the lens. Study the appearance of the bubble when the centre and then the upper surface are in focus (B, C).

The student will have an opportunity by-and-by of observing the appearance of a bubble of air in Canada balsam (fig. 55, 2), and an oil globule in water (fig. 55, 3).

8. Cotton Fibres (H).—Place a few fibres of cotton-wool in water, cover, and examine.

(*a.*) Observe the fine translucent flattened threads, which are really tubes, each looking as if twisted on itself at intervals (fig. 56).

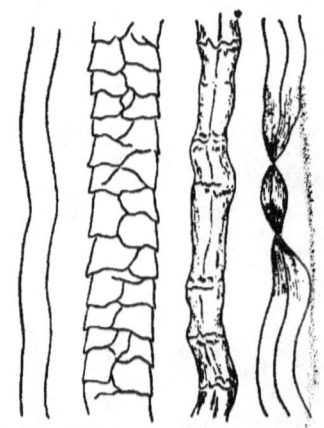

Silk. Wool. Linen. Cotton.
FIG. 56.—Fibres of Silk, Wool, Cotton, and Linen.

(*b.*) Remove the cover-glass and the water, add a drop of iodine; cover and examine; note that the fibres are yellowish; add a drop of strong sulphuric acid or a drop of the following mixture:—Glycerine 2, water 1, and sulphuric acid 3 parts. At the edge of the cover-glass suck it through with blotting-paper, and note that the fibres become *blue*. They are composed of cellulose, which does not give a blue with iodine alone, but with iodine and sulphuric acid.

9. Linen Fibres (H).—Examine in water.

(*a.*) Observe the cylindrical or flattened translucent fibres with no twist; but they have a few markings on them here and there, and at these points the fibre is generally slightly thicker. They are in reality tubes with thick walls (fig. 56).

10. Wool (H).—Examine in water.

(*a.*) Observe the cylindrical fibres with numerous zigzag transverse lines due to the epithelial covering of the fibre (fig. 56). It is convenient to examine dyed wools also, as some of these exhibit the zigzag imbricate scales even better than undyed wool.

11. Vegetable Cells forming a Membrane.

—With a pair of forceps peel off a thin layer of the covering of a fresh onion bulb. Examine it either in water or in water after staining with a solution of iodine (H). Observe the cells (fig. 57) with well-defined walls,

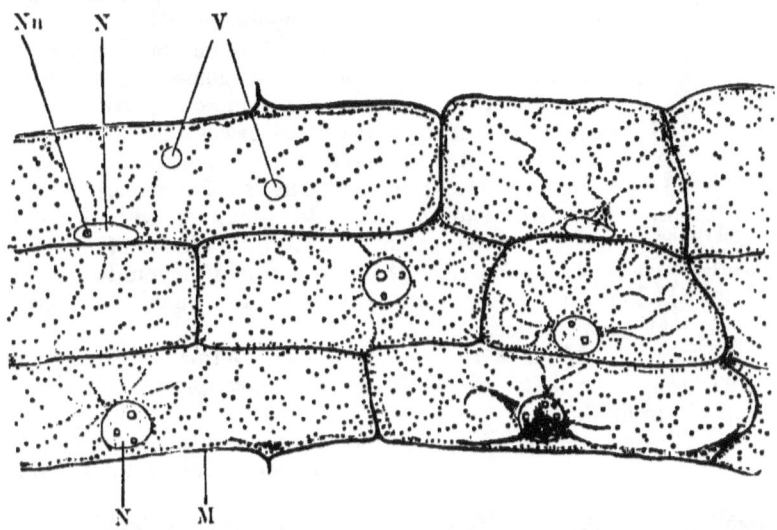

FIG. 57.—Cells from bulb of a fresh onion forming a membrane. M, cell membrane; N, nucleus seen from the surface; Nu, nucleolus; V, vacuoles, × 240.

and united to each other by the edges to form a membrane—the excentrically placed spherical nucleus and the granular cell-contents.

12. Cells with sinuous margins, and containing chlorophyll granules, may be studied in a leaf of the Duckweed or *Lemna minor*, also the reaction for cellulose (sulphuric acid and then tincture of iodine = blue) in the same plant.

13. The currents in protoplasm are well seen (H) in *Chara vulgaris*, which is so common in our streams.

Vegetable Micro-Organisms.—These are frequently found in microscopic preparations, while others are the cause of various diseases. They are usually classified as follows:—

1. *Moulds*.
2. *Saccharomycetes* (yeast-like organisms).
3. *Schizomycetes* (bacteria-like organisms).

14. Penicillium (H).

—This mould is readily found on starch-paste or on Müller's fluid preparations which have been left standing uncovered. Place a little on a slide, add a drop of water, cover, and examine.

(a.) Observe the stems, consisting of narrow, clear, oblong cells, joined end to end, and on the summit of each are several rows of small spores. The rootlets or mycelium consist of elongated chains of narrow oblong cells.

FIG. 58.—Development of a Gonidiophore bearing Stylogonidia (*st.g*) on the Hyphæ of Penicillium.

The *mycelium* is composed of much-branched *hyphæ*. The cells, elongated and narrow, composing these are separated from each other by numerous partition walls (fig. 60, *e*). The walls consist of cellulose and contain protoplasm with vacuoles.

(b.) Search for one of the hyphæ which bear on their free ends a brush-like group of cells, some of which become constricted to form chains of spherical *stylogonidia* (fig. 58). The ripe stylogonidia have a green colour, and they give the mould its green appearance.

15. Yeast (H).—Place some German yeast in sugar and water; keep it for several hours in a warm place. Examine a drop of the fluid.

(a.) Observe the small oval yeast-cells, each with an envelope, a large clear vacuole, and granular protoplasm.

FIG. 59.—Yeast-Cells, *a. b* and *c* Budding.

(b.) Search for one budding, and notice the small faintly granular bud adhering to and projecting from the mother-cell (figs. 59, 60, *f*).

Each yeast plant is a single cell or morphological unit. It is composed of a transparent thin delicate envelope or cell-wall, composed of cellulose. When the cell is ruptured the empty cell-envelope may be seen in the field. Within is *granular protoplasm*, containing, as a rule, a clear space or *vacuole*. Some cells contain a nucleus.

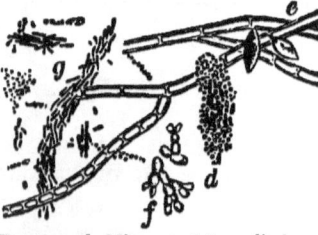

FIG. 60.—*d.* Micrococci in cylinders; *e.* Mycelium of Penicillium; *f.* Yeast-Cells; *g.* Bacilli and Micrococci.

(c.) Stain the preparation with a watery solution of magenta. Note that all the cells do not stain equally well. The buds on the side are stained of a deep red, and to a less extent the protoplasm of most of the cells.

16. Micrococci and Bacteria (H).—Set aside a solution of peptone or a watery extract of a piece of flesh for some time, until a scum forms on the surface due to putrefaction. Place a little of the scum on a slide, cover, and examine.

(*a.*) Observe groups of small round specks, often held together by a homogeneous medium. These are micrococci (fig. 60, *d*).

(*b.*) Small elongated rod-like bodies, each moving across the field in a zigzag like manner. These are bacteria (fig. 60, *g*).

N.B.—*In all cases sketches must be made of the objects examined. This holds good for this and all succeeding lessons.*

17. Determine the magnifying power of the microscope.—Do this according to the method described at p. 19.

ADDITIONAL EXERCISES.

18. Staining of Fission Fungi or Schizomycetes.—They stain readily with aniline dyes, especially basic aniline colours. Scrape off a little of the coating which accumulates on the surface of the molar teeth, press it between two cover-glasses, so as to make cover-glass preparations. Glide the one glass off the other. Place the cover-glasses, film surface downwards, in a 1 per cent. solution of methyl-violet, gentian-violet, or methylene-blue contained in a watch-glass. Heat the watch-glass and its contents over a gas-flame until a faint cloud of vapour rises; allow it to cool, and in 5-10 minutes or so the coloration is complete. Remove the cover-glasses, wash them in water, and place them in absolute alcohol. Remove the cover-glasses and allow them to dry in the air and mount in xylol balsam. A moist cover-glass preparation from absolute alcohol may be clarified with xylol.

(*a.*) Observe many epithelial cells coloured, but so are the organisms.

If it is desired to have only the organisms coloured, the stained cover-glasses are taken direct from the staining solution, or laved in absolute alcohol and placed in a solution of iodine composed of 1 part iodine, 2 parts potassic iodide, and 300 water, in which they are kept for a few minutes. Transfer them to and wash them in absolute alcohol until nearly all colour is gone, and clarify with xylol. Only the organisms are of a dark-blue tint; the other tissues are decolorised or nearly so. This is **Gram's method**.

Stain bacteria from a putrefying proteid fluid, *e.g.*, peptones or meat-extract, in the same way.

LESSON II.

THE BLOOD.

UNDER the microscope blood is seen to consist of a clear, transparent fluid, the **plasma** or **liquor sanguinis**, in which are suspended the **blood-corpuscles**. The blood-corpuscles are of two kinds, the **red** or **coloured**, and the **white, pale**, or **colourless**. Besides these, there fall to be examined the **blood-plates**, or, as they are also called, **blood-tablets** or **platelets**.

$$\text{Blood} \begin{cases} \text{Blood-plasma.} \\ \text{Corpuscles} \begin{cases} \text{Red.} \\ \text{White.} \\ \text{Platelets.} \end{cases} \end{cases}$$

BLOOD-CORPUSCLES OF AMPHIBIANS.

(A.) **Coloured Corpuscles of Amphibians (Newt or Frog)**.—After destroying the brain of a newt or frog, a drop of blood may be obtained from the cut end of the tail of the former, or from the cut surface after amputation of the foot of the latter.

1. **Red Corpuscles (H)**.—Place a small drop of blood in the centre of a perfectly clean slide and cover it at once with a coverglass. Examine it with a high power. To avoid the pressure of the cover-glass, a short length of a hair may be placed in the blood droplet before the cover-glass is applied.

(*a.*) Observe the coloured and colourless corpuscles, the former much more numerous than the latter (fig. 61).

(*b.*) *Study the red corpuscles.* Observe that they are very numerous, elliptical in outline when seen on the flat, slightly yellowish in colour; their border or contour is even and well defined. Select one seen on edge, and note that it is a thin ellipse, pointed at the ends, becoming gradually thicker in the centre, so that it is a *bi-convex elliptical disc*. Sometimes one corpuscle can be seen overlying a subjacent one, in which case the outline of the latter can be distinctly seen through the former, indicating that the corpuscles are transparent (fig. 61). Notice within each corpuscle a lighter oval, central area, indicating the existence of an elliptical, colourless, elongated, granular-looking included body—the **nucleus**. The long axis of the colourless nucleus coincides with the long axis of the corpuscle. At first, the nucleus may not be very distinct, but after a time it becomes distinctly visible, and it can be readily made so by the action of certain reagents, especially weak acids. Small vacuoles frequently appear in the body of the corpuscle, more especially in frogs that have been kept some time. In others there may be seen a faint radiate striping of the corpuscles as if there were fine folds in it due to partial drying.

The yellow colour is due to the presence of **hæmoglobin**, which is enclosed within the meshes of a colourless **stroma** or delicate framework.

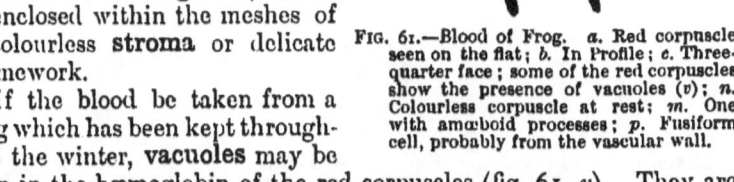

FIG. 61.—Blood of Frog. *a.* Red corpuscle seen on the flat; *b.* In Profile; *c.* Three-quarter face; some of the red corpuscles show the presence of vacuoles (*v*); *n.* Colourless corpuscle at rest; *m.* One with amœboid processes; *p.* Fusiform cell, probably from the vascular wall.

If the blood be taken from a frog which has been kept throughout the winter, **vacuoles** may be seen in the hæmoglobin of the red corpuscles (fig. 61, *v*). They are rapidly produced in frogs after the injection of ammonium chloride.

Sketch two or three corpuscles, both red and white; the red corpuscles both on the flat and on edge, and two overlapping each other.

2. Acetic Acid.—To the edge of the cover-glass of the same preparation apply a drop of dilute acetic acid (1 per cent.—1 cc. glacial acetic acid to 99 cc. of normal saline solution). The acid runs in under the cover-glass, but if it does not, apply a small triangular piece of blotting-paper to the opposite edge of the cover-glass. Let one of the angles of the blotting-paper touch the fluid under the cover-glass, and it will suck up some of the blood, and thus cause the acetic acid to run in at the opposite side. This is known as the process of **irrigation**. Move the slide, and find a part of the object which has been acted on by the dilute acid.

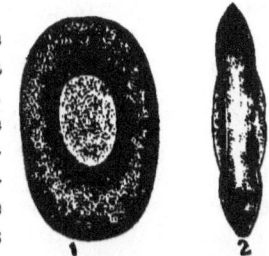

FIG. 62.—Amphibian Coloured Blood-Corpuscles seen on the Flat and on Edge, × 1000.

(*a.*) Observe that the red corpuscles, or at least most of them, become spherical and decolorised, the nucleus becomes very distinct and granular, while the outline of the corpuscle may become very indistinct (fig. 63). The nucleus appears somewhat shrunken or shrivelled. If the corpuscles are roughly treated, the nucleus may be seen placed excentrically, or even extruded from the corpuscle. The acid acts on the hæmoglobin, forming a new compound, which diffuses out of the corpuscles and stains the surrounding plasma a faint yellow. The nucleus in some of the corpuscles may absorb some of the yellow pigment, and become stained thereby, especially if a strong solution of acid has been used. The action on the colourless corpuscles is referred to at p. 112.

3. Dilute Hydrochloric Acid, 1 per cent. (H).

(*a.*) To a fresh drop of blood add, as before, a drop or two of dilute hydrochloric acid. Watch diligently one or two of the red corpuscles. They gradually enlarge, become spherical, and may all of a sudden burst and discharge their contents, the nucleus coming clearly into view during the process. After the rupture, the residue of the stroma of the corpuscles may be seen in the field. In other cases the corpuscles become clear, globular, and transparent, with

FIG. 63.—Frog's Red Blood - Corpuscle acted on by Dilute Acetic Acid, × 300.

FIG. 64.—Action of Water on an Amphibian Coloured Blood-Corpuscle.

FIG. 65.—Action of Syrup on Frog's Red Blood - Corpuscle, × 300.

here and there fine shreds stretching between the nucleus and the surface of the spherical corpuscle.

4. Water (H).

—To a fresh preparation of blood apply a drop of water to the edge of the cover-glass, and notice its effects upon the corpuscles.

(*a.*) The water rapidly diffuses into the corpuscles and renders them spherical, while at the same time it decolorises them, the hæmoglobin diffusing outwards into the plasma, and staining it slightly yellow. The nucleus also becomes spherical. Thus the outline of the corpuscles becomes very faint in the field of the microscope, the corpuscles themselves now almost consisting of a nucleated stroma (fig. 64).

5. Strong Syrup (H).

—Place a small drop of blood on a slide and near it a drop of syrup; mix the two with a needle, and apply a cover-glass.

(*a.*) Observe that some of the red corpuscles are rapidly shrivelled and puckered, especially when seen on edge, owing to fluid passing out of them by exosmosis (fig. 65). Some of them may present here and there a reddish tinge. All the corpuscles are not affected equally or at the same time. The same effects are produced by strong saline solutions.

6. Tannic Acid (H).—On a slide mix a drop of blood with a drop of tannic acid, using a relatively large drop of the acid fluid; wait for about a minute, and then apply a cover-glass.

(*a.*) Observe that some of the corpuscles become globular, while the hæmoglobin passes out of the corpuscles at one or more spots, and appears on the surface in the form of one or more small granular buds (fig. 66, *c*). Sometimes the bud or buds are small. At others the bud may be as large as the remainder of the corpuscle, which has become smaller and partly decolorised. In other cases it may be collected around the nucleus (*Roberts*). The tannic acid causes a separation of the hæmoglobin from the stroma.

The solution of tannic acid is made by dissolving 2 grains of tannic acid in 1 oz. of boiling water, and allowing it to cool.

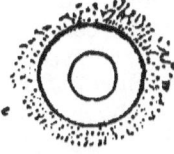

FIG. 66.—Action of Tannic Acid on the Red Blood-Corpuscles of Man (*a*, *b*) and Frog (*c*).

FIG. 67.—Action of Boracic Acid on a Frog's Red Blood-Corpuscle.

7. Boracic Acid.—Mix a drop of newt's or salamander's blood with a 2 per cent. solution of boracic acid, which takes a short time to act on the corpuscles and produce its effects.

(*a.*) Observe that the hæmoglobin is collected around the nucleus, so that the presence of the latter is obscured, but in many of the corpuscles fine threads of hæmoglobin remain attached to the circumference of the corpuscle, so that the retracted hæmoglobin may have a stellate form, while the rest of the corpuscle is colourless. To the latter Brücke gave the name *Oikoid*, to the former *Zooid* (fig. 67).

If these corpuscles are acted on for twenty-four hours with 1 per cent. osmic acid, they are "fixed," and may be mounted permanently in glycerine-jelly.

8. Magenta.—Mix a drop of blood and a drop of the special magenta fluid (p. 74) on a slide, cover, and examine.

(*a.*) Observe that the nuclei of the corpuscles, both red and white, are stained of a brilliant red, although the surrounding part of the corpuscle is not so stained, unless the magenta be in great excess. All the corpuscles are not stained equally brightly. On the edge of some of the corpuscles at one or more points will be found small coloured spots or thickenings.

9. Osmic Acid and Picrocarmine (H).—This preparation is best made by mixing a few drops of blood with an equal volume of 1 per cent. osmic acid in a small tightly-corked tube, or by exposing a thin film of blood to the vapour of a 2 per cent. solution of osmic acid. After two to four hours pour off the supernatant fluid and cover the residue of corpuscles with picrocarmine. After twenty-four hours the picrocarmine can be poured off, and a little of the deposit placed on a slide and mixed with glycerine-jelly dissolved by heat, and covered (p. 85).

(a.) Observe that the nuclei of the corpuscles are bright-red, and the perinuclear part of the corpuscles yellow. Within the granular-looking nucleus, with a good lens, may be seen a network of fibrils.

This preparation is permanent, and must be sealed up or "ringed" after the manner described at p. 88.

10. Blood of Bird (H).—Mount a drop of fresh blood.

(a.) Observe the red corpuscles, which are elliptical, biconvex, nucleated bodies, but smaller in size and more pointed than amphibian corpuscles (fig. 68).

These corpuscles behave towards reagents as those of amphibians.

11. Blood of Fish (H).—Mount a drop of fresh blood, which is readily obtained from a gold-fish or salmon, but the blood coagulates rapidly.

(a.) Observe the red corpuscles, which are elliptical, biconvex, and nucleated, but the ends are not so pointed as in the bird, while, like the bird's corpuscles, they are smaller than those of amphibians (fig. 68, F). The blood must be quite fresh.

In both cases a few colourless corpuscles may be noticed.

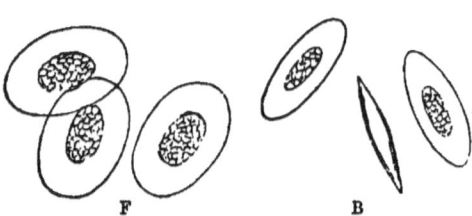

FIG. 68.—Red Blood-Corpuscles of Fish (F) and Bird (B), × 450.

(B.) **The White Corpuscles of the Frog or Newt.**

12. White Corpuscles.—In a fresh preparation of blood (taking care to place a hair under the cover-glass) search for the colourless corpuscles, of which there are several varieties. They are much less numerous than the red.

(a.) Observe that there is: (i.) The **finely granular form**, consisting of a nucleated mass of protoplasm larger than a red corpuscle (fig. 61, *m*). (ii.) The **coarsely granular** variety may be found. In it the granules are large and refractive, and often lying at one side of the corpuscles. (iii.) A third variety, much smaller than the others, may be found (fig. 61, *k*).

(*b.*) In all the varieties, by careful observation, may be detected a nucleus, which in some is irregular or subdivided. The surface of these corpuscles is sticky, as can readily be shown by giving the cover a push with a needle, when the coloured corpuscles will be seen to glide over each other, while some of the colourless ones will be seen adhering to the glass; and even if a coloured one impinges on them, they are rarely displaced by the impact, so firmly do they adhere to the glass.

13. Amœboid Movements of the White Corpuscles.—Select one of the large finely granular corpuscles, and at once make a sketch of its outline; in a minute or two make another sketch, and do this every few minutes. The sketches will vary, because the corpuscle has slowly

FIG. 69.—Showing the Amœboid Movements of Colourless Blood Corpuscles.

changed its shape, even at the ordinary temperature. A process of its protoplasm may be extruded on one side, while a part of the corpuscle may be drawn in at another. The corpuscle, therefore, exhibits spontaneous *irregular and indefinite movements*, constituting what is known as *amœboid movements*. It may even be seen to change its place, and thus exhibit *locomotion*.

In the *coarsely granular* form, if one be found, the processes are not so pointed, while the granules may be seen to pass suddenly

FIG. 70.—Ranvier's Moist Chamber. *s.* Disc on which the fluid to be examined is placed; *d.* Air-Chamber; *c.* should be placed on the square part on which the cover-glass rests.

from one side of the corpuscle to the other. The granules seem to be passive, and their motion is due to movements of the protoplasm.

These movements may be watched for a long time if the preparation be sealed up either with melted paraffin wax or with oil. The former is to be preferred. When sealed up in paraffin wax, the preparation may be kept without evaporation for several days. The margins of the cover-glass—preferably a square one—are sealed down with paraffin wax by heating a wire bent at right angles to melt the paraffin and applying it along the edges of the cover-glass.

Ranvier uses a moist air-chamber (fig. 70). In the centre of the chamber is a small piece of glass, less in height by $\frac{1}{10}$th of a millimetre than the height of the walls of the cavity. The fluid containing the corpuscles is placed on this, covered with a cover-glass which is sealed down by means of paraffin wax. The object is thus protected from evaporation, while it has in relation to it a layer of air.

14. Acetic Acid (1 per cent.).—On irrigating a fresh preparation with dilute acetic acid, the protoplasm is made clear and transparent, and the complex nucleus—usually consisting of three parts, and hence called tripartite—is distinctly revealed (fig. 71, a). If the acid act vigorously, it may be almost impossible to see the outline of the now clear protoplasm.

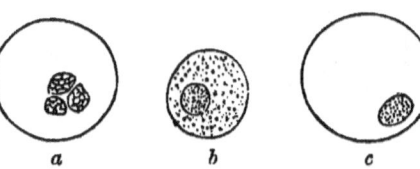

FIG. 71.—*a*. Action of Acetic Acid on the Colourless Blood-Corpuscles of a Frog. Action of Water on the Colourless Blood-Corpuscles of a Frog. *b*. Early, and *c*. Later Stages.

15. Water.—On irrigating a fresh preparation with water, the colourless corpuscles are killed, and they assume a globular form, the protoplasm becoming at first more granular (fig. 71, *b*), and subsequently clear and transparent, thus distinctly revealing the presence of the nucleus. The granules may exhibit Brownian-movements, while each cell has an outline round it as if a membrane had been formed round it. So that by the addition of water these corpuscles may be changed so as to resemble salivary corpuscles (Lesson IV.).

16. Magenta.—This stains the nucleus deeply and the protoplasm to a less degree.

For **Fibrin** (Lesson III.

Lymph.

17. (*a*.) Pith a **frog**, and with a fine pipette withdraw a little lymph from the dorsal lymph sac, or, better still, curarise a pithed frog, and next day remove the lymph from its sub-lingual lymph sac. The lymph accumulates there in large amount.

Lymph (H).

(*a*.) Observe numerous leucocytes, mixed perhaps with a few red blood-corpuscles, exactly like the three varieties of leucocytes found in the blood. A coagulum of fibrin will ultimately be formed.

(*b*.) The **toad** yields a very large quantity of lymph. Destroy its brain, wipe one leg dry, and cut off the projecting toe. At once large drops of lymph flow out, which soon coagulates (*S. Mayer*).

ADDITIONAL EXERCISES.

18. Migration of Colourless Corpuscles (H).—Heat and draw out a glass tube so as to form an excessively fine capillary tube. Pith a frog, expose its heart, make a cut into the latter, and as the blood flows suck up blood into the capillary tube. Seal the tube at both ends by holding it for a second or two in a gas-flame.

Place the tube on a slide in a drop of glycerine or clove-oil, cover it and examine.

(*a.*) (L) Observe that the blood-clot shrinks, squeezes out serum from the clot, and by-and-by some of the colourless corpuscles migrate from or are squeezed out of the red clot into the serum, where they may be seen exhibiting amœboid movements. This shows that the vitality of the white corpuscles is not abolished by the coagulation of the blood.

19. Feeding White Corpuscles (H).—Get some blood from the heart of a frog into a capillary pipette, as described in 18. Allow the blood to clot and exude serum, which it does in about an hour.

Rub up a little Indian-ink in a few drops of normal saline in a watch-glass till a greyish fluid is obtained. Blow the contents of the capillary tube upon a clean slide, remove the clot; the serum contains numerous colourless corpuscles. Mix a little of the greyish Indian-ink fluid with the serum, put a hair under the cover-glass, and seal up the latter with melted paraffin to prevent evaporation. After a time, observe that the minute particles of Indian-ink are seen included in the protoplasm of these cells. The cells throw out processes which surround a particle, meet and coalesce, and thus the particles come to be included in the corpuscle (*Schäfer*).

20. Movements and Division of White Corpuscles.—Ranvier[1] seals up the preparation on the slide devised by him, this being done by paraffin wax. A water immersion lens is placed on the microscope, and then the microscope with the slide is placed in a glass vessel filled with water at the required temperature. In this way the movements of the corpuscles, and, as I have myself seen, even the division of leucocytes, can be readily studied.

21. Glycogen in White Corpuscles (H).—Irrigate a preparation of frog's or newt's blood with 1 per cent. solution of iodine containing 2 grams of potassic iodide. The red corpuscles are stained yellow, the white ones are killed; many of them are also stained yellow, but in some of them may be seen mahogany-coloured granules of stained glycogen.

22. Elder Pith Preparation (H).—Into the dorsal sac of a pithed frog introduce a small piece of elder pith soaked in normal saline, and leave it there for twenty-four hours. Withdraw it, make a thin section, and examine it in normal saline solution.

(*a.*) Observe the large polygonal cells of the elder pith crowded with lymph corpuscles—several varieties—which have, in virtue of their amœboid movements, "wandered" into and permeated the cellular pith.

The elder pith may be hardened in Flemming's fluid (p. 32), and after thoroughly washing it, sections are made and mounted in Farrant's solution.

23. Serous Fluids of Mammals.—A small mammal, *e.g.*, mouse, rabbit, is killed by decapitation. Open its peritoneal cavity by means of a sharp red-hot knife, *i.e.*, by a thermo-cautery. This is to avoid any blood being shed into the cavity. By means of a fine pipette withdraw some of the "lymph" or "serosity." It will be found not to be clear like whey, but slightly opalescent.

(*a.*) **Corpuscles** (H).—Some are like ordinary lymph corpuscles, many are very granular, and there are always red blood-corpuscles also present (*Ranvier*).[2]

---

[1] *Comptes Rendus*, vol. 110, p. 686, 1890.
[2] *Ibid.* vol. 110, p. 768, 1890.

24. Dry Cover-Glass Preparation of Blood.—Place a drop of frog's blood on a cover-glass, and to it apply another cover-glass. Press the two glasses together, and then slip them asunder. There will be a thin film of blood on both glasses. Allow the films to dry.

In a watch-glass place a dilute watery solution of methyl-green, and on this float the cover-glasses, with the blood-surface next the staining reagent, as in the method of staining bacteria.

After ten minutes remove the cover-glass; move it in water; touch the edge of it on blotting-paper to remove the surplus green fluid, and allow the green stain on the glass to dry. After it is dry, add a drop of xylol balsam and place the cover-glass on a slide. This forms a permanent preparation of blood-corpuscles, whose nuclei are stained green.

The same process may be practised with other aniline dyes, or eosin may be combined with methyl-green, as the former stains the hæmoglobin, and the latter the nucleus. A very good stain is eosin-hæmatoxylin (p. 70).

25. Double-Staining of Blood-Corpuscles (*Methyl-green and Eosin*).—Diffuse a thin layer of frog's blood on a cover-glass, allow it to dry, and pour on the dry residue a 1 per cent. watery solution of methyl-green. Leave it on for ten minutes, and then wash off the surplus stain. Pour on a weak watery solution of eosin, and after five minutes wash it off also by rinsing the slide gently in water; dry, and add xylol balsam; cover.

Observe that some of the corpuscles have the nucleus green and the surrounding part coppery-red. Eosin is almost a specific reagent for detecting hæmoglobin.

Numerous combinations of this kind may be made, such as magenta and iodine-green, fuchsin and methylene-blue.

26. A better plan is to place a drop of blood on a cover-glass, and to this apply another cover-glass, press the glasses together, and then separate them so that a thin film of blood adheres to each. Allow them to dry. After they are dry, place them for two hours in a mixture of equal parts of absolute alcohol and ether, which coagulates the proteids of the corpuscles. Float the cover-glasses blood-surface downwards upon a saturated solution of methylene-blue. After an hour or more, wash the cover-glasses in water, then in absolute alcohol, and clear them up with clove-oil in which a little eosin is dissolved. Remove the clove-oil by immersing the cover-glasses in xylol, and mount in xylol balsam. A beautiful preparation is obtained. The nuclei are blue, and the protoplasm of the colourless corpuscles of pale-rose colour.

27. The advanced student should also study the effect of **sulpho-cyanide of potassium** (5 per cent.), **urea, ammonium chromate, pyrogallic acid, heat, carbolic acid** (1 to 1000 of normal saline), and other agents on the coloured corpuscles. **Dilute alcohol** reveals a nucleolus within the nucleus. This preparation may be subsequently stained with magenta.

28. Urea.—A strong watery solution causes the corpuscles to assume irregular forms and to send out processes, which may separate from the corpuscle. Changes not unlike these are produced by neutral ammonium chromate, very bizarre forms being thus produced.

29. Bile, *e.g.*, of mammal or frog, dissolves human red blood-corpuscles.

30. Zinc Sulphate (.25–.5 per cent.) causes the hæmoglobin to separate from the stroma. One may see it inside the corpuscle, or on the surface as small buds.

31. Pseudo-Membrane of the Corpuscles.—Irrigate a preparation with dilute alcohol. This decolorises the corpuscles. Then stain with magenta or Spiller's purple. The nucleus and so-called membrane of the corpuscles are thereby stained.

32. Preservation of Blood-Corpuscles by Hayem's Fluid (p. 122).—This is an excellent method. The corpuscles retain their form, and can be subsequently stained, e.g., by picro-carmine, and keep well when mounted in glycerine jelly.

33. Leucocytes of Crayfish Blood or Hæmolymph.—The colourless blood of the crayfish does very well for the study of colourless corpuscles. Make a slit in the ventral surface of the abdomen between the rings and the blood flows freely. Receive it in normal saline solution. It clots quickly. Mount some on a slide, and if it be desired to fix the corpuscles, allow a jet of steam to play on the cover-glass for a few seconds. If the fluid contain too much blood there is so much coagulable proteid, that on its coagulation by the steam a white film obscuring the leucocytes is formed. The corpuscles can be subsequently stained with picro-carmine and mounted in glycerine. The blood contains two forms of corpuscles—one with well-marked amœboid movements and provided with a large spherical nucleus; the other filled with highly refractile granules, which stain red with picro-carmine.

The best method is to allow a drop of blood to fall into a large drop of 1 per cent. osmic acid previously placed on a slide. This at once kills and "fixes" the corpuscles, which can then be stained with picro-carmine, or 1 per cent. watery solution of eosin. The latter stains their protoplasma and its expansions a rose-pink.

LESSON III.

HUMAN BLOOD—CRYSTALS FROM BLOOD— BLOOD PLATELETS.

WRAP a twisted handkerchief round the ring-finger of the left hand, and begin at the base of the finger, gradually constricting the finger from the base towards the nail. The end phalanx will thereby become greatly congested. With a sharp clean sewing-needle prick the skin at the root of the nail; a drop of blood will exude, to which rapidly apply a slide; cover the drop of blood on the slide with a cover-glass, and examine it as quickly as possible.

Observe various kinds of **corpuscles** floating in a fluid, the **blood-plasma** or **liquor sanguinis**. Note the red and white corpuscles, the former being far more numerous. Blood platelets are also present; but it requires special precautions in order to preserve them.

1. Human Coloured Blood-Corpuscles (H).—(a.) Observe the field of the microscope crowded with the red or coloured disc-shaped corpuscles much smaller than those of the newt; they are only $\frac{1}{3200}$th of an inch or 7.7 μ (7.2 μ–7.8 μ) in breadth (p. 21), and $\frac{1}{12000}$ of an inch or 2 μ in thickness. The observer may notice that when the corpuscles cease to move, after a time corpuscles may be seen adhering to each other by their flat surfaces, until a chain of these bodies resembling a pile of coins is produced (fig. 72). This is the so-called **formation of rouleaux** (fig. 72, *c*). In order to obtain *rouleaux* a fairly large drop of blood must be taken. These chains by-and-by increase in number and intersect each other, so as to produce a network-like appearance in the field. In the *rouleaux* all the corpuscles are seen edgeways.

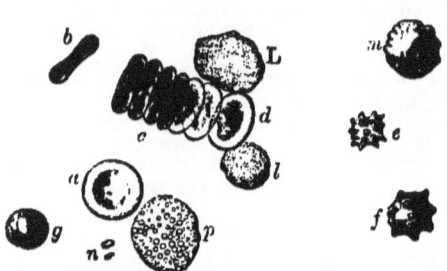

FIG. 72.—Human Red and White Blood-Corpuscles. *a.* Red corpuscle seen on the flat; *b.* In profile; *c.* A *rouleau;* *d.* Three-quarter face; *e, f.* Crenated corpuscles; *g.* Spherical; *m.* Slightly crenated; *L.* Large white corpuscle; *l.* Small white corpuscle; *p.* Granular leucocyte; *n.* Free granulations, × 1000.

(*b.*) Move the preparation until a part of it is found where the red corpuscles are to be seen not in *rouleaux*, but isolated and lying on the flat.

Study a Single Red Corpuscle.—Observe its *shape;* when seen on the flat it is circular in outline, and on bringing its edge sharply into focus, a darker area is seen in its centre (fig. 74, *a*). If the fine adjustment be used, so as to bring the lens nearer the corpuscle, the dark centre is replaced by a lighter area, while the rim becomes darker (fig. 74, *b*).

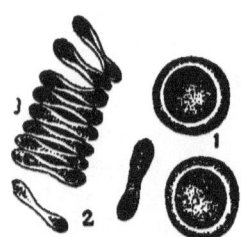

FIG. 73.—Red Blood-Corpuscles. Human; 1. Seen on their surface; 2. Seen edgeways; 3. United into a *rouleau.*

Sketch the two appearances.

This is due to the fact that these bodies are *bi-concave circular discs.* The dark area in the centre is not due to the presence of a nucleus, as they are *non-nucleated.*

This is confirmed by examining a corpuscle seen on edge, when it appears somewhat dumb-bell shaped (figs. 72, *b,* 73, 2).

In *size* they are much smaller than those of Amphibia, being only $\frac{1}{3200}$th of an inch (7.7 μ) in their greatest diameter. Their *border* is smooth, rounded, and regular, their *colour* is pale straw-yellow, their *surfaces* are smooth, and they are homogeneous throughout. As to *transparency*, the outline of one corpuscle can be seen through one overlapping it. They are soft and flexible, so that if they impinge on other objects they may change their form, which they rapidly regain, so that they are elastic.

FIG. 74.—Red Corpuscle seen on the flat. *a.* On raising, and *b.* On depressing the objective, × 1000.

(*c.*) Measure the actual size of several corpuscles by the method described at p. 20. It will be found that all the corpuscles are not of absolutely the same size. This variation in size becomes more marked in some diseases.

2. **Colourless Corpuscles.**—With a little care in observing, especially when the red corpuscles are in *rouleaux*, here and there in the field a few colourless corpuscles will be found. They remain isolated and do not form groups. They are few in number, only three to ten being found in one field. The proportion is about 3 to 10 per 1000 of red. They are usually spherical, and although they exhibit amœboid movements at 40° C., at the ordinary temperature they do not do so, and appear as nucleated finely granular masses of protoplasm. They rapidly alter after they are shed. Some of them are larger ($\frac{1}{2500}$ inch or 10 μ in diameter), others smaller than the red corpuscles (fig. 72, L, *l*).

Move the cover-glass so as to cause a current in the preparation. Note that the colourless corpuscles adhere to the glass,—they are sticky and adhesive,—while the coloured ones being smooth, and polished, glide over each other, and frequently impinge on the colourless ones without displacing the latter.

3. **Blood-Plates** (see p. 123).

4. **Acetic Acid, Tannic Acid** (fig. 66, *a*, *b*), **Water, Syrup, and Magenta** act in great part in the same manner on the coloured and colourless corpuscles of man and mammalia as on the corpuscles of Amphibia. The differences are due to structural differences in the corpuscles.

Thus in the coloured mammalian corpuscles **acetic acid** decolorises them and renders them spherical (but of course no nucleus is revealed), leaving a hull or stroma, almost invisible in the field. **Syrup** shrivels them, **magenta** reveals no nucleus, but on the side of some of the corpuscles a little spot may be observed. **Bile** makes them pale, and finally dissolves them. **Water** decolorises the red corpuscles, and makes the blood "laky."

5. **Amœboid Movements of White Blood-Corpuscles.**—In order to study these movements in the white blood-corpuscles of man and warm-blooded animals generally, the temperature of the blood must be near the temperature of the body. To keep the preparation warm, some form of **warm** or **hot stage** is necessary.

For the purposes of the student the following simple contrivance is sufficient. Take an oblong copper plate, 76 mm. long, 25 mm. broad, and 1.5 mm. thick, with a rod 100 mm. long projecting from one side of it, as shown in fig. 75. In the centre of the oblong plate of copper is a hole 15 mm. in diameter. Fix the oblong plate to an ordinary glass slide by means of sealing-wax.

Make a preparation of blood. Take a cover-glass 1 inch square, and on it place a drop of normal saline solution, and to the latter add a drop of blood. Mix them. Apply a $\frac{3}{4}$-inch cover-glass, so as to have a layer of diluted blood in a thin film between the two cover-glasses, and exactly filling the space between them. Any surplus fluid at the edge of the cover-glasses must be removed by blotting-paper. With a camel's-hair pencil then run a layer of oil round the edge of the smaller cover-glass, or the preparation may be sealed up with melted paraffin wax. This will prevent evaporation. The larger cover-glass acts the part of a slide.

Fix the warm stage on the stage of the microscope by means of the clips on the microscope, and over the aperture in the warm stage place the cover-glasses. Focus the corpuscle between the two cover-glasses, and notice that the colourless ones do not exhibit amœboid movements.

Heat the projecting rod by means of a spirit-lamp. The heat travels along the copper, and finally warms the cover-glasses and the layer of blood between them. It must, however, not be overheated. In order to prevent overheating, make previously a mixture of cacao-butter and white wax, which melts at 38° C., and place a fragment of this on the copper stem near the copper plate. Heat should be applied until the wax just begins to melt.

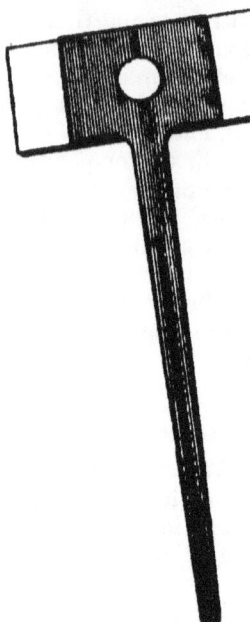

FIG. 75.—Simple Copper Warm Stage.

Observe the colourless corpuscles when heated beginning to

FIG. 76.—Warm Stage made by Reichert of Vienna. It costs £1. A, A', Screws to fix it to the stage of a microscope; B, Inflow, and B', Outflow of water.

exhibit amœboid movement. Sketch a corpuscle, and in a few minutes make another sketch and compare the two.

In the more expensive forms of **warm stages**, as those of M. Schultze and others, warm water at a known temperature is passed through a brass box which rests on the stage of the microscope, and the exact temperature is determined by means of a delicate thermometer. Fig. 76 shows a convenient form of hot stage, which can be clamped to the stage of any microscope. It is heated by means of warm water which passes in at B, and, after traversing a system of tubes, out at B'. It is provided with a thermometer.

6. Crenation of Coloured Corpuscles (H).—Mix a drop of human blood with a 2 per cent. solution of common salt. Note the change of colour. (*a.*) Observe that some corpuscles shrink in part, and become crenate or beset with short spines (fig. 77). This is due to exosmosis of fluid from the corpuscles. The colour becomes slightly deeper than in normal corpuscles. All the corpuscles are not affected simultaneously or to the same extent.

FIG. 77.—Crenation of Human Red Blood-Corpuscles, × 300.

In some individuals, merely exposing the blood to the air for a few minutes before applying a cover-glass suffices to produce this condition (fig. 72, *e*, *f*); but in any specimen of blood it may be readily produced in the majority of red corpuscles by acting on them with a saline solution of appropriate concentration.

It has been observed that in the blood of a mammal poisoned by Calabar-bean the blood-corpuscles are crenated.

7. Dilute Alkalies.—Use a 0.2 per cent. solution of caustic potash (*i.e.*, 2 grams in 1000 cc. of normal saline). It dissolves both the red and white corpuscles.

8. Fibrin (H).—Make a preparation of human blood, using a large drop. Cover it and put it aside for half-an-hour or longer until the blood coagulates.

(*a.*) Observe carefully, and in the meshes between the *rouleaux* fine threads forming a delicate network will be seen. They are fibrils of fibrin.

(*b.*) A better method, however, is to mix on a slide a drop of blood with a drop of normal saline solution. Cover and put it aside for an hour or so to clot, and then irrigate with water or dilute alcohol (p. 25), which rapidly decolorises and washes away the red corpuscles, and thus brings into view a fine fibrillar network of fibrin in the field. Irrigate with a watery solution of Spiller's purple (1 per cent.), which stains the network of fibrin a purplish tinge (fig. 78). Raise the cover-glass, and to it will be found adhering a thin film of fibrin. Dry the film, apply a drop of balsam, and mount the specimen as a permanent preparation. Numerous purple stained threads of fibrin are seen stretching from colourless corpuscles, the latter having their nuclei stained purple.

Crystals from Blood.—The hæmoglobin of certain animals crystallises very readily, *e.g.*, rat, guinea-pig.

9. Hæmoglobin Crystals of Rat's Blood (H).—Place a drop of defibrinated rat's blood on a slide, add two drops of water, and mix. Apply a cover-glass. After a few minutes, near the edge of the cover-glass oblique rhombic crystals of hæmoglobin will be found. At first they are small, but they gradually become larger. They may be single, or arranged in rosettes, or crossing each other (fig. 79).

10. Crystals from Guinea-Pig's Blood (H).—To a drop of the

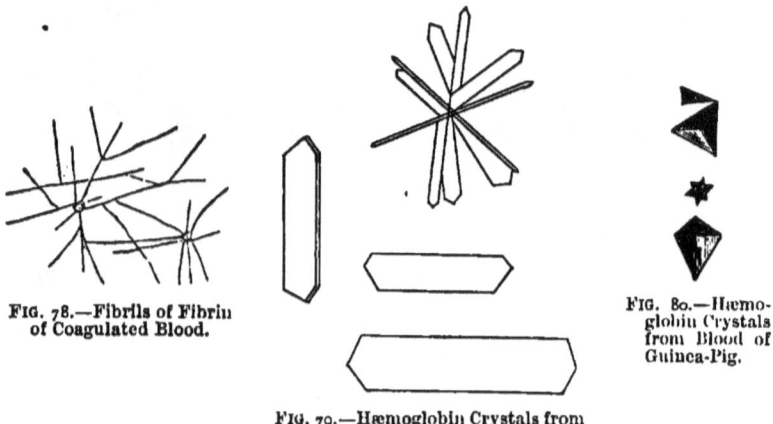

FIG. 78.—Fibrils of Fibrin of Coagulated Blood.

FIG. 79.—Hæmoglobin Crystals from Rat's Blood, × 300.

FIG. 80.—Hæmoglobin Crystals from Blood of Guinea-Pig.

defibrinated blood add a drop of Canada balsam or clove-oil, mix, and apply a cover-glass. It is perhaps better to place the balsam or clove-oil on the slide first, then to place the drop of blood on the top of the oil or balsam, and then to mix them; very soon (about 10 minutes) large red tetrahedral crystals are formed. These crystals cannot be preserved for any length of time (fig. 80).

Sometimes very good crystals of hæmoglobin from human blood are obtained from leeches which have sucked blood some weeks previously.

11. Hæmin Crystals (H).—Place a few particles of dried blood on a slide, add a small crystal of common salt, and two drops of *glacial* acetic acid. Cover. Heat over the flame of a spirit-lamp until bubbles of gas are given off, *i.e.*, until it boils. Allow it to cool. Or mix fresh blood on a slide with a minute quantity of $\frac{1}{2}$ per cent. salt solution. Heat the mixture until it becomes brownish in tint, add glacial acetic acid and heat. On cooling, crystals of hæmin are obtained. Should there be any cubes of common salt present irrigate with water, which soon removes the latter. Hæmin crystals are insoluble in water.

(*a.*) Observe the small brownish or black rhombic crystals, either singly or in rosettes, scattered over the field or on the surface of the larger masses of blood (fig. 81).

FIG. 81.—Hæmin or Teichmann's Crystals.

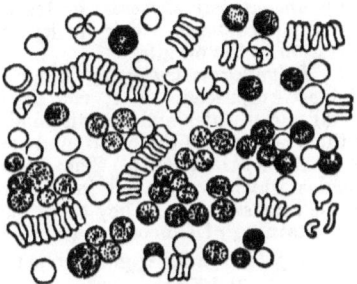

FIG. 82.—Leukæmic Blood.

(*b.*) To preserve them, remove the acid, raise the cover-glass, dry the preparation, and mount them in balsam.

12. Enumeration of the Blood Corpuscles.—See the author's *Outlines of Practical Physiology*, Lesson VI., p. 42.

13. Leukæmic Blood (H).—If a small quantity can be obtained from a patient in the hospital, examine it.

(*a.*) Observe the great excess of colourless corpuscles. According to the variety of the leukæmia, they may be larger or smaller in size than most of the coloured corpuscles (fig. 82).

ADDITIONAL EXERCISES.

14. Human Blood and Aniline Dyes (H).—Make a cover-glass preparation of human blood (Lesson II. 24). Ehrlich heated the cover-glass in an air-oven to a temperature of 120° for several hours to coagulate the proteids. To avoid this, place the covers for two hours in the alcohol and ether mixture (Lesson II. 26), and then stain them in methylene-blue as directed for frog's blood-corpuscles, or stain in 1 per cent. Spiller's purple or a weak alcoholic solution of rosein. After washing and drying, mount them in xylol balsam.

In the methylene-blue preparation the coloured corpuscles are unstained, but some of the colourless ones have their nuclei stained blue, while the surrounding protoplasm is unaffected; in other colourless corpuscles the granules in the protoplasm are stained. These corpuscles correspond to Ehrlich's basophile corpuscles. In the other preparations both the coloured and colourless corpuscles are stained.

15. Varieties of Leucocytes in Blood.—According to Ehrlich, the following varieties of colourless corpuscles are present in the blood. (Also Lesson XXXVIII.)

(*a.*) *Small lymphocytes.* They are slightly smaller than the red corpuscles, and possess a large spherical readily-stained nucleus, which almost fills the cell, being surrounded only by a small quantity of protoplasm.

(*b.*) *Large lymphocytes* are said to represent an advanced stage of (*a*). They are twice as large as (*a*), have a large nucleus surrounded by a well-defined zone of protoplasm. (*a*) and (*b*) together make up 25 per cent. of the leucocytes in blood.

(*c.*) *Mononuclear elements*, or transition forms, are distinguished from the large lymphocytes by their nucleus not being quite spherical, and having a depression in the middle.

(*d.*) *Polynuclear leucocytes*, which are smaller than (*c*), but larger than red blood-corpuscles. They contain a nucleus composed of several lobes, or several nuclei which stain readily and deeply. They represent 70 per cent. of all the leucocytes of the blood, and can migrate from the vessels.

(*e.*) *Eosinophilous cells.* The nucleus stains less deeply than (*d*). The granules which are present in the protoplasm stain deeply with eosin, *i.e.*, become intensely red. They occur but sparsely in normal blood.

16. Staining of Leucocytes (*Ehrlich*).

(1.) Make a cover-glass preparation of blood and dry it for several hours at 120° C.

(2.) Stain it for several hours in Ehrlich's acid-hæmatoxylin, eosin solution, or in a strong glycerine solution of eosin.

(3.) Wash in water,—dry and mount in xylol-balsam.

The nuclei of the white blood-corpuscles, as well as the lymphocytes and polynuclear forms, are deeply coloured; the nuclei of the mononuclear forms are bluish-gray, the red corpuscles copper-red, and the eosinophile granules red (*Kahlden*).

17. Eosinophilous Cells.—If it be desired to stain only these cells, stain a cover-glass preparation with a strong glycerine solution of eosin. The results of Ehrlich and his pupils, will be found in his pamphlet.[1]

18. Action of Hayem's Fluid.—This is an extremely useful fluid for preserving and fixing the blood-corpuscles, and can be employed both for the blood of animals and of man. It consists of

Sodic chloride	1 grm.
Sodic sulphate	5 grms.
Corrosive sublimate	0.5 grm.
Distilled water	200 cc.

The blood is run direct from a blood-vessel into this fluid in the proportion of 1 of blood to 100 of the fluid. It takes several hours to harden and fix the corpuscles, but twenty-four hours is not too long. By-and-by the corpuscles subside, and the supernatant fluid can then be decanted, and the deposit of blood-corpuscles well washed with water to get rid of the salts of the mixture. These corpuscles can then be stained with various reagents, or eosin may be added to the fluid.

(*a.*) Make an experiment with frog's blood, and stain the corpuscles for a day or so with borax-carmine and mount the stained corpuscles in glycerine or glycerine-jelly.

(*b.*) Stain another specimen with very dilute eosin-hæmatoxylin (p. 70). The hæmoglobin is stained by the eosin, and the nuclei by the hæmatoxylin. It is preserved as (*a*). Other combinations of dyes will suggest themselves.

(*c.*) Make similar preparations of mammalian blood.

[1] *Farbenanaly. Unters. z. Histologie und Klinik d. Blutes,* Berlin, 1891.

19. Blood-Plates or Platelets (H).—Wrap a handkerchief round a finger to obtain some blood (Lesson III., p. 115). On the skin at the root of the nail place a drop of normal saline containing methyl-violet (.75 gram in 1000 cc.). *Through* this drop prick the finger, and blood runs into it. Place a little of the mixed blood and methyl-violet solution on a slide; cover at once and examine.

It requires a good microscope and careful observation to see the platelets. The red corpuscles are unstained while the colourless corpuscles are stained. In the field are to be seen small oval, refractive, very delicate, non-nucleated bodies, much smaller than the red corpuscles—these are the blood-plates (fig. 83, 3). They are about 2.5 μ in diameter. They undergo changes exceedingly rapidly in shed blood.

Instead of methyl-violet solution, the skin may be pricked through a drop of the following mixture:—1 part of 1 per cent. osmic acid, and 2 parts .75 per cent. sodic chloride.

By far the best method of obtaining blood-platelets in large quantity, and in a condition in which they do not disappear, is to allow blood to flow into a solution of oxalate of potash until the mixture contains at least 1 per 1000 of the salt. This prevents coagulation of the blood. On placing such blood in a

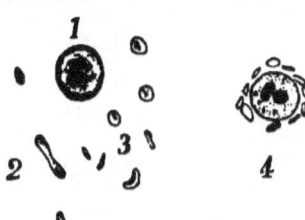

FIG. 83.—1, 2. Coloured corpuscles with 3. Blood-plates; 4. Lymph corpuscle surrounded by blood-plates.

centrifuge, when the corpuscles subside a film of grayish material accumulates on their surface, which consists chiefly of blood-plates. A drop of this spread on a slide, dried, and stained with a watery solution of methyl-violet 5 B and mounted in balsam, yields permanent preparations of these bodies (*Moser*).

20. Blood-Platelets—Dry Method (H).—Clean a slide thoroughly; sterilise it in the flame of a Bunsen-burner, and allow it to cool. Obtain a drop of blood from the finger in the usual way; get a drop on the slide, and with the edge of another slide rapidly spread the drop of blood as a thin film on the sterilised slide. Move the slide to and fro in the air until the film of blood dries. The whole process should not occupy more than four seconds. Cover the dry film with a cover-glass and seal the latter at the corners with paraffin.

(*a.*) Observe the coloured blood-corpuscles, which for the most part retain their shape, and between them the blood-platelets or, as Hayem calls them, the hæmatoblasts. They are readily seen thus in human blood. The white corpuscles are somewhat altered in shape.

21. Weigert's Method of Staining Fibrin.—Embed a thrombus or a piece of lung affected with acute croupous pneumonia in celloidin, and make a section. Float the section from water on to a slide. Stain it for ten minutes or so with a drop of the following fluid:—

Gentian violet (5 per cent.) . .	44 cc.
Aniline oil	1 ,,
Alcohol (96 per cent.) . . .	6 ,,

Remove the stain by pressing on the section two or three plies of blotting-paper, or, better, unsized printing-paper. Add a drop of iodine in iodide of potassium (iodide of potassium 5 per cent. and saturated with iodine). This quickly decolorises most of the stained parts. Remove the iodine with blotting-paper in the same way as before.

Pour on the section a mixture of equal volumes of xylol and aniline oil, moving the mixture over the preparation. Remove this and apply a fresh supply of the xylol-aniline oil mixture. This removes all the water. As long as there is a trace of whiteness in the section it still contains water. After all the water is removed, dry the preparation with paper as before, to remove as much aniline oil as possible. Wash the preparation several times with xylol to remove the last traces of the aniline oil and mount in xylol-balsam.

The stages are as follows:—

1. Harden in alcohol.
2. Stain 5–15 minutes in concentrated aniline water solution of gentian violet.
3. Dry with blotting-paper.
4. Apply iodine solution (2–3 mins.).
5. Dry with blotting-paper.
6. Decolorise and wash in aniline-xylol.
7. Remove the latter and mount in xylol-balsam.

(*a.*) Observe the threads of fibrin—very fine and numerous—stained a beautiful violet.

22. Solvent Action of Serum.—Place some blood of a rabbit or guinea-pig in a drop of blood serum of a dog. The red corpuscles are completely dissolved in a few minutes. The blood-corpuscles of a pigeon or frog are similarly but more slowly dissolved, except the nuclei. This property of dog's serum is set aside by previously heating the serum to 50°–60° C. for about half-an-hour.

LESSON IV.

EPITHELIUM (STRATIFIED) AND ENDOTHELIUM.

Epithelium presents the following general characters:—

1. It is always disposed on surfaces.
2. The cells are united by cement.
3. There are no blood-vessels within the cells.

Varieties of Epithelium.

1. **Squamous.**
2. **Columnar.**
3. **Secretory.**
4. **Transitional.**
5. **Ciliated.**

Squamous Epithelium may occur either in a *single layer*, or in *several layers*; in the former case it is sometimes called endothelium, in the latter it is said to be stratified.

(A.) In a **single layer** it lines serous and synovial membranes, heart, blood- and lymph-vessels, air-cells of lung, pigmentary layer of retina, posterior surface of cornea, anterior surface of iris, membranes of brain and spinal cord, surfaces of tendons and tendon sheaths, &c.

As *endothelium*, it consists of a single layer of flattened squamous transparent cells united to each other at their edges by means of a cement substance.

(B.) As **stratified squamous epithelium**, it covers the skin (epidermis), and lines the following cavities and surfaces:—mouth, pharynx (lower half), œsophagus, conjunctiva, over anterior surface of cornea, vagina, and lower half of cervix uteri, and entrance of urethra.

I. Isolated Squamous or Scaly Epithelium Cells.—With the finger-nail or a small section-lifter gently scrape the inner surface of the lip, place the scrapings on a slide, add a drop of saliva, skimming off any air-bubbles with a needle, cover, and examine.

1. Squames (H).—(*a.*) Observe large flat plates or squames floating in the field, either singly or in groups; the cells of the latter may be united by their edges, or by their edges and surfaces. Select a single squame seen on the flat. Note its large *size*, being five to ten times broader than a red blood-corpuscle; its polygonal shape; the colourless, transparent, and, it may be, slightly granular body of the cell, and the small oval excentrically-placed nucleus.

(*b.*) On some of the cells fine lines may be seen, some of them due to folds, but most of them to facets, indicating that the cell has been overlapped and slightly indented by its neighbour. Select a group of cells where the cells can be seen adhering to each other by their edges and surfaces, as the squames occur in several layers. It is well also to look for a cell seen edgeways, to observe that it is really a flat plate (fig. 84).

(*c.*) Not unfrequently fungi, such as bacteria, may be seen adhering to the squames.

(*d.*) **Salivary Corpuscles** may be seen. They are sharply-defined spherical cells, provided with a membrane, and about the size of a colourless blood-corpuscle. They contain fine granules, which in the fresh condition of the corpuscles may be seen to exhibit Brownian movement (Lesson I. **6**). Each cell may contain one small, spherical, excentrically-placed nucleus, or sometimes two nuclei may be present. They seem to be leucocytes distended with fluid.

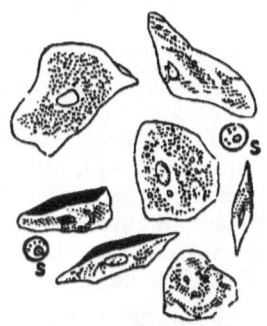

FIG. 84.—Cells of Stratified Squamous Epithelium detached from the Mouth. *s.* Salivary corpuscles.

2. Magenta.—Irrigate the preparation with a watery solution of magenta (p. 74), which stains deeply the nuclei of the squames and salivary corpuscles, while the peri-nuclear parts of these cells are less deeply stained. As the magenta contains alcohol, the latter precipitates the **mucin** of the saliva either in the form of fine threads or in membranes, which are stained red by the magenta.

3. **Epidermis of Newt—Superficial Layers of Stratified Epithelium (L and H).**—Keep a newt or frog for a day or two in a small quantity of water, and do not change the water. Very soon the superficial layers of squames will be "cast" as thin membranes. Take these, and harden them in absolute alcohol. Stain a thin piece in hæmatoxylin, and mount it in balsam.

(a.) Try to get a layer sufficiently thin, so that the cells are only one layer thick. Note the polygonal, large, nucleated cells united to each other by their edges by means of a clear cement substance (fig. 85).

Fig. 85.—Superficial Layer of Squames cast from the Epidermis of a Newt. Alcohol and hæmatoxylin, × 300.

(b.) The nucleus is usually excentric, and surrounded by slightly granular material. In the nucleus are usually two or three nucleoli, and sometimes an intra-nuclear plexus of fibrils is visible, especially in the cells of the deeper layers. Sometimes granules of the pigment melanin are seen in the cells, especially from a dark-pigmented newt.

4. **V.S. of Stratified Epithelium.**—This may be conveniently examined either in a vertical section of the skin of the palm of the hand, or a similar section of the mucous membrane covering the hard palate of a cat or the conjunctiva on the cornea. The skin must have been previously prepared by being hardened in absolute alcohol or chromic acid and spirit mixture (p. 29), while the mucous membrane may be hardened in the chromic acid and spirit or Müller's fluid.

If the mucous membrane of the hard palate be taken, stain it with picro-carmine and mount in Farrant's solution. Or the mucous membrane may be stained in bulk in borax carmine, embedded and cut in paraffin. In the latter case the sections are best mounted in balsam.

(a.) Examine it first with (L). Observe the epithelium arranged in many layers, covering a connective-tissue basis—the latter stained red—and projecting in the form of fine papillæ into the epithelial layer (fig. 86). Neglect the connective-tissue basis meantime.

(b.) (H) Select a thin part of the deeper layers of the epithelium near the papillæ, and note that the cells there are somewhat small and cylindrical, with their nuclei stained red. Study the change in the form of the cells towards the free surface of the epidermis (fig. 86), where they become corneous and less granular.

(*c.*) Note in the deeper layers of the epithelium **prickle-cells**. Adjacent cells are connected by very fine processes or "**intercellular bridges**." The fine spaces between the bridges are called "**intercellular channels**." When these are broken across and the cells isolated, the cells present the appearance of being beset by very fine processes, and hence they are called "prickle-cells."

(*d.*) The cells vary in their shape and characters from below upwards. It will be easy to detect the *horny layer* above, composed of many layers of flattened, hardened cells. This forms a fairly well-marked layer, the cells being clearer than those situated more deeply; the nuclei are less conspicuous, and the cells generally stain yellowish with picro-carmine.

FIG. 86.—V.S. Mucous Membrane of Hard Palate of Cat. Epidermis with corneous and deeper layers with prickle cells. Below, connective tissue of the mucous membrane with a papilla. Chromic acid and spirit, picro-carmine, Farrant's solution.

5. Non-Corneous Stratified Epithelium from the Œsophagus (H).—Macerate the mucous membrane of the œsophagus of a calf or other animal for a week in $\frac{1}{10}$ per cent. bichromate of potash. This "dissociates" the epithelial cells, so that when the surface is scraped, one obtains isolated cells. Make a cover-glass preparation (p. 114), and stain it in aniline-water-gentian-violet, *i.e.*, gentian-violet solution dropped into aniline water (p. 73). Dry and mount in balsam.

(*a.*) Observe the isolated squamous cells, each with a small excentrically-placed nucleus stained of a violet tint. Numerous facets are seen in the cells, and their shape both on the flat and on edge can be carefully studied.

(*b.*) If desired, stain some in picro-carmine. This is best done by keeping them in the picro-carmine on a slide placed under a bell-jar, with water to prevent evaporation, *i.e.*, in a moist chamber.

6. Horny Epidermis.—Macerate a shred of epidermis in 35 per cent. caustic potash. After a time it is softened and can be broken up with needles. The cells fall asunder and swell up in the fluid, and appear as spheroidal cells with a membrane, but no nucleus is visible. Examine them in the potash solution. No water must be added, else the cells are dissolved.

7. Prickle Cells (H).—Place for twenty-four hours in 1 per cent. osmic acid a small piece of the palmar surface of the skin —less than one-eighth of an inch cube—from a freshly-amputated finger. Make vertical sections by freezing, or after embedding in paraffin, and mount the former in Farrant's solution and the latter in balsam.

(*a.*) Observe the prickle-cells *in situ*, *i.e.*, polygonal cells in the deeper layers of the rete mucosum, with fine processes connecting adjoining cells, leaving thus a system of fine spaces between the cells (fig. 87). The fine fibres which pass from cell to cell form "**intercellular bridges**," and when these bridges are broken across they give the appearance as if the cells were beset with fine prickles.

8. Isolated Cells from the Different Layers of the Epidermis (H).—It is usual to macerate very small pieces of any membrane covered with stratified epithelium in $\frac{1}{8}$ or $\frac{1}{10}$ per cent. of potassic bichromate, which usually takes more than a week to dissociate the cells.

FIG. 87.—Prickle-Cells from the Deeper Layers of the Epidermis of the Palm, showing intercellular bridges and channels. Osmic acid.

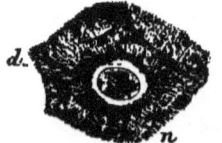

FIG. 88.—Prickle Cells Isolated from the Human Epidermis by means of Iodised Serum. *n.* Prickles; *d.* Space between nucleus and cell-body, × 800.

A much speedier method is that of Schiefferdecker. Make a watery extract of "pancreaticum siccum" of Dr. Witte of Rostock. Filter, and in the filtrate place a small fragment of fresh skin or the pad from the upper jaw of a sheep. Place the fluid near the fire or in an oven at 40° C. Within four hours, the epidermis can be detached, and its cells fall readily apart. Preserve it in a test-tube in a mixture of equal parts of water, alcohol, and glycerine. It forms a deposit at the bottom of the tube. A little of the deposit is mounted in glycerine or Farrant's solution, covered and examined. It may be stained with picro-carmine or methylene-blue.

(*a.*) Observe numerous cells of different shapes, some flattened, others cubical, and many "prickle-cells" (fig. 88). Many of the cells exhibit facets where they have been pressed against each other.

9. Isolated Prickles and other Cells from the Pad of a Sheep's Mouth (H).—Isolated by means of "pancreaticum siccum" (*v. supra*). Many of the somewhat cubical-shaped cells show the "prickles" beautifully. These cells may be stained with an aniline dye or picro-carmine, and some of them show two nuclei.

10. Endothelium of Central Tendon of Diaphragm.

—Mount in balsam a small piece of the central tendon of the diaphragm of a rabbit stained in silver nitrate (Method, p. 77).

(*a.*) Observe the polygonal areas which map out the outlines of the single layer of squames covering the surface of the tendon. These areas are bounded by brown or black lines, the so-called "**silver lines**" (fig. 89). As a rule, no nucleus is visible in the cells, but it may be brought into view by staining with logwood and mounting the preparation in Canada balsam.

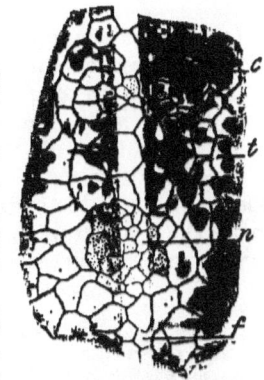

(*b.*) Amongst these may be seen small groups or islands of smaller granular cells. In some specimens stomata surrounded by granular or germinal cells may be seen. These stomata open into a plexus of lymphatic capillaries in the substance of the tendon (Lesson on Lymphatics).

FIG. 89.—Endothelium of the Peritoneal Surface of the Central Tendon of the Diaphragm of a Rabbit. Silvered. *f.* Lymphatic slit; *t.* Tendon; *c.* Ordinary endothelial cells; *n.* Islands of small cells.

11. Omentum of Young Rabbit (L and H).

—Mount in balsam a small piece of the silvered omentum of a young rabbit (one week old) (Method, p. 77).

(*a.*) Observe a thin fibrous membrane mapped into polygonal areas—some of them with more or less sinuous outlines—by means of black "silver lines" (fig. 90).

The omental membrane composed of connective tissue is covered on both surfaces with a continuous layer of endothelium.

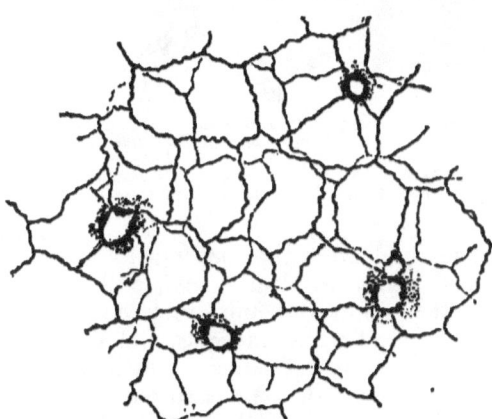

(*b.*) Raise the lens by means of the fine adjustment until the upper layer of squames comes distinctly into view; then depress the lens and focus through the thickness of the membrane to bring into view the layer of squames covering the deeper surface, and note that the outlines of

FIG. 90.—Omentum of a Young Rabbit stained with Silver Nitrate (× 300), showing the epithelium on the upper and under surfaces; the outlines of the latter faintly indicated. Some holes are also seen in it.

the cells do not correspond with those of the upper layer. In focussing through the membrane, note the plexus of elastic fibres in it.

(c.) To see the nuclei of the endothelial cells, stain another piece of the silvered omentum in logwood and mount it in balsam.

The omentum of a *young* rabbit is chosen because it is nearly a complete membrane with few fenestræ or holes in it.

12. Omentum of Cat (L and H).—Mount in balsam a small piece of the silvered omentum of a cat (Method, p. 77). In cutting the omentum into small pieces, the easiest way is to spread it out, or rather float it out, on a sheet of paper, and then cut the paper and omentum into pieces of the necessary size. The pieces are thus less liable to fold up, and are more readily manipulated on the slide.

(a.) (L) Observe a mesh work of trabeculæ (T) bounding open polygonal spaces (fig. 91, *m*). In the larger trabeculæ may be seen blood-vessels (*c*), an artery or a vein, or both, surrounded here and there by groups of large clear cells—fat-cells (*f*). All the trabeculæ are completely covered with endothelial cells, whose outlines, mapped out by silver lines, can just be recognised.

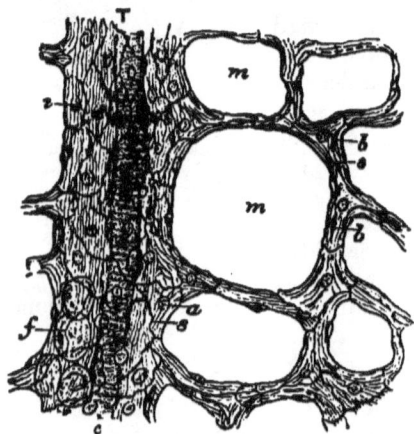

FIG. 91.—Omentum of Cat Silvered. T = Trabecula, with *c*. Blood-vessel; *f*. Fat-cells; *s*. Silver lines, and *a*. Nuclei of the endothelium; *m*. Meshes; *b*. Nuclei of the connective-tissue corpuscles. Silver nitrate and hæmatoxylin, × 100.

(b.) (H) Select a large strand with a blood-vessel in it (T). Focus the silver lines (*s*) on its upper surface, and gradually depress the lens until the fibrous tissue composing the strand comes into view, and, still lowering the lens, bring the endothelium on the under surface into view.

(c.) Select a fine trabecula, and note the silver lines on it; also observe the fibrous tissue of which it is composed.

(d.) If desired, a preparation may be stained with logwood to reveal the nuclei of the cells of the endothelium (*a*), as well as those of the fibrous tissue composing the membrane (*b*). The nuclei of the endothelial cells are superficial, and usually spherical (*a*); those of the connective-tissue corpuscles are in the substance of the membrane, and are usually more flattened and somewhat oval (*b*).

ADDITIONAL EXERCISES.

13. Dogiel's Methylene-Blue Method.—Place any fresh membrane, *e.g.*, omentum, mesentery, capsule of kidney, in a 4 per cent. solution of methylene-blue in normal saline. Allow it to stain for ten minutes or so. Wash it twice in a saturated watery solution of picrate of ammonia and examine in glycerine. The outlines of the cells are stained of a purplish colour. This method may be used instead of the silver method to demonstrate the existence of endothelium on any surface, *e.g.*, on tendons (rat) in blood- or lymph-vessels. It is also used to demonstrate lymph-spaces.

14. Sections covered by or consisting of epithelial cell structures may be stained in a weak watery solution of **benzo-azurin**. This stain is not removed by alcohol, and the sections can be mounted in balsam. Or **benzo-purpurin (B.)** may be used. This gives a reddish stain, any excess being removed by alcohol rendered feebly alkaline, *e.g.*, by lithium carbonate. Benzo-azurin stains connective tissue of a bright blue.

LESSON V.

COLUMNAR, SECRETORY, AND TRANSITIONAL EPITHELIUM.

Columnar Epithelium lines the mucous membrane of the alimentary canal from the cardiac orifice of the stomach onwards; the greater part of the ducts of the glands opening into it; other gland ducts.

II. Columnar Epithelium.—Slit open the small intestine of a rabbit or a cat *just killed*. Wash the mucous surface with normal saline to remove any adherent particles.

1. Fresh Condition (H).—With a scalpel gently scrape the mucous surface and transfer what is on the scalpel to a drop of normal saline solution on a slide. Diffuse the scrapings in this fluid, and to prevent the pressure of the cover-glass, place in the fluid a hair half-an-inch in length. This preparation is not to be preserved.

(*a.*) Observe numerous columnar epithelial cells, a few isolated, but most of them adhering together. **Side View of the Cells.**—Select an isolated cell, notice its columnar form, usually tapering to a blunt point at one end, while the body of the cell is faintly granular, and contains a clear oval nucleus. The free broad end is covered with a highly **refractile disc**—seen on edge as a narrow refractile band—with fine vertical striæ in it (fig. 92).

Many more or less complete villi may be seen. Along the edge of the villus, covered by its layer of columnar cells, the clear disc is readily seen, and by focussing the surface of a villus the nucleated mosaic formed by the ends of the cells, with the rounded mouths of the goblet-cells, come into view.

(*b.*) If the animal was killed whilst its food was undergoing digestion, refractive fatty granules may be seen in the protoplasm of the cells.

(*c.*) **End View of the Cells.**—Move the preparation until a fragment of detached epithelium is seen showing the *free* ends of the cells directed towards the observer (fig. 92, *d*).

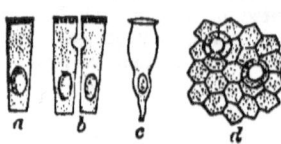

FIG. 92.—*a, b.* Isolated Columnar Epithelial Cells from the Small Intestine of Cat; *c.* Goblet-cell; *d.* Ends of columnar cells and open mouths of goblet-cells directed towards observer.

(*d.*) Note in such a group of cells the polygonal outlines of the ends of the cells, and in each polygonal area a large spherical nucleus, which appears almost to fill the area. A nucleolus can often be seen. Here and there may be seen the rounded opening of a goblet-cell. Focus carefully and notice the appearance of the goblet-cell. When the open mouth is in focus, it is seen as a circle of small diameter, and on depressing the tube the broad part of the cell comes into view (fig. 92, *d*).

(*e.*) Amongst the cells may be found isolated goblet-cells (fig. 92, *c*). Each cell has an open mouth, while the lower part of the cell contains a nucleus embedded in a small quantity of protoplasm.

2. **Isolated Columnar Epithelium of Newt (H).**—Mount in glycerine a small quantity of isolated columnar cells which have been dissociated by maceration in dilute alcohol and subsequently stained with picro-carmine. Place a short length of hair under the cover-glass.

The small intestine of a rabbit may be used, but far larger cells are obtained from the intestine of a newt. Macerate the whole intestine in dilute alcohol for twenty-four hours and stain in bulk in picro-carmine for at least another twenty-four hours. On scraping the mucous surface, the cells are detached and diffused in formic glycerine.

(*a.*) Observe the large, tall, columnar cells, often tapering to a point at their lower end, the red-stained nucleus, and the clear striated disc. The nucleus usually exhibits a distinct nucleolus (fig. 93), and sometimes two nuclei are present. Sometimes two nucleoli are seen in a nucleus, and the cell itself may be branched at its fixed extremity.

If the intestine of a newt be macerated for twenty-four hours in

5 per cent. ammonium chromate, then the nuclei and cell-body show a distinct fibrillar structure.

III. **Glandular Epithelium** occurs in the secretory glands, *e.g.*, liver, pancreas, salivary, gastric, intestinal, and other glands. Necessarily it must vary very greatly in shape and in its functions.

3. **Secretory or Glandular Epithelium of Liver.**—With a clean scalpel scrape the cut surface of the liver of an animal just killed, *e.g.*, a rat. Place the scrapings in dilute alcohol (twenty-four hours), pour off the alcohol and cover it with picro-carmine (twenty-four hours).

Liver-Cells (H).—(*a.*) Examine the isolated cells in glycerine with the usual precautions. Observe the cubical cells, which may be isolated or adhering in groups of two or three. The granular

FIG. 93.—Isolated Columnar Epithelial Cells from the Newt's Intestine. Dilute alcohol and picro-carmine, × 300.

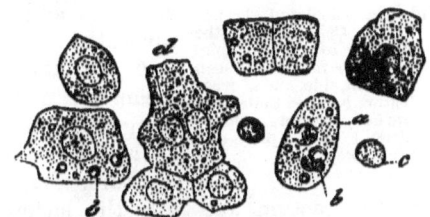

FIG. 94.—Isolated Hepatic Cells. *d.* With two nuclei; *b.* Oil-drops; *c.* Isolated nucleus of a cell. Teased fresh.

protoplasm is stained yellowish, and each cell has a spherical bright-red nucleus. The protoplasm may contain globules of oil, and occasionally two nuclei may be seen in a cell, especially in the liver-cells of a young rat (fig. 94).

(*b.*) If the cells be much broken up, liberated nuclei and granules of oil, and sometimes red blood-corpuscles, are seen in the field.

(*c.*) *Acetic acid* clears up the protoplasm, makes the nucleus more distinct; it does not affect any fatty particles, but merely makes them more evident.

Sulpho-cyanide of potassium (5 per cent.) is an excellent medium for maceration of the liver-cells. The plexus of fibrils in the nuclei is thus rendered visible.

IV. **Transitional Epithelium** occurs in the urethra, urinary bladder, ureters, and pelvis of the kidney. It is confined to the genito-urinary mucous membrane. It consists of several (2-3-4) layers of cells. The superficial cells are large and flattened (especially if the bladder has been kept distended), often with two nuclei, and with depressions on their under surface produced by the large rounded ends of the cells of the next layer. The cells of the next layer are somewhat pyriform (fig. 95, *b*), while the deepest layers are composed of smaller polyhedral cells.

4. Transitional Epithelium of Bladder (H).—Place the distended bladder of a frog or cat in dilute alcohol for twenty-four hours, stain it *en masse* for the same time in picro-carmine, scrape off a little of the mucous surface and diffuse it in glycerine, add a hair and cover. The cells may also be macerated by using instead $\frac{1}{8}$ per cent. bichromate of potash solution.

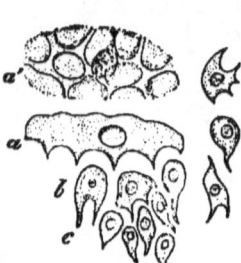

FIG. 95.—Isolated Transitional Cells from the Bladder of a Guinea-pig. *a*. A superficial cell seen from the side, and *a'* from below; *b* and *c*. Cells from the deeper layers. Dilute alcohol and picro-carmine, × 300.

(*a*.) Observe various forms of cells, some of them more or less flattened with facets on their surfaces (fig. 95, *a*, *a'*), others elongated with finger-shaped processes (*b*), some pear-shaped, and others cubical. All are nucleated.

5. To Distend a Frog's Bladder.—By means of a pin transfix the skin at the margins of the anus, and tie round the pin a thread so as to completely occlude the aperture. Open the abdomen, make a slit into the rectum, and from the latter, after removing its contents, inject dilute alcohol (p. 25) into the bladder. When the bladder is distended, ligature it, cut it out, and suspend it in its inflated condition in dilute alcohol for twenty-four hours.

If the bladder of a cat or guinea-pig be used, it is distended from the urethra with dilute alcohol, and suspended for twenty-four hours in a large volume of the same liquid.

ADDITIONAL EXERCISES.

6. To Silver the Free Ends of Columnar Epithelium.—A small piece of the mucous surface of the small intestine of a cat is washed in distilled water, and then placed for ten minutes in a $\frac{1}{2}$ per cent. silver nitrate solution, and silvered in the usual way (Method, p. 77). After hardening in alcohol, if the epithelium be detached and mounted in glycerine, it is easy to obtain a view of the free ends of the epithelial cells, with the cement substance between them indicated by "silver lines," and also to see the open mouths of the goblet-cells. The view obtained is that shown in fig. 92, *d*.

N.B.—Other preparations of these forms of epithelium will be obtained in sections of the organs in which they occur.

LESSON VI.

CILIATED EPITHELIUM.

Ciliated Epithelium occurs in the nasal mucous membrane (except that of the olfactory region), the cavities accessory to the nose, the upper half of the pharynx, the Eustachian tube, larynx, trachea, and bronchi, the uterus (except the lower half of the cervix), Fallopian tubes, vasa efferentia to lower end of epididymis, the ventricles of the brain, and the central canal of the spinal cord.

V. Ciliated Epithelium.—A ciliated cell may be any shape, but usually it is more or less columnar. Only the cells of the superficial layer bear cilia. The bunch of cilia are directly continuous with the protoplasm of the interior of the cell, and are planted on a clear disc, which is said to be composed of small "knobs" placed side by side to form a bright refractile disc; a cilium being attached to each knob. The deeper cells may be pyriform or polyhedral, according to the situation from which the epithelium is obtained. Goblet-cells may be found between the superficial ciliated cells.

1. Ciliary Motion in the Frog (H).—After pithing a frog, gently scrape the roof of its mouth with a scalpel, and diffuse the scraping in a drop of normal saline, add a short piece of hair, and cover.

(*a.*) Notice groups of cells with cilia on their free surface. The cilia bend quickly, at the rate of ten to twelve times per second, with a whip-like motion in one direction, and then rapidly unbend, thus creating currents in the liquid, and thereby moving the corpuscles, granules, or other free particles that may be present. They bend more rapidly in one direction than the other. All the cilia covering one surface are not in motion at one time, but the movement passes from cell to cell in a wave-like form. If a portion of cell with cilia attached is in view, the fragment of the cell may be seen to be moved in a definite direction by the vibratile motion of its own cilia. Cilia detached from a cell cease to move.

2. Ciliary Motion in the Mussel (L and H).—Open a salt-water mussel, collect the salt water which escapes, cut out a fragment of one of the flattened yellow gills, and place it in a drop of salt water. By means of two needles separate slightly the bars which compose the gills, cover and examine.

(*a.*) (L) Observe the bars with their free rounded ends and their surfaces beset with a fringe of moving cilia, which cause the particles suspended in the fluid to be carried along in a definite direction.

(*b.*) (**H**) Select a single, *cilium* observe that it is a clear homogeneous tapering filament, placed on a clear band which covers the cells on the surface of the bar of the gill. Notice how the cilium bends more at the top than base, and how it straightens itself again. The movement may go on for several hours. The backward movement is less rapid than the forward stroke.

3. Heat (**H**).—By means of a camel-hair brush run a ring of oil

FIG. 96.—Slide with a Ring of Glass Tube fixed to it, for Studying the Action of Chloroform on Cilia.

round the preparation (2), and put it aside for an hour or so, until the ciliary motion becomes slower.

(*a.*) Place the slide on a hot stage (fig. 75) and gradually apply heat. As the cilia are warmed they move more quickly; but if the temperature be too high, of course the proteids are coagulated and the cells killed. If, while the cilia are moving rapidly, the source of heat be removed, as the preparation cools the cilia gradually move more and more slowly.

Use the hot stage described in Lesson III. **5.** But in using it, put the preparation of cilia on a cover-glass moistened with a drop of sea-water, and invert the cover-glass over the aperture in the hot stage, so that the drop of fluid and cilia hang in the little circular cavity.

4. Weak Alkalies.—To a preparation in which the cilia move languidly, apply a drop of $\frac{1}{5}$ per cent. solution of caustic potash. This immediately revives their action for a short time; but as the alkali penetrates into the cells, it ultimately kills them.

5. Chloroform (**L** and **H**).—Place a fragment of a mussel's gill in a drop of salt water on a cover-glass. Put a small drop of chloroform in a glass cell and place the cover-glass on the cell, with the drop of fluid hanging into the latter, as shown in fig. 96.

(*a.*) (**H**) Observe the movement of the cilia, and, as the chloroform vapour diffuses into the drop of water and acts on the cilia, how they move slower and slower. If the action of the chloroform be pushed too far, their movement will be arrested. If the action of the chloroform be not too prolonged, and the preparation removed and freely exposed to the air, the cilia may begin to move again.

6. Action of Gases on Ciliary Motion, *e.g.*, **Carbon Dioxide.**—Carbonic acid is generated in the usual way in a flask containing marble and dilute hydrochloric acid, and by means of a caoutchouc tube it is conducted to a glass gas-chamber (fig. 97, C), over which the preparation of cilia on a cover-glass is inverted.

If it be preferred, the following moist chamber, by Ranvier, for studying the action of gases may be used. It consists of a brass box

FIG. 97.—Gas-Chamber for Studying the Action of Gases on Cilia. A. Inlet; B. Outlet-tube; C. Glass gas-chamber.

about the size of a microscopical slide, and perforated at the centre by an aperture 2 cm. wide, which is closed below by a plate of glass.

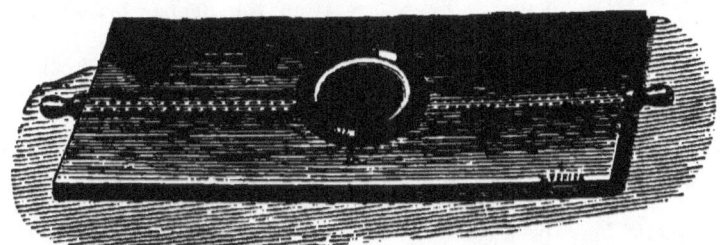

FIG. 98.—Ranvier's Moist Chamber for Applying Gases to a Preparation.

In the centre of the aperture is fixed a plate of glass (fig. 98, a), less in diameter, thus leaving a circular trench all round (b). Moreover, the height of this circular plate of glass is less than the height of the brass box by at least $\frac{1}{10}$th mm. The box is perforated by two tubes, through which the gases can be conducted to the preparation, which is placed between the top of the circular glass disc and the cover-glass which covers in completely the aperture in the brass box. In this way the gas or vapour can be applied to a preparation still in a normal fluid medium.

FIG. 99.—Various Forms of Ciliated Cells from the Mucous Membrane of the Hard Palate and Œsophagus of the Frog. Dilute alcohol and picro-carmine, × 300.

(*a.*) If carbon dioxide be used, observe that it rapidly arrests the movement of the cilia and renders the cells granular, probably from the precipitation in them of paraglobulin.

7. Isolated Ciliated Epithelium and Goblet-Cells (Frog) (H).—

Scrape off a little of the epithelium from the mucous membrane of the palate of a frog, which has been macerated for twenty-four hours in dilute alcohol and afterwards stained by picro-carmine. Before staining, it is advisable to place the isolated cells for several hours in $\frac{1}{2}$ per cent. osmic acid. Diffuse the cells in glycerine, put in a hair, cover, and examine.

(*a.*) Observe an isolated ciliated cell; it is short and columnar, perhaps tapering or divided at one end, while the other end is beset with cilia, resting on a clear, transparent, **refractile disc**; the protoplasm is granular, and encloses an oval red-stained nucleus with one or two bright excentrically-placed nucleoli (fig. 99).

Besides the ciliated cells, there are others without cilia, oval or elongated, pointed at one end. These are from the deeper layers of the ciliated surface (fig. 102).

FIG. 100.—Isolated Ciliated Cell from the Œsophagus of a Frog. *c.* Cilia; *p.* Clear disc; *n.* Nucleus; *m.* Irregular extremity. Iodised serum, × 1000.

(*b.*) Numerous **goblet** or **chalice cells** will also be seen. They are cup-shaped cells, with an open mouth, and containing mucigen. There is a small amount of protoplasm at one end of the cell—the end by which it is fixed—which encloses a spherical, oval, or compressed nucleus. The goblet-cells may be seen in two conditions, some clear with their mucigen discharged, and others full of granules or "loaded" with mucigen.

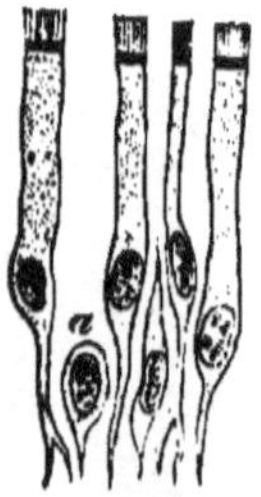

The cells may also be isolated or dissociated by macerating the membrane in iodised serum (p. 25). A cell isolated in this way, and magnified 1000 diameters, is shown in fig. 100.

8. Isolated Ciliated Cells (Mammal).—

Use the trachea of a cat or other mammal, and macerate small pieces in dilute alcohol (twenty-four hours). Stain it in bulk in picro-carmine. The isolated cells are examined in glycerine or glycerine-jelly. If the trachea of the ox be used, observe

(*a.*) The tall narrow ciliated cells (fig. 101), each with its cilia and clear disc. The ends of the cells may be pointed or branched. Amongst these may be seen oval or battledore-shaped cells (*a*), the younger cells which exist in the deeper layers of the mucous membrane.

FIG. 101.—Ciliated Cells from Trachea of Ox. *a.* Cell from the deeper layers. Dilute alcohol and picro-carmine, × 300.

9. Ciliated Epithelium (L and H).—Mount a vertical section of the respiratory mucous membrane of the septum of the nose of a cat or other animal. The tissue has been previously hardened in Müller's fluid, or in chromic and spirit fluid, or, what is better, a saturated watery solution of corrosive sublimate for two or three hours. In the last case, every trace of the metallic salt must be removed by prolonged and frequent washing with alcohol. Stain the section with logwood and mount it in balsam; or picro-carmine can be used.

(*a*.) Observe several layers of cells (fig. 102), but only the superficial layer of cells is furnished with cilia, which are placed on a clear disc on the free end of the cell. Notice the shape of the cells in the subjacent layers. Those of the lowest layers are nearly spherical, while in the intermediate layers they are more elongated,

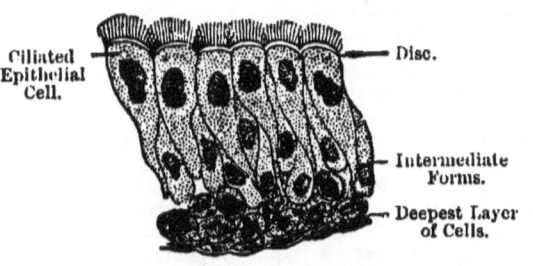

FIG. 102.—V.S. of Ciliated Epithelium.

FIG. 103.—Various Forms of Goblet-Cells from the Mucous Membrane of the Hard Palate and Œsophagus of the Frog. One of the cells shows mucus exuding from the open mouth of the cell. Dilute alcohol and picro-carmine, × 300.

and are described as battledore-cells, arranged in between the others. They replace the ciliated cells when the latter are shed.

10. Isolated Goblet-Cells (H).—These are readily obtained by macerating the stomach of a frog in dilute alcohol for twenty-four hours. Scrape the surface and diffuse the cells in glycerine. They may be stained with picro-carmine, or they may be diffused in salt solution and stained with methyl-violet, but the latter preparation cannot be preserved in glycerine. Numerous goblet-cells will be found in **7**.

(*a*.) Observe the isolated cells (fig. 103). Each cell is filled for more than three-fourths of its capacity with mucus, while at the lower tapering end there is a nucleus embedded in a small quantity of protoplasm. Sometimes a plug of mucus may be seen exuding from the open mouth of a cell.

(*b*.) With a high power the interior of the upper part of these cells may be seen to contain a fine network of fibrils. In the meshes is a substance, **mucigen**, and when this is acted on by water it yields **mucin**. In a certain sense these bodies are unicellular muciparous glands.

ADDITIONAL EXERCISES.

11. Cover-Glass Preparation of Goblet-Cells.—Place the œsophagus or stomach of a frog in dilute alcohol for twenty-four hours. Scrape the mucous surface and compress the scrapings between two cover-glasses. Separate the cover-glasses, allow the film adhering to each glass to dry, and then stain it with eosin or aniline-water-methyl-violet, or safranin-O. Wash off the surplus stain with absolute alcohol, allow the film to dry, and mount it in xylol-balsam.

Perhaps a better plan still is to stain the cover-glass preparations for twenty-four hours in Ehrlich-Biondi fluid. It is prepared thus:—

Ehrlich-Biondi Fluid.

Saturated watery solution of orange	. . .	100 cc.
,,	,,	acid fuchsin . . .	20 ,,
,,	,,	methyl-green . . .	50 ,,

The solutions used, however, must be *saturated*. When used as a staining agent, this strong fluid is diluted with about forty volumes of water.

(*a.*) Observe the goblet-cells with their characters retained intact. In the Ehrlich-Biondi preparation the protoplasm is stained red, the nuclei and nucleoli bluish.

12. Cover-Glass Preparation of Ciliated Epithelium.—The mucous membrane of the œsophagus of a frog is placed for twenty-four hours in dilute alcohol. A cover-glass preparation is made of the epithelium, and stained as described under 11, with methylene-blue, safranin-O, gentian-violet, or Ehrlich-Biondi fluid, and mounted in xylol-balsam.

13. T.S. Tongue of Frog (H).—By means of hedgehog-spines, pin out the tongue of a frog on a thin layer of cork with a small hole in it. Harden it for two hours in a saturated watery solution of corrosive sublimate; remove every trace of the sublimate by prolonged washing in alcohol—not water. Stain in bulk in picro-carmine or borax-carmine. Make transverse sections—best by the paraffin infiltration embedding method (p. 41)—and mount the sections in balsam.

FIG. 104—V.S. Ciliated Epithelium of Frog's Tongue. *m.* Muscular fibres. Corrosive sublimate and picro-carmine, × 250.

(*a.*) Observe the fibro-muscular basis of the tongue, covered on the surface with ciliated epithelium cells; between the ciliated cells the goblet-cells, each with an open mouth and its plexus of fibrils with mucigen in its meshes (fig. 104).

(*b.*) Observe, too, how the expanded ovoid goblet-cells compress the ciliated cells and cause the latter to have a peculiar shape, a broad expanded top and a narrow body.

(*c.*) The young cells at the base of the ciliated cells.

If the specimen be stained in picro-carmine and mounted in balsam, the yellow colour of the picric acid can be retained by putting a little picric acid into the alcohol used to dehydrate it, or by picric acid placed in the clove-oil or xylol used to clear up the section.

LESSON VII.

STRUCTURE OF CELLS—MITOSIS OR KARYO-KINESIS.

Structure of the Animal Cell.—To see all the structures in an animal cell is by no means easy. Speaking generally, the tissues of the articulata, amphibians, and reptiles yield the largest tissue elements for examination. The cells may be examined in the fresh condition or after "fixing," hardening, and staining.

In Fresh Condition.—Examine a teased preparation in an indifferent fluid, *e.g.*, normal saline, the liquid of Ripart and Petit (p. 24), or in solution of methyl-green (p. 74). The preparation may be sealed up with paraffin wax (p. 111) and examined after some time.

After Hardening.—The best osmic acid "fixing" fluids for this purpose are the fluids of Flemming, Rabl, and Fol, and the best stains safranin and gentian-violet. Mount in xylol-balsam.

To see the finer details an oil-immersion lens and Abbe's condenser must be used.

Mitosis or Karyokinesis.—By these terms is meant the remarkable series of phenomena which take place in cells—animal and vegetable—when they undergo a process of indirect division. In this connection it is important to remember the constitution of a cell and some of the terms which have been applied by different authors to its several parts. A cell may or may not possess a distinct **cell-wall**, but the **cell-body** appears to be made up of two substances, which Flemming names as follows :—One composed of threads, seldom forming a network, and called by him **cyto-mitoma** or **mitoma** (μίτος, thread), also called **spongioplasm**, and the other, homogeneous and lying in the meshes of the latter, is the **paramitoma** or **hyaloplasm**. The cell-contents are generally described as consisting of a finely-granular soft substance, the so-called *protoplasm* (fig. 105). This protoplasm consists of a network, sometimes called a "*filar mass*" or *spongioplasm*, which lies embedded in a homogeneous ground-substance or "*interfilar mass*" or *hyaloplasm*. The filar mass corresponds to the *mitoma* of Flemming and to the *spongioplasm* of some other authors, while the interfilar mass corresponds to the *paramitoma*, or the *hyaloplasm*, or *paraplasma* of some authors.

The **nucleus** (fig. 105), bounded by a **nuclear membrane**, composed of two layers, an outer one, which does not stain (achromatic), and an inner one, which does (chromatic). Within the membrane is an **intranuclear network** or **karyomiton** (κάρυον, a kernel) or **karyomitoma**, consisting of a **reticulum** of threads or fine fibres, arranged sometimes in the form of a regular network. As these threads, or at least particles in them, stain readily with certain dyes, *e.g.*, safranin, they have been called **chromatin** or composed of chromoplasm.

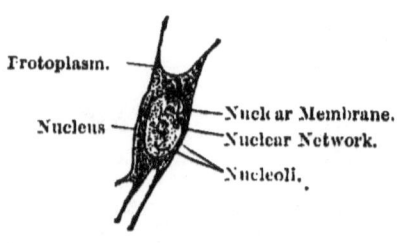

FIG. 105.—Connective-Tissue Corpuscle from the Skin of a Salamander.

In the meshes of this more or less perfect network lies the **nuclear fluid**, which, however, does not stain with certain pigments: it has been called **achromatin**.

In the meshes of the reticulum lies one—usually more than one—**nucleolus**. It is more refractile than the rest of the nucleus. Generally, however, as stated, two or more nucleoli are present, and they seem to differ in their chemical constitution, so that Flemming speaks of *principal* and *accessory* nucleoli. Many of the bodies described as nucleoli are really parts of the intranuclear fibrillar network seen in optical section. Other observers have applied different terms to these structures, but here it is not necessary to multiply terms. The "**attraction sphere**" existing in the protoplasm of some cells seems to exercise some influence on the dividing nucleus.

The great majority of cells reproduce themselves by **indirect cell-division** or **mitosis**, and in this process the network within the nucleus plays a most remarkable part. The division of a cell is always preceded by the division of the nucleus. Starting from the **resting nucleus**, where the threads are not well developed, soon two poles appear in the nucleus, and then the threads grow thicker, more numerous and tortuous, forming the convolution stage. The various stages are indicated in the following scheme :—

Mitosis.

1. *Resting nucleus.*   The mother nucleus showing the fibrils in the reticulum or network stage.
2. *Skein or spirem.*   Close skein of fine convoluted fibrils, then thicker loops running from polar to antipolar regions, the nucleoli disappear.
3. *Cleavage of fibrils.*   Each loop (usually V-shaped) splits longitudinally into two and the achromatic spindle appears.

4. *Monaster, Star, or equatorial stage.*	The achromatic spindle distinct with two poles, terminating in two polar corpuscles (with cytasters), nuclear membrane lost. The V-shaped chromatin fibres arrange themselves in the equator of the spindle. Cytoplasm separates into a clear and a granular zone.
5. *Metakinesis or Divergence.*	Sister threads separate and move towards poles along fibres of the spindle.
6. *Dyaster or Double Star.*	Complete separation of the two sets of V-shaped sister threads towards the poles of the spindle.
7. *Dispirem or Double Skein.*	Open skein in daughter nuclei passing into a close skein. Nuclear membrane forming. Cell itself divides.
8. *Network or Reticulum.*	Resting condition of daughter nuclei. Cytoplasm divided, remains of spindle disappear. Chromatic fibres more twisted.

In the preparations one readily finds examples of these and the other stages of mitosis.

Mitosis.—By far the best animals to use for studying the process of mitosis are the larvæ of the water-salamander. The young

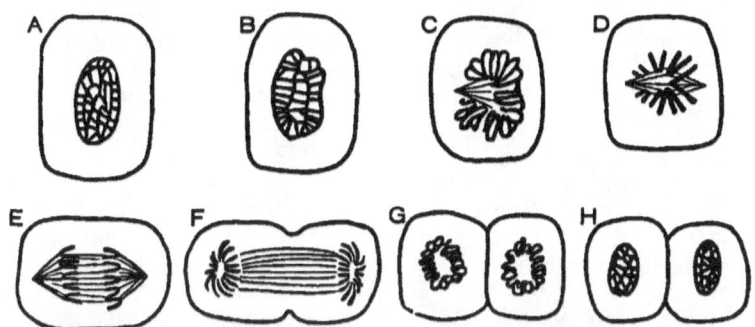

FIG. 106.—Mitosis. A. Nuclear reticulum, resting-stage; B. Preparing for division; C. Wreath stage; D. Monaster with achromatic spindle; E. Barrel or pithode stage or metakinesis, *i.e.*, chromatin fibrils travelling along the achromatic spindle towards the poles; F. Diaster; G. Daughter wreath stage; H. Daughter cells passing to resting-stage.

animals must be carefully fed, else the mitotic figures are not well seen. They are killed at various stages, when they vary in length from ¾ to 1 inch in length. They may be hardened in ⅙ per cent. chromic acid.

A very good hardening reagent is one-sixth p.c. chromic acid. From a week to ten days is sufficient. If Flemming's mixture (p. 32) be used, the tissues must remain in it twenty-four hours or less. Some prefer Rabl's mixture (12–24 hours) (p. 31), others absolute alcohol or picric acid. After being "fixed" the tissues are placed in

30 to 50 per cent. alcohol, the alcohol being renewed frequently until no colouring matter is given off. They can then be preserved in strong alcohol, but on prolonged keeping the nuclei change somewhat.

1. T.S. Tail of Larva of Salamander (H).—Stain a section for 12–24 hours, or even longer, in a solution of safranin (p. 75)—a saturated alcoholic solution diluted with half its volume of water. In dealing with such delicate sections, it is well to "fix" the section on a slide beforehand, especially if it be cut in paraffin. The section may be fixed by albumin and glycerine, the paraffin removed by turpentine or naphtha. After staining, wash in spirit and place in acid alcohol (100 cc. absolute alcohol to 3–5 drops of hydrochloric acid). This rapidly decolorises it ($\frac{1}{2}$–1 min.). Instead of acid alcohol the absolute alcohol may be used. It removes the surplus stain more slowly. The difficulty is just to hit the moment when the dye is washed out of the nuclear matrix, the fibrils being still stained. The section is then transferred to absolute alcohol and clarified, and mounted in xylol-balsam. The best agent to clarify the section is cedar-wood oil, as it does not dissolve the safranin, which clove-oil does.

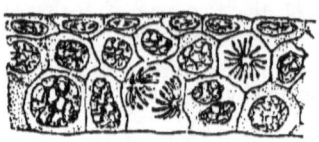

FIG. 107.—V.S. Epidermis of a Young Salamander with Resting Nuclei, Monaster and Diaster Stages of Dividing Nuclei. Chromic acid and safranin, × 300.

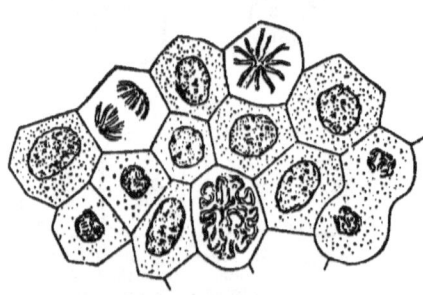

FIG. 108.—Mitotic Figures from the Epidermis of a Young Salamander. Chromic acid and safranin, × 300.

(*a.*) Observe the layers of epithelium of the epidermis (fig. 107). In several of the nuclei the characteristic mitotic figures are to be seen, and in one or two sections it is not difficult to pick out examples of nearly all the stages of nuclear division. While the nuclear fibres are well seen in safranin-stained specimens, the nuclear spindle is not usually well seen.

2. Surface-Scraping of the Epidermis, Cornea, or External Gills.—Instead of making a section, scrape the surface of the skin of the tail, or break up the external gills in water, or stain (safranin) and mount the cornea in xylol-balsam. Stain either with safranin or logwood, mount in balsam, and numerous mitotic figures will be found (fig. 108).

Some prefer to stain the sections,—*e.g.*, after hardening in picric acid—with Kleinenberg's logwood diluted with the alcoholic solution of alum and calcium chloride (p. 68), allowing the sections to stain for 12 hours or longer. They are mounted in balsam.

N.B.—It is important to note that tissues must be treated differently according as one wishes to see the chromatic fibres or the achromatic spindle. To see the ordinary mitotic figures, osmic acid, or any fluid containing it, is good (12–24 hours); but in order to see the achromatic spindle, it is better to use a chromo-acetic mixture (p. 31) for 12 hours.

ADDITIONAL EXERCISES.

Mitosis, however, can also be studied in mammalian tissues.

3. Mitosis in Omentum of New-Born Rabbit.—Orth recommends the omentum of a new-born rabbit. Harden it for twenty-four hours in Flemming's fluid (p. 25); wash it thoroughly, and stain it in safranin-O; wash in water, and remove the surplus dye, if necessary, by means of alcohol acidulated with hydrochloric acid (p. 144). In the cells of the milk-spots and in the walls of the blood-vessels it is easy to detect mitotic figures, but they are much smaller than in the salamander.

4. Mitosis in the Amnion.—One of the readiest sources is the amnion of a pregnant rat, as recommended by Solger. After the rat is killed, the uterus is excised and placed in a saturated watery solution of picric acid. The uterus and the membranes round each fœtus are opened under the picric fluid. Harden for twenty-four hours; wash well in water, and harden in the various strengths of alcohol, beginning with 70 per cent. Better results are, I think, obtained by removing the picric acid by washing in alcohol instead of water. Select the amniotic membrane and tinge a small part of it in Ehrlich's acid hæmatoxylin (p. 69) diluted one-half. The membrane may also be hardened in Flemming's fluid and stained with safranin.

5. Method of Martinotti and Resegotti.—Small pieces of the tissue, *e.g.*, a rapidly growing tumour, are hardened in absolute alcohol. Sections are made and coloured in a watery solution of safranin-O. The decolorisation, however, is obtained by a hydro-alcoholic solution of chromic acid. Take one to two parts of a 1 per cent. solution of chromic acid to eight or nine of alcohol. After it is sufficiently decolorised—*i.e.*, the colour is removed from every part except the fibrils of the nuclei—wash the section in absolute alcohol, clarify in oil of bergamot, and mount in balsam. This method yields excellent results.

6. Mitosis in Plants.—Various plants have been recommended, but my own experience is limited to the following:—

(*a.*) Take the fruit of *Fritillaria imperialis* when they are 30–40 mm. in length, and place them in absolute alcohol for a week. Then in equal parts of glycerine and absolute alcohol for 24 hours. Then cut the fruit in two; with a dissecting microscope search for the embryo-sac. It shows various stages of mitosis after staining for 12–24 hours in safranin. Mount in xylol-balsam.

(*b.*) A transverse section of the fruit of *Lilium candidum* also does very well.

Harden the buds in absolute alcohol when they are about 1-1½ cm. long. Stain in safranin and mount in xylol-balsam.

(c.) I have also tried the young growing shoots of an onion bulb placed in water. The tips of the growing rootlets were hardened in Fol's or Rabl's fluid and stained with safranin. Fairly good specimens were thus obtained.

LESSON VIII.

CELLULAR AND HYALINE CARTILAGE.

Cartilage.—The varieties of cartilage are classified as follows:—

1. **Cellular** or **Parenchymatous** occurs in the chorda dorsalis, ear of mouse and rat.
2. **Hyaline** encrusts the articular ends of bones, occurs in costal, tracheal, bronchial, laryngeal (except epiglottis and cornicula laryngis), nasal cartilages, external auditory meatus, and in the "temporary" cartilages of the fœtus.
3. **Fibrous.** { (a.) White fibro-cartilage.
              { (b.) Yellow fibro-cartilage.

(a.) *White fibro-cartilage* occurs in the intervertebral discs, interarticular fibro-cartilages, as marginal cartilages on the edge of joints (hip, shoulder), lining tendon grooves, in sesamoid bones, the sacro-iliac synchondrosis and symphysis pubis.

(b.) *Yellow fibro-cartilage* occurs in the epiglottis, cartilages of Wrisberg and Santorini, external ear, and Eustachian tube.

I. Cellular Cartilage (H).—Kill a rat or mouse; snip off the ear. With a stout pair of forceps remove the skin and the other tissues from the ear until the thin lamella of cartilage which forms the basis of the ear is exposed. Harden the cartilage in absolute alcohol. Mount a thin part of the cartilage in Farrant's solution.

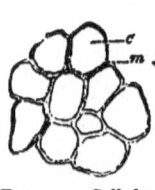

FIG. 109.—Cellular Cartilage from the Ear of a Rat. Absolute alcohol, × 250.

1. **Cellular Cartilage (H).**

(a.) Observe the clear *cells* (fig. 109, c); some of them may be spherical, but most are polygonal in shape, closely pressed together, and united by a very small amount of matrix or *intercellular substance* (m). By focussing, rows of them in several planes may be seen. Usually no nucleus is visible in the cells.

If desired, a section can be stained with hæmatoxylin and mounted in balsam.

VIII.] CELLULAR AND HYALINE CARTILAGE. 147

II. Hyaline Cartilage.—This consists of **cells** or **corpuscles** embedded in a **hyaline matrix**.

2. Cartilage of Newt (L and H).—Snip off a small piece of the thin cartilage of the sternum of a freshly-killed newt, and with a scalpel scrape away any fibrous tissue or muscle adhering to it. Mount it in normal saline solution or ½ per cent. solution of alum.

(*a.*) Observe a homogeneous **matrix** (fig. 110, *m*), like ground glass, in which are embedded here and there **cartilage-cells** or **corpuscles** (*r*). The matrix is comparatively small in amount, and hyaline.

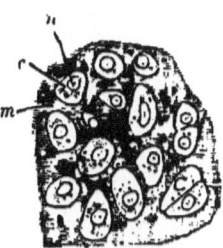

FIG. 110.—Hyaline Cartilage. *m.* Matrix; *c.* Body of cartilage-cell; *n.* Nucleus. × 250.

(*b.*) Each corpuscle consists of a spherical mass of transparent, finely-granular **cell-substance** (*c*). Sometimes the protoplasm contains refractile granules of oil, and in it is placed a spherical, clear, granular **nucleus** (*n*). Near the margin of the preparation may be seen cavities or capsules from which the cell-contents have fallen out; others where the cell-contents have shrunk from their capsule; while at other places the cells completely fill the spaces in which they lie. On focussing through the thickness of the tissue, the cells are seen to be two or more layers deep, *i.e.*, in a section of moderate thickness they lie in several planes. The cells may lie singly or in groups.

3. Effect of Acetic Acid (H).—Irrigate with a 2 per cent. solution of acetic acid.

(*a.*) Observe that the nucleus becomes more distinct and granular, the cell-contents clearer, and the cell shrinks from its capsule, so that a space is left between the capsule and the irregular shrunken cell-contents.

4. Action of Gold Chloride (H).—Mount in Farrant's solution a section of articular cartilage from the head of the femur of a freshly-killed frog which has been stained by the gold chloride method (p. 78). If the gold be reduced by formic acid the bone is thereby softened, so that both bone and cartilage can be cut together in a freezing microtome.

(*a.*) Observe the matrix faintly stained, and the corpuscles or **cell-contents**, but not nucleus, stained of a purple hue. The cell-contents have shrunk very little. Here and there an empty cartilage-capsule may be seen. The gold chloride has a special affinity for the protoplasm of the cartilage-cells, so that they stand out distinctly in contrast to the less-stained matrix in which they lie embedded. This preparation represents, as it were, the "positive picture," in contrast to that obtained by the use of silver nitrate, which yields the "negative picture" (p. 77).

As a general rule, it may be stated that gold chloride stains the cellular elements, whilst silver nitrate stains the cement substance, *e.g.*, connective tissue, cornea, &c.

5. Costal or Tracheal Cartilage (L and H).—With a razor make a thin transverse section of a fresh rib cartilage or tracheal cartilage, *e.g.*, of a dog, cat, or rabbit, and examine the section in normal saline.

(*a.*) (L) Observe the circular or oval outline of the section surrounded by the **perichondrium** firmly adherent to the cartilage, which consists of **cells** embedded in a hyaline **matrix** (fig. 111, Pch).

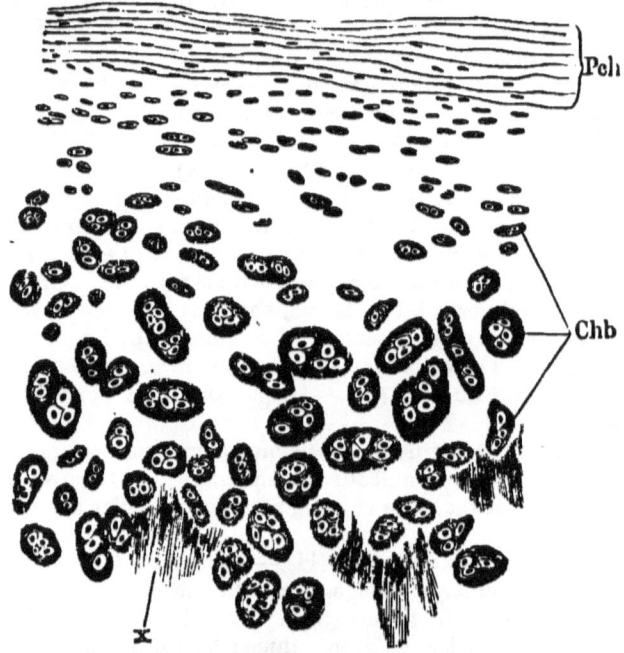

FIG. 111.—Hyaline Cartilage. T.S. Human thyroid hardened in alcohol. x. Fibrous matrix; Chb. Cartilage-capsules; Pch. Perichondrium.

(*b.*) (H) Observe that the cells are smaller and flattened near the periphery, fusiform farther in, oval or spherical nearer the centre of the section. They may lie singly, or in groups, or in rows.

(*c.*) The matrix is generally hyaline, but in some places it may be fibrous (figs. 111, x, and 112, *f*).

(*d.*) Around each cell, or, it may be, each group of cells (fig. 112, 3 or 2), look for a *cartilage-capsule* (fig. 111). It is firmly united to and continuous with the matrix; but by tilting the mirror slightly, so as to modify the light, it may be seen distinctly as a

VIII.] CELLULAR AND HYALINE CARTILAGE.

well-defined membrane bounding the cell-cavity. Not unfrequently several so-called "daughter-cells" may be seen within one capsule.

6. Hardened Costal Cartilage.—Mount a transverse section of costal cartilage which has been hardened in a saturated solution of picric acid. The picric acid is removed by washing in alcohol.

(*a.*) Stain a section with picro-carmine and mount it in Farrant's solution.

(*b.*) Stain a section in hæmatoxylin and mount it in balsam. The matrix is stained of a light blue, and the corpuscles of a deeper tone.

(*c.*) A very good stain for hyaline cartilage is Merkel's indigo-carmine stain (p. 67). The preparation can be mounted in balsam, and is not too transparent.

(*d.*) **Carmine.** — Place a similar section for twenty-four hours in a strong solution of ammoniacal carmine (p. 63). Wash away the surplus carmine, and allow a drop of strong glacial acetic acid to fall on the section. After a minute or so, wash the section thoroughly in water to remove all the acid, and mount it in Farrant's solution.

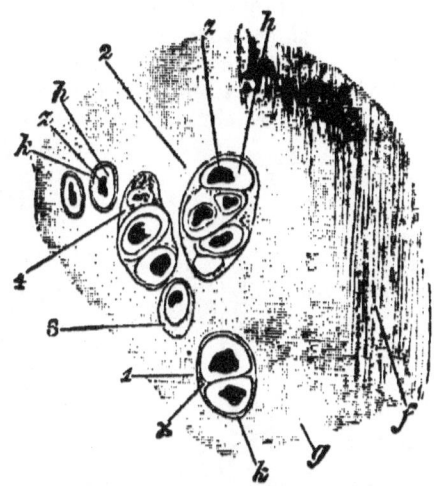

FIG. 112.—T.S. Human Costal Cartilage. *z.* Cell shrunk from the wall of its cavity, *h*; 1. Two cells in one cartilage-capsule, *k*; at *x* is the commencement of a separation wall; 2. Five cartilage-cells within one capsule, but the lowest cell has fallen out of its cavity; 3. Cartilage-capsule cut obliquely, so that it appears thicker at one side; 4. Cartilage-capsule not opened into; *g.* Hyaline matrix; *f.* Fibrous matrix, × 300.

Observe the same arrangement of the cells as before, but the cells are stained red while the matrix is colourless. The connective-tissue corpuscles of the perichondrium are also red. Round each cartilage-cell is a thin outline deeper stained than the rest, indicating the presence of a **cartilage-capsule**.

7. Fat in Cartilage-Cells (H).—Place a section of costal cartilage (preferably made from a piece of costal cartilage taken from a person over fifty years of age) in 1 per cent. osmic acid for an hour, wash it in water, and mount in Farrant's solution.

(*a.*) Observe in some of the cells small and somewhat larger black spots, which are globules of oil blackened by the osmic acid.

8. Eosin (H).—Stain with a watery solution of eosin a section of human costal cartilage from an adult. The section becomes uniformly red. Wash it in dilute acetic acid and mount it in Farrant's solution.

(*a.*) The cells are more deeply stained than the matrix, and numerous cells will be seen in groups or in rows due to the proliferation of cartilage-cells. The *cartilage-capsules* are usually more deeply stained than the surrounding matrix. Look for a part of the matrix which has become *fibrous*. It is deeply stained. If the mirror be slightly tilted, or the light shaded from the preparation by the hand, the cartilage-capsules are usually distinctly seen.

9. Articular Cartilage.—Decalcify the head of a long bone (*e.g.*, the femur) of a cat or other animal in picric acid or chromic and nitric acid, with the precautions indicated at p. 37. When it is thoroughly decalcified, make—by freezing—vertical sections, so as to include the encrusting cartilage and the subjacent cancellous bone. Place some sections in 1 per cent. osmic acid for twenty-four hours, and stain others in picro-carmine. Mount examples of both in glycerine-jelly, as glycerine or Farrant's solution makes the tissues rather too transparent.

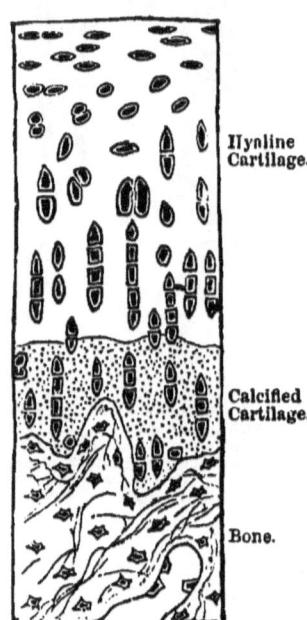

FIG. 113.—V.S. Articular Cartilage. Chromic and nitric fluid. Picro-carmine.

(*a.*) (**L**) Observe the layer of encrusting cartilage fixed upon the cancellated bone beneath (fig. 113), a bold, irregular, wavy line separating the cartilage from the bone, but the one dovetails into the other. In the cartilage notice two areas, an upper and larger one, with a hyaline matrix; and a lower, narrower one, with a more granular matrix. The latter is the zone of **calcified cartilage**. A fine wavy delicate line indicates where the **hyaline matrix** ends and the calcified matrix begins. In the matrix note the cartilage-cells, flattened at the circumference—*i.e.*, next the joint cavity—in small groups deeper down, and in vertical rows in the substance of the cartilage.

(*b.*) (**H**) Study the shape of the cells from the free or joint surface downwards. At the circumference they are flattened or fusiform, and deeper down they are more or less polyhedral and arranged in vertical rows. Some of the cells may be somewhat shrunk within their capsules. A row of cells may be seen partly in the hyaline and partly in the calcified matrix. Note the finely granular character of the calcified matrix.

(*c.*) Observe the bone with its lamellæ and bone-corpuscles, and its open meshes containing bone-marrow.

10. Cartilage of Cuttlefish (H).—Mount in Farrant's solution a section of the cephalic cartilage of a cuttlefish. The cartilage must have been hardened previously in picric acid, alcohol, or osmic acid.

(*a.*) Stain a section in picro-carmine. Observe that the cells lie in groups of three or four, and from the periphery of the group processes are given off which anastomose with processes from adjoining groups of cells (fig. 114).

FIG. 114.—Branched Cartilage-Cells of the Cartilage of Loligo. Picric acid, eosin, and logwood.

Eosin and Hæmatoxylin.—Stain a section slightly with a dilute solution of eosin and afterwards with dilute hæmatoxylin. Mount in Farrant's solution. The matrix is reddish, and the cells and their processes purplish in hue.

ADDITIONAL EXERCISE.

11. Silver Nitrate and Cartilage Matrix (H).—Rub a piece of solid silver nitrate upon the cartilaginous end of the freshly-excised femur of a frog. Expose the cartilage in water to sunlight. It rapidly becomes brown. Make a thin surface section with a razor, and mount it in Farrant's solution.

(*a.*) Observe the matrix stained brown, and a large number of unstained spaces apparently empty. The latter are the cavities in which the cells lie, but the cells themselves are too transparent to be readily seen. This picture is the reverse of that obtained with gold chloride (p. 147).

LESSON IX.

THE FIBRO-CARTILAGES (WHITE AND YELLOW).

III. Fibro-Cartilages.
A. White Fibro-Cartilage.
1. **Intervertebral Disc.**—Decalcify in chromic and nitric acid fluid (p. 37) or picro-sulphuric acid (24 hours) an intervertebral disc

and its adjacent pieces of bone (rabbit or cat). By freezing make vertical sections to include the disc and its adjacent bones; place the sections for twenty-four hours in 1 per cent. osmic acid, wash them thoroughly in water, and mount in Farrant's solution or glycerine-jelly.

(*a.*) (L) Observe the decalcified bones, and between them the disc (fig. 115). The bone consists, for the most part, of cancellated bone, and the disc stretches between the two plates of denser bone which cover the ends of the vertebræ. A thin layer of hyaline cartilage exists on the surface of the bodies of the vertebræ.

(*b.*) The disc itself is of considerable thickness, and consists of many parallel bundles, between which, and at right angles to them, are other bundles cut transversely, the fibres of adjoining bundles being arranged sometimes in a zigzag fashion (fig. 115). In these bundles are cartilage-cells, more numerous in the central bundles and fewer in the outer ones. In the centre of the disc may be seen a more pulpy tissue, the remains of the chorda dorsalis.

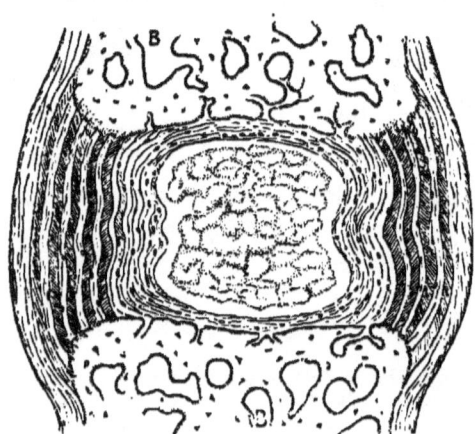

FIG. 115.—V.S. Intervertebral Disc of Cat. B. Bone; D. Disc. Chromo-nitric acid fluid and osmic acid, × 15.

(*c.*) Externally on both sides is a ligament of connective tissue which passes from one vertebra to the other. It gradually shades into the fibres of the disc.

(*d.*) (H) Observe the fibres, with a greater or less number of large oval cells lying between them. The fibres can perhaps be traced into the matrix of the bone.

(*e.*) The cells—oval and with a hyaline capsule—are most numerous in the central part of the disc, and usually there is no difficulty in seeing groups of cells in the hyaline cartilage forming a thin coating on the bone. The boundary-line between the disc and the bone is never straight, but wavy. This can readily be made out by tilting the mirror slightly.

2. **White Fibro-Cartilage (H).**—Snip off a small piece of the intervertebral disc of an ox or sheep or man, after hardening a small portion for a day in a saturated solution of picric acid or spirit, or

picro-sulphuric acid (twenty-four hours). The hardening is completed in the various strengths of alcohol. Tease it with needles in Farrant's solution.

(a) Observe the fibrous **matrix**, consisting of very fine, wavy unbranched **fibrils** (fig. 116), and between them oval or spherical nucleated **cells**, each one with a distinct thick hyaline **capsule**. In some of the latter, concentric rings indicating the deposition of successive capsules, may be seen.

(b.) The bundles of fibres run in various directions, and each fibril is unbranched. The cells —with thick capsules—are not very numerous, and lie either singly or in groups of two or three between the fibres.

3. **Another Method.** — The cells of this cartilage do not stain very readily, but the following method gives good results: —Harden the cartilage—small pieces—for a day or so in Kleinenberg's fluid (p. 30), and after washing the pieces free from the picro-sulphuric acid, place them for several days in borax-carmine, and stain them *en masse*. Concentrate the pigment in the cells by placing the pieces in acid alcohol for twenty-four hours (p. 65). Sections can be made either by freezing or embedding in paraffin. The cells are stained bright red.

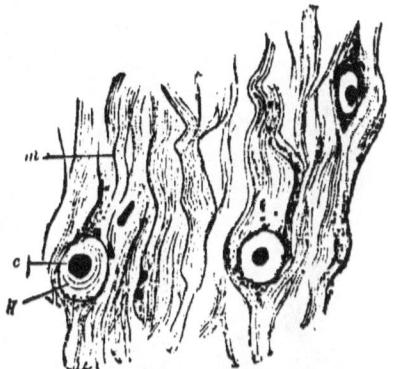

FIG. 116.—From Human Intervertebral Disc. *m.* Matrix or fibrous ground-substance; *c.* Cartilage-cell; *k.* Capsule surrounded with calcareous particles. Kleinenberg's fluid and borax-carmine, × 250.

B. **Yellow Fibro-Cartilage.**—(i.) Harden the epiglottis of a sheep, dog, or cat for forty-eight hours in absolute alcohol.

(ii.) Harden a part of the ear of a pig in a saturated solution of picric acid for twenty-four hours. Wash away the picric acid with alcohol, and in the various strengths of alcohol complete the hardening.

By freezing make sections of the epiglottis and ear.

4. **Epiglottis.**—Stain a section in picro-carmine and mount it in Farrant's solution.

(a.) (L) Neglecting the stratified epithelium and glands which are present, observe the **perichondrium** (fig. 117, *c, f*), embracing the mass of cartilage, and firmly adherent to the latter, which has a fibrous, yellow-stained **matrix**, studded with **cells**—stained red—embedded in it. The mass of cartilage may appear to be interrupted, or it may even be perforated.

(b.) (H) Observe the **perichondrium**, composed of connective tissue, with numerous elastic fibres; the latter can be traced into, and become continuous with, the elastic fibres of the matrix (fig. 117, *e*, *f*). The **matrix** consists of fine branched and anastomosing fibres of elastic tissue, stained yellow with picric acid. Where the fibres are cut transversely, they appear as yellow dots or granules. In addition to these, however, there are numerous granules of elastin scattered in the matrix. In this meshwork notice the nucleated **cells** stained red. Each cell has a capsule, but near the perichondrium they are smaller and flattened (*i*), while in the substance of the cartilage they are larger, oval, or spherical (*r*).

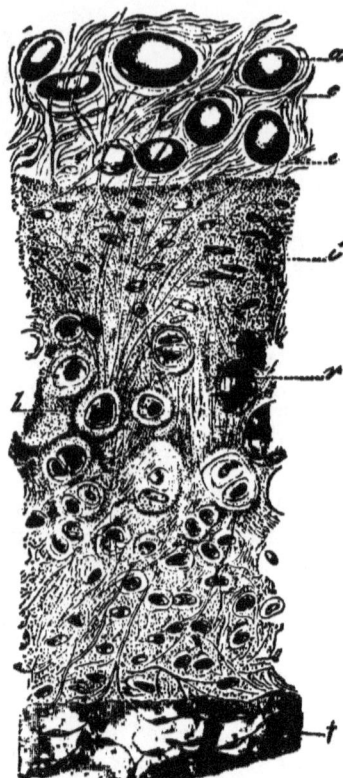

FIG. 117.—T S. Epiglottis of a Dog. *a.* Fat-cells in perichondrium, *c*; *e.* Elastic fibres; *i.* Superficial layers of smaller cells; *r.* Layer of larger cells with elastic granules, *l.*; *f.* Perichondrium. Alcohol and picro-carmine.

5. Acid Fuchsin Method.—Stain a section with a watery solution of acid fuchsin. Wash the section for a long time in absolute alcohol and mount in balsam. The network is intensely red, and the other parts uncoloured.

6. Double Staining of the Epiglottis.—(i.) Stain a section with picro-carmine, and then faintly with logwood. Mount in balsam. To preserve the yellow colour of the fibrils, the clove-oil, with which the section is cleared up, must be made yellowish by dissolving in it a little picric acid. The fibres are yellow, the cells red, and the nuclei purplish.

(ii.) Stain another section with dilute eosin-hæmatoxylin, and mount it in Farrant's solution or balsam. The cells take the logwood tint, and the fibres the colour of eosin.

7. Ear of Pig or Horse.—Stain a section in picro-carmine, and mount it in Farrant's solution.

(**L** and **H**) Observe the skin, its glands and muscles. Neglect these, and note the perichondrium enclosing the cartilage with a characteristic arrangement of the cells. The cells near the surface are small, flattened, and parallel to it, while those in the centre are larger and arranged across the long axis of the section.

THE FIBRO-CARTILAGES.

Between the cells is a matrix, which may be partly hyaline and partly yellow fibrous.

8. Transition of Hyaline to Elastic Cartilage (H).—Dissect out the arytenoid cartilage of an ox or sheep. Harden and preserve it in alcohol. Cut sections through the part where the hyaline cartilage merges into the elastic variety. This is quite visible to the naked eye, the elastic part being more opaque and yellowish white in tint. Stain a section with picro-carmine, and

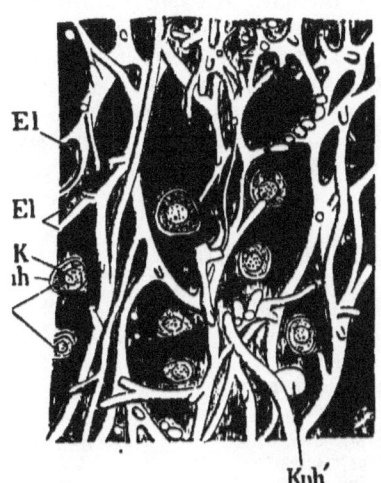

118.—Elastic Cartilage Ear of Horse, hardened in alcohol. El. Elastic fibres cut in various directions; K. Nucleus of cartilage-cell; Knh. Contour of cartilage-cell cavity; Knh'. Empty cartilage capsule; Knz. Cartilage-cell.

FIG. 119.—Elastic Cartilage developing in Hyaline Cartilage in Arytenoid Cartilage of a Calf, × 100. The clear spaces indicate the position of the cells, the shadow part the hyaline matrix.

mount it in Farrant's solution. On making a section of such a cartilage in the fresh condition, one part has the pale-bluish colour of hyaline cartilage, and the other part is very faintly yellow.

(*a.*) At one part observe hyaline cartilage, whose matrix gradually becomes fibrous. At first only a few scattered granules of elastin are seen, then the hyaline matrix is traversed by elastic fibres, and gradually the matrix loses its hyaline character, and becomes distinctly fibrous. Around each cell there is a clear area —hyaline—devoid of fibres (fig. 119).

LESSON X.

CONNECTIVE TISSUE.

THE group of **Connective Tissues** includes cartilage, ordinary connective tissue (with adipose tissue), adenoid or retiform tissue, mucous tissue, bone and dentine. (1) These all subserve more or less mechanical functions in the organism; (2) they all have much in common in structure, *i.e.*, they are composed of cells, and an intercellular matrix, but usually the development of the matrix exceeds that of the cells; and (3) they are all developed from the mesoblast of the embryo.

ORDINARY CONNECTIVE TISSUE.

It consists of the following structural elements :—

A. *Structural Elements.*

Fibres. { White or gelatinous. Yellow or elastic.

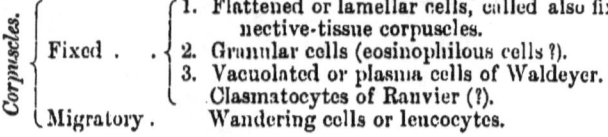

Corpuscles.
- Fixed.
  1. Flattened or lamellar cells, called also fixed connective-tissue corpuscles.
  2. Granular cells (eosinophilous cells?).
  3. Vacuolated or plasma cells of Waldeyer. Clasmatocytes of Ranvier (?).
- Migratory. Wandering cells or leucocytes.

B. *Arrangement of these Elements.*

(*a.*) Areolar, *e.g.*, subcutaneous and submucous tissues.
(*b.*) Bundles in parallel groups, *e.g.*, tendon (with parallel fibres) and fasciæ (fibres crossing at right angles).
(*c.*) Fenestrated fibrous membranes, *e.g.*, omentum.
(*d.*) Compact bundles crossing in all directions, *e.g.*, skin.

The **lamellar cells** are flattened or winged plates which lie on the bundles of fibrils. They have a large oval nucleus lying in a clear plate.

The **granular cells**, or "Mastzellen" of Ehrlich, are often found near blood-vessels, and in the fat present in areolar tissue, in the submucous tissue of the intestine, and in Glisson's capsule. The cells are often spherical, and the granules are numerous and proteid in nature, and stain with aniline dyes, *e.g.*, eosin, hence the term sometimes applied to them "eosinophilous cells."

The **plasma cells** were formerly confused with the foregoing, but in the plasma cells the protoplasm is vacuolated, and the vacuoles contain fluid. They sometimes have short processes. What the relation of the clasmatocytes to these other cells may be is so far not determined; nor, indeed, do we know the relation between the granular and the plasma cells.

The **migratory cells** are identical with the white blood-corpuscles or lymph-corpuscles, and may therefore be regarded as an adventitious element.

Yellow or Elastic Fibres occur in the ligamentum nuchæ of animals (large fibres); lig. subflava; stylo-hyoid ligament; connective tissue generally; in the walls of the air-tubes and lungs; the larger blood-vessels, especially arteries; the vocal cords and some ligaments of the larynx; many organs, *e.g.*, spleen.

1. Yellow or Elastic Fibres—Thick Fibres (H).—Tease out in water a fragment of the ligamentum nuchæ of an ox; cover and examine it. It can be mounted in Farrant's solution.

(*a.*) Observe the broad fibres with a definite outline, yellow in colour, refracting the light strongly, branching and anastomosing, and sometimes curling at their ends where they are broken across (fig. 120, *f*). A small quantity of white fibrous tissue will be found between and supporting the fibres (*b*).

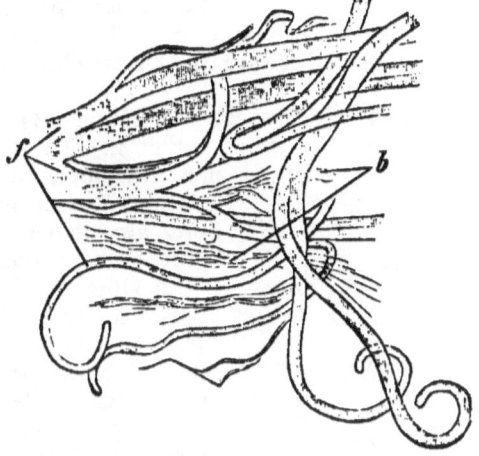

FIG. 120.—*f.* Elastic fibres from the ligamentum nuchæ; *b.* Fine white fibrous tissue. × 300.

(*b.*) Measure the size of one of the larger fibres. They are about 7–8 μ ($\frac{1}{3100}$th inch) in diameter.

(*c.*) Irrigate with acetic acid. The fibres are not affected, and no nuclei are revealed in them. They consist of the substance **elastin**, which is unaffected by acetic acid.

Make longitudinal and transverse sections of the ligamentum nuchæ (hardened in alcohol). Stain both in **picro-carmine** and mount in Farrant's solution. The connective tissue is thereby stained red, and the elastic fibres yellow.

2. L. S. Ligamentum Nuchæ (H).—Observe the fibres (yellowish), with a small amount of connective tissue (red) between them. The fibres are broad with well-defined margins, have a feeble yellow tint, and are transparent. They branch and anastomose, and where ruptured curl up at their ends.

3. T.S. Ligamentum Nuchæ of Ox (H).—Observe the polygonal ends of the broad fibres—yellow—and nearly as broad as, or broader than, a coloured blood-corpuscle, sometimes single, mostly in groups of three or more (fig. 121, *a*)—homogeneous throughout. A small amount of connective tissue (*c*) (red) between the groups.

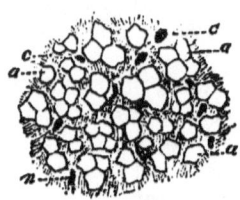

FIG. 121.—T.S. Ligamentum Nuchæ of Ox. *a*. Elastic fibres; *c*. Connective-tissue between them; *n* Nuclei of connective-tissue corpuscles. Alcohol and borax-carmine, × 300.

4. Another section may be stained with a watery solution of **magenta** and mounted in Farrant's solution. The fibres are stained red, but the pigment is apt to diffuse into the Farrant's solution.

5. A good plan is after hardening the ligamentum nuchæ in alcohol to stain it in **borax-carmine** for several days, with the precautions stated at p. 65. Transverse sections show the white fibrous tissue between the elastic fibres, with its nuclei stained red (fig. 121, *n*).

6. Fine Yellow Elastic Fibres (H).—Harden the mesocolon or mesentery of a young rabbit in Flemming's fluid, and stain it in methyl-violet as directed under Lesson X. 14, or stain it with magenta, when the elastic fibres are stained red; or with safranin after hardening in chromic acid.

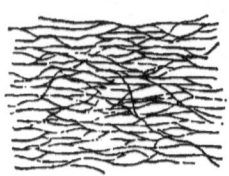

FIG. 122.—Fine Network of Elastic Fibres from the Mesocolon of Rabbit. Flemming's fluid and safranin.

(*a*.) Observe the network of fine elastic fibres. Many of the fibres have a diameter equal to one-sixth, or less, of that of a coloured blood-corpuscle ($1\ \mu$ or $\frac{1}{25,000}$ inch in diameter). The fibres branch and anastomose, and by carefully focussing, one can observe that the fibres do not all lie in the same plane (fig. 122).

7. Fenestrated Membranes (H).—Sometimes the elastin is so arranged as to form sheets or plates of elastic tissue, *e.g.*, in the large arteries; at other times these are perforated with holes, and are called fenestrated elastic membranes.

With a pair of forceps tear off a little of the endocardium from a sheep's heart, spread it on a slide, and treat it with caustic potash. Or use the basilar artery, slit it up and scrape away the outer coats, and use caustic potash as before.

(*a*.) Observe near the margin of the preparation the elastic membrane with holes in it (fig. 123).

8. White Fibres of Areolar Tissue (H).

Dissect off a thin lamella from an intermuscular septum, or remove a little of the subcutaneous tissue of a rabbit or rat. Place it on a *dry* slide, and rapidly spread it out into a thin film, but do not let it become dry, which can easily be avoided by breathing on the preparation. This is known as the "**half-drying**" or "**semi-desiccation method**," and is a very useful one, especially for sections containing much connective-tissue. Place a drop of normal saline solution on the cover-glass and apply it to the preparation.

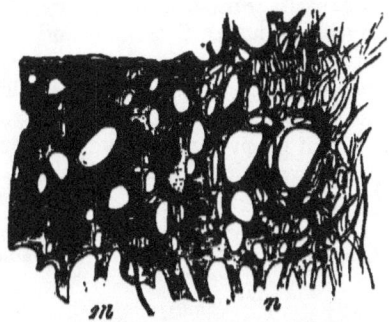

FIG. 123.—Network of Thick Elastic Fibres. *n.* Passing into a fenestrated membrane. *m.* Human endocardium. Fresh and caustic potash.

(*a.*) Observe the unbranched **white fibres**, wavy in their course,

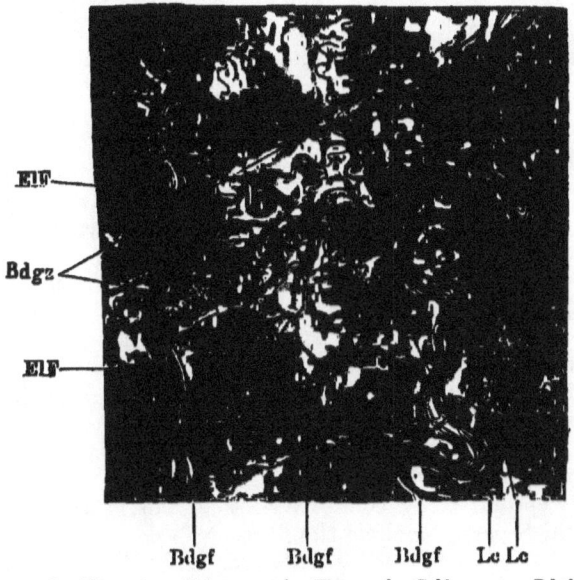

FIG. 124.—Areolar Tissue from Intermuscular Tissue of a Calf, × 200. Bdgf. Connective-tissue fibres, *i.e.*, bundles of fibres; Bdgz. Connective-tissue cells; ElF. Elastic fibres; Lc. Leucocytes.

with a faint, ill-defined outline, crossing each other in various directions. They are colourless, of feeble refractive power, and transparent. The fibres are striated longitudinally, and are seen to be made up of excessively delicate fine unbranched **fibrils**. The fibres vary from 6 μ to 8 μ in diameter ($\frac{1}{4200}$–$\frac{1}{3100}$ inch). They may be round or flattened, and are of indefinite length. Amongst the white fibres may be seen a few fine **elastic fibres** (ElF), recognised by their sharper contour, and by the fact that they branch and anastomose. . They run between, but never in the white fibres. It is not often that the **corpuscles** are visible without the action of special reagents. The corpuscles are best seen in young animals (fig. 124, Bdgz).

(*b.*) Irrigate with a 2 per cent. solution of glacial acetic acid; observe that the white fibres swell up, become clear, gelatinous, and homogeneous; the elastic fibres being unaffected, come clearly into view. The latter have a sharply-defined outline, branch and anastomose, and sometimes curl at the ends.

The **corpuscles**, or at least their nuclei, come into view. Observe the oval or fusiform nuclei of the *fixed connective - tissue corpuscles* (Bdgz)—they may be surrounded with some soft protoplasm—and the much smaller compound nuclei of the *wandering cells* or *leucocytes* (Lc). In the rat especially one is very likely to find the very granular nucleated cells known as **granular cells**. They frequently lie along the course of the small blood-vessels. Do not preserve this specimen.

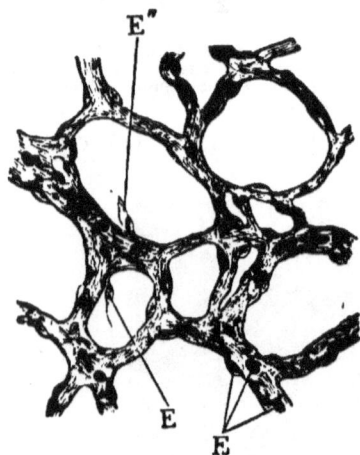

FIG. 125.—Omentum of Dog. E′, E″. Partially detached endothelial cells. E. Nuclei of endothelial cells, × 130.

If the areolar tissue be taken from the sub-arachnoid space of the brain and treated with dilute acetic acid, the fibres lose their fibrillated structure, nuclei appear, and the fibres themselves may be seen to swell up here and there, while they are constricted at irregular intervals by a thin fibre. This is due to these fibres being partly embraced by connective-tissue cells, which have long branches which partly encircle the fibre.

9. Fenestrated Fibrous Tissue (L and H).— Harden the omentum of a dog or cat in Müller's fluid. Stain a piece in logwood, and mount in balsam.

Note the meshes bounded by areolar tissue. The fibrils which compose the fibres are readily seen, and sometimes an endothelial cell may be seen partially detached (fig. **125**, E′), for the omentum lying in a serous cavity is covered with endothelium. The nuclei on the surface are the nuclei of endothelial cells, and those in the substances of the trabeculæ belong to connective-tissue cells.

ADDITIONAL EXERCISES.

10. Martinotti's Reaction for Elastic Fibres (H).—Harden elastic tissue, e.g., ligamentum nuchæ, or an organ containing elastic fibres, e.g., skin, artery, lung, trachea, in .2 per cent. chromic acid for three weeks. Cut sections and place them for twenty-four to forty-eight hours in a saturated alcoholic solution of safranin-O. Wash them in acid alcohol (p. 65), and then in absolute alcohol, to remove the surplus dye; clear in xylol, and mount in xylol-balsam.

(a.) All the elastic fibres, and they alone, are now either purplish, or, if the fibres be fine, black. This is a most excellent method for differentiating elastic fibres. The one thing of importance is to secure a good sample of safranin; some samples are quite inactive.

11. Elastic Fibres (*Herxheimer's Method*).—Place the sections containing elastic fibres in an alcoholic solution of hæmatoxylin, to which is added a few drops of a saturated solution of lithium carbonate. Stain them for a few minutes. Place them for five to twenty seconds in tincture of perchloride of iron, which rapidly decolorises all except the elastic fibres, which remain bluish or blackish. Wash in water and mount in balsam. This method is admirably adapted for demonstrating the longitudinal layer of elastic fibres in the trachea and bronchi.

12. Violet-B Method.—Cut out the hyaloid membrane of a frog's eye, or a piece of the omentum of a young rabbit, or the suspensory ligament of the liver of a rabbit. In normal saline pencil away the epithelium covering the membrane. Stain the section with violet-B (1 gram violet-B in 300 cc. of normal saline). This stains the cells and the elastic fibres. The preparation cannot be mounted in glycerine or Farrant's solution, as these dissolve out the dye, but a strong solution of common salt may be used (*S. Mayer*).

13. Areolar Tissue—Permanent Preparations.

(i.) By means of a hypodermic syringe (fig. 126) make an interstitial injection of silver nitrate (1 : 1000) into the subcutaneous tissue of a dog or rabbit. In this way an artificial œdema is produced and the tissues are "fixed." With a pair of scissors curved on the flat, snip out a little of the now œdematous connective tissue and stain it with picro-carmine. It requires some time to stain, and the preparation should be left for ten to twelve hours in a moist chamber (fig. 70), and then the picro-carmine is slowly displaced by acid glycerine, i.e., glycerine slightly acidulated with formic acid. In this way the

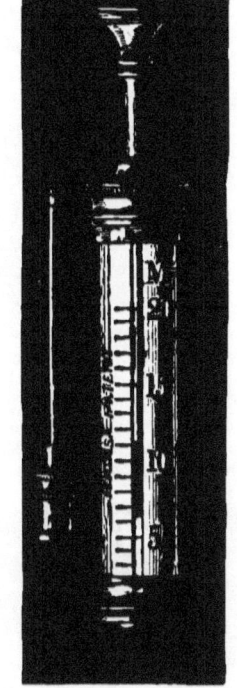

FIG. 126.—Hypodermic Syringe for making a Subcutaneous or Interstitial Injection.

L

various elements—fibrous and cellular—are usually brought distinctly into view (*Ranvier*).

(ii.) An excellent plan is to inject picro-carmine interstitially, and to leave the bulla several hours before snipping out a small part of it and mounting it in formic glycerine. In this preparation connective-tissue fibres with constrictions at intervals are frequently seen.

(iii.) A fibre may be stained with acid hæmatoxylin and mounted in glycerine (p. 69).

14. Coarsely Granular Cells ("*Mastzellen*" *of Ehrlich*).—Place part of the omentum of a young rabbit—or the fat from around the kidney of a rat or rabbit—in a watery solution of gentian-violet, to which a filtered watery solution of aniline-oil has been added. Heat the whole in a capsule until the vapour begins to rise, and allow it to cool.

After staining for twenty-four hours, remove the tissue and wash it in acid alcohol until most of the blue is gone. Dehydrate it in absolute alcohol, clear with xylol, and mount it in xylol-balsam.

(H) Search for a blood-vessel, and along its course will be found large oval cells crowded with numerous granules stained blue (fig. 127). These cells are found also apart from the blood-vessels.

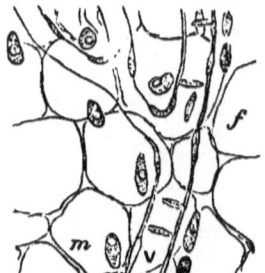

FIG. 127.—Coarsely Granular Cells, the "Mastzellen" of German authors, from Rat. *f*. Fat cells ; *v*. Vein ; *m*. "Mastzellen."

15. Gentian-Violet and Carmine Preparation.—The preparation may be double stained, thus : After washing in acid alcohol, stain the preparation in lithium-carmine for a few minutes, and again extract with acid alcohol. Mount as before in balsam. Observe the granules of the cells, blue as before, the nucleus red. All the other nuclei in the field are now red.

16. Clasmatocytes.—Ranvier[1] has given the name clasmatocyte (κλάσμα, fragment, κύτος, cell) to cells which can be seen in thin connective-tissue membranes of vertebrates. In *Triton cristatus* these cells may be 1 mm. in length. Stretch a membrane, *e.g.*, the omentum of a mammal or mesentery of a small reptile on a slide. Fix its elements by dropping on it a drop of 1 per cent. osmic acid, and then stain it with methyl-violet-BBBBB (1 part) dissolved in distilled water (10 parts). Examine the preparation in the fluid or in water. Large branched cells are seen stained of a bluish tint. At the extremities of the branches are small islands or granulations similarly tinted, and it is for this reason Ranvier has given them this name.

FIG. 128.—Cell-Spaces in Areolar Tissue. Silver nitrate.

17. Cell-Spaces in Areolar Tissue (H).—From a freshly-killed rabbit snip out a small piece of the subcutaneous tissue as free from fat as possible. Spread it upon a *dry* slide, and drop on it from a pipette a half per cent. solution of silver nitrate. Allow the silver to act for ten to twelve minutes, remove it, cover the film with glycerine and expose it to light. It rapidly becomes brown. It is better to use connective-tissue from a calf, but the layer used must not be too thin.

(*a*.) If successful, note a brownish ground—the cement substance—and in it clear branched spaces corresponding in shape to the fixed connective-tissue corpuscles. These are the cell-spaces (fig. 128).

[1] *Comptes Rendus*, vol. 110, p. 165, 1890.

18. Pericœsophageal Membrane of Frog.—To Ranvier[1] we owe our knowledge of the value of this membrane. The œsophagus is surrounded by a lymph-sac, which is separated from the pleuro-peritoneal cavity by an excessively thin membrane. The sac is readily distended by insufflation, after pulling out the œsophagus through the mouth by means of a hook. This membrane is covered on both surfaces by epithelium which can be stained with silver nitrate; its texture consists of fine connective-tissue with elastic fibres, but containing clasmatocytes, and in it are also to be found blood-vessels and non-medullated nerves. Treat it as described in **16**, to see the clasmatocytes. By the same process the non-medullated nerve fibres and elastic fibres will also be stained.

LESSON XI.

TENDON.

Tendon is composed of *white fibres* arranged longitudinally and parallel to each other. The fibres are arranged in bundles, the *tendon-bundles*, which are held together by a *sheath* and *septa* of connective-tissue. The fibres are united to each other by a *cement substance*, and on the primary bundles of the fibres are placed the *tendon-cells*, which vary in their shape and arrangement in different tendons. It is supplied by few blood-vessels, and contains only a few elastic fibres.

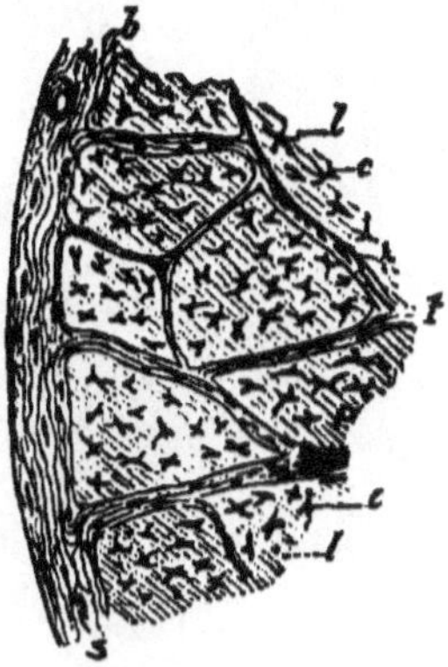

Tendon.—Harden a small tendon of a man, calf, dog, or cat, in Müller's fluid, alcohol, bichromate of potash, and complete the hardening in alcohol. By freezing, make transverse and longitudinal sections, or use the celloidin method. Tendons cannot be cut after being embedded in paraffin, they become too hard.

1. T.S. Tendon (L).—(*a.*) Stain a section in picro-carmine, and mount it in Farrant's solution. Observe the **sheath** (fig. 129, *s*), composed of connective tissue arranged circularly, sending **septa** (*t*) into the substance of the tendon, thus breaking it up into polygonal areas of different sizes, which are filled by the cut ends of the longitudinally-arranged fibres.

FIG. 129.—T.S. Tendon. *s.* Sheath, with *b.* Blood-vessel; *t.* Trabeculæ or septa; *c.* Branched spaces in the tendon for tendon-cells; *l.* Matrix or cut ends of the fibres, × 50.

[1] *Comptes Rendus*, vol. 111, p. 863, 1890.

(*b.*) The branching stellate spaces, **interfascicular spaces** (fig. 129, *c*), between the fasciculi or bundles of fibres. If these spaces contain air, they appear somewhat dark. These spaces can readily be seen as branched dark spaces if a transverse section is made— by means of a knife, not a razor—of a small tendon dried at the ordinary temperature.

(*c.*) (H) Observe the cut ends of the fibres (*l*), which appear almost homogeneous, but amongst them here and there may be seen a few dots, which are the transverse sections of elastic fibres, and the branched **tendon-spaces** (*c*), some of them with a nucleated branched cell.

FIG 130.—L.S. Human Tendon (*Tibialis anticus*).
SZ. Rows of nuclei of tendon-cells.

2. L.S. Tendon.—Stain a section with logwood, and mount it in balsam. (H) Observe the longitudinal arrangement of the *fibres*, and between them rows of fusiform *tendon-cells*, or rather the long fusiform nuclei of the tendon cells arranged between the fibres. The other parts of the cell become too transparent to be seen (fig. 130). Sometimes a L.S. of one of the septa may be seen.

3. Fibrils in Tendon (H).—Macerate a tendon from the tail of a rat for twenty-four hours in a saturated solution of picric acid, or for 3–4 hours in baryta water. Tease a small part, and examine it first in water. Mount it in Farrant's solution. This is apt, however, to render the fibrils too transparent. Perhaps a better method is to place the fine tendons for twenty-four hours in equal parts of 1 per cent. osmic acid and 1 per cent. silver nitrate. Mount a teased preparation in Farrant's solution.

(*a.*) Observe the isolated *fibrils* ($\frac{1}{50000}$ inch in diameter), excessively fine; wavy, and unbranched (fig. 131).

4. Tendon of Rat (*Gold Chloride Method*).—Kill a rat, cut off its tail, forcibly rupture the tail, when a long leash of fine white glistening threads or tendons will be obtained. Prepare the tendons by one of the gold chloride methods. One of the best methods is the lemon-juice method of Ranvier (p. 79), but the boiled formic acid and gold method also yields excellent results. It is to be remembered that it is not necessary to use gold chloride to demonstrate the tendon-cells; this can be done by hæmatoxylin.

With regard to the action of gold chloride, my experience leads me to believe, that in order to see the rows of tendon-cells with their lateral protoplasm expansions, the lemon-juice method is very good; while the old acetic acid method makes the fibres less swollen up, and on teasing they are readily isolated, thus enabling one to see cells either singly or in rows clasping them. Not unfrequently isolated tendon-cells are to be seen in the field of the microscope.

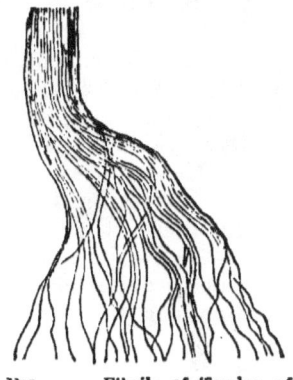

FIG. 131.—Fibrils of Tendon of Rat Isolated by Picric Acid, × 300.

Tease a small part of the gold tendon in Farrant's solution.

(*a*.) (**H**) The fibres are swollen up and transparent, and lying on them are rows of **tendon-cells** (fig. 132, *b*, *b*) stained of a violet tint. Each cell is somewhat oblong with a distinct nucleus, and

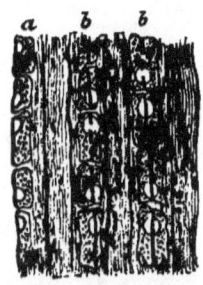

FIG. 132.—Gold Chloride, Tendon, Tail of Rat. *a.* Tendon - cells seen on edge and embracing a fibre; *bb.* On the flat, the cells with a ridge.

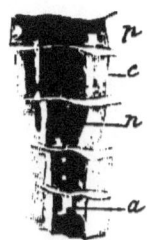

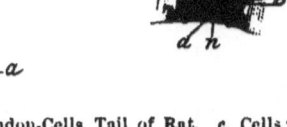

FIG. 133.—Tendon-Cells, Tail of Rat. *c.* Cells; *p.* Lateral prolongation or expansion of the cell protoplasm; *n.* Nucleus; *a.* Stripe or ridge.

bears a flattened wing-shaped expansion (fig. 133). Along the cells is usually to be seen a **stripe** or **ridge** (fig. 133), produced by the cells being compressed between several adjoining fibres. This ridge may be seen to be interrupted in some of the cells. The nuclei of the adjacent cells may be seen to be close together.

(*b.*) If a side view of the cells is obtained (fig. 132, *a*), they partially clasp the fibre, but never envelope it completely; in this respect these cells differ from endothelium.

5. T.S. of Gold Tendon (H).—Remove the skin from the tail of a young rat, cut out a piece of the tail a quarter of an inch in length with its tendons, and subject it to the gold chloride process (p. 79). When, after reduction, it has become purple or brownish, decalcify the bone, harden it in alcohol, and make transverse sections. Mount one in balsam.

(*a.*) Many tissues will be seen, including muscle, nerve, fat, and bone. Neglecting these, observe the small rounded areas at the circumference, the transverse sections of the small tendons, each surrounded by its own sheath of connective tissue (fig. 134, *t*). In each observe the branched stellate spaces (fig. 134, *c*), frequently anastomosing with each other. These **interfascicular spaces** are

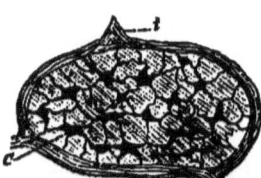

FIG. 134.—T.S. of a small Tendon, Tail of Rat. *t.* Sheath; *c.* Interfascicular spaces with tendon-cells. Gold chloride.

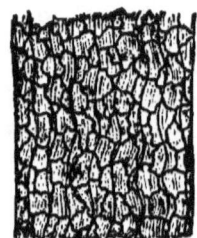

FIG. 135.—Layer of Endothelial Cells on the Surface of a Tendon, Tail of Rat. Silver nitrate.

purplish in colour; they contain the **tendon-cells**, and also a purplish deposit due to the gold chloride acting on the lymph which they contain in the fresh condition.

6. Endothelial Sheath of Tendon (L and H)—Silver a leash of the fine tendons from the tail of a rat. Mount a short length of one of them in balsam.

(*a.*) Observe the tendon made up of parallel fibres, and note on their surface a single layer of endothelium. The squames are large, polygonal, and mapped out by "silver lines," but no nuclei are visible.

7. Fresh Tendons and Acid Logwood (H).—Place three or four tendons (rat's) 1¼ inch long, on a *dry* slide, and fix their ends with paraffin, so as to keep them extended. Make a solution of acid logwood by adding one part of 1 per cent. glacial acetic acid to three parts of logwood solution. This solution is red. Place a drop of it on a cover-glass, and lay it on the tendons. The acid brings into view rows of narrow, granular, nucleated cells between the fibres of the tendon, while at the same time the logwood stains them. Instead of acid logwood use picro-carmine or acid hæmatoxylin (p. 69). Displace the dye with water and mount in glycerine. The tendons are purposely taken longer than the breadth of the cover-glass, so that they may remain stretched.

8. Fresh Tendon.—On a black surface, tease in normal saline a small piece of any tendon of a calf. Observe the fibres and fibrils, but no cells are visible. Irrigate the preparation with 2 per cent. acetic acid. The mass becomes clear and transparent to the naked eye, and now under the microscope one sees the fusiform nucleated cells singly or in line in order between the swollen-up fibres. The preparation may be stained with magenta, which brings into clearer view the cells, and any elastic fibres present.

ADDITIONAL EXERCISES.

9. Tendon of Rat (*Dogiel's Method*).—Very good preparations are obtained by placing the fresh tendons for several days—the longer the better—in Grenacher's alum-carmine (p. 65). This fluid stains but slowly. The cells, however, are stained, and if a tendon be teased, isolated cells, and cells on the fibres are easily seen. It is a good method for showing the relations of the cells to the fibres.

10. T.S. Tail of Rat (*Corrosive Sublimate and Borax-Carmine*).—Harden short lengths of the tail of a rat, the skin being first removed, in corrosive sublimate for three hours or so. Remove every trace of the mercuric salt by prolonged washing in alcohol. Stain the tissue in bulk in borax-carmine, and then decalcify it in dilute hydrochloric acid. Make transverse sections, after embedding it by the interstitial method in paraffin. Sections may also be made by freezing, but they are apt to fall asunder. This method also yields beautiful preparations, comparable to those by the gold chloride methods. The transverse sections of the tendons are very characteristic.

11. Dried Rat's Tendons.—A very convenient method is to dry the tendons of a rat's tail, keeping them extended during the process. After drying, they can then be used at any time. By acting on them with dilute acetic acid they swell up slowly, and the rows of cells are thereby revealed. The cells—after washing away the acetic acid—can be stained with picro-carmine, and the preparation mounted in dilute glycerine.

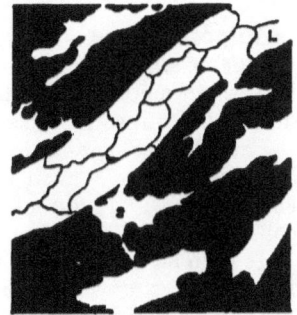

FIG. 136.—Cell-Spaces in the Central Tendon of the Diaphragm of Rabbit. *l* Lymphatic; *s.* Cell-spaces. Silver nitrate.

12. Cell-Spaces (*Saft-Canälchen*) **in Central Tendon** (*Silver Method*).—Place the central tendon or the whole diaphragm for five minutes in a ¼ per cent. silver nitrate. Remove it, and with a camel's-hair pencil brush both surfaces of the tendon to remove the endothelium. Replace it in the silver solution for fifteen minutes. Remove it; wash it in water, and expose it to light to reduce the silver. One piece may be mounted in balsam; another piece should be stained with acid logwood or picro-lithium carmine (in this case use dilute hydrochloric acid), and mounted in balsam.

(*a.*) Observe a large number of clear, branched anastomosing spaces, surrounded by brown areas of ground-substance. The former are the cell-spaces and *Saft-Canälchen*, or juice-canals, and some of the latter may be seen to communicate with the lymphatics (fig. 136).

(*b.*) In the stained specimen, stained nuclei are seen in the spaces, *i.e.*, the nuclei of the cells which occupy these spaces.

13. Cell-Spaces (*Iron Sulphate Method*).—Using the fresh central tendon of the diaphragm of a mouse or guinea-pig or rat, place it for a few minutes in 1 per cent. sulphate of iron. Pencil away the surface endothelium, and leave it in the iron solution for five to seven minutes. Remove it, wash it, and place it in 1 per cent. ferricyanide of potash, in which it becomes blue. Mount it in Farrant's solution or balsam. In this preparation the cell-spaces and juice-canals are again clear, but the ground-substance is blue.

14. Cell-Spaces in Rat's Tendon.—The fine tendons are placed in silver nitrate ($\frac{1}{4}$ per cent.) for two minutes, and then the epithelium is brushed off by a camel's-hair pencil. Five or six sweeps of the brush usually suffice. The tendons are stained for other ten minutes in silver, washed, and exposed to light in alcohol. Rows of clear, somewhat quadrangular spaces in a brownish matrix are obtained.

LESSON XII.

ADIPOSE, MUCOUS, AND ADENOID TISSUES— PIGMENT CELLS.

ADIPOSE TISSUE (FATTY TISSUE).

Adipose Tissue.—A fat-cell consists of a *membrane* enclosing a *globule of oil*, which pushes the oval flattened *nucleus* (surrounded by a small amount of protoplasm) to one side, so that it lies close under the cell-wall. Size, 40 μ to 80 μ ($\frac{1}{500} - \frac{1}{300}$ inch).

Fat-cells are arranged in groups, which form *lobules*, and these again form *lobes*. Each lobule has an afferent *artery*, one or two efferent *veins*, and a dense network of *capillaries* between the fat-cells, each capillary surrounding one or more fat-cells.

It is to be remembered that cells in connective tissue containing fat may have a two-fold origin. Fat may be formed in ordinary connective-tissue cells, but there are other cells of a connective-tissue nature, which seem to be more specifically fat-cells. During development this tissue is formed at certain parts, *e.g.*, in the groin, axilla, and neck, and presents a grayish-yellow appearance in the form of lobules, surrounded by connective-tissue—readily seen in a young animal. The cells at first contain granules. The

XII.] ADIPOSE TISSUE. 169

fat-cells derived from this tissue form lobules, and are supplied with blood-vessels very much as a gland is supplied with blood; so that each lobule is provided with an artery, vein, and capillaries.

1. Fat-Cells.—Cut out a small piece of the omentum of a cat, selecting a piece that contains a little fat, and mount it in normal saline.

(*a.*) (**L**) Observe the large, highly refractive fat-cells arranged singly or in groups (fig. 137).

(*b.*) (**H**) The large fat-cells of variable size —some of them polygonal—highly refractive contents, but no nucleus visible; connective tissue passing between some of the cells.

FIG. 137.—Fat-Cells, some showing a Nucleus. The central one shows crystals of margarine. × 100.

2. Action of Osmic Acid (**H**).—Place a small part of the omentum of a cat or embryo ox in ¼ per cent. osmic acid for an hour. Wash it thoroughly, and mount in Farrant's solution.

(*a.*) (**L**) Observe the fat-cells, which first become brown and ultimately quite black. The fat-cells are in groups, and their relations to the blood-vessels can also be seen in fig. 138.

3. Membrane and Nucleus of Fat-Cells (**H**).—Place a piece of the omentum or subcutaneous adipose tissue in absolute alcohol for several days, and afterwards in ether for a day or two; or the tissue may be boiled for a few minutes first in alcohol and then in ether. Transfer a small piece to hæmatoxylin and allow it to stain for several hours. Wash it in water and place it in

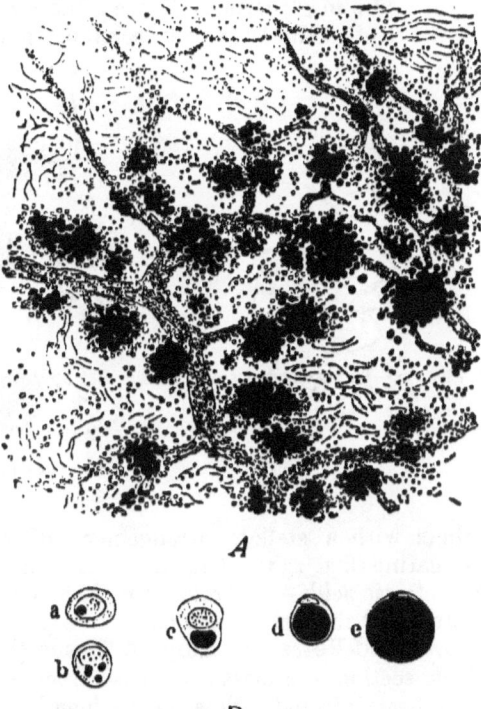

FIG. 138.—Fat-Cells stained with Osmic Acid from the Omentum of an Embryo Ox. *A.* The fat-cells in groups or lobules, blackened by the osmic acid, and showing their relation to the blood-vessels; *B.* a-e fat-cells in different stages of development.

absolute alcohol. Extract it with turpentine, clear it up with clove-oil, and mount it in balsam.

(*a.*) Observe the collapsed membrane of the fat-cells, with a small oval blue-stained nucleus immediately under the cell-wall. A small quantity of protoplasm surrounding the nucleus may be visible.

4. Or harden a small piece of the omentum of a rat or other animal in absolute alcohol. Select a piece which contains some fat. Stain it for twenty minutes in lithium-carmine, then place it in acid alcohol (1 per cent. hydrochloric acid in 70 per cent. alcohol); place it in absolute alcohol; clear it up with clove-oil or xylol, and mount in Canada balsam.

(*a.*) Observe the envelopes of the fat-cells and the nuclei of the fat-cells, the latter stained bright red (fig. 139). Other red-stained nuclei are visible, but they are the nuclei of blood-capillaries between the fat-cells.

5. **Margarine Crystals (H).**—Place a small piece of fat for forty-eight hours in glycerine. Tease a piece in Farrant's solution.

(*a.*) Notice the large cells some of them with granular contents,

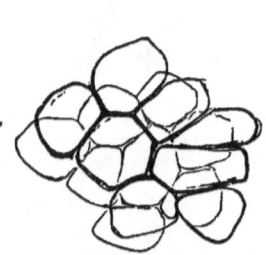

FIG. 139.—Empty envelopes of Fat-Cells. Alcohol and ether.

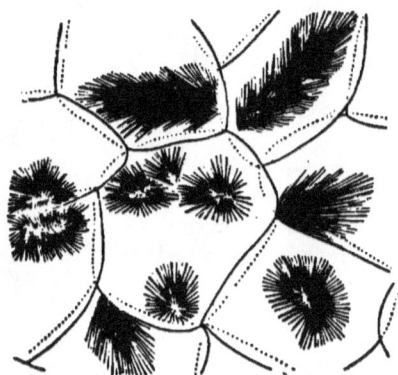

FIG. 140.—Fat-Cells containing Crystals of Margarine.

others with a stellate arrangement of needle-shaped crystals of margarine (figs. 137 and 140). If the star of crystals—really palmitic and stearic acids—be broken up, then the needle-shaped crystals are distributed throughout the cell.

6. **Blood-Vessels of Adipose Tissue (L and H).**—Make a rather thick section of a mass of adipose tissue in which the blood-vessels have been injected. This can be done by injecting the blood-vessels of a rabbit with a carmine-gelatine mass. As the fat is very soft, the best method of obtaining such sections is to saturate the tissue with paraffin and cut it in paraffin. Mount sections—not too thin—in balsam. A section which has been saturated with paraffin and cut in this substance must have the paraffin removed by soaking in

ADIPOSE TISSUE.

turpentine or xylol, which dissolves out the paraffin; the section is afterwards mounted in balsam.

(L) Observe the very vascular small *lobules*, composed of fat-cells. To each lobule there passes one artery, and from it emerge one or two veins.

(H) Observe the loop of capillaries round each fat-cell or around several fat-cells.

7. Development of Fat-Cells.—These may be studied in the subcutaneous tissue of a newly-born rat, or in the omentum of a newly-born rabbit. Stain a small piece of any of these tissues in osmic acid.

(*a.*) (L) Observe the fat-cells in groups surrounded by connective tissue.

(*b.*) (H) Observe the shape of the cells, with small globules of oil—black—scattered throughout the protoplasm, the nucleus pushed to one side in the more developed fatty cells by a globule of oil, which is stained black (figs. 138, 141).

FIG. 141.—Developing Fat-Cells from the Subcutaneous Tissue of a Fœtus. Osmic acid.

8. Atrophic Fat-Cells.—These are readily obtained from the yellow bone-marrow of an old person who has died from some wasting disease, or from the sub-pericardial fat of a person who has died from phthisis.

(H) Observe the envelope of the fat-cell, now no longer completely filled with fat, but containing a little protoplasm and some serous fluid.

9. Injection Method.—By means of a hypodermic syringe, inject under the skin of a dog or cat silver nitrate (1 in 1000), which causes a local œdema and isolates the elements of the areolar tissue, including the fat-cells (*Ranvier*).

MUCOUS TISSUE.

Mucous Tissue.—In the embryo it exists under the skin; it forms Wharton's jelly of the umbilical cord, and in the adult it forms the vitreous humour. It is essentially an embryonic tissue.

10. Mucous Tissue (H).—Harden the umbilical cord of a three-months fœtus in Müller's fluid and then in alcohol. Make transverse sections by freezing, and stain them in picro-carmine or hæmatoxylin or gentian-violet. Mount in Farrant's solution.

(*a.*) Observe the large, branched, granular, nucleated cells, which anastomose with each other. Between the cells is a fluid which contains mucin, and according to the stage of development of the cord, there is a greater or less number of fibres. The older the cord, the more the fibres increase in number, and its characters approach those of ordinary connective-tissue. A better view of the finer processes is obtained by examining the tissue in normal saline (fig. 142).

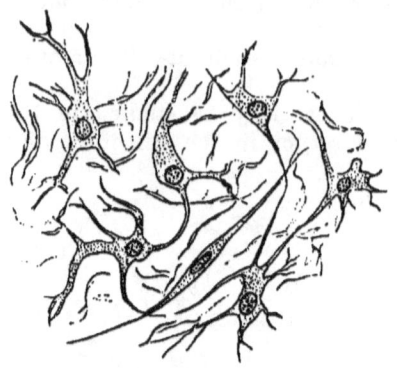

FIG. 142.—Mucous Tissue of Umbilical Cord of Fœtus. Müller's fluid and logwood, × 300.

ADENOID TISSUE.

Adenoid, Retiform, or Reticular Tissue consists of a reticulum or network of fine fibrils, which run in all directions, forming a meshwork in several planes. Some regard it as made up of branched corpuscles, the processes of which anastomose. In the meshes are leucocytes or lymph-cells, which usually occur in such numbers as to obscure the presence of the fine meshwork in which they lie.

It is very widely distributed, e.g., in lymphatic glands, simple and compound, tonsils, solitary glands, and Peyer's glands; in the bronchial, pharyngeal, nasal, intestinal mucous membrane, spleen, thymus, and a few other situations.

11. Adenoid Tissue of Lymphatic Glands.—This may be prepared in several ways.

(i.) Harden an abdominal lymphatic gland of a calf or kitten for two weeks in Müller's fluid. Make sections, and shake up one in a test-tube with some water; this dislodges the lymph corpuscles, and in places leaves the fine reticulum visible.

(ii.) A better plan is to inject into a fresh lymph gland a $\frac{1}{4}$ per cent. solution of osmic acid, or $\frac{1}{2}$ per cent. solution of silver nitrate. In either case an œdema is produced which separates the parts and reveals the network. The injection is made by means of a hypodermic syringe (fig. 126). The syringe is filled with the solution, and the sharp nozzle is thrust into the gland, and the contents of the syringe rapidly injected haphazard into the organ. It passes in, and forms, as it were, an œdema, and separates, and at the same time hardens, the constituent parts of the organ. This is the method of **interstitial injection**, one which is frequently employed.

(iii.) The gland may be hardened for twenty-four hours in picric

XII.] ADENOID TISSUE. 173

acid, and the sections stained with eosin-hæmatoxylin and mounted in balsam.

(H) Observe some parts crowded with lymph corpuscles, but where these are wanting, note the very fine network of fibres, with nuclei here and there at the points of intersection (fig. 143).

The lymph corpuscles may be got rid of by applying to a fresh preparation a dilute solution of caustic potash which dissolves them.

12. Pigment-Cells and Guanin-Cells.—These may be studied by pinning out on a frog-board one of the webs between the toes of a frog (Lesson XIX. 11, *e*).

(*a.*) (**L** and **H**) Observe large, branched, corpuscles loaded with

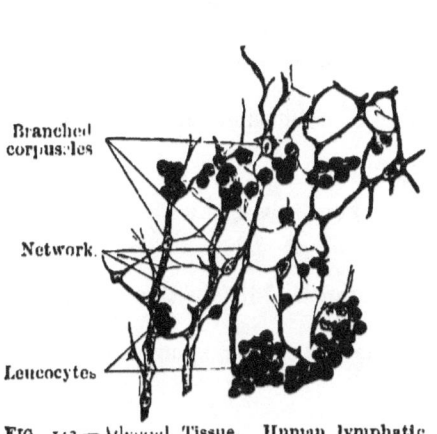

FIG. 143.—Adenoid Tissue. Human lymphatic gland. Picric acid and eosin-hæmatoxylin.

FIG. 144.— Pigment and Guanin-Cells of Frog. A Contracted; B, C. Partially relaxed pigment-cells. G Guanin cells.

black granules of melanin (fig. 144, B), also smaller black spots, which are cells with their processes retracted. Every intermediate stage between these two states may be seen.

A permanent preparation is readily made by stripping off the skin from the web of the toe of a dead frog, hardening it in alcohol, and mounting in balsam. The web should be fixed in an extended position before it is placed in the alcohol.

In such a preparation, not only will pigment-cells be found, but **guanin-cells** also, *i.e.*, small oval cells filled with white refractive granules of guanin (fig. 144, G). To see the guanin-cells turn off the light reflection from the mirror, when the granules in the guanin-cells will appear bright and refractile on a black ground.

For pigmented connective-tissue corpuscles from the choroid, see Lesson on Eye.

ADDITIONAL EXERCISE.

13. Mucous Tissue.—Many branched cells are seen in a T.S. of the tail of a tadpole, young triton, or salamander, hardened in 1 per cent. osmic acid and cut in paraffin.

It is also to be found under the skin of the flanks in frogs at the breeding season. It gives good preparations when stained with methyl-violet-5B (*S. Mayer*).

LESSON XIII.
BONE, OSSEOUS TISSUE, &c.

THE essential elements of **osseous tissue,** of which bone consists, are a calcified fibrous *matrix* or ground-substance, with cells or *bone-corpuscles* embedded in it; the latter are lodged in spaces called *lacunæ*. A **bone,** however, is a complex organ. The following scheme may facilitate the comprehension of its minute structure:—

BONE.

In a longitudinal section of a long bone, observe with the naked eye—

Periosteum covering the bone.

Compact or dense bone, the substantia dura (with Haversian canals).

Cancellated or spongy bone, the substantia spongiosa (with Haversian spaces and cancelli).

Medullary cavity with marrow.

Histologically dry compact bone shows—

Lamellæ
- Peripheric or circumferential.
- Haversian or concentric.
- Intermediary, interstitial or ground.
- Perimedullary.

Sharpey's perforating fibres { white. yellow.

HAVERSIAN SYSTEM.

Dry bone.	Recent bone.
Haversian canal	Blood-vessels, connective tissue, lymphatics, osteoblasts.
,, lamellæ	Between the lamellæ, branched bone-corpuscles.
Lacunæ	Processes of bone-corpuscles and lymph.
Canaliculi	Lymph.

XIII.] BONE, OSSEOUS TISSUE, ETC. 175

Periosteum
1. External layer, fibrous, with the larger blood-vessels.
2. Internal or osteogenic layer, with finer blood-vessels, numerous elastic fibres, and *osteo-blasts*, and sometimes *osteoclasts*.

Osteoclasts or myeloplaxes of Robin.
Blood-vessels, nerves, and lymphatics.

Development
1. Endochondral in cartilage.
2. Intra-membranous or periosteal.

1. T.S. of Dense or Compact Bone.—It is better to buy a prepared transverse section of the shaft of a small, dry, long bone. In it most of the smaller spaces, being filled with air, appear black.

(*a.*) (**L**) Observe the **medullary cavity**, bounded by a ring of bone, less dense internally.

(*b.*) Sections of tubes which appear round, or oval—**Haversian canals**—surrounded by **concentric lamellæ** (fig. 145, *c*), and between these lamellæ are **lacunæ** (*e*), with fine channels passing from them—the **canaliculi**. Some lamellæ are arranged parallel to the circumference of the bone—the **peripheric** (*a*)—while others of larger radius are incomplete, and jammed in between the Haversian system: they are **intermediary** (*d*). Around the medullary canal are the **perimedullary lamellæ** (*b*).

(*c.*) (**H**) Observe the shape of the lacunæ—flattened branched spaces — with their numerous wavy branching canaliculi, and how adjoining canaliculi anastomose by traversing the lamellæ (figs. 145, 147, *e*). At the outer part of each Haversian system, some of the canaliculi of the

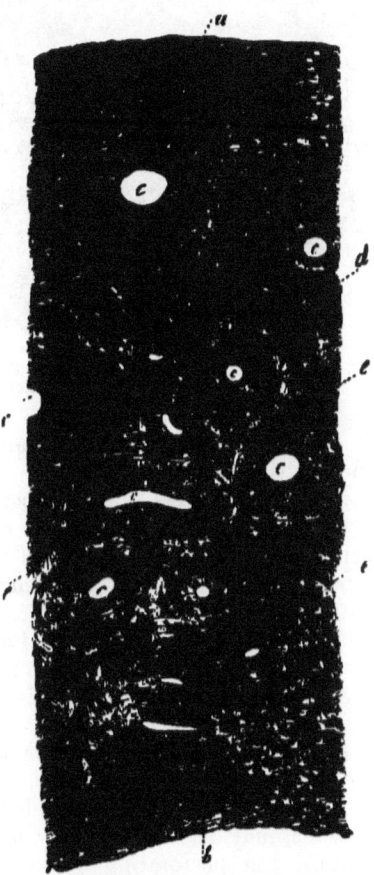

FIG. 145.—T.S. Human Metacarpal Bone. *a.* Peripheric lamellæ; *b.* Perimedullary lamellæ; *c.* Haversian canals surrounded by their Haversian lamellæ; *d.* Intermediary lamellæ; *e.* Lacunæ, × 20.

outermost row of lacunæ will be found to form loops and open into the lacunæ from which they arose: these are **recurrent canaliculi** (fig. 147, *a*). The canalicular system is for the distribution of lymph to all the parts of the calcified fibrous matrix. Notice also that the intermediary lamellæ are parts of circles with a much

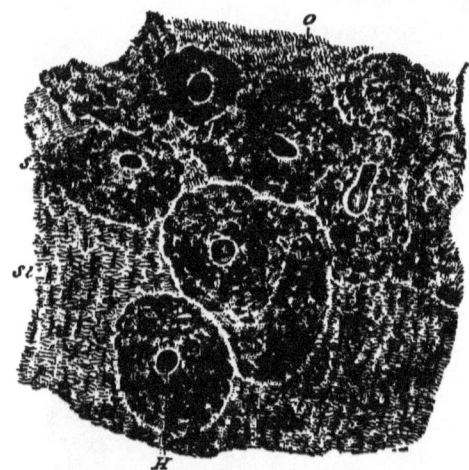

FIG. 146.—T.S. The Shaft of a Human Femur. *H.* Haversian canals; *s.* Haversian lamellæ; *si.* Intermediary lamellæ, × 40.

larger radius than those of the Haversian system. Each Haversian canal, with its system of lamellæ, lacunæ, and canaliculi, forms a **Haversian System**. The greatest diameter of the lacunæ is about 14 μ ($\frac{1}{1800}$ inch).

(*d.*) The **lamellæ** in a Haversian system on transverse section appear as thin concentric bands, a clear transparent one alternating with one which looks more granular. These are not due to different kinds of lamellæ, but in the clear ones one looks upon the long axis of the fibrils composing the lamella, and in the others upon the ends of the fibrils.

2. **L.S. of Dense Bone**, prepared in the same way.

(*a.*) (**L**) Observe Haversian canals running chiefly in the long axis of the section, with here and there oblique, short, junction canals. Near the surfaces some open externally, and others communicate with the medullary cavity. The lamellæ of any Haversian system run parallel to its own Haversian canal. In the system of canals—each 20–100 μ wide—the canals frequently divide dichotomously, and ultimately form a network in the compact bony substance.

(*b.*) If the section be near the surface of the bone, so as to include the peripheric lamellæ, canals for blood-vessels, perforating the lamellæ and not surrounded by lamellæ as in the Haversian systems, may be seen. They are called **Volkmann's canals**, and contain the perforating vessels. They are connected with the Haversian canals proper. They are well seen in sections of the femur of a guinea-pig, but, unlike Haversian canals, they are not surrounded by Haversian or concentric lamellæ.

(*c.*) (**H**) Observe the flattened oval lacunæ with their canaliculi;

XIII.] BONE, OSSEOUS TISSUE, ETC. 177

their arrangement, as well as that of the lamellæ (fig. 148). If a
canal be viewed carefully many fine dots will be seen in it. These
are the openings of the canaliculi.

3. **T.S. Decalcified Shaft of a Bone.**—The bone, cut into short
lengths, must be decalcified in picric acid or chromic and nitric
fluid, with the precautions laid down at p. 38. When sufficiently

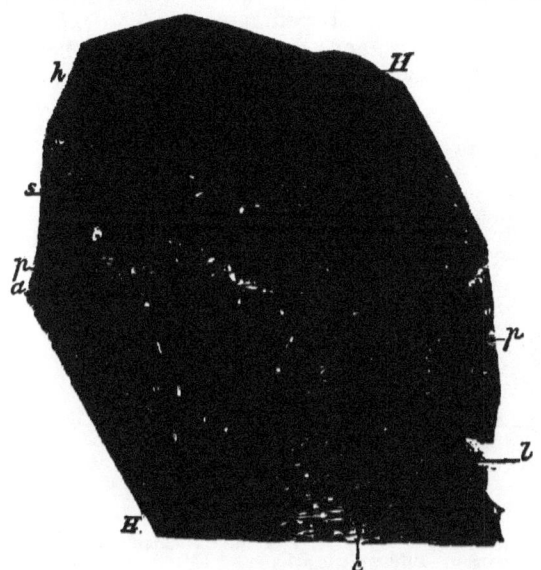

FIG. 147.—T.S. Shaft of Human Femur. *H.* Haversian
canals; *c.* Lacunæ with bone-corpuscles; *a.* Lacunæ
with recurrent canaliculi; *s.* Intermediary lamellæ with
Sharpey's fibres, *h*; *p.* Large fibres of Sharpey in intermediary lamellæ; *l.* Confluent lacunæ. These Ranvier
supposes are bone-corpuscles and lacunæ undergoing
atrophy, × 300.

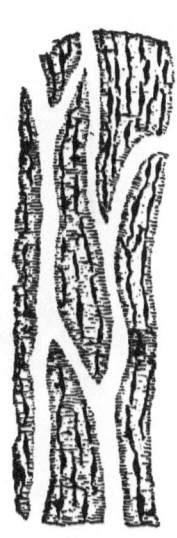

FIG. 148.—L.S. Dense
dry bone, × 40.

soft—ascertained by pricking it with a pin—it is hardened in
alcohol in the usual way. Sections are best made by freezing.

(i.) Stain a section in picro-carmine and mount it in glycerine-
jelly. Glycerine or Farrant's solution tends to make the preparation rather too transparent.

(ii.) Place some sections in 1 per cent. osmic acid for twenty-four
hours and mount them in glycerine-jelly.

(*a.*) (L) Observe the **periosteum** (fig. 149), embracing and
adherent to the bone. In the bone itself the lacunæ, and especially
the canaliculi, are no longer black, and are not so visible as in the
non-decalcified bone. Each lacuna contains a highly refractive,
branched, nucleated, and stained corpuscle.

Bone-Corpuscles.—Observe their arrangement following that of the lamellæ, but the latter are not so distinct as in dry bone. Several lamellæ lie between two consecutive rows of bone-corpuscles. The Haversian canals contain blood-vessels, connective tissue, and other cells, or **osteoblasts**.

(b.) (H) The **periosteum** consists of an *external layer* stained red, and composed chiefly of white fibrous tissue. Attached to it may be found small fragments of striped muscle. The *internal layer* contains many elastic fibres, and, especially in young bones, there may be seen one layer or more of flattened or cubical cells,

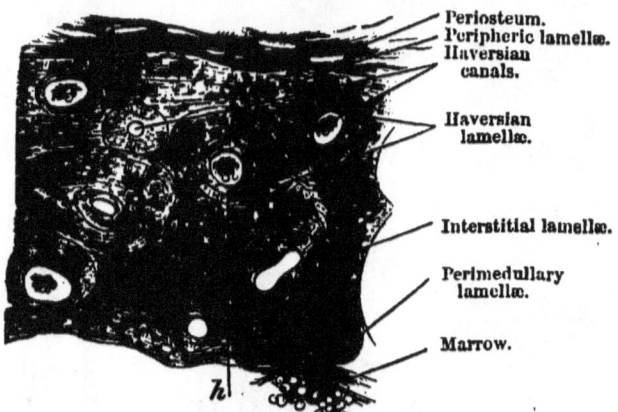

FIG. 149.—T.S. Part of a Human Metacarpal Bone, × 50. *h.* Haversian space with marrow. Dilute nitric acid.

called **osteoblasts** (fig. 156, c). The latter may be seen not only under the periosteum, but also passing along with blood-vessels into the Haversian canals.

Fibres may be seen passing from the deep surface of the periosteum into the bone—the **perforating fibres of Sharpey**.

(c.) Observe the **lamellæ**, but their outline is not very distinct, while the **canaliculi** will not be distinctly visible, being indicated by fine lines traversing the lamellæ. The **bone corpuscles** are nucleated refractive cells, each lying in a lacuna. In such a preparation, one cannot make out that they send processes into the canaliculi.

(d.) Select a large Haversian canal and study its contents. Note the presence of an artery and vein with very thin walls, and the cavity lined by osteoblasts (fig. 150, Obl.), and the remainder filled up with medullary tissue.

4. **Perforating Fibres.**—(i.) From a membrane bone of the skull (*e.g.*, the parietal or frontal bone, which has been softened in 2 per cent. hydrochloric acid or in v. Ebner's fluid (p. 37)

and from which the acid has been removed by steeping in water, and subsequently in spirit) with forceps remove the periosteum, and tear off a thin lamella of osseous tissue. Place its under-surface uppermost on a slide in water or v. Ebner's fluid.

(H) Observe fine tapering fibres like nails—perforating fibres—projecting from the surface. Some apertures may be found from which corresponding fibres have been withdrawn.

(ii.) These are far better developed in the bones of the skull of birds. Soften the vault of the skull of a

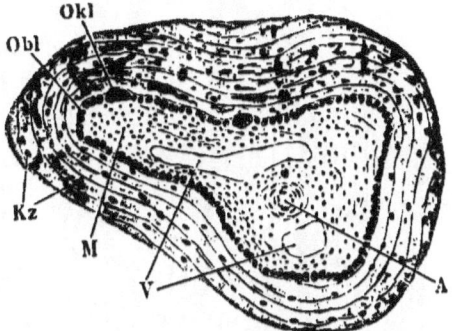

FIG. 150.—T.S. of a large Haversian Canal with its Soft Parts. Femur of a Dog, ×130. A. Artery; V. Vein; Kz. Bone-cells; M. Medullary tissue; Obl. Osteoblast; Okl. Osteoclast.

fowl in v. Ebner's fluid and harden in alcohol. Make not too thin sections, and with needles tear the lamellæ asunder. Examine it in v. Ebner's fluid. Or make sections of a human frontal bone softened in dilute hydrochloric acid, and examine it in water.

(H) Numerous perforating fibres passing between the separated lamellæ, and, it may be, the sockets from which they have been withdrawn, will be seen (fig. 151). In some of the sections branched perforating fibres are visible. The important point is to make the sections as nearly as possible parallel to the course of the fibres. The preparation is apt to be made too transparent by Farrant's solution, so that the fibres are not so distinctly seen in this medium as in water or v.

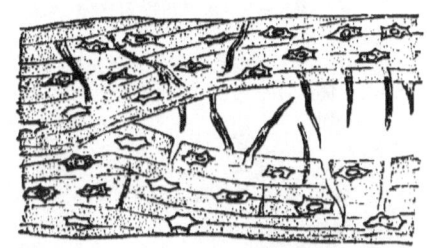

FIG. 151.—Sharpey's Perforating Fibres.

Ebner's fluid. Observe that there are no Sharpey's fibres in the Haversian systems.

5. **Blood-Vessels of Bone** (L).—These are not easily injected. Inject with a fluid carmine mass (p. 89) the posterior half of the body of a rabbit. Do this from the abdominal aorta. Or use the injection fluid mentioned at p. 181. It requires considerable pressure to cause the injection to traverse the blood-vessels of bone. Therefore clamp the inferior vena cava to prevent the exit of the injection mass. Decalcify the injected bone, and afterwards make

T.S. and L.S. The sections, however, must not be too thin. L.S. are the most instructive, and are mounted in glycerine-jelly.

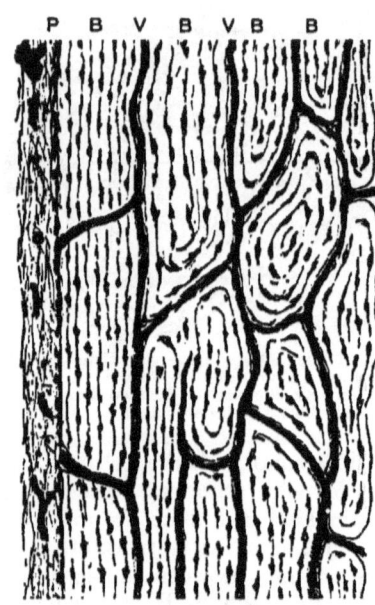

FIG. 152.—L.S. Injected Bone. P. Periosteum; B. Bone; V. Blood-vessels.

(*a.*) Observe that the blood-vessels lie in the Haversian canals, and follow the arrangement of the latter (fig. 152). If the marrow of the bone be preserved, fine blood-vessels may be seen in it. Perhaps "perforating vessels" lying in Volkmann's canals may be noticed, especially in transverse sections.

6. **Cancellated Bone (L** and **H).**—In the vertical section of the head of a long bone showing articular cartilage a view of the open lattice-work will be obtained (Lesson XIV. **3**). Or a T.S. may be made across the head of a long bone, *e.g.*, a femur, preferably of a young animal. Stain a section in hæmatoxylin, picro-carmine, or eosin-logwood, and mount it in glycerine-jelly.

(*a.*) (**L**) Note the network of osseous trabeculæ (fig. 153) bounding the spaces or **cancelli**. In the latter lies red marrow, and on their walls are osteoblasts. In

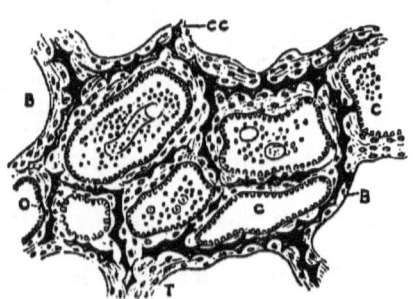

FIG. 153.—Cancellated Bone. C. Cancellus; T. Trabecula; CC. Calcified cartilage; B. Bone; O. Osteoblast.

the centre of some of the trabeculæ may be seen a deeper stained irregular bar, the remains of **calcified** cartilage (fig. 153, CC). On the calcified cartilage is deposited osseous tissue.

(*b.*) (**H**) In each trabecula note the fibrous matrix and the bone-corpuscles. In the interior of some of the trabeculæ the remains of unabsorbed calcified cartilage. Note also how the osseous tissue with spherical bulgings advances upon the calcified cartilage. By carefully shading the light, it will be seen that a more or less spherical mass of osseous tissue surrounds, and in fact is formed by, each bone-corpuscle (fig. 154). There are thus spherical masses—cell-territories, as it were—and in the centre of each a bone-corpuscle.

7. Fibrillar Structure of Lamellæ (H).—Decalcify a bone in v. Ebner's fluid (10-15 per cent. sodic chloride and 1-3 per cent. hydrochloric acid). Either this fluid or that given at p. 37 may be used. Scrape off a thin lamella and examine it in water.

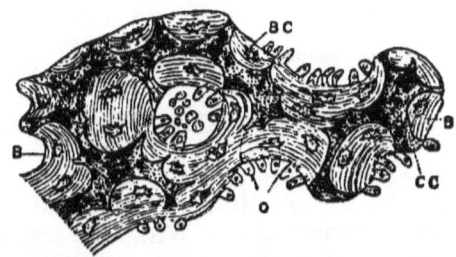

FIG. 154.—Small Part of Fig. 153, × 300. CC. Calcified cartilage; B. Bone; O. Osteoblast; BC. Bone-corpuscles.

(H) Observe the fibres of which it consists. They are composed of fibrils arranged in bundles. They are best seen near the edge. Fibres in different planes cross each other at a right or obtuse angle.

ADDITIONAL EXERCISES.

8. Blood-Vessels of Bone.—These may be injected from the descending aorta of a rabbit with a saturated and filtered watery solution of Brücke's Berlin blue. One must remember that the injection often fails. Inject from the abdominal aorta, after a time using pretty high pressure. It is well to ligature the inferior vena cava after some injection has flowed from it. Decalcify the bone in chromic acid. The sections may be stained with very dilute fuchsin.

9. Decalcified Bone.—Make a T.S. of dense bone—rather a thick section—steep it in alcohol, transfer it to a slide, and mount it in glycerine-jelly, or in a morsel of dry Canada balsam. In the latter case the slide must be heated to melt the balsam. Examine it at once, when the lacunæ and canaliculi will be seen black on a clear ground. They still contain air, hence their black appearance. Gradually, however, the jelly penetrates into the canalicular system, and this characteristic appearance vanishes, but it is an instructive exercise for students to perform this experiment. Afterwards the lamellæ remain quite distinct in the preparation, there are also faint indications of the canaliculi.

10. Bone in Polarised Light.—Examine a non-decalcified transverse section of the shaft of a long bone in polarised light (Lesson XVI. 16). When the Nicols are crossed, each Haversian system has a bright cross on a dark ground, while the lamellæ are alternately bright and dark. The crosses are also seen in decalcified bone, so that they are produced by the organic basis or ossein of the bone.

LESSON XIV.

BONE AND ITS DEVELOPMENT.

BONE is developed either in connection with cartilage or membrane. The former is called **endochondral** and the latter **intramembranous ossification**. With the exception of a part of the skull—its sides and vault—and nearly all the facial bones, all the bones are laid down in hyaline cartilage.

1. Development of Bone—T.S. of Fœtal Bone.—Decalcify the shaft of the femur or other long bone, *e.g.*, the radius and ulna,

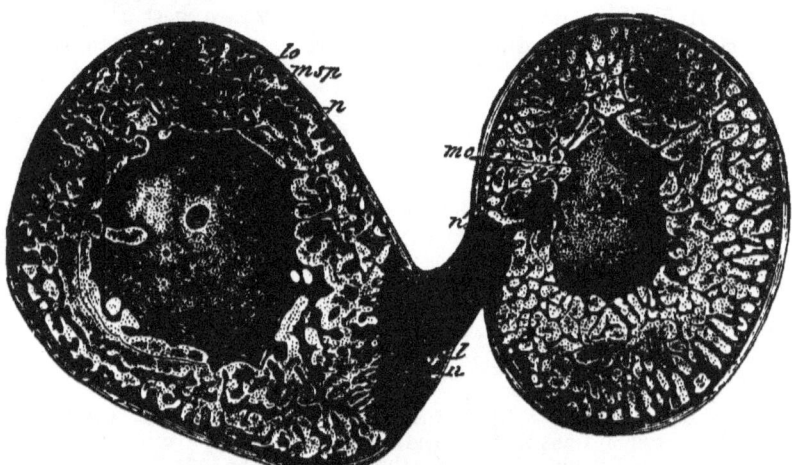

FIG. 155.—T.S. Radius and Ulna of an Embryo Dog. *l.* Interosseous ligament; *p.* Periosteum; *mc.* Medullary cavity; *msp.* Subperiosteal tissue; *lo.* Osseous trabeculæ; *n.* Point where the ligament enters the bone; *n'.* Union of ligament with the periosteum. Observe that at first the interosseous membrane is inserted into a depression in the bone; when the membrane becomes ossified a ridge is formed.

of a newly-born kitten in picric acid (p. 37). Make transverse sections, and stain them with picro-carmine. Mount one in Farrant's solution.

(*a.*) (L) Observe the **periosteum** (fig. 156, *a*, *b*), composed externally of connective tissue, with fusiform corpuscles stained red. Under this, one or more layers of cubical or somewhat flattened nucleated cells, **osteoblasts** (*c*). They pass into and line the **Haversian spaces**, thus reaching the cancelli and medullary cavity, which they also line, so that every spicule or surface of young bone is covered by them. From them the bone-corpuscles are formed.

(*b.*) The osseous tissue is stained red, and forms an anastomosing series of *trabeculæ* bounding large spaces—**Haversian spaces**—containing blood-vessels and marrow, and lined by osteoblasts. Processes from the trabeculæ project into the deeper layer of the periosteum. At this stage a concentric arrangement of the lamellæ leading to the formation of Haversian systems has not yet taken place. In the bone matrix, the **bone-corpuscles** (*l*), irregular, refractive, nucleated cells, each lying in a lacuna.

(*c.*) (H) Under the periosteum and in the spaces may be seen larger multinucleated cells.

Osteoclasts or **Myeloplaxes** (fig. 156, *k*).—The cells are much larger than the osteoblasts, contain many nuclei, and lie in little depressions of the bone eroded by themselves. These depressions are called **Howship's lacunæ**. These cells are concerned in the removal of bone.

2. **Intra-Cartilaginous Formation of Bone (L and H).**—Decalcify in picric acid the phalangeal bones of a four-months fœtus. Make longitudinal vertical sections, stain in picrocarmine, and mount in Farrant's solution.

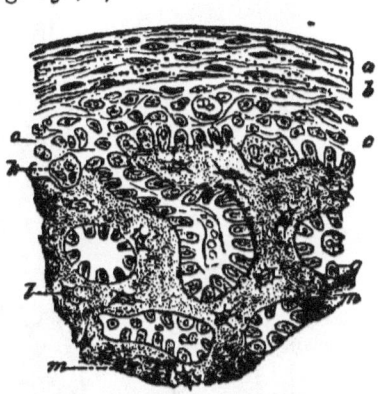

FIG. 156.—T.S. Fœtal Bone of Kitten. *a, b.* Superficial and deep layers of periosteum; *c.* Layers of osteoblasts, with *k.* Osteoclasts; *m.* Matrix of bone; *l.* Lacunæ, with bone-corpuscles.

(*a.*) (L) At the head of the bone observe a mass of **hyaline cartilage** (fig. 157), and lower down bone, and where the two are continuous an irregular festooned margin, the **line of ossification**.

(*b.*) The cartilage at the upper part is hyaline, with the cells small, and arranged singly or in groups, while deeper down the cartilage-capsules are beginning to be arranged in rows, and below this are larger cartilage-capsules with clearer contents, and a somewhat refractive matrix between them.

(*c.*) Under this, the line of ossification, with its festooned margin, are spicules of **calcified cartilage** passing downwards towards the medullary cavity. In the bony part, the **primary medullary spaces**, with their osteoblasts, blood-vessels, osteoclasts, and the newly-formed bone deposited on the calcified cartilage.

(*d.*) The first bone is formed under the periosteum (fig. 157), by means of the subperiosteal osteoblasts. As these osteoblasts become embedded in the bone matrix they become bone-corpuscles. This piece of bone is perforated by blood-vessels, which pass into the primary medullary spaces in the hollowed-out cartilage.

(*e.*) (**H**) Examine the several parts. Search for an osteoclast lying in a little cavity—**Howship's lacuna**—and notice that while the margins of the trabeculæ, covered by recently-formed bone, are stained of a deep red colour by the carmine, no such red stain

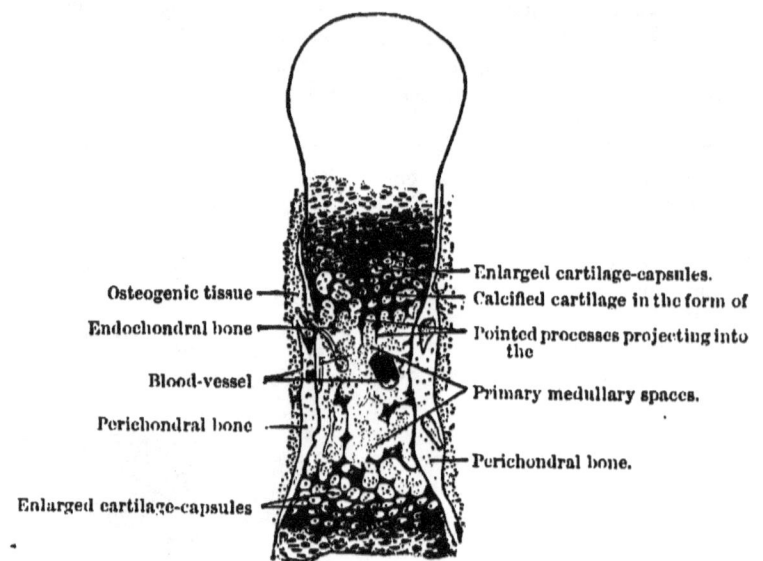

Fig. 157.—Dorso-palmar Longitudinal Section of the Second Phalanx of the Finger of a Four-Months Fœtus.

is seen where the osteoclast is embedded. It, in fact, is eroding or eating away bone.

3. Epiphysis and Epiphysial Cartilage (**H** and **L**).—Make longitudinal vertical sections of a young rabbit's femur or tibia to show the epiphysial cartilage. Stain it in picro-carmine and mount in glycerine-jelly; or, better still, double stain it in hæmatoxylin and picro-carmine or hæmatoxylin and eosin. In the last case the cartilage will be blue, the rest red or copper-red. The method of double staining is particularly valuable for the study of bone development.

(*a.*) (**L**) Observe the thin layer of **encrusting cartilage** on the head of the bone (fig. 158, *C*), and under it the cancellated bone of the head of the tibia. This cartilage is continuous below with—

(*b.*) A broad layer of cartilage between what is to be the head of the bone—the **epiphysis** (*E*)—and the future shaft or diaphysis of the bone (*D*).

(*c.*) In the epiphysial cartilage (*EC*) the vertical rows of cartilage-cells, smaller above, and larger and more quadrilateral below (fig. 158 and fig. 159, *C*).

(*d.*) The shaft with longitudinal spaces—the **primary medullary spaces** (fig. 158, *MS*)—bounded by trabeculæ of calcified cartilage partly covered by bone (fig. 159, *b*). The spaces contain young marrow (*c*).

(*e.*) (**H**) Study specially the cells of the epiphysial cartilage (fig. 159, *C*), and notice that the cells and capsules are smaller above and larger below—zone of enlarged cartilage-capsules. Some of the enlarged cartilage-capsules of the lowest row may be seen opening into the primary medullary spaces. The line of termination of these spaces is called the *line of ossification*. The bone grows in length by the proliferation of the epiphysial cartilage-cells.

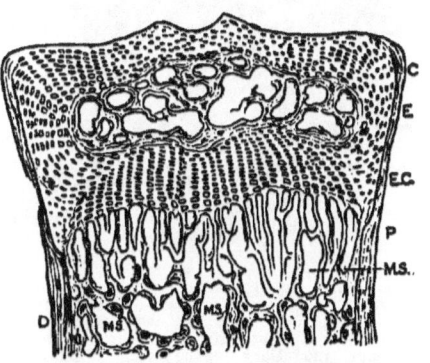

FIG. 158.—V.S. Head of Tibia of a Young Rabbit, greatly reduced to show details. *C.* Encrusting cartilage; *E.* Epiphysis; *EC.* Epiphysial cartilage; *D.* Diaphysis; *p.* Periosteum; *MS.* Primary medullary spaces.

(*f.*) Bounding the primary medullary spaces, **directive trabeculæ** of calcified cartilage (*b'*) with a deposit of bony matter on them. Note specially the osteoblasts (*p*) partially embedded in the osseous matter which they themselves secrete or form. When they become embedded in the osseous products of their own activity, they are then called **bone-corpuscles**. Note also that the bone on the cartilage is bounded by convex surfaces which fit into corresponding depressions in the cartilage. On the larger trabeculæ may be found osteoclasts lying in little cavities—**Howship's lacunæ**—which they have eroded. The spaces themselves are filled with red marrow (*c*) and blood-vessels (*v*).

4. **Intra-Membranous Formation of Bone (H and L).**—Take the parietal bone of a fœtus when the parietal bone is only partially ossified; scrape off the periosteum from a part near the periphery of the ossified part; stain in picro-carmine and mount in Farrant's solution.

(*a.*) Observe at one part the fibrous matrix, and shooting from it calcified fibres of connective tissue, these covered with osteoblasts (fig. 160).

Marrow of Bone.—It fills the medullary cavity of long bones, the cancelli of spongy bone, and it occurs in some of the larger Haversian canals. There are two varieties, **yellow** and **red**, the difference in colour being due to the former containing a large amount of fat-cells.

Yellow marrow occurs in the medullary cavity or canal and the larger cancelli of long bones. Besides a small amount of connective tissue and blood-vessels, it consists principally of fat-cells.

Red marrow (H) occurs in the spongy tissue at the ends of long bones, in the short bones of the hands and feet, flat bones of the skull, epiphyses of long bones, clavicle, ribs, and in the medullary canal of the long bones of some animals, *e.g.*, guinea-pig, rat, rabbit. It consists of delicate connective tissue with blood-vessels and numerous cells of several varieties.

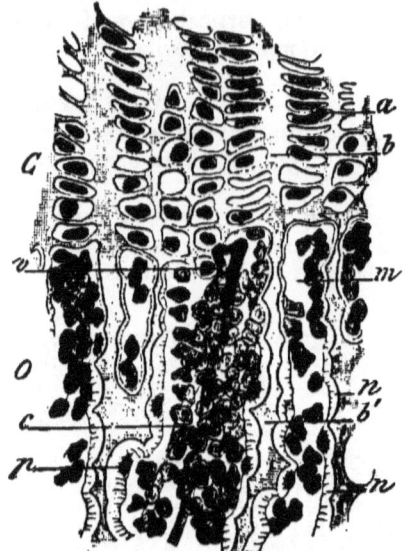

FIG. 159.—L.S. Head of a Metacarpal Bone of a Rabbit, aged three months. *C.* Epiphysial-cartilage; *O.* Bone; *a.* Row of cartilage-cells; *b.* Cartilaginous trabeculæ; *m.* Primary medullary space; *n.* Osseous deposit; *b'.* Directive cartilaginous trabeculæ; *p.* Bone-corpuscle being embedded in the osseous matrix; *c.* Marrow-cells; *v.* Injected blood-vessel, × 240.

(1.) **Medullary** or **Marrow Cells** (fig. 161, *a*, *b*, *c*).—They are the most numerous, and are nucleated cells, not unlike large leucocytes. They are spherical, with finely granular protoplasm and a spherical pale nucleus. Sometimes they contain two nuclei. Examined quite fresh in serum, no nucleus is visible, but it is revealed by the action of acetic acid or dilute alcohol. The protoplasm of some of them contains numerous highly refractile granules, and that of others some brownish granules. There is also a smaller

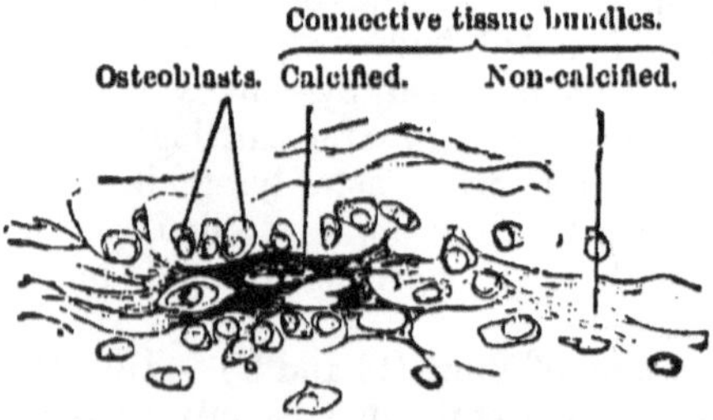

FIG. 160.—Surface Section of a Parietal Bone of a Human Embryo × 240.

variety of corpuscle. They exhibit amœboid movements under proper conditions.

(2.) **Cells with Budding Nuclei** (fig. 161).—Far less numerous, but easily distinguished, are large finely granular cells, each with a single large, often twisted, nucleus, which in some looks as if it were composite, in others it would seem to consist of several parts united by some substance like the nucleus itself. There is very great variety in the shape of those nuclei, which are visible in the fresh condition of the cell. These cells are not amœboid (fig. 161, h, i).

(3.) **Myeloplaxes.** — These are larger than the foregoing, consisting of a finely granular protoplasm with numerous nuclei (fig. 161, m).

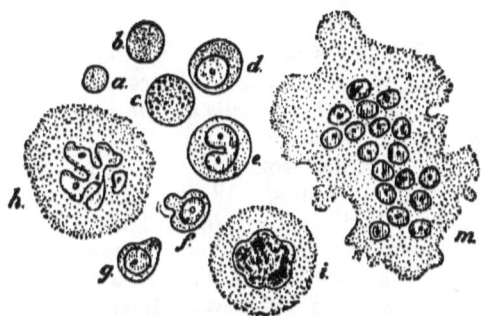

FIG. 161.—Cells from the Red Marrow of the Tibia of a Young Rabbit. a, b, c. Marrow-cells examined in normal saline; d, e. Marrow-cells after dilute alcohol; f, g. After dilute alcohol; h, i. Large cells with budding nuclei; m. Myeloplaxes, × 300.

What relation there is between (2) and (3), or if there is any relation at all, is entirely unknown.

(4.) There are to be found cells smaller than but not unlike (1), with a homogeneous protoplasm which has a reddish tint and a spherical nucleus. Some of them have a small bud at the side (fig. 161, f, g). They are regarded as cells from which coloured blood-corpuscles are formed. They are regarded by Bizzozero as similar to the nucleated red blood-corpuscles of the embryo.

(5.) Always a few fat cells.

(6.) Numerous red blood-corpuscles from the blood-vessels in the red marrow.

(7.) Sometimes osteoblasts may be detached along with the other constituents of the marrow.

Lay open a rib or a long bone of a guinea-pig or rabbit; remove a little of the red marrow and diffuse it in blood-serum or normal saline. With a high power search for examples of each of the foregoing kinds of cells. The vertebra of a calf may be used, and from it the red marrow is readily expressed by squeezing it in a vice.

The nuclei in some of the cells are best revealed by the action of dilute alcohol.

5. Red Marrow.—(i.) Squeeze out some of the red marrow from a rib of a guinea-pig or rat; shake it in a test-tube containing normal saline tinged with methyl-green, aldehyd-green, or aniline-green. The marrow will fall as a precipitate. Examine a little of this in normal saline. Stain some in picro-carmine, and mount in dilute glycerine.

(H) Observe the various forms of marrow and other cells met with, the nuclei tinged red. The large myeloplaxes with numerous nuclei have their protoplasm green and the nuclei red. The fat-cells are quite green.

(ii.) Place some of the red marrow for 12-24 hours in *dilute alcohol*, mount and stain the cells with picro-carmine. In this way the nuclei of all the cells are stained red, their protoplasm yellowish, while nucleoli are revealed.

(iii.) Harden some of the red marrow in Hayem's fluid for twenty-four hours. Wash the deposit, stain (picro-carmine) and mount a little in glycerine. This is a good hardening medium for this purpose; the red blood-corpuscles can be readily distinguished, while the nuclei in all the cells are distinct.

6. T.S. Red Marrow.—Small pieces of the ribs of a young rabbit are placed in the following fluid for one day:—

Sodic sulphate	2.5 grams.
Mercuric chloride	0.25 ,,
Water	500 cc.

Rinse in water, and transfer to picric acid to decalcify the bone. The tissue can then be soaked in gum, then hardened in alcohol and sections cut. Sections may be stained with eosin and logwood. The eosin stains any cells containing hæmoglobin of a reddish-orange tint.

ADDITIONAL EXERCISES.

7. Cover-Glass Preparation of Red Marrow (H).—As described for blood, get a thin layer of red marrow on a cover-glass and stain it for twenty-four hours in Biondi's fluid (p. 140). Dry cover-glass preparations may also be stained with methylene-blue, eosin-hæmatoxylin, or methyl-green. Mount the cover in xylol-balsam. If desired, clear it up with oil of cloves in which a little eosin is dissolved, and remove the clove-oil by xylol. This method yields excellent preparations.

8. Squeeze on a slide a little of the red marrow from the rib of a young rabbit. Squeeze a little between two cover-glasses to get in a thin film, and expose the latter for a minute or two to the vapour of osmic acid. Stain it with picro-carmine and mount it in glycerine.

LESSON XV.

MUSCULAR TISSUE.

Muscle histologically occurs in two varieties—(1.) Non-striped; (2.) Striped. Non-striped muscles are involuntary, while striped muscles, as a rule, are voluntary; but the heart-muscle is an exception, for though striped it is involuntary.

Non-Striped (smooth, involuntary). Occurs in the outer coats of the lower half of œsophagus; muscular coat and muscularis mucosæ of stomach and intestines; villi; ureter, bladder, and urethra; pelvis and capsule of kidney; trachea (trachealis muscle); bronchi; oviduct; uterus; iris, ciliary muscle; erector pili muscles of skin, sweat glands; coats of blood- and lymph-vessels; capsule and trabeculæ (lymphatics, spleen), and ducts of some glands, salivary, bile-ducts, &c.

It consists of nucleated elongated fusiform contractile cells, held together by a clear cement. The cells taper towards their extremities, and although they appear homogeneous in reality they seem to consist of longitudinally arranged fibrils held together by a sarcoplasm, or at least a transparent material. The nucleus is oval or rod-shaped. Some state that the cells have an elastic sheath. Each cell is from 4 to 10 μ ($\frac{1}{6000} - \frac{1}{2500}$ inch) in breadth and 40–200 μ ($\frac{1}{600} - \frac{1}{120}$ inch) in length.

1. Non-Striped Muscle (H).—Place thin strips of the muscular coat of the intestine for forty-eight hours in a 25 per cent. solution of nitric acid, which softens the connective tissue and renders the tissue yellow. Wash it thoroughly in water. Tease a small part in glycerine or Farrant's solution. Or macerate a small piece of intestine in $\frac{1}{8}$ per cent. bichromate of potash for 48 hours. Cells can then be readily isolated.

(*a.*) Select an isolated fibre. It is spindle-shaped, elongated, or fusiform, tapering to both ends, and in its centre there is an oval nucleus, distinguished by its being rather more refractive than the rest of the cell (fig. 162). At the poles of the nucleus there may be a few granules.

It is very difficult to stain these cells after the action of nitric acid, but this may be done with magenta, provided the nitric acid be entirely washed out of the tissue beforehand. Each cell is said to have a sheath, but that cannot be seen in fibres prepared in this way.

2. Muscle-Cells from the Frog's Bladder (L and H).—Distend the frog's bladder with dilute alcohol thus. Transfix the skin on each side of the anus with two pins, and tie round them a thread so as to occlude the anus; open the abdomen, make a slit into the large intestine, clear out any residues it may contain, and inject dilute alcohol through the intestine into the bladder. When the latter is full, ligature it at the neck, and suspend it for twenty-four hours in a large quantity of dilute alcohol. Then open the bladder, and with a camel's-hair brush pencil away all the lining epithelium. Stain a portion of the bladder in logwood and mount it in balsam, or stain in picro-carmine and mount it in Farrant's solution.

(*a.*) Observe the thin membrane, traversed in every direction by thicker or thinner trabeculæ of smooth muscle (fig. 163, *a*). The trabeculæ consist of numerous long fusiform nucleated cells (*c*)—the nuclei, long, narrow, and oval. Some of the cells are triradiate (*b*). These are what S. Mayer has called atypical cells. Oval nuclei with blunter ends are seen in the fibrous covering of the bladder (*d*). They are the nuclei of connective-tissue cells.

FIG. 162.—Isolated Smooth Muscular Fibres. Nitric acid, × 300.

3. T.S. Non-Striped Muscle (H).—This is obtained by making transverse sections of the circular muscular coat of the small intestine (cat), previously hardened in chromic acid and spirit, or Müller's fluid. Stain a section in hæmatoxylin and mount in balsam.

(*a.*) Observe polygonal areas of *unequal size*, mapped out from each other by a refractive (fig. 164) **cement substance.** Each area corresponds to the transverse section of a fibre. Some of the areas contain a nucleus (*n*), others not. Surrounding groups of these areas are fine septa of connective tissue (*s*), which map out the fasciculi or bundles of cells. The fibres are arranged in bundles or **fasciculi,** each surrounded by an envelope of

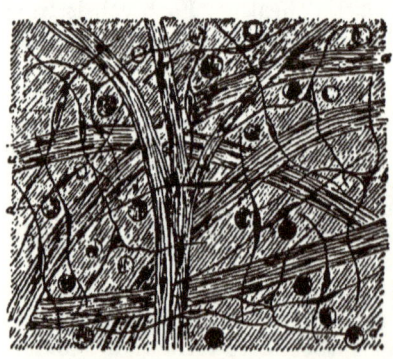

FIG. 163.—Bladder of Frog. *a.* Large strands of smooth muscle; *b.* Triradiate, and *c.* Fusiform muscle-cells; *d.* Nuclei of connective-tissue corpuscles.

connective tissue. The fasciculi are large or small according to the number of fibres entering into their composition.

(b.) In some preparations one can see intercellular bridges between adjacent cells, like those that occur in squamous epithelium.

4. **L.S. of Non-Striped Muscle (H).**—This may be obtained by making a longitudinal section of the longitudinal coat of the intestine, but it is better to strip off a thin lamella of this coat from the intestine of a rabbit hardened in spirit or Müller's fluid. Stain in hæmatoxylin and mount in balsam.

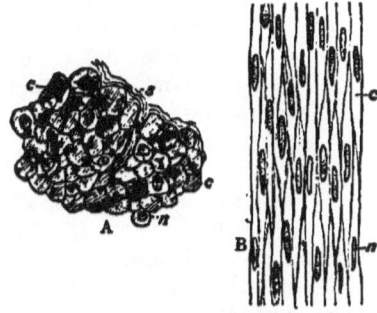

FIG. 164.—A. T.S. non-striped muscle, intestine of cat; B. Longitudinal strip of intestine of rabbit; c. Cell; n. Nucleus; s. Septum of connective tissue. Chromic and b'chromate fluid, hæmatoxylin.

(a.) Observe the oval fusiform nuclei (fig. 164, n) lying in narrow fusiform areas—the cells. The boundary-lines between the cells are usually not well defined.

5. **Cement Substance of Non-Striped Muscle (H).**—With distilled water wash out the contents of the small intestine of a freshly-killed rabbit, or the large intestine of a frog. Tie one end of the gut, and fill it with .5 per cent. solution of silver nitrate, and tie the other end of the gut. Suspend the whole in ¼ per cent. solution of silver nitrate for ten minutes or so. Slit up the gut along the line of attachment of the mesentery. Wash it in water and expose it to light. It soon becomes brown. Lay it on a glass plate, mucous surface uppermost, and with a scalpel scrape away all the mucous and submucous coat, which is very easily done, especially if the intestine has been macerated for about twenty-four hours in water. There remain only the muscular and thin serous coats. Harden in alcohol. Snip out a piece, dehydrate completely in absolute alcohol, and mount in balsam.

FIG. 165.—Cement Substance of Smooth Muscle, Intestine of Rabbit. Silver nitrate.

(a.) Observe the narrow elongated fusiform areas bounded by silver lines (fig. 165); they indicate the outline of the fusiform cells. On focussing upwards and downwards, notice the longitudinal and circular direction of the fibres crossing each other. In this preparation also many lymphatic paths lined by sinuous epithelium may be seen. They are recognised by the dilatations in their course, and by the character of the endothelial cells lining them. By focussing deeply, the silver lines of the endothelium of the serous membrane may be seen.

6. **Methylene-Blue Method** (*S. Mayer*).—Into the blood-vessels of a cat—*e.g.*, from the aorta—inject the following solution:—

Methylene-blue (S. Mayer's)[1] . . . . 1 gram.
Normal saline . . . . 300 cc.

After an hour or so, open the abdomen; the intestines appear blue; cut out a portion of the muscular coat and place it in the following mixture:—

Picro-Glycerine Mixture.

Saturated watery solution ammonium picrate . 100 cc.
Glycerine . . . . 100 ,,

On teasing a small piece of the muscular coat in the same mixture, isolated muscle cells are readily found, the nucleus being stained of a faint rosy-pink.

ADDITIONAL EXERCISES.

7. **Fibrillar Plexus in Muscle-Cells** (H).—Kill a newt, open its abdomen, and pin out its intestine and mesentery on a thin piece of cork with a hole in it corresponding to the mesentery. Place it for twenty-four hours in a 5 per cent. solution of ammonium chromate. After this wash away all the chromate, stain a piece of the mesentery in logwood, and mount it in balsam.

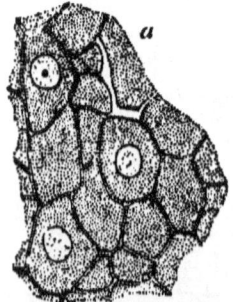

FIG. 166.— T.S. Smooth Muscle of a Cat's Intestine. Chromo-aceto-osmic acid. Shows intercellular bridges, and at *a* an artificial slit, × 1000.

(*a.*) In the membrane observe narrow strands of non-striped muscle composed of very large fusiform cells. In the large nuclei a plexus of fibrils, while a leash of fine fibrils will be seen in the perinuclear part stretching from the poles of the nucleus to the ends of the fibre.

8. **Fibrils in Smooth Muscle.**—These may be seen for a short time by macerating small pieces of the stomach of a frog for twenty-four hours in dilute alcohol, or in 8-10 per cent. of sodic chloride (*Engelmann*).[2]

9. **Isolated Smooth Muscle-Cells.**—To get these cells quickly macerate a small piece of the muscular coat of the stomach, intestine, or bladder in 33 per cent. caustic potash for 15-20 mins. Tease and observe in the same solution. Water must not be added, else the effect of weak caustic alkalies is produced, viz., solution of the cells.

[1] To be obtained from Dr. Grubler, Leipzig, or Bindscheidler & Busch, Basel.
[2] *Pflüger's Arch.*, xxv.

10. Grooving on Smooth Muscular Fibres.—Harden the muscular coat of the small intestine of a cat in chromic and bichromate fluid or in Flemming's fluid. Stain and cut sections in paraffin. (H) Observe the polygonal areas, but note that fine intercellular bridges connect adjacent cells (fig. 166). The surface of each cell seems to be grooved with canals, but what these canals contain is not known. In each cell the cut ends of fine fibrils are seen.

LESSON XVI.

STRIPED OR STRIATED MUSCLE.

Striped Muscle—sometimes called voluntary or skeletal muscle—occurs in the muscles of the skeleton, pharynx, upper half of the œsophagus, diaphragm, the sphincter of the bladder, external anal sphincter, and the muscles of the outer and middle ear. The muscular fibres of the heart are also striped, but they are involuntary.

A Muscular Fibre is cylindrical in form, and tapers at its extremities. They vary in breadth from 10 to 50 μ ($\frac{1}{2500} - \frac{1}{500}$ inch), but they are broadest in the muscles of the extremities. The fibres are 1 to $1\frac{1}{2}$ inches in length.

A muscular fibre consists of the following parts:—
(1.) **Sarcolemma**, or sheath.
(2.) **Sarcous substance**, which is transversely striated.
(3.) **Muscle-corpuscles.**

1. The Sarcolemma (H).—(i.) To avoid the effect of the contractility of the muscle, kill a frog several hours before it is required. Dissect out the sartorius muscle, because it is composed of parallel fibres. Tear off a thin strip, and with needles tease it in *distilled* water.

(*a.*) Observe the cylindrical shape of the fibres, marked transversely by alternate light and dim stripes. Run the eye along the edge of a fibre, and perhaps a clear, transparent bulla or bleb will be seen. If so, it is the **sarcolemma** raised from the subjacent sarcous substance by water diffusing into the fibre (fig. 167, A).

(*b.*) The sarcolemma is a clear, transparent, colourless, homogeneous elastic membrane, forming a tubular sheath for the sarcous substance. It is allied to, but not identical with, elastic tissue. It is much tougher and less easily ruptured than the sarcous substance which it contains.

(ii.) A much better plan is the following:—Tease out a few

fresh muscular fibres on a slide in normal saline, and across the direction of the fibres place a hair. Cover, and press the cover-glass down firmly on the fibres. The hair ruptures the sarcous substance, which retracts and leaves the tougher unbroken sarcolemma between the ends of the ruptured fibre. Remove the hair, and the now empty sarcolemma will be seen (fig. 167, C).

(iii.) Another excellent method is to leave the muscle in a saturated solution of ammonium carbonate. The sarcolemma will

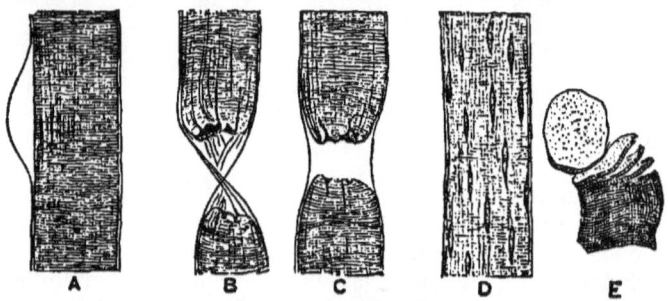

FIG. 167.—A. Striped muscle of frog, sarcolemma raised in the form of a bleb; B. Ruptured fibre with sarcolemma; C. Fibre ruptured by a hair; D. Effect of acetic acid on a muscle-fibre; E. Muscle-discs. Ammonium carbonate.

be seen raised for long distances from some of the fibres, with a little of the sarcoglia sometimes adhering here and there to its under-surface.

2. **Muscle-Corpuscles, or Nuclei (H).**—Tease a piece of a fresh frog's muscle, and irrigate it with 2 per cent. acetic acid. I have often found the nuclei beautifully stained in a piece of muscle that has been in ammonium carbonate and then in picro-carmine.

(a.) Observe the sarcous substance becoming swollen up and more homogeneous, while a number of fusiform, somewhat shrivelled or shrunken nuclei, with their long axis in the long axis of the fibre, come distinctly into view. They are now slightly more refractive than the altered sarcous substance, hence they are seen with greater distinctness. Focus carefully, and note that these nuclei lie not only under the sarcolemma, but also in the substance of the fibres. Had a mammalian muscle been used instead of one from an amphibian—in the case of most muscles—the nuclei would have been found directly under the sarcolemma only. The presence of the nuclei is merely revealed by the action of the acid, which alters the refractive index of the sarcous substance, and thus brings the nuclei into view (fig. 167, D). Sometimes faint longitudinal striation is exhibited by such a fibre.

3. **Isolated Muscular Fibres.**—(i.) Pith a frog, and plunge it in a beaker of water at 55° C. Leave it in the water, and allow the

XVI.] STRIPED OR STRIATED MUSCLE.

water to cool gradually (*Rancier*). It will now be found that the fibres of any muscle can be dissociated with great ease; the muscles to be preserved in 70 per cent. alcohol until they are required. By careful manipulation very long fibres may be isolated from the sartorius. These muscular fibres exhibit the ordinary characters of striped muscle. This is by far the easiest method of obtaining isolated fibres.

(ii.) Place small pieces of a fresh muscle in the following mixture. Nitric acid saturated with potassic chlorate. There must be crystals of the latter in the fluid. The tube or vessel is speedily filled with yellow nitrous fumes. It is usually advised to leave the muscle several hours in this fluid. I find, however, if this be done, that the muscle is dissolved. Half an hour is usually sufficient. With glass rods remove the now softened orange-coloured muscle, and place it in water. It becomes whitish. Shake it in a tube with water. The fibres fall asunder quite readily, and after having all the acid removed by prolonged washing, they can be stained and mounted.

4. Fibrillæ of a Muscular Fibre (H).—Place a frog's or mammal's muscle in water (two hours), and afterwards in dilute alcohol for twenty-four hours. Tease a small fragment of the now softened muscle in glycerine, or tease a fresh muscle of a calf in white of egg. If a fibre be split up, bundles of fibrils may be seen as in fig. 168.

FIG. 168.—Part of a Striped Muscular Fibre of a Calf, teased in white of egg, and showing isolated bundles of fibrils, × 200.

Select a fibre, and note that at its free end it splits up longitudinally into a large number of very fine **fibrils** or **fibrillæ**, each of which is transversely striated like the original muscular fibre. Much larger fibrils are obtained from the muscles of insects, *e.g.*, Hydrophilus or Dytiscus (fig. 171).

5. Muscle-Discs (H).—The usual directions for obtaining these are to place dead muscle for several days in dilute (.2 per cent.) hydrochloric acid. The muscular fibre then cleaves transversely. I have not found this to be a very satisfactory method. A much better plan is to place small pieces of the muscle in a saturated solution of ammonium carbonate for several hours.

(*a.*) Not only will the sarcolemma be seen, but inside it, here and there, the sarcous substance will be seen cleft transversely into discs (fig. 167, E).

6. Ending of Muscle in Tendon.—This is readily seen by taking the lower end of the sartorius with its tendon from a frog treated as in Lesson XVI. 3, (i.), and which has been afterwards placed in 70 per cent. alcohol. Tease out the muscular fibres, when their conical ends will be seen ending abruptly, and the small tendons beginning as abruptly (fig. 169).

7. Crab's Muscle (Sarcous Substance) (H).— Plunge the living muscles of a crab—or better still, use a stag-beetle—into absolute alcohol for twenty-four hours. Stain a fragment in *dilute eosin-hæmatoxylin*, and mount it in Farrant's solution or balsam. In the latter case the clarifying reagent must contain a little eosin to restore the eosin colour to the preparation.

FIG. 169.— Relation of a Tendon T. to its Muscular Fibre, the latter with a Conical Termination.

(*a.*) Observe the fibre striped as in fig. 170, with alternate **light** and **dim discs**, the *dim discs* (*a*) stained of a logwood tint, while the light discs are of a faint eosin tint.

(*b.*) In the **light disc** (*b*) observe a fine line or series of dots running transversely, sometimes called **Dobie's line** or **intermediate line** (figs. 170, *b*, 171, *c*), dividing the light disc into two equal parts, these adhering to the ends of the dim disc. They are then called **lateral discs**.

(*c.*) The dim disc may exhibit slight vertical striation, indicating a tendency to cleave longitudinally. If transverse cleavage be associated with simultaneous longitudinal cleavage, small "elements" are obtained, which were called "**sarcous elements**" by Bowman.

(*d.*) **Nuclei** may be seen.

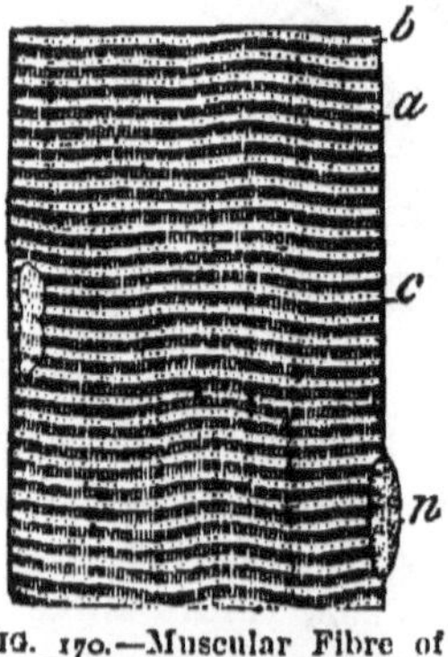

FIG. 170.—Muscular Fibre of Great Adductor of Rabbit. Living and Extended. *a.* Dim disc; *b.* Light disc; *c.* Intermediate or Dobie's line; *n.* Nucleus seen in profile. Examined in its own juice, × 300.

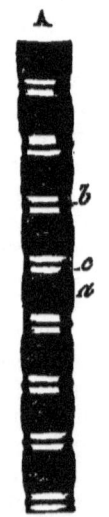

FIG 171. — Fibril of Muscle of Hydrophilus. *a.* Dim, *c.* Light disc; *b.* Intermediate line, × 2000. Picrocarmine and formic glycerine.

A similar preparation of an insect's muscle may be stained in picro-carmine. Tease it so as, if possible, to isolate a fibril. Mount in formic glycerine. The

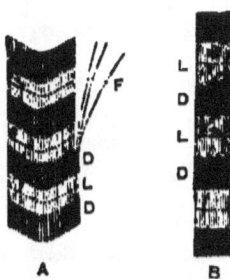

FIG. 172.—Crab's Muscles, partly schematic. *A.* Non-stretched; *B.* Extended; *F.* Fibrils; *D.* Dim; *L.* Light stripes.

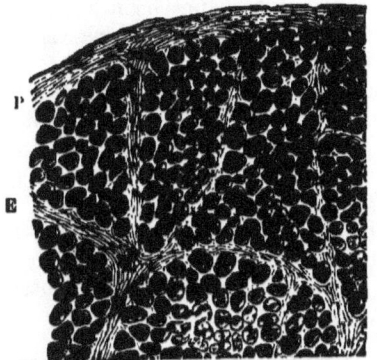

FIG. 173.—T.S. Muscle, with its Sheaths. P. Perimysium; E. Endomysium.

dim disc is stained red (fig. 170, *a*), and if the fibre be stretched

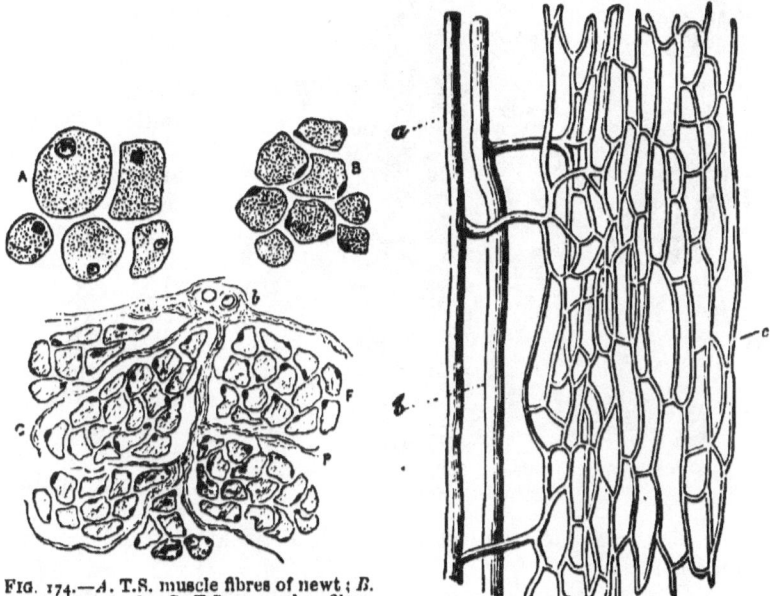

FIG. 174.—*A.* T.S. muscle fibres of newt; *B.* Of mammal; *C.* T.S. muscular fibres (mammal), with endomysium; *b.* Bloodvessel; *F.* Fasciculus; *P.* Perimysium.

FIG. 175.—Capillary Plexus in Muscles. *a.* Artery; *v* Vein; *c.* Capillaries, × 250.

the details of its structure can be better seen, especially if a very high objective be used (fig. 171).

8. T.S. of Muscle.—Make transverse sections of a small mammalian muscle which has been kept stretched and hardened in 0.5 per cent. chromic acid and afterwards in alcohol. Stain one in logwood and mount it in Canada balsam.

(*a.*) (L) Observe the sheath of connective tissue or **perimysium** surrounding the whole muscle, and that from it septa pass between groups or fasciculi of the muscle-fibres, and also a small amount between the muscle-fibres, forming the **endomysium** (figs. 173, 174).

The ends of the muscular fibres somewhat polygonal or rounded, with stained nuclei (one, two, or three) immediately under the sarcolemma. In amphibian muscles (fig. 174, A) and in a few mammalian muscles, *e.g.*, the semi-tendinosus of the rabbit, nuclei also occur within the sarcous substance.

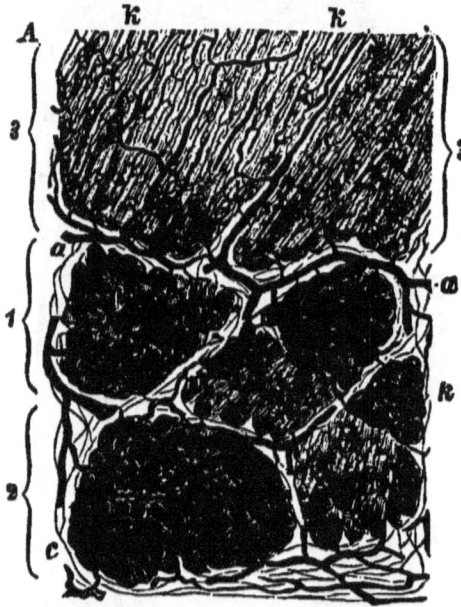

FIG. 176.—T.S. and L.S. Injected Striped Muscle. 1. and 2. T.S. Muscular fibres; 3. L.S.; *a.* Arteries.

(*b.*) (H) The ends of the fibres appear finely dotted with clear interspaces, the dots corresponding with the ends of the bundles of fibrils or **muscle-prisms** or **sarcostyles**, while the clear areas are due to what is called **sarcoglia**.

9. L.S. Injected Muscle (L).—Make a longitudinal section of an injected muscle, *i.e.*, parallel to the direction of its fibres. It is better to inject the whole of the posterior half of the body, *e.g.*, of a rabbit, from the aorta. The preparation made for injected bone will yield injected muscle. Mount it in balsam.

(*a.*) Observe the elongated quadrilateral meshes of capillaries between the muscular fibres, but outside the sarcolemma, and that capillaries run between the fibres with short transverse connecting branches. Trace their origin from an artery and their termination in a vein (fig. 175).

10. T.S. Injected Muscle.—Mount, either stained or unstained,

XVI.] STRIPED OF STRIATED MUSCLE.

in balsam. In this the cut ends of the capillaries between the fibres will be seen (figs. 176, 177).

It is to be noted that there is a difference in the T.S. of contracted and uncontracted muscles (*Spatleholz*). Beautiful figures are given in.[1]

11. L.S. Red Muscle of Rabbit Injected (L).—Use the semitendinosus or soleus of a rabbit. This shows the same general arrangement of the blood-vessels, but some of the transverse branches and some of the veins have small dilatations or ampullæ upon them, while the capillaries are usually more tortuous than those of the pale muscles.

CARDIAC MUSCLE.

12.—Harden small pieces of the heart in 20 per cent. nitric acid (forty-eight hours), or 2 per cent. potassic bichromate or ammonium chromate for thirty-six to forty-eight hours. Tease a small piece in Farrant's solution. Small pieces of muscle cardiac placed fresh in picro-carmine for several days show the structure well when mounted in glycerine.

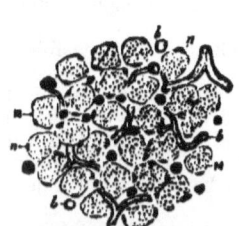
FIG. 177.—T.S. Muscle Injected. *M.* Muscle, with *n.* Nuclei, *b.* Blood-vessel (capillaries).

FIG. 178.—Muscular Fibres of the Human Heart.

FIG. 179.—T.S. of a Fresh Frozen Muscular Fibre, showing Cohnheim's areas.

(*a.*) Observe the faintly transversely-striated fibres made up of short quadrilateral pieces with short oblique processes, which join other muscular fibres. The muscle-cells branch and anastomose. A rather indistinct line of clear cement joins the ends of adjacent cells.

(*b.*) There is no sarcolemma, but a well-defined nucleus lies in the substance of the fibre, while the transverse striation is much less

---

[1] "Die Vertheilung d. Blutgefässe im Muskel," *Abhand. d. math. phys. slassed. k. Sächsig. Gesell. d. Wissensch.*, 1888.

distinct than in skeletal muscles (fig. 178). In a transverse section of a cardiac muscle the nucleus lies in the centre of the fibre, and radiating from it are fine lines (fig. 207); an appearance somewhat similar to this is shown in T.S. of insects muscles.

ADDITIONAL EXERCISES.

13. T.S. Frozen Muscle (H).—With a freezing microtome make a transverse section of a muscle taken from a recently-killed animal.

(*a.*) The ends of the fibres are mapped out into a large number of small polygonal areas—**Cohnheim's areas**—separated from each other by a clear network of lines. The darker areas correspond to the ends of a bundle of fibrils—the so-called **muscle-prisms**, while the clear material between them is the **sarcoglia** (fig. 179).

14. Living Muscle (H).—Remove, with as little injury as possible, some of the muscular fibre from the leg of a water-beetle (Dytiscus or Hydrophilus), place the muscle on a slide *without the addition of any other fluid*, and cover it. Examine it as quickly as possible, and with the highest available objective. In insects' muscles there is far more protoplasmic matter or sarcoplasm between the muscle prisms or sarcostyles than there is in vertebrate muscle.

(*a.*) Observe the alternate cross stripes, some of which will be distinctly seen, while at other parts they may be very close together, or the fibre may exhibit contraction waves.

The dim disc or band may exhibit slight longitudinal striation, while a dotted line—*Dobie's line*—will be seen running across the bright or light disc, dividing it into two so-called *lateral discs*.

The nuclei, surrounded by a small quantity of protoplasm, may be visible.

15. Crab's Muscle (Methyl-Violet).—Stain a fragment of a crab's muscular fibre (hardened in alcohol or Müller's fluid and spirit, p. 29) with methyl-violet as described for fibrin (Lesson III. 18). Decolorise it with Lugol's solution of iodine in iodide of potassium, clarify it in aniline-oil and xylol, and mount it in balsam. Use all the precautions detailed under Weigert's method for fibrin. In successful portions, the dim disc, and it alone, will be obtained of a deep violet.

Do this with a contracted fibre, one extended, and one relaxed. In the extended fibre observe that it is chiefly the light disc which has been elongated by the extension (fig. 172, B). In the contracted muscle, the discs are closer together and narrower, while the fibre is broader at the contracted part.

16. Polariscope.—An ordinary microscope can be fitted with a polariscope, which consists of two Nicol's prisms; one is placed below the object, and is called the **polariser** (fig. 180), while the other, the **analyser**, is placed above the ocular.

The light reflected from the mirror as it passes through the polariser is polarised. With the analyser in position, look into the ocular, and slowly turn the analyser. The best forms are provided with a graduated circle to indicate the extent of the rotation. There are two positions of the analyser in which the field is quite dark, caused by the polarised rays being cut off. This occurs when the planes of polarisation of the two prisms are at right angles to each other, *i.e.*, when the Nicols are crossed. Between these two positions of the analyser a greater or less amount of polarised light is transmitted. Certain transparent histological preparations when placed on the stage of the microscope are dark when the Nicols are crossed, others under the same conditions cause the light to reappear, and appear bright on a dark field. They are said to be doubly refractive. The dim disc is doubly refractive or *anisotropous*, while the light disc is singly refracted, and is *isotropous*.

Either a preparation of fresh muscle or a balsam preparation may be used. It is well to take a muscle which has broad and distinct stripes to see the phenomena, one set of bands bright and refractive, and the others dark on a dark ground, *i.e.*, with crossed Nicols.

Instead of a glass-cover slip, cover the preparation with a thin slip of mica, or place a thin plate of gypsum under a preparation of striped muscle. The field shows various colours, red, pink, green, &c., according to the thickness of the mica plate, the position of the Nicols, and the relation of the axis of the mica to that of the Nicols. Suppose the general tint of the field to be pink, then any doubly refractive substance assumes a tint complementary to the pink, *i.e.*, of a greenish hue. On turning the analyser, the tint of the field varies,

FIG. 180.—A. Polariser to fit into Zeiss's large stand in a frame under Abbe's condenser; B. Section of A showing arrangement of the prisms.

and with it the colour of the anisotropous substance, while the isotropous substance, being singly refractive, and having no effect on the direction of the polarised ray, has the same colour as the field.

The doubly refractive property is possessed by bone (p. 181), smooth muscle, and the white fibres of connective tissue.

17. **Obliquely striated Muscle of Anodon.**—Place a wedge between the partially opened valves of Anodon, the fresh-water mussel, so as to put the fibres of the posterior adductor on the stretch. The whole animal may then be placed in dilute alcohol, or 1 per cent. potassium bichromate, for two days. On teasing a portion, isolated fibres showing oblique striation are obtained. Or the muscle may be examined fresh in sea-water or in the blood of the animal.

Retro-lingual Membrane of Frog (*Ranvier*).[1]—Under the tongue of the frog is a lymph-sac which is separated from the buccal cavity by a thin membrane. It contains striped muscular fibres which anastomose with each other, thus forming a plexus—in this respect presenting a peculiarity. Cut off the head, pull out the tongue, and remove the membrane. It can be floated in salt solution to see it. Place the membrane for twenty-four to forty-eight hours in dilute alcohol, then pencil away the epithelium, and place it for twenty-

[1] *Comptes Rendus*, 1890.

four hours in dilute methyl-violet-5B. Wash and mount in glycerine. The elastic fibres are stained of a bright blue, and they seem to spring from the ends and sides of the muscular plexus—also stained blue. The elastic fibres are fixed to the sarcolemma—which, however, is not stained by the methyl-violet—and the union is a very firm one. Thus the muscular fibres divide, and are in elastic sheaths connected with elastic fibres.

Ranvier has used the same membrane for studying the changes which take place in a striped muscular fibre during contraction.

LESSON XVII.

NERVE-FIBRES.

NERVE-FIBRES are of two kinds, **medullated** and **non-medullated**; the former are found chiefly in the white matter of the nerve-centres and the cerebro-spinal nerves, while the latter occur in large numbers in the sympathetic system.

I. **Medullated Nerve-Fibre.**—The essential part is the **axis-cylinder**, a soft, transparent rod or thread running from end to end of the fibre, and composed of **primitive fibrils**. It is covered by the **myelin**, or **white substance of Schwann**, or **medullary sheath**, which envelops the axis-cylinder everywhere except at the termination of the fibre and at the nodes of Ranvier (fig. 181). The myelin gives the nerve-fibre its highly refractive appearance and its double contour, and it can be shown to consist of a stroma or network of fibrils of a peculiar chemical substance called **neurokeratin**, enclosing a semifluid fatty-like substance, containing, amongst other chemical substances, protagon, a complex phosphorised fat. Histologically it consists of **cylinder-cones** or **medullary segments**, whose ends are bevelled and fit one into the other, but separated from each other by oblique clefts or **incisures**.

Outside the medullary sheath is a thin, transparent, tough elastic sheath, the **primitive sheath, sheath of Schwann**, or **neurilemma**. It is not present in all nerve-fibres, being absent from the fibres of the central nervous system.

Between the axis-cylinder and the myelin is a thin layer of matter, called by Kühne **axilemma**. By others it is regarded as an albuminous cement.

At fairly regular intervals—about 1 mm.—along the course of a fibre are constrictions, the **nodes of Ranvier**, where the myelin is absent, so that the neurilemma appears to produce a constriction at these points. The part between any two successive nodes of

Ranvier is an **interannular segment**, or internode, and about the centre of this, under the neurilemma, is a flattened oval nucleus—**nerve-corpuscle**—surrounded by a small quantity of protoplasm, and lying in a slight depression of the myelin. Nodes of Ranvier are absent from the nerve-fibres of the brain and spinal cord. Osmic acid blackens the myelin; silver nitrate produces the so-called **Ranvier's crosses**. The manner of their production is given in the text. Nerve-fibres do not, as a rule, branch except towards their terminations.

The fibres vary greatly in diameter, some being only half as broad as a red blood-corpuscle ($4\,\mu$, $\frac{1}{8400}$ inch), others as broad ($8\,\mu$, $\frac{1}{3200}$ inch), and others broader still; so that they vary in diameter from $2\,\mu$ to $20\,\mu$.

II. **Non-Medullated Nerve-Fibres.**—They occur specially in the sympathetic system, but are also present in the cerebro-spinal nerves. Each fibre consists of a bundle of fibrils enclosed in a transparent structureless sheath. Some observers doubt the existence of this sheath. The fibres are somewhat flattened; they branch and anastomose, and in their course are oval nuclei (fig. 182). As they have no myelin, they are not blackened by osmic acid, so that this reagent serves to distinguish the two kinds of fibres.

Nerve-trunks consist of bundles or funiculi of nerve-fibres, each bundle containing a greater or less number of fibres. Several bundles are held together by a common connective-tissue sheath—the **epineurium**. The sheath around each funiculus is composed of lamellated connective tissue, covered on both surfaces by endothelial cells, and is called the **perineurium** or lamellated sheath. Lymph spaces exist between the lamellæ. Delicate fibrils of connective tissue lie between the nerve-

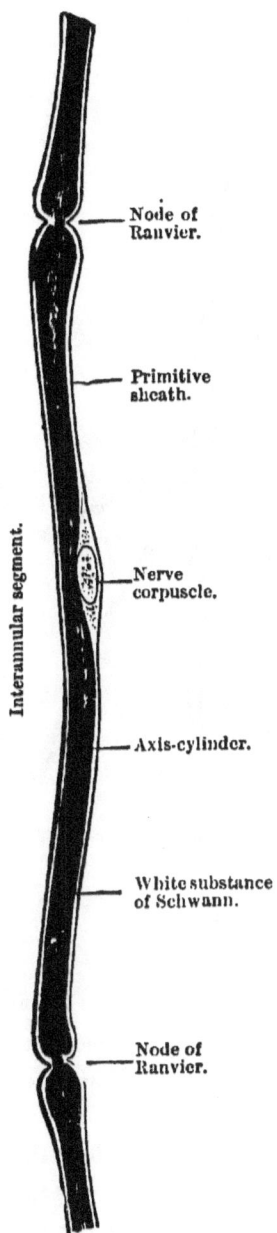

FIG. 181.—Medullated Nerve-Fibre. Osmic acid.

fibres, and constitute the **endoneurium**. The larger blood- and lymph-vessels lie in the epineurium and perineurium, while the endoneurium supports the few capillaries which are distributed to the nerve-fibres (fig. 190).

The term **Sheath of Henle** is applied to the prolongation of the perineurial sheath—usually a single lamella—around a small branch, or even one or two nerve-fibres.

The following statement may facilitate the study of the parts to be investigated:—

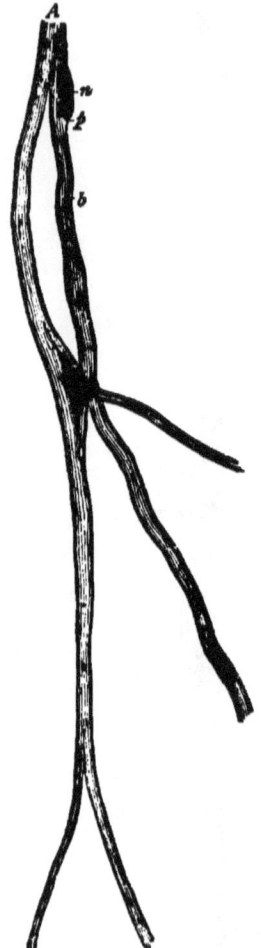

FIG. 182.—Non-Medullated Nerve-Fibre, Vagus of Dog. *b*. Fibrils; *n*. Nucleus; *p*. Protoplasm surrounding it.

NERVE-FIBRES.

I. **Medullated** (chiefly in cerebro-spinal system).

II. **Non-Medullated** (sympathetic or fibres of Remak).

A. **Medullated Nerve Fibre** consists of—

(1.) *Primitive sheath*, sheath of Schwann (or neurilemma).
(2.) *Nerve-corpuscles* occur under the primitive sheath in each inter-annular segment.
(3.) *White substance of Schwann*, myelin, or medullary sheath (with cylinder cones and incisures). It contains a net-work of *neurokeratin*.
(4.) *Axis-cylinder*, composed of primitive fibrils (surrounded by a sheath called the axilemma).
(5.) *Nodes of Ranvier* and internodal or interannular segments between two successive nodes.

Ranvier's crosses and *Frommann's lines*, obtained by using nitrate of silver.

Sheaths of a Nerve-Trunk.—Epineurium, perineurium, and endoneurium.

B. **Non-Medullated Nerve-Fibres** consist of—

(1.) A bundle of fibrils, usually enclosed in
(2.) A transparent sheath or neurilemma (?).
(3.) On the fibres are oval nuclei.

1. Medullated Nerve-Fibres (H).

—Select a small nerve, *e.g.*, the sciatic or one of its branches, of a frog. Cut out half an inch of it, and place it on a dry slide, but add no fluid. Fix one end of the thread by pressing on it with any thin blunt object, *e.g.*, the flat surface of a mounted needle, and fray out the opposite end in a fan-shaped manner with a mounted needle, so as to isolate some nerve-fibres. To prevent it from drying, breathe on the specimen from time to time. Add a drop of normal saline, cover, and examine (fig. 183).

(*a.*) Observe highly refractive **medullated nerve-fibres** of variable size, some as broad as a red blood-corpuscle and others narrower. Each fibre has a *double contour*, *i.e.*, two thin lines on each side of the centre. The double contour may be interrupted here and there. The double contour is due to the **white substance of Schwann, medullary sheath,** or **myelin.**

(*b.*) In the centre a clear bright rod, the **axis-cylinder.**

(*c.*) Outside the myelin is a thin transparent sheath, **primitive sheath** or **neurilemma,** scarcely to be detected as such unless the fibre is ruptured or the sheath raised from the myelin, or stretching as a funnel-shaped prolongation from the end of a torn fibre.

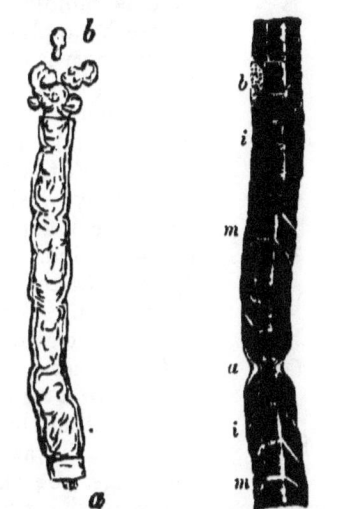

FIG. 183. — Fresh Nerve-Fibre Examined in Normal Saline. *a.* A short piece of the axis-cylinder projecting; *b.* Myelin drops.

FIG. 184. — Nerve-Fibre of Frog. *a.* Ranvier's node; *b.* Nucleus; *i.* Incisures; *m.* Myelin blackened. Osmic acid, × 400.

(*d.*) Selecting a fibre isolated for a considerable distance, trace its outline, and observe at intervals slight constrictions—**nodes of Ranvier**—where the myelin is absent.

(*e.*) A small quantity of delicate connective tissue—**endoneurium** —with, perhaps, a capillary and a few blood-corpuscles between the fibres.

(*f.*) The myelin tends to exude from the ruptured ends of the fibres, and appears as highly refractive spherical droplets—**myelin drops**—often with concentric markings, but there is no nucleus in them (fig. 183, *b*). The myelin drops exude more rapidly in a preparation irrigated with 1 per cent. acetic acid, or 2 per cent. caustic soda, or even with distilled water alone.

Stain the preparation with picro-carmine, and note that this reagent diffuses into the fibres at their cut ends and at the nodes of Ranvier. It stains the axis-cylinder red, and also the **nerve-nuclei** or corpuscles which lie just under the neurilemma.

2. **Nerve-Fibres in Osmic Acid (H).**—A nerve-fibre is rapidly blackened by osmic acid, as can be shown by applying a drop of 1 per cent. solution to a fresh nerve, but for good permanent preparations it is well to stain the nerve after the action of the osmic acid. Place a small piece of nerve in a small glass thimble along with 2 cc. of .5 per cent. osmic acid. Cork the thimble, and after twenty-four hours thoroughly wash the preparation; tease it a little, and place it for twenty-four hours in a solution of picro-carmine. In fact, if it be left for days in this dye it is better, as the fibres can then be more readily dissociated. Tease a small fragment, and mount it in glycerine acidulated with formic acid.

(a.) Observe in each fibre the myelin stained black. Search for a **node of Ranvier**, a narrow constriction, and note that the myelin is absent at the constriction, although the axis-cylinder and neurilemma are present (figs. 184, a, 185). Find the next node, and, between the two adjoining nodes, the stretch of nerve—the **internodal** or **interannular segment** (fig. 181).

(b.) In the interannular segment, just under the neurilemma, and lying in a slight depression of the myelin (fig. 184, b), a red-stained oval **nucleus** surrounded by a small quantity of protoplasm, and about midway between the two nodes (fig. 185, n). The **axis-cylinder** stained red, and continuous throughout the fibre.

(c.) In the myelin what look like oblique slits—**incisures**—running obliquely outwards from the axis-cylinder to the neurilemma (fig. 185, i). They correspond on each side, and break up the myelin into a number of short lengths—**cylinder cones**—the bevelled end of one cylinder-cone fitting into the oppositely bevelled end of the next cylinder-cone. Many cones lie in an internode.

(d.) Some of the fibres are broad, and others narrow, about half the breadth of the others.

(e.) Some fibres are not blackened by osmic acid at all. They

FIG. 185.—Nerve-Fibre (Osmic Acid). a. Axis-cylinder; r. Node of Ranvier; i. Incisures; n. Nucleus; p. Protoplasm; s. Neurilemma.

appear as flattened bands with oval nuclei at intervals in their course. They are **non-medullated nerve-fibres**, which, as they have no myelin, are not blackened by the osmic acid.

(*f*.) A small quantity of connective tissue and capillaries. To get a good view of the incisures and nodes, the best plan is to stretch the sciatic or other nerve of a frog on a match before placing it in osmic acid. In this way the fibres are kept straight, and the cylinder-cones pulled asunder as far as possible, thus making the incisures wide and distinct.

3. **Ranvier's Crosses (H).**—Rapidly tease out on a dry slide one of the branches of the sciatic nerve (frog), and stain it for five minutes with .3 per cent. solution of silver nitrate. Wash off the silver, apply a drop of glycerine, and expose it to daylight. It rapidly becomes brown.

(*a*.) Observe the fibres, but at each node will be seen a brownish-black cross (fig. 186); the silver nitrate diffuses into the nerve-fibre only at the nodes, stains the cement joining one internode with another, thus making the transverse bar of the cross, and as it diffuses along the axis-cylinder it stains some cement substance on the latter, and thus makes the vertical bar of the cross.

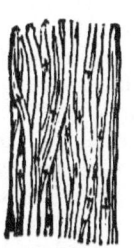

FIG. 186.—Nerve with Ranvier's Crosses. Silver nitrate, × 30.

FIG. 187.—Nerve-Fibre. *a*. Node and cross of Ranvier with Frommann's lines. × 200. Silver nitrate.

Occasionally a number of transverse (fig. 187) lines—**Frommann's lines**—are seen on the axis-cylinder, *i.e.*, on the vertical bar of the cross (p. 210). Place a sciatic nerve of a frog for twelve hours in 1 per cent. AgNO$_3$, and keep it in the dark. Wash and harden in absolute alcohol, and mount in balsam. The crosses and lines are then seen with great distinctness.

4. **Endothelial Cells of the Perineurium. Intercostal, or other Small Nerve.**—(i.) It is well also to stain with silver nitrate one of the small intercostal nerves of a rat or some other small nerve. Stain the whole nerve in silver nitrate.

(ii.) Open the abdomen of a frog, remove the abdominal viscera, so as to expose the nerves coming from the vertebral canal. Pour on the nerves .3 per cent. silver nitrate. After three minutes, cut out the nerves and place them for half an hour in fresh .3 per cent. AgNO$_3$. Wash in distilled water, tease a piece in glycerine, and expose it to light.

(*a*.) The crosses are seen as before, but above them is the

endothelial sheath of the nerve-fibre, composed of polygonal squames (fig. 189).

5. Axis-Cylinder.—Harden for two or three days a nerve in chromate of potash, tease a piece, and stain it in carmine. Observe the axis-cylinder stained red (fig. 188). Or treat a fine nerve for four days to a week in $\frac{1}{10}$ per cent. chromic acid or $\frac{1}{5}$ per cent. bichromate of potash. On teasing long stretches of isolated nerve axis-cylinders are readily found.

6. T.S. of a Nerve.—Select a rather large nerve, e.g., the human sciatic, and harden about an inch of it in picric acid for forty-eight hours, or in 2 per cent. ammonium bichromate for two weeks. Wash out the bichromate. Complete the hardening in alcohol. Sections may be made, and stained with logwood or carmine, but they are very apt to fall to pieces. It is preferable, therefore, to stain the hardened tissue "in bulk." Place it in borax-carmine for three days, then transfer it to acid alcohol, and pass it through absolute alcohol and turpentine, and embed it in paraffin. Cut sections, when the paraffin keeps all the parts in their places.

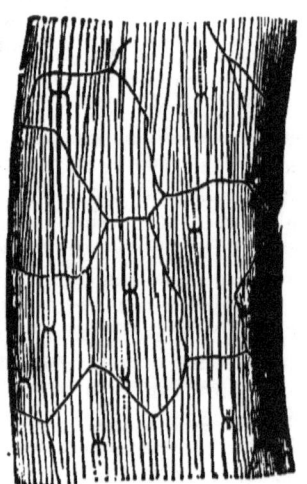

FIG. 189.—Intercostal Nerve of Mouse, Ranvier's Crosses and Endothelial Covering. AgNO$_3$, × 300.

FIG. 188.—Peripheral Nerve-Fibre. *a.* Axis-cylinder; *b.* Ranvier's node; *c.* Nucleus. Chromate of potash and carmine, × 200.

Fix a section on a slide with white of egg, remove all the paraffin by placing the slide in turpentine, clear it up in clove-oil, and mount in balsam.

(*n.*) (**L**) Observe the connective-tissue sheath—**epineurium**—or sheath surrounding the whole nerve, sending processes into the nerve,—numerous bundles—some large, others smaller—or **funiculi** of nerve-fibres, each surrounded by a lamellated sheath—**perineurium**—which sends fine septa—**endoneurium**—into each funiculus. In each bundle the cut ends of the fibres are directed towards the observer (fig. 190).

The large *blood-vessels* are in the epineurium, and a few in the endoneurium.

(*b.*) (**H**) Select a bundle. Observe the *perineurium*, made up of several concentric lamellæ with nuclei between them. The cut ends of the fibres varying in diameter.

(*c.*) Note in each fibre the section of the stained axis-cylinder. Surrounding this a clear transparent ring, indicating the position of the myelin, which has been dissolved out in the process of preparation.

(*d.*) Outside this a thin circle — the *primitive sheath*. Between the nerve-fibres a small quantity of connective tissue or **endoneurium**.

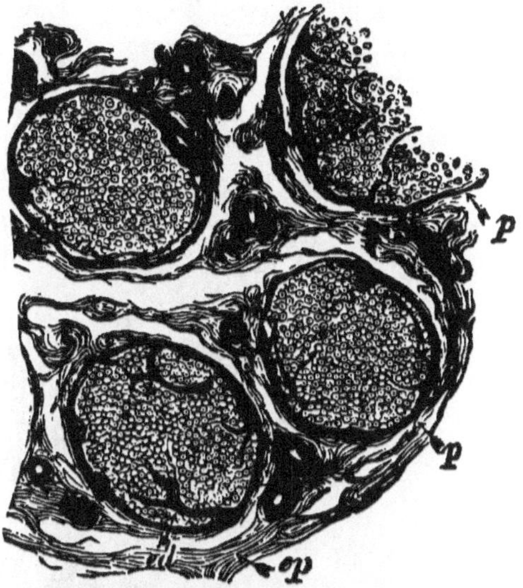

FIG. 190.—T.S. of Several Funiculi of the Median Nerve. *p.* Perineurium; *ep.* Epineurium; *ed.* Endoneurium.

Non-Medullated Nerve-Fibres.

7. Non-Medullated or Sympathetic Nerve-Fibres.—(i.) These are readily found in the large splenic nerve of the ox, or in a portion of the sympathetic nerve of the thoracic or abdominal chain. The nerve has a pretty thick sheath. Cut this open, and cut off a small piece. Tease it in normal saline. The process of teasing is greatly facilitated by placing the nerve for twenty-four hours in dilute acetic acid (5 drops in 100 cc. water).

(**H**) Observe that there are very few medullated fibres, the great majority being non-medullated (fig. 192). Their outlines are not very distinct; they are faintly striated longitudinally, and have oval nuclei at intervals. They may be stained with picro-carmine.

FIG. 191.—Non-Medullater Nerve-Fibres, Sympathetic Nerve of Rabbit, × 200.

(ii.) Place the cervical sympathetic nerve of a rabbit in .25 per cent. osmic acid for twenty-four hours. Wash it in water, and stain it for several hours in picro-carmine. Tease a fragment in glycerine.

(iii.) Or tease the vagus of a rabbit upon a *dry* slide, taking care that the nerve does not become dry. Cover it for 5–10 minutes with 1 per cent. osmic acid, wash away the osmic acid, add picro-carmine, and place the whole for 24-48 hours in a moist chamber

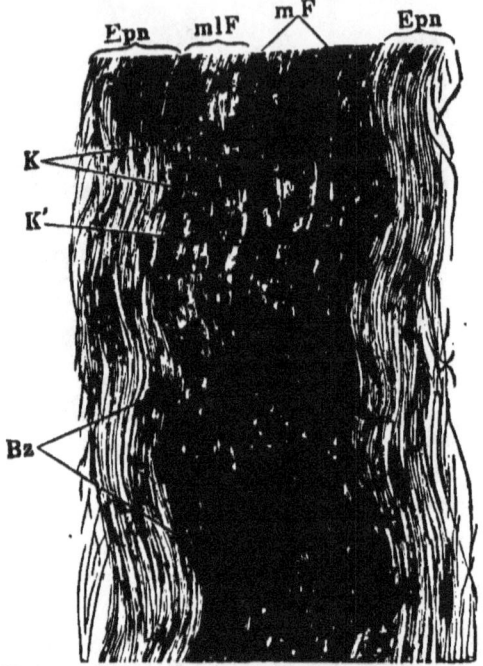
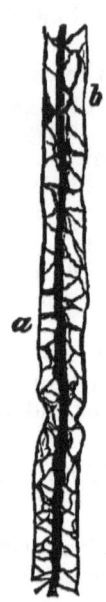

FIG. 192.—Part of the Human Sympathetic Nerve in Osmic Acid. There are two medullated fibres amongst Remak's fibres, × 350. Externally the epineurium, *Epn*; *mF*. Medullated fibres with Ranvier's nodes; *K*. Nuclei of Remak's fibres; *mlF*. Remak's fibres; *K'*. Nucleus of medullated fibre; *Bz*. Connective-tissue cells.

FIG. 193.—Neurokeratin Network in a Medullated Nerve-Fibre.

(fig. 47), and afterwards displace the picro-carmine by glycerine, as recommended at p. 82.

(H) Observe a few small medullated fibres (blackened by OsO_4), and numerous non-medullated fibres, some of which may be seen to branch. Note the oval nuclei on the fibres (fig. 191). If the part be taken from near a ganglion, often nerve-cells may be seen.

ADDITIONAL EXERCISES.

8. Neurokeratin Network and Axis-Cylinder.—Place the fresh sciatic nerve of a frog in a dilute solution of ferric chloride consisting of—

Liquor ferri perchloridi	1 part
Distilled water or spirit	3-4 parts.

XVII.] NERVE-FIBRES. 211

Leave it in this fluid for three or four days. Wash every trace of the iron salt out of the preparation, and preserve it in alcohol. Place small pieces of the nerve several days in a saturated solution of dinitrosoresorcin in 75 per cent. alcohol. Tease a fragment, dehydrate it with alcohol, clarify with xylol, and mount in balsam (*Platner*). This is an excellent method.

(*a*.) Observe the axis-cylinder stained green. It can be seen with the utmost distinctness passing from one internode to the next one, and across the nodes of Ranvier, which are particularly sharply defined. In the myelin, a network of fibres—the neurokeratin network—stained green. The axis-cylinder appears as distinct as in fig. 193.

(*b*.) **Kühne and Ewald's Method.**[1]—Harden a nerve for twenty-four hours in absolute alcohol. Boil it in absolute alcohol, and extract it with ether. To remove everything except the network, a teased preparation is digested in pancreatic juice.

(*c*.) Heidenhain's hæmatoxylin (p. 70) may be used to stain nerve-fibres. It stains the axis-cylinder and the neurokeratin network.

9. Frommann's Lines and Ranvier's Crosses (H).—Place a fresh nerve in .5-1 per cent. silver nitrate for forty-eight hours, and keep it in the dark. Wash it in water, and expose it to light for 2-3 days in equal parts of formic acid, water, and glycerine, and preserve it in glycerine. Tease a piece in glycerine.

FIG. 194.—Frommann's Lines on an Axial Cylinder.

(*a*.) Observe the crosses of Ranvier sharply defined, and on the axis-cylinder well-defined transverse markings, extending for a long distance along the axis-cylinder. **Frommann's lines.**—If an axis-cylinder be dislodged from its fibre, a *biconical swelling* may be seen (*a*). It corresponds to that part of the axis-cylinder opposite a node of Ranvier (fig. 194).

10. Axis-Cylinder—(*a*.) **Action of Collodion (H).**—Tease a fresh sciatic nerve of a frog without adding any fluid. Add a large drop of collodion and apply a cover-glass, or tease a fresh nerve in chloroform. Examine quickly, as the preparation soon spoils. The neurilemma is distinct, the myelin is transparent and finely granular, while the axis-cylinder appears as a dark cylindrical rod—often with a curved course—in the centre of the fibre, and it may even project beyond the end of the fibre.

(*b*.) Isolated axis-cylinders are readily obtained from the central nervous system after maceration of the white matter of the cord in methyl-mixture (p. 26), or Müller's fluid or ammonium chromate.

11. Schwann's Sheath.—Macerate a peripheral nerve for several days in ammonium chromate (1 : 3000). The myelin is dissolved while the neurilemma and axis-cylinder remain (*Schiefferdecker*).

FIG. 195.—Peripheral Nerve-Fibre of Frog. *a*. Longitudinal fibrillæ in axis-cylinder; *b*. T.S. of nerve-fibre. Osmic acid and Bismarck brown, × 1000.

12. T.S. Nerve, Osmic Acid (H).—Stretch a nerve on a piece of wood, and place it—wood and all—for two days in .5 per cent. osmic acid, or, better still, in Flemming's mixture for one day. On the second day, add a little more osmic acid to Flemming's mixture, and harden the nerve for another day. It is better to embed the nerve in paraffin and make transverse sections. The sections are fixed on a slide by a fixative,

[1] *Kühne's Untersuch.*, Heidelberg, 1878.

the paraffin is extracted by turpentine, and the sections mounted in balsam. They may be stained with a watery solution of Bismarck brown. Instead of this, after fixing the nerve in osmic acid, and hardening in 90 per cent. alcohol, place it for 24–48 hours in a strong saturated watery solution of acid fuchsin (S.N. 30), wash in alcohol, embed and cut in paraffin. The ends of the fibrils on the axis-cylinder are stained red (fig. 195).

(*a.*) Note the axis-cylinder in the centre surrounded by a dark ring (figs. 195, 196, *b*), the myelin blackened by the OsO_4. If the section of a nerve-fibre is through incisures, the double contour of the myelin may be seen sometimes with a narrow black edge, at others with a broad black edge externally.

(*b.*) If a piece of the nerve be placed for forty-eight hours in a solution of Bismarck brown, and then teased, the appearance shown in fig. 195, *a*, is obtained, when the axis-cylinder presents a longitudinally striated appearance.

13. Size of Nerve-Fibres.—The osmic acid method has yielded the best results. Some of the fibres are broad, and others are narrow or fine. Thus the anterior roots of the upper cervical nerves, and the third cranial nerve, contain only broad nerve-fibres, while the second and succeeding thoracic nerves contain broad and fine fibres (fig. 196). In nerves going to muscles there are many large and few small medullated fibres, while in nerves going to viscera the fine medullated fibres are far more abundant than the broad fibres.

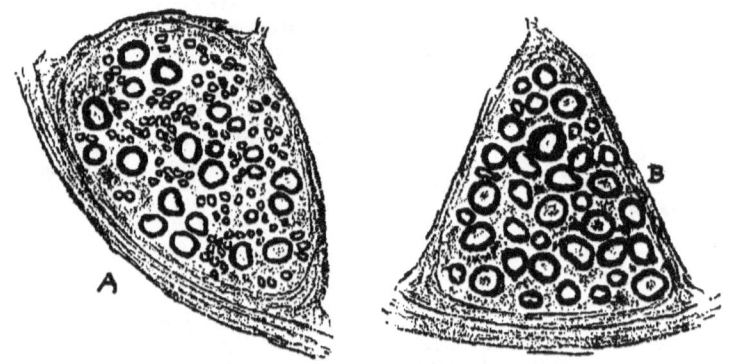

FIG. 196.—*A.* T.S. of the anterior root of a spinal nerve below the first dorsal nerve. *B.* T.S. of a part of a cervical nerve. Osmic acid.

14. Living Nerve-Fibres are readily studied in the inflated lungs of a newt or frog. The frog's lung is best kept inflated by Holmgren's apparatus.[1] Nerve-fibres are also readily seen in the tongue of a frog arranged as for studying the circulation of the blood (Lesson XIX.).

15. Marchi's Method for Degenerated Fibres.—Harden a nerve in Müller's fluid for eight days and then in the following fluid:—

Marchi's Fluid.

Müller's fluid . . . 2 parts.
Osmic acid (1 per cent) . . 1 part.

Embed in celloidin and mount in warmed balsam. It is well not to employ balsam dissolved in chloroform, as then the darkened parts lose their dark colour. The degenerated parts of nerve-fibres are black. This method is particularly useful for degenerations in the nerve-centres before sclerosis has

[1] *Beiträge z. Anat. u. Phys.*, *C. Ludwig, gewidmet*, Leipzig, 1874, p. cxvi.

set in, but it is also applicable to nerve-fibres that have undergone degeneration, e.g., after section of a nerve, constituting Wallerian degeneration. It will also detect any degenerated fibres in an ordinary nerve.

16. Isolated Schwann's Sheath.—Place a stretched nerve of a frog in the following fluid for twenty-four hours in the dark:—

Boveri's Fluid.

Silver nitrate (1 per cent.) . . . 10 cc.
Osmic acid (1 per cent.) . . . 10 ,,

Wash it in water, and place for twenty-four hours in very dilute caustic potash (2-3 drops of a concentrated solution in 15 cc. water). Tease in glycerine. The axis-cylinder shrinks and Schwann's sheath may be traced as a continuous sheath without any interruptions at the nodes.

17. Nerve-Fibres of the Spinal Cord.—These are devoid of Schwann's sheath, but they possess both Ranvier's nodes and incisures of **Lantermann.** Boveri's fluid may stain both. It is evident then that these two structures have no relation to Schwann's sheath, but are related entirely to the myelin. The cylinder-cones are readily isolated by Schiefferdecker's methyl-mixture (p. 26).

18. Degeneration of Nerve-Fibres.—This is readily studied in the rabbit. The skin is first disinfected with solution of corrosive sublimate, and then the median and ulnar nerves are exposed on the inner aspect of the upper arm, the nerves divided, and the wound sealed with collodion (C. Huber).[1] In different animals the nerves are excised 2, 3, 4, . . . 8 or 10 days after the operation. The excised nerves are kept extended on wood and fixed for twenty-four hours either in Hermann's fluid (Lesson XXXV.) or the picro-osmium mixture of Benda, prepared by saturating a 1 per cent. solution of osmic acid with picric acid and filtering. They are then washed in water and hardened in alcohol. The sections are stained with safranin and light-green. Besides showing the usual degeneration phenomena, they show mitotic division of the nuclei of Schwann's sheath, showing that these proliferate.

LESSON XVIII.

NERVE-GANGLIA, NERVE-CELLS, AND PERIPHERAL TERMINATIONS OF MOTOR NERVES.

Spinal Ganglia.—Harden a spinal ganglion of a cat or dog in 2 per cent. ammonium bichromate for three weeks, and subsequently in alcohol. Make transverse and longitudinal sections of the ganglion, stain with logwood or carmine, and mount in balsam.

1. L.S. Mammalian Spinal Ganglion (L).—(*a.*) Note the cap-

[1] *Archiv f. mik. Anat.*, xl. p. 409, 1892.

sule (fig. 197, c) surrounding the ganglion; nerve-fibres (a) enter the ganglion at one end and leave it at the other. They run in groups, chiefly through the central part of the ganglion, so that they are cut in different planes.

(b.) Numerous spherical cells (fig. 197, b) lying singly or in groups between the nerve-fibres, but chiefly towards the surface.

(c.) (**H**) Select a single **ganglion-cell**; note its spherical shape,

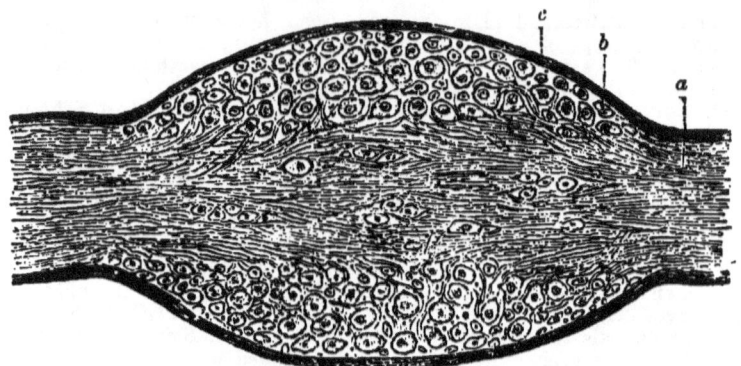

FIG. 197.—L.S. Spinal Ganglion. *a.* Nerve-fibres; *b.* Nerve-cells; *c.* Capsule of the ganglion.

its granular contents, and single, large, distinct, excentrically-placed nucleus, often with one or more distinct nucleoli (fig. 198). The nucleus has a well-defined nuclear membrane.

(d.) Around each cell is a **capsule**, which is lined by a single layer of flattened cells, but only the nuclei of these cells are seen.

FIG. 198.—Two Cells in a Spinal Ganglion (Human), the Protoplasm shrunk from the Capsule, × 200.

The cell-substance is frequently somewhat retracted from the capsule, so that a space may intervene between the two.

In a T.S. of such a ganglion, notice the capsule of the ganglion sending in coarse septa, the nerve-cells near the circumference, and the nerve-fibres chiefly in the centre.

2. **Isolated Cells of a Spinal Ganglion (Mammal) (H)**.—Into a dorsal ganglion of a young rabbit make an interstitial injection of osmic acid (2 per cent.). Tease a small piece in picro-carmine and mount the preparation in glycerine. Sometimes a cell with its single process may be found. The cells are *unipolar*. It is more difficult to find the connection of the issuing axis-cylinder with a nerve-fibre, forming what Ranvier has described as T-shaped nerve-fibres, but with care such processes can be found.

3. Spinal Ganglion of Frog.—These ganglia lie under cover of the small white calcareous sacs situated on each side of the vertebral column, which are seen at once when the abdominal cavity is opened and the abdominal viscera removed. Remove the white chalky mass, and the greyish semi-transparent small ganglion will be seen. With sharp-pointed forceps it is not difficult to tear away the capsule of the white calcareous mass. These sacs contain arragonite; some of the crystals are large, but the smaller arragonite particles when examined in water exhibit Brownian movement. Treat it in the same way as directed for the frog's Gasserian ganglion.

(H) In a carefully-teased specimen (use a dissecting microscope, p. 22), it is by no means difficult to find large unipolar cells, each cell with a distinct hyaline capsule, and the cell itself with a relatively large nucleus and well-defined nucleolus. Moreover, the continuation of the body of the cell with a nerve-fibre is not difficult to establish. The methylene-blue method may be used (p. 222).

4. Spinal Ganglion of a Skate.—Make an interstitial injection of osmic acid (2 per cent.) into such a ganglion. Stain a piece in picro-carmine and tease it in glycerine. Bipolar cells are readily found. Each cell shows a distinct capsule enclosing a nucleated cell with a pole at either end continuous with a nerve-fibre (fig. 199).

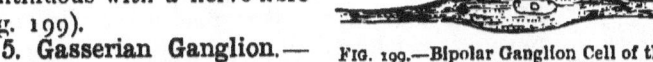

FIG. 199.—Bipolar Ganglion Cell of the Spinal Ganglion of a Skate.

5. Gasserian Ganglion.—(*a.*) The Gasserian ganglion of a sheep does very well; harden it in the same way as for spinal ganglia or in Müller's fluid. The same general arrangement of fibres and cells is seen, only the cells are larger, and their protoplasm frequently contains granules of a yellow pigment. Very instructive results are obtained by double-staining it with eosin and hæmatoxylin, first with hæmatoxylin and then with eosin, or use eosin-hæmatoxylin, and mount in balsam. The nuclei are blue, the other parts reddish. If a cell be isolated after interstitial injection of osmic acid, as recommended for spinal ganglia, the cells have the form shown in fig. 200.

(*b.*) A fresh ganglion may be teased in salt solution. The large spherical cells are readily isolated, but they usually shell out of their capsule. Stain them with magenta solution.

6. Gasserian Ganglion of Frog.—Destroy the brain and spinal cord of a frog, remove the lower jaw, divide the skull into two longitudinally by a vertical incision. Tear off the mucous membrane covering the roof of the mouth. From a foramen just behind the eyeball there issue a few fine threads, branches of the fifth nerve. Scoop out the brain. These threads are readily recognised by being usually somewhat pigmented. With a pair of scissors make a snip in the base of the skull at right angles to the cut already made. Turn up the bone, and on the fifth nerve, which runs towards the foramen behind the eyeball, will be found a small oval swelling surrounded by a tough capsule. Divide the latter and remove the ganglion. I have usually found that the little ganglionic swelling is somewhat pigmented. At any rate, it is easily found by tracing the fifth nerve backwards. The nerve is accompanied by an artery.

Tease the ganglion in .25 per cent. osmic acid, and let it stain in this fluid for two hours. Stain it for several hours in picro-carmine (under a moist-chamber, p. 82), tease a fragment in glycerine. Try to find a cell with its single process prolonged into a nerve-fibre. In most of the cells, however, the process is apt to be detached.

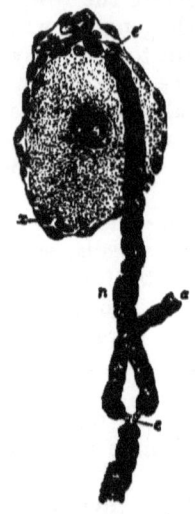

FIG. 200.—Nerve-Cell from Rabbit's Spinal Ganglion. *x.* Nuclei of the cell-capsule; *n.* Nucleus of nerve-fibre; *a.* Nerve-fibre; Fibre dividing at *e* at *a* node of Ranvier, T-shaped fibre.

7. Sympathetic Ganglia (Frog).—Lying in contact with the spinal column of the frog is a row of small semi-transparent ganglia, the sympathetic chain, and between them and the roots of the spinal nerves pass fine nerve filaments. Open the abdomen of a freshly-killed frog, remove the intestinal tract and liver, cut through the peritoneum above the kidneys, raise the kidneys and excise them. There will be seen the white nerves issuing from the cord. Between these and the sympathetic ganglia fine nerves run transversely. Cut out the sympathetic ganglia and treat them with chloride of gold by the method (p. 79, 3); or cut out the aorta and the adjacent tissues, and subject them to the gold chloride method. After the piece of tissue has acquired a purplish colour, examine it with a low power to find nerve-cells. The nerve-cells may be isolated or arranged in groups. It requires great care to get a satisfactory preparation. Note the pyriform shape of the cell, each with a large nucleus, the cell-substance continued into a **straight process**, which may be seen to be encircled by a **spiral process**. The body of the cell is surrounded by a well-marked capsule, which is continued over the cell-processes, and has nuclei on its inner surface.

8. Sympathetic Ganglion (Mammal) (H).—Harden the first thoracic sympathetic ganglion or the superior cervical ganglion of a **man or rabbit in 2 per cent. ammonium bichromate (2–3 weeks)**, and subsequently in alcohol. Make transverse sections. Stain in picro-carmine or a watery solution of nigrosin (several hours), and mount the former in Farrant's solution, and the latter in balsam.

(*a.*) Observe the fibrous **capsule** of the ganglion ending in septa, and numerous bundles of non-medullated nerve-fibres cut obliquely or transversely. A few blood-vessels.

(*b.*) The **nerve-cells**, each with a capsule, showing nuclei. The nucleated cell-substance is frequently somewhat shrunk from its capsule, and at one side it usually contains some yellowish-brown pigment granules, especially if the human cervical ganglion be used. It is difficult to see the process, which becomes continuous with a nerve-fibre, but with care it may be seen passing out of one or more of the cells (fig. 201).

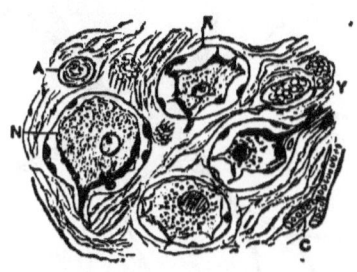

FIG. 201.—Human Superior Cervical Sympathetic Ganglion. *A.* Small artery; *C* Capillary; *V.* Vein; *K.* Capsule; *N.* Nerve-cell, × 300.

(*c.*) The veins have long fusiform **dilatations** upon them; this is not unfrequently seen in teased preparations of a human ganglion, but can only be fully demonstrated in an injected specimen of the ganglion.

9. Isolated Multipolar Nerve-Cells of the Spinal Cord.— There are several ways of preparing these.

(i.) Cut out a small part of the anterior cornu of the spinal cord of an ox, calf, sheep, or other animal, and place it in very dilute chromic acid (.01 per cent.) or .2 per cent. potassium bichromate for a few days, and do not change the fluid. Wash, and place it for twenty-four hours in strong carmine solution (p. 63). Place a little of the red pulp on a slide, and, with the aid of a dissecting microscope, try to isolate one or more multipolar nerve-cells.

(ii.) Or, what is a better method, take small fragments of the anterior cornu of the spinal cord of an ox or calf, and place them in dilute alcohol for forty-eight hours or longer. After this time we can see better the distinction between the grey and the white matter. Shake the fragments in the dilute alcohol, and allow the debris to subside. Pour off the alcohol, and "fix" the cells with .25 per cent. osmic acid (one hour); pour this off, and stain the cells for forty-eight hours with picro-carmine. Pour off the picro-carmine and replace it by glycerine-jelly. When the glycerine-jelly is warmed, a drop of the fluid placed on a slide is almost certain to contain one or more isolated multipolar nerve-cells.

With a low power find a cell.

(*a.*) (**H**) Observe the large size of the cell (100 μ, $\frac{1}{250}$ inch, and therefore visible to the naked eye), with numerous **processes—branched processes**—which run in all directions (fig. 202). The processes branch; and then branch again and again to form a fine protoplasmic system of processes — the **protoplasmic processes**. One process is always unbranched, it is by no means difficult to see—the **axis-cylinder process**—which becomes continuous with, or in fact is, or becomes, the axis-cylinder of a nerve-fibre (fig. 202, *a*).

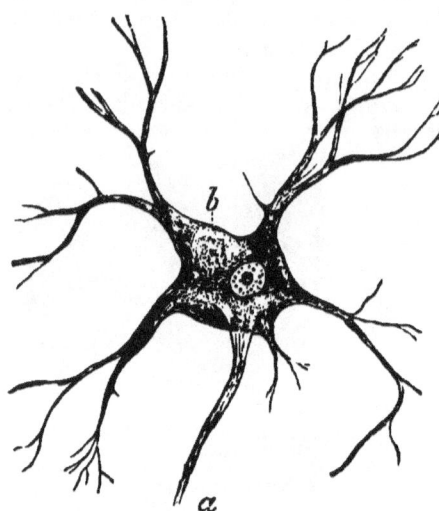

FIG. 202.—A Multipolar Nerve-Cell from the Anterior Cornu of the Grey Matter of the Human Spinal Cord. *a.* Axis-cylinder process; *b.* Pigment. × 150.

(*b.*) The multipolar cell itself has no cell-wall, and it contains a large, conspicuous, spherical, nucleolated nucleus, the latter with a distinct envelope. The protoplasm is fibrillated, and the fibrils may be seen to stretch into the branched processes. Sometimes the cells contain pigment (fig. 202, *b*).

10. Cover-Glass Preparation of multipolar Nerve-Cells.—(i.) From a perfectly fresh cord of a sheep or ox snip off a small piece of the anterior cornu; press it between two cover-glasses, so as to form a thin film. Separate the cover-glasses and allow the film adhering to each to dry. Float the cover-glass—film surface downwards—on a concentrated watery solution of methylene-blue for several hours. Wash the cover-glass in water mixed with alcohol, drain it, allow it to dry, and mount it in xylol balsam; the cover-glass can be passed through absolute alcohol, cleared with xylol, and mounted in xylol balsam. The multipolar nerve-cells are all deeply stained blue (*Thanhoffer*).

(ii.) If the use of aniline colours be objected to, the following method gives good results:—For three or four days macerate a small part of the grey matter of the anterior cornu in 20 cc. of water, containing 1 gram of each of the following: Neutral ammonium chromate, potassic phosphate and sodic sulphate (Landois' fluid, p. 26), and then stain it in bulk for 24-48 hours in equal parts of the above solution and strong ammoniacal carmine.

Squeeze a little of the red pulp between two cover-glasses, and treat it as recommended for the methylene-blue preparation. It would be difficult to get preparations that surpass in beauty those prepared by the methylene-blue method of Thanhoffer.

(iii.) Make a cover-glass preparation from a fresh spinal cord. Heat one of them by passing it two or three times through the flame of a Bunsen-burner. Thereby the proteids are coagulated and partially charred. In some of these preparations good views of the blood-vessels may also be obtained.

(*a.*) Observe—especially in the methylene-blue preparation—the large cells with numerous branched processes; some of them are very long, and each shows distinct fibrillation.

(*b.*) The unbranched axis-cylinder process.

(*c.*) The body of the cell, nucleated and nucleolated, with its cell-contents, traversed by blue-stained fibrils, running in certain definite directions through the cell.

Other forms of nerve-cells are referred to under Cerebrum and Cerebellum.

ADDITIONAL EXERCISES.

11. Motor Nerves to Muscles.—(i.) If the skin over the sternum of a small frog be divided longitudinally, on raising the skin a small thin muscle—musculus cutaneus pectoris—will be seen running from the skin to the sternum. Keep the muscle stretched and "fix" it by pouring on it a little osmic acid. Cut out the muscle, and after dehydrating, mount it in balsam. It is apt to darken on exposure to light.

(L) Observe the nerve is black—sending branches over the muscular fibres; trace these onwards over the muscular fibres until a single nerve-fibre is found.

(H) Note that when a nerve-fibre divides, it does so always at a node of Ranvier. The nerve-fibre can be traced to a muscular fibre, but it apparently stops abruptly, because the myelin stops where the nerve pierces the sarcolemma. Other methods are required to see the termination within the sarcolemma.

(ii.) **May's Method.**—Select a thin muscle, *e.g.*, the cutaneus pectoris, sartorius, mylo-hyoid, &c., and place it in water containing 2 per cent. glacial acetic acid for twelve hours. Make—fresh—the following mixture:—

½ per cent. potassio-gold chloride	. . .	1 cc.
2 ,, osmic acid	. . .	1 ,,
2 ,, glacial acetic acid	. . .	50 ,,

and place the muscles in it for 2–3 hours. Then transfer them to the following mixture:—

Glycerine	40 cc.
Water	20 ,,
Hydrochloric acid (25 per cent.)	1 ,,

for several hours. They become very transparent, and can be investigated in glycerine or Farrant's solution.

12. Nerves of Frog's Sartorius.—Suppose the sartorius to be selected. A beautiful view of the distribution of the motor nerves—black—is obtained (fig. 203).

The single nerve-trunk enters the muscle on the median aspect and on its under surface about the level between the middle and lowest thirds. Several large branches—usually two—run nearly parallel towards both ends of the muscle—two longer, towards the upper end of the muscle *A*—and two or three shorter, towards its lower end, the latter following a slightly more oblique course. Numerous branches form elongated quadrilateral meshes. At two points in the muscle, towards its ends, there are more fine branches than elsewhere. The fibres form **plexuses** and divide. Note specially that the knee and pelvic ends are *devoid of nerve-fibres.*

13. Motor Nerve-Endings.—(i.) Take a thin muscle, *e.g.*, the eye-muscles or intercostal muscles of a small mammal, the thin leg muscles of a lizard, or the thin cutaneous muscles which pass between the skin and the wall of the chest in snakes, and stain them with gold chloride by the formic acid gold chloride method (p. 79). They must remain in the gold solution about one hour. The gold may be reduced, either in water acidulated with acetic acid, by exposure to the light, or in the dark, in 25 per cent. formic acid.

(H) Tease a piece of the purplish-violet muscle in glycerine, and search for a purple nerve-fibre termination in an arborescent branched **end-plate** lying on the sarcous substance of the muscle (fig. 204). Nuclei are present in the protoplasm of the end-plate.

(ii.) **Golgi's Method.**—Place the muscles of a newly-killed lizard for a minute or two in a .5 per cent. solution of arsenic acid, and directly afterwards in a solution of .5 per cent. solution of chloride of gold and potassium for 15-20 minutes, and reduce the tissue in sunlight in a 1 per cent. solution of arsenic acid. Instead of the above gold solution, use the following mixture, devised by Kühne :—

FIG. 203.—Distribution of Nerve-Fibres in the Frog's Sartorius. *A.* Upper, *B.* Lower end; *aa* and *bb.* Numerous fine branches; *P.* Pelvic end, and *K.* Knee end, with no nerve-fibres.

Arsenic acid (.5 per cent.)	60 cc.
Osmic acid (2 per cent.)	3 ,,
Chloride of gold and potassium (1 per cent.).	12 ,,

The tissue is then placed in 1 per cent. arsenic acid, and reduced by exposure to sunlight. The process may be greatly hastened by doing the reduction process at a temperature of 50° C., but it must be done in the direct rays of the sun. The pieces of tissue can be preserved in the following fluid, devised by Mays :—

Glycerine	60 cc.
Arsenic acid (1 per cent.)	10 ,,
Methylic alcohol	10 ,,
Water	20 ,,

In working with solutions of gold, do not use steel instruments. They must be either glass, platinum-iridium, or the substance known as "nickeline."

In birds, reptiles, and mammals the nerve-fibres terminate in "**end-plates**," which are disc-shaped bodies lying under the sarcolemma, *i.e.*, they are hypolemmal in position, 40-60 μ long and 40 μ broad. They consist of a finely granular protoplasm with nuclei. As the gold chloride stains only the axis-cylinder, one sees its branched arborescent terminations in the protoplasm (fig. 205). A good plan to see the unaltered end-plates is to examine the

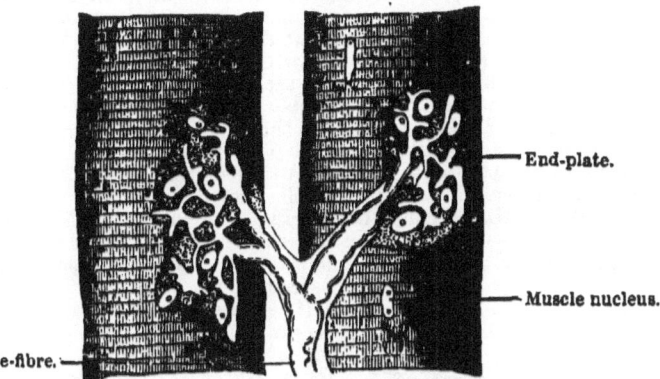

FIG. 204.—Termination of a Nerve-Fibre in End-Plate of a Lizard's Muscle.

muscle in a freshly-prepared 1 per cent. solution of sulphate of iron or ammonio-sulphate of iron (*Kühne, Mays*).

14. Frog's Motor Nerve-Endings.—There is no "end-plate," as in mammalian muscles, but the axis-cylinder splits up into bayonet-shaped branches under the sarcolemma. The myelin is continued up nearly to the sarcolemma, but as the nerve-fibre perforates the sarcolemma and becomes hypolemmal, the myelin stops, so that within the sarcolemma the branches are pale, and consist of branches of the axial-cylinder only. The gold method (p. 79) may be used. A simple method is to stain a small piece of a fresh muscle (*e.g.*, sartorius, near its middle, not at the ends) with Delafield's logwood. It stains the hypo-sarcolemmal nerve-terminations. The mode of termination of nerves in sensory surfaces will be found under Skin, Eye, and the sense-organs generally.

15. Terminations of Nerves in Tendon.—This is best shown by one of the gold-chloride methods. Perhaps the following by Manfredi is the best:—

Place the tendinous ends of the gemelli muscles of a rabbit (or the enucleated eyeball with its muscles attached) in

FIG. 205.—End-Plate of a Lizard's Muscle. One end-plate seen in profile, the other from above, with a nerve-fibre axis-cylinder terminating in it. Gold chloride, Golgi's method.

Potass bichromate (2 per cent.) . . for 3 days.
Acetic or arsenic acid (1 per cent.) . ,, 30 mins.
Gold chloride (1 per cent.) . . . ,, 30 ,,

Wash and leave exposed in sunlight in 1 per cent. arsenic acid until it assumes a violet-blue colour.

16. Pyriform Nerve-Cells (Frog).—These are most readily found in the ganglion of the vagus as it issues from the skull. Pith a frog, distend the œsophagus by pushing a small test-tube into it, place the frog on its belly, reflect the skin over the shoulder-blade, divide the trapezius and remove the fore-limb. The vagus will be seen coming out, along with the glosso-pharyngeal, through a large foramen immediately in front of the occipital condyle. Clear away the muscles from the region of this foramen, snip and excise a small part of the bone, so as to trace the nerve as far back as possible. Its greyish gelatinous semi-transparent ganglion is seen. Remove the nerve with the ganglion and place it for twenty minutes in 1 per cent. osmic acid and then stain it in picro-carmine. Tease a small piece in glycerine, and it is by no means difficult to find pyriform cells, each with a large nucleus, usually near the broad end of the cell. The protoplasm frequently contains large refractile granules. The straight process from the cells is readily seen, and with care the spiral process also can be seen.

The methylene-blue method (p. 192) shows very well the straight and the spiral fibre. Place the fresh tissue in the methylene fluid.

17. Isolated Cells of Sympathetic Ganglion (Mammal).—(*a.*) Tease a sympathetic ganglion—best done after maceration for 24 hours in weak acetic acid (2 drops to 100 cc. water). To isolate a cell showing its connection with a nerve-fibre requires much patience.

(*b.*) Place a small piece of the superior cervical ganglion of a rabbit in ½ per cent. osmic acid for 2-3 hours. Leave it in water for a day or two to macerate, and then tease it in diluted glycerine. A better plan is to make an interstitial injection of the osmic acid. It requires to be carefully teased to get isolated cells showing several nerve-fibres passing off from them.

(H) Note the spherical cell, with a nucleated capsule. It gives off many processes, each of which becomes continuous with a nerve-fibre, *i.e.*, it is *multipolar*. These cells in the rabbit usually contain two nuclei (fig. 206).

(*c.*) **Double Impregnation Method of Ramón y Cajal.**—It is better to use the sympathetic ganglia, *e.g.*, G. stellatum of embryos (dog, rabbit), or those of new-born animals. The fresh ganglion is placed in Golgi's bichromate-osmic acid fluid (3 days), then wash it with distilled water and afterwards with .75 per cent. silver nitrate. Leave it for 1-2 days in fresh silver nitrate (.75 per cent.). Wash it again, and place it again in osmico-bichromate mixture (4-4½ days), and then again in silver nitrate. Sometimes even a "treble impregnation" is useful. This process— "*intensivo*" or "impregnacion doble"—we owe to R. y Cajal, who has obtained good results with it, and so has L. Sala,[1] whose paper[1] contains figures of his results and a resumé of the literature. He finds that the cells are multipolar with a single unbranched process and numerous branched protoplasmic processes.

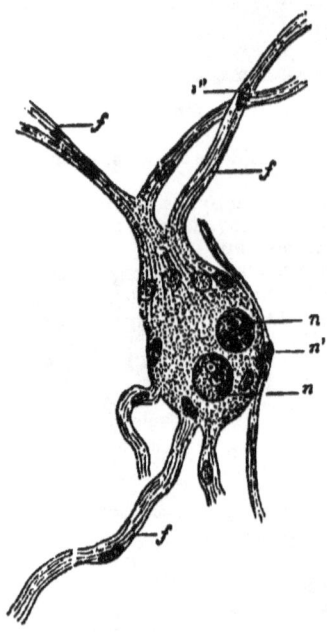

FIG. 206.—Isolated Nerve-Cell from Superior Sympathetic Ganglion of a Rabbit. *f.* Remak's fibres; *n'n'*. Nuclei of these fibres; *nn*. Nuclei of cell.

18. Methylene-Blue Method.—This is an admirable method, and depends on the fact that this substance stains blue the axis-cylinders or fibrils of nerves *in vivo*. It is applicable for studying the terminations of nerve-fibres in any tissue where they terminate, and also for the connection between nerve-fibres and nerve-cells. When injected into the blood-vessels of an animal, it

---

[1] *Archiv ital. de Biologie*, 1893, p. 439.

acts on the nerves, and on the latter being exposed to the air they become blue. It may also be applied to fresh tissues.

Inject some of the following solution into the blood-vessels :—

| Methylene-blue | . | . | . | 1 gram. |
| Normal saline | . | . | . | 300 cc. |

Or introduce a 3 per cent. solution, or even the solid substance, into the lymph-sac of a frog. After an hour or two, expose a muscle to the air, or use the cornea, or any other tissue with nerves in it, and on examining it under the microscope the nerves will be found stained blue. To preserve such specimens, mount them in a solution of picrate of ammonia and glycerine (p. 192). Fresh tissues, *e.g.*, cornea or a thin muscle, may be immersed in a weak solution of methylene-blue with the same result.[1]

The methylene-blue method may be used for the study of nerve terminations in any organ, *e.g.*, the sense organs, and in arteries one can see most beautifully the plexus of non-medullated fibres in the muscular coat.

19. Nerve-Cells of Crayfish.—Select a small individual and inject into its abdominal cavity 1 to 2 cc. of a 0.2 per cent. solution of methylene-blue. After 8 or 10 hours, remove the chitinous covering over the gangliated nerve-cord and expose the latter in a vessel, which admits air and yet prevents evaporation. In 24 hours or so excise a ganglion and observe it in a drop of glycerine tinged with picrate of ammonia (*Retzius*).

LESSON XIX.

THE HEART AND BLOOD-VESSELS.

Heart.—The wall of the heart consists of—(1.) **Pericardium**; (2.) **Myocardium**; (3.) **Endocardium**.

The **pericardium** covering the heart is a serous membrane composed of fibrous tissue, with numerous elastic fibres, and covered on its free surface by serous endothelium. It is sometimes called epicardium. The fibrous tissue is continuous with that which invests the bundles of muscles of the myocardium itself. Underneath the epicardium are the blood-vessels, nerves (ganglia), and the lymphatics.

The **myocardium** is composed of striated muscular fibres, whose characters have been described already (Lesson XVI. **12**). The fibres are arranged in bundles separated from each other by a greater or less amount of connective tissue, in which run the blood-vessels and nerves.

The **endocardium** in structure resembles the pericardium, but it is thinner. It consists of a fibrous basis, with elastic fibres covered by a single layer of endothelium. It contains a few smooth muscular fibres.

An **artery** consists of three coats :—

(1.) **Tunica intima**, or inner coat, composed of a single layer

---

[1] S. Mayer, *Zeitsch. f. wiss. Mikros.*, vi. 422, 1889, gives numerous references.

of endothelium resting on an **elastic lamina** composed of elastic networks, or an elastic membrane. In many arteries, however, there is a layer of connective tissue between the epithelium and the elastic lamina—the *sub-epithelial layer*.

(2.) **Tunica media**, or middle coat, consists of a varying number of layers of circularly-disposed, short, smooth, muscular fibres; but in most arteries, and chiefly in the large ones, it is intermixed with elastic fibres or laminæ.

(3.) **Tunica adventitia**, composed of fibrous tissue with elastic fibres, the latter especially numerous near the middle coat.

There are, however, great variations in structure in arteries, according to their size and other conditions.

Veins.—Speaking broadly, the veins have the same general structure as the arteries. They are, however, much thinner, and some of them have **valves**. They consist of three coats; the **inner coat** is thinner than in arteries, and the elastic lamina thinner

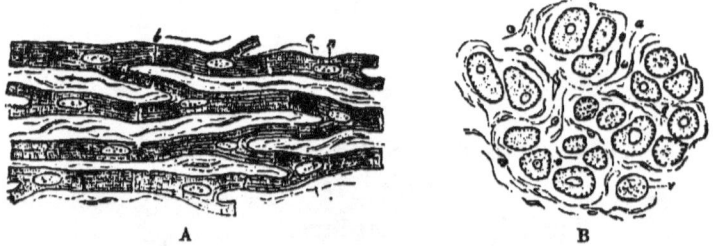

FIG. 207.—A. Fibres of the heart cut longitudinally; B. Transverse sections of the heart-fibres; *c.* Cell; *n.* Nucleus; *a.* Connective tissue; *v* Vein.

and often incomplete. The shape of the endothelial cells is different (fig. 213, V). The **middle coat** is also thinner, and has less muscular and elastic tissue, and relatively more connective tissue. The **outer coat** is relatively very strong, and is composed of fibrous tissue, which sends processes into the middle coat. There are, however, great variations in the structure of veins.

Capillaries.—They form networks of fine tubes of uniform diameter, sufficient to allow blood-corpuscles to pass along them freely in single file. The arrangement of the network varies in different tissues. The walls, when examined fresh, appear to be homogeneous, but they are composed of flattened epithelial cells or endothelial cells united to each other by their edges by cement substance, which is blackened by silver nitrate.

1. **Heart.**—Harden small pieces of the heart (human) in alcohol. Stain a small piece in bulk in borax-carmine, and then place it for twenty-four hours in acid alcohol. It is best to cut sections by the paraffin method and mount them in balsam. A very instructive

preparation is to make a transverse section across both ventricles of the heart of a small mammal, e.g., guinea-pig. This is best done in a heart stained in bulk.

Sections may be cut by means of a freezing microtome, and then stained with picro-carmine, and mounted either in Farrant's solution or balsam. In both cases it is well to include a section of the pericardium and also of the endocardium. Transverse sections of the papillary muscles are very instructive.

(*a.*) (**L**) Observe the branched and anastomosing fibres, but in addition some of them will be cut obliquely, and others transversely, with the nuclei stained (comp. p. 199).

(*b.*) (**H**) The faint transverse striation, short branches, absence of sarcolemma, the nucleus in the substance of the fibre, and the indistinct cement substance (fig. 207).

(*c.*) The fibrous character of the pericardium, which sends fine septa between the bundles of muscular fibres. If the pericardium be not included in the section, still connective tissue will be seen between the bundles of fibres, especially in transverse sections of these (fig. 207, B).

2. **Purkinje's Fibres** (**H**) occur under the endocardium in the heart of the sheep and some other animals. Open a ventricle of a sheep's heart, observe the network of fine glistening lines; strip off the endocardium, snip out a little piece of the heart-muscle, and place it for thirty-six hours in dilute alcohol or 5 per cent. ammonium chromate for two days. Tease a very small piece in picro-carmine, and mount in glycerine.

(*a.*) Search for isolated polygonal cells, each with usually two nuclei, and the edges only of the cells striated. These are heart-cells apparently arrested in the process of striation (fig. 208).

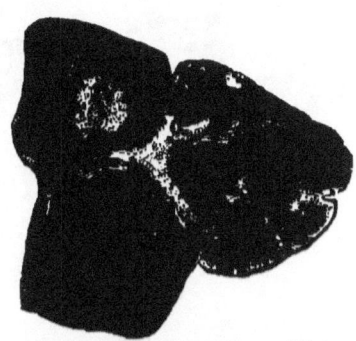

FIG. 208.—Purkinje's Fibres. Dilute alcohol, × 300.

3. **Endocardium** (**H**).—Harden a part of the ventricle of the human heart in alcohol or potassic bichromate. Make sections to include the endocardium.

(*a.*) Observe oval nuclei on the surface, the nuclei of the endothelial cells. Under this a superficial layer of fibrous tissue (fig. 209, *a*), with a few smooth muscle-cells (*ml*), and underneath this fibrous tissue the basis of the membrane (*tc*).

(*b.*) Outside this is the myocardium (*mc*).

4. **T.S. Heart-Valve** (**H**).—Harden a cusp of a human tricuspid

valve in chromic and spirit fluid (two weeks), make transverse sections, stain in logwood, and mount in balsam. Alcohol does well as a hardening reagent, and the sections can then be readily stained in picro-carmine.

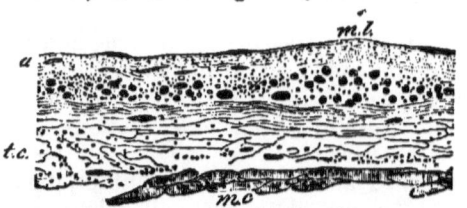

FIG. 209.—Endocardium of Left Ventricle (Human). a. Superficial layer; ml. Smooth muscular fibres; tc. Fasciculated fibrous tissue; mc. Muscle of heart, × 150.

(a.) On the surface, toward the auricle (A), note a superficial layer of lamellated connective tissue, which is covered with endothelium (fig. 210, a,) and underneath this a fibrous basis with elastic fibres (re).

(b.) If the section passes through the insertion of one of the chordæ tendinæ, it presents the appearance shown in fig. 210, ct.

5. Aorta.—(i.) Make transverse (and longitudinal) sections of the human aorta, or of that of an ox, which has been hardened in alcohol, or, preferably, in 2 per cent. potassic bichromate (ten days). Stain a section in picro-carmine, and mount it in glycerine.

(ii.) Another good method is to slit up any large artery, pin

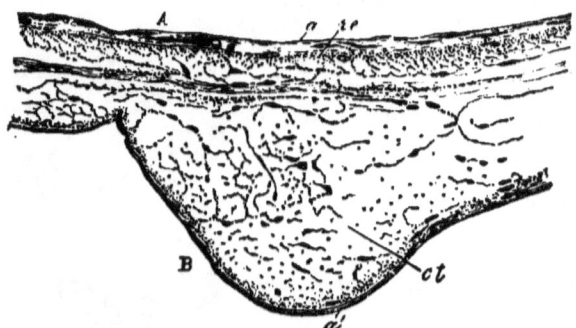

FIG. 210.—T.S. of the Cusp of the Human Tricuspid Valve, vertical to the axis of the cusp. A. Auricular, B. Ventricular surface; a. Superficial lamellated layer of the auricular, and a'. of the ventricular surface; ct. T.S. of one of the chordæ tendineæ, where it is inserted into the valve; re. Fibrous tissue, the basis of the valve, × 100.

it, inner surface upwards, upon wood. The pins must be close together to prevent too great shrinking of the tissue. Place it in a dry, well-aired place, say near a fire, so that it dries within a few hours. Make transverse sections with a sharp razor. This is best done by making a slit in a cork and clamping the dried membrane in the slit. Place the sections in water; they swell up greatly. Remove them, stain with picro-carmine, and mount in formic glycerine.

(a.) (L) Observe the subdivision into three coats (fig. 211), the tunica intima (inner), media (middle), and adventitia (outer).

(b.) (H) The **inner coat** is lined by a layer of squames, whose nuclei may be detected as slight oval swellings (not seen in the dried specimen). Under this several layers of yellow elastic membrane, with a small amount of pink-stained connective tissue between them. The outermost layer of elastic membrane is generally thicker than the others, and marks the outer limit of this coat.

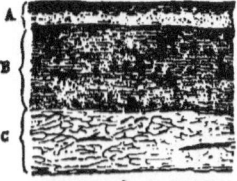

FIG. 211.—L.S. Human Thoracic Aorta. A. Internal, B. Middle, and C. External coat, × 20. Drying, picro-carmine, and acid glycerine.

(c.) The **middle coat**, composed also of numerous elastic laminæ, stained in this case bright yellow, and between them patches of smooth muscle with a somewhat brownish tint, and some connective tissue stained pink. The colour of these two tissues is quite distinct.

(d.) The **outer coat**, composed of white fibrous tissue (pink), some elastic laminæ, and a few smooth muscular fibres.

It is obvious, therefore, that elastic tissue enters largely into the structure of the larger arteries.

6. **Fenestrated Membrane of Henle.**—Tear off a thin lamella from the inner surface of a large artery, e.g., the aorta of a sheep. Irrigate it with acetic acid, or place it in 35 per cent. caustic potash, and mount it in Farrant's solution.

(H) Observe the elastic laminæ, some of them with holes in them (fig. 123). These laminæ tend to curl up at their edges. (See also Lesson X. 7.)

7. **T.S. Medium-Sized Artery**, e.g., the femoral artery of a child, prepared and stained as the aorta.

(a.) (L) Note the three coats.

(b.) (H) The **inner coat**, with its endothelium and **internal elastic lamina**. In many arteries a layer of sub-epithelial connective tissue lies between the endothelium and the elastic lamina. The elastic lamina is thrown into folds in an empty artery owing to the contraction of the middle coat (fig. 212).

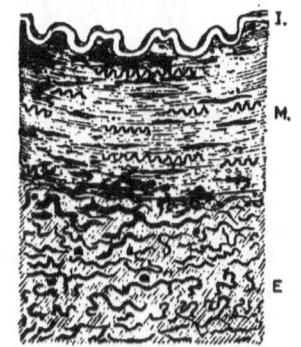

FIG. 212.—T.S. Artery. I. Tunica intima; M. Media; E. Externa.

(c.) The **middle coat** is composed of several layers of smooth muscle arranged circularly. In a Canada balsam preparation the nuclei stand out distinctly. Amongst the muscle-fibres are twisted elastic fibres (fig. 212).

(*d.*) The **outer coat**, chiefly composed of white fibrous tissue intermingled with elastic fibres, some of which are cut transversely. The elastic fibres are more numerous towards the inner part of this coat. Here and there a section of a blood-vessel may be seen.

8. **Endothelial Lining of Veins and Arteries.**—Cut open the external jugular vein of a rabbit just killed. Pin it to a piece of cork—inner surface uppermost—by means of hedgehog-spines. Wash the internal surface with distilled water, and then apply to it for five minutes ½ per cent. solution of silver nitrate. Wash off the silver, place the vein in water or alcohol and water, and expose it to light. Do the same with any large artery.

(**A.**) Snip out a small portion of the **vein**, and mount it in balsam, inner surface uppermost.

(**H**) Observe the "silver lines," indicating the existence of a layer of polygonal squames composing part of the inner coat. The

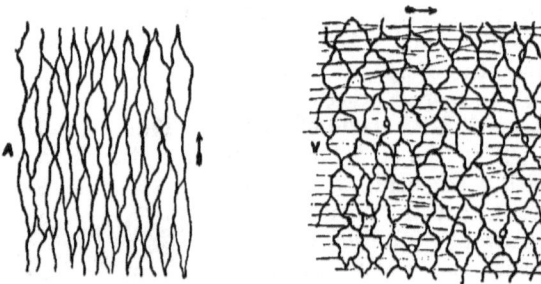

FIG. 213.—*A*. Epithelial lining of an artery of a calf, and *V*. of the jugular vein of a rabbit. The arrows show the direction of the blood-stream. Silver nitrate.

long axis of the squames lies across the long axis of the vein itself (fig. 213, V).

By focussing through the thickness of the wall, narrow fusiform areas, bounded by black lines, may be seen, indicating the existence of the smooth muscular fibres in the middle coat.

(**B.**) With a razor shave off a thin layer from the brown inner coat of the **artery**. Mount it in balsam.

(**H**) Observe the elongated lancet-shaped endothelial cells (fig. 213, A); the long axis of each cell in the long axis of the tube. The variation in the shape of the epithelial lining of vessels seems to have relation to the velocity of the blood-stream in these vessels.

9. **Pia Mater, Capillaries, Small Arteries, and Veins (H).**—Carefully remove a small piece of the pia mater from the brain of a sheep recently killed. Lay it on a glass plate, outer surface

THE HEART AND BLOOD-VESSELS.

lowest. With a camel's-hair pencil dipped in normal saline solution brush away all the brain matter, leaving a somewhat villous-looking surface. Cut out a small part of the membrane, and mount it in normal saline.

(*a.*) Find a **capillary**, note its diameter, uniform calibre, and oval nuclei bulging slightly into the lumen of the tube. The rest of the wall is homogeneous (fig. 214). Trace it backwards until the small artery or *arteriole* with which it is continuous is found.

(*b.*) A **small artery or arteriole** with a thin outer coat, and a middle coat composed of a single layer of smooth muscular fibres arranged circularly (fig. 216).

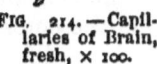

FIG. 214.—Capillaries of Brain, fresh, × 100.

(*c.*) Select a small **vein** which is somewhat thinner than the corresponding artery; the muscular coat is very imperfect. Irrigate with 2 per cent. acetic acid. Nuclei wherever present come distinctly into view.

(*d.*) The oval nuclei of the lining endothelium, the long axis of

FIG. 215.—*A*. A small artery with the lumen in focus. *B*. Small arteriole just before it passes into a capillary, × 300.

FIG. 216.—Small Artery from Human Brain, Pigment Granules in the Adventitia, × 150.

the nuclei in the long axis of the vessel. Outside this the **elastic lamina** appears as a somewhat refractive membrane with longitudinal folds.

(*e.*) The vessel crossed transversely by the long, oval nuclei of the muscular fibres of the middle coat (fig. 216). Note in some cases they are not distributed at equal distances, but in groups (fig. 215). One point must be particularly studied, viz., to focus through the thickness of a small vessel, and observe carefully that the appearance of the vessel varies with the position of the lens, *i.e.*, whether the upper surface, lumen, or deeper part of the artery is in focus.

(*f.*) In the outer coat the elongated fusiform nuclei of the connective tissue cells are arranged longitudinally.

For a permanent preparation, the pia mater after being removed is hardened in 2 per cent. potassic bichromate and preserved in alcohol. A thin piece is selected and stained with logwood or with eosin-logwood, and mounted in balsam.

10. T.S. Small Artery and Vein.—Select a small artery and vein, harden in Müller's fluid, stain in bulk with hæmateïn, and cut in paraffin.

(**H**) Observe the three coats in each, but they are much thinner in the vein than the artery. In the former the intima is very thin and the outer coat relatively thicker (fig. 217).

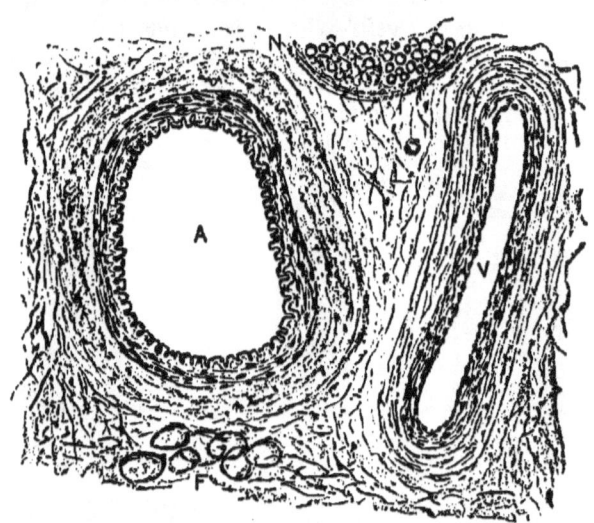

FIG. 217.—T.S. Small Artery and Vein. *A.* Artery; *V.* Vein; *N.* Nerve.

11. Injection of Silver Nitrate into Blood-Vessels.—(i.) From the aorta inject the blood-vessels of a **rabbit** with .25 per cent. silver nitrate. Before doing so wash out the blood-vessels with normal saline to remove all the blood, and then with distilled water. Slit up the intestine, wash out its contents, and expose it to light in alcohol and water. Scrape away the mucous membrane, leaving only the muscular coats. Dehydrate a small piece and mount it in balsam. It is easy to find large and small vessels as well as capillaries.

(*a.*) (**H**) Select a small artery, and note in it the endothelial lining (fig. 219, E) and the circular muscular fibres mapped out from each other by silver lines (*m*).

Arterioles and Small Arteries.—(*a.*) **(H)** Select a small artery or arteriole. Note the layers already described. If the circular muscular fibres be arranged in one layer, note that the fibres are arranged in alternate groups on opposite sides of the vessel (fig. 215, A).

(*b.*) Trace an arteriole into a capillary, and note the change in structure, especially how the muscle disappears (fig. 215, B).

(*c.*) Select a larger vessel, and in it observe the structure already described and shown in fig. 218.

(*d.*) Find **capillaries**, and note the endothelium of which they

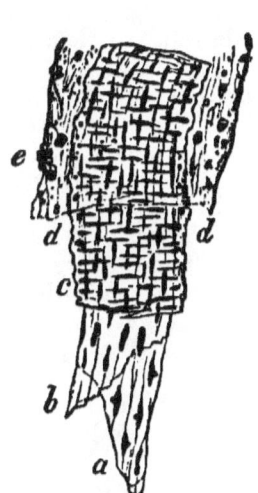

FIG. 218.—Middle-sized Artery of Brain. *a.* Endothelium; *b.* Fenestrated membrane; *c.* Middle or muscular coat; *d.* Adventitia; *e.* Pigment. × 300.

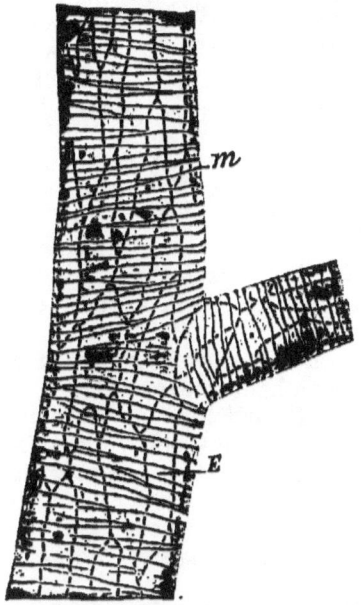

FIG. 219.—Arteriole of the Rabbit's Small Intestine. *E.* Endothelial cells of the intima; *m*, Circular muscular fibres; indicated by silver lines, × 200. Nitrate of silver.

are composed (fig. 220). In order to see the nuclei of these cells, the preparation should be stained with logwood.

(ii.) Instead of a rabbit, the blood-vessels of a **frog** may be injected in the same way from the aorta or ventricle. Use a glass syringe. Wash out the vessels first with normal saline, then with distilled water, and, finally, inject the silver nitrate solution. The bladder, intestine, and mesentery are particularly serviceable for obtaining small vessels and capillaries.

12. Circulation of Blood.—A day before the frog is required, let the brain of the animal be destroyed. An hour before the web is to be examined, place two or three drops of a .5 per cent. watery solution of curare in the lymph-sac under the skin of the animal's back. The drug should not act too rapidly. After a hour or so it paralyses the extremities of the motor nerves, and thus makes the frog motionless. A small dose of the drug is given in order not to affect the calibre of the blood-vessels.

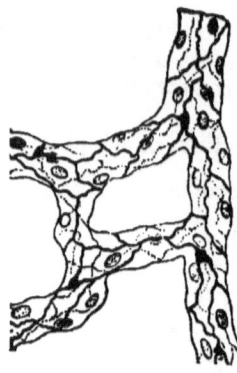

FIG. 220.—Capillaries Injected with Silver Nitrate.

Make a frog-plate by taking a piece of stout cardboard or thin slip of wood 15 cm. long (6 inches) and 5 . cm. (2 inches) broad. At one end of it cut a triangular slit whose base is 2 cm. or less in width. Tie a thread round the tip of, *e.g.*, the third and fourth toes of the hind-limb, place the frog on its belly on the board, and by means of the two threads gently stretch the web across the triangular slit. It must not be drawn too tightly. The threads can be fixed in slits made in the horns bounding the triangular aperture. To prevent evaporation from the frog, it had better be placed in a moist cotton rag or surrounded with moist blotting-paper. Moisten the web with a drop of water, and cover it with a narrow fragment of a cover-glass. In selecting a frog, choose a light-coloured one.

(*a.*) (L and H) Find an **artery**, and note that the blood flows from larger into smaller vessels with what appears to be considerable velocity. Contrast it with a **vein**, in which the blood flows in an opposite direction, *i.e.*, from smaller vessels—capillaries—to larger ones, but the current is slower in the veins than in the arteries. The walls of the vein are slightly thinner than those of the artery.

(*b.*) The **capillaries**, small and of uniform diameter, with the corpuscles moving in single file. The flow is uniform.

(*c.*) In the arteries and veins note the rapid red central stream, or axial zone of coloured corpuscles, and next the wall, on either side, the peripheral zone or space of Poiseuille, narrow and free from red corpuscles, but containing a few white ones rolling lazily along the vascular wall (**H**).

(*d.*) If a red corpuscle happen to be arrested at the bifurcation of a capillary, other corpuscles impinge on it and bend it. As soon, however, as it is dislodged it regains its shape, so that the red corpuscles are highly elastic.

(*e.*) Numerous pigment-cells, some of them contracted, others expanded, are visible (fig. 144).

If desired, the phenomena of **inflammation** can readily be studied by applying some irritant to the web, e.g., mustard or creosote (one minute).

ADDITIONAL EXERCISES.

13. Elastic Fibres in Arteries (*Martinotti*).—Harden the blood-vessel in chromic acid, make sections, and stain them with safranin as directed in Lesson X. 10. All the elastic fibres are purplish or black. **Herxheimer's Method** (p. 161) also yields good results.

14. Development of Blood-Vessels.—(1.) Harden the omentum of a newly-born rabbit in Flemming's fluid for twenty-four hours. Wash it thoroughly to remove all the hardening solution. Stain a piece for 24-36 hours in safranin. Remove the surplus stain in the usual way with acid alcohol. Mount in balsam.

(ii.) Kill a rabbit five days old with chloroform; do not bleed it. Open the abdomen, remove the stomach and spleen, and attached to them the omentum. Place all in a saturated watery solution of picric acid for one hour; wash away all the picric acid, cut out a small piece, stain it in logwood and then with eosin, or double stain it at once in eosin-hæmatoxylin. Mount in Farrant's solution.

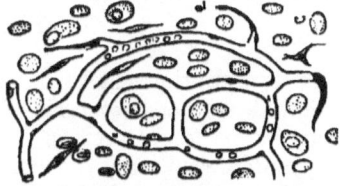

FIG. 221.—Capillaries and Developing Blood-Vessels from the Omentum of a New-Born Rabbit. Flemming's fluid and safranin.

(a.) (H) Search for a network of capillaries, which is easily found. Try to find one of them which gives off a long, narrow, blunt process. The process may be found partially channelled (fig. 221). By the union of two such processes, which ultimately become hollow, new capillary arches are formed. The blood-corpuscles in a preparation of (ii.) are stained with eosin.

15. Nerves and Nerve-Cells in a Frog's Heart.—Pith a frog, expose its heart, cut away the pericardium, divide the frænum which connects the posterior surface of the ventricle to the pericardium; raise the heart, find the sinus venosus, ligature the inferior and two superior venæ cavæ which open into the latter, make an incision into one of the aortæ, and into it tie a fine glass cannula. Inject normal saline so as to wash out the cavities of the heart. Distend the heart-cavities with the following mixture:—Four parts of gold chloride (2 per cent.) and one of formic acid boiled together and allowed to cool. Ligature the other aorta, so as to get the heart-cavities fully distended. Place a ligature below the cannula, cut out the heart and place it for $\frac{1}{2}$-1 hour in 5 cc. of the gold mixture. Open the auricles, wash out the heart in water, and expose it to light in distilled water—50 cc.—containing three drops of acetic acid. Reduction of the gold takes place slowly in 3-4 days. Cut out the auricular septum and examine it in glycerine.

Pyriform nerve-cells, each with a straight and a spiral process, will be found along the course of the nerves in and near the auricular septum.

The nerve-fibres in the auricular septum are readily found by using instead of the gold a .2 per cent. solution of osmic acid.

The **Methylene-blue** method yields excellent results, if the fresh auricular septum be placed in a weak solution of the blue, and then mounted in ammonium picrate glycerine (p. 192).

16. Circulation in the Tongue of Frog.—Destroy the brain of a frog, and after a time curarise it and fix it on its back on a frog-plate of cork with a hole cut in it just in front of the head of the animal, the hole corresponding in size to that of the tongue. The tongue is attached in front, and it can thus readily be pulled out of the mouth so as to display its blood-vessels from the under-surface. Pin out the tongue over the hole in the cork, when the under-surface of the tongue will be uppermost. Fix the tongue under the microscope and examine it (L). One can study the circulation in its vessels, and also observe nerve-fibres in their normal condition.

LESSON XX.

THE LYMPHATIC SYSTEM—SPLEEN—TONSILS—THYMUS GLAND.

THE **lymphatic vessels** have thin translucent walls, and in all essential respects resemble veins in structure. There are three coats in the larger vessels. The muscular fibres are abundant in the middle coat, and the epithelium of the inner coat is in some situations sinuous in outline. **Valves** are numerous in some situations. The larger vessels spring from so-called **lymph capillaries**, which are usually wider than blood-vessels, and they unite with each other and form an irregular plexus. They often present dilations and constrictions, and consist of a single layer of endothelium with sinuous outlines. They are without valves, and open into the smallest regular lymphatic vessels.

Lymphatic Glands.—These vary much in shape, size, and colour, but they are frequently oval or kidney-shaped. At one part is a depression—the **hilum**—where the medullary part of the gland comes to the surface, and where the blood-vessels enter the gland and the efferent lymphatic vessel leaves it. There is usually only one efferent vessel and several afferent lymphatics; the latter perforate the capsule and enter the gland on its convex side. On making a section of a gland, with the naked eye one can see that it is divided into a **cortical** and a **medullary part**. The gland is invested by a fibrous **capsule**, which in some animals contains smooth muscular fibres. It consists of two layers, the outer of coarser and the inner of finer connective tissue. It sends somewhat flattened, large, usually unbranched, septa or **trabeculæ** into the **cortex**, thus dividing it into a series of compartments or **alveoli**

(fig. 222). These trabeculæ are continued into the medulla, where they branch, become finer, and anastomose to form an irregular network. The lymphoid tissue lies in the meshes of this trabecular framework, but everywhere separated from it by a **lymph sinus** or **lymph channel.**

In the **cortex** the alveoli are arranged in a regular manner, and the greater part of each alveolus is occupied by a mass of adenoid tissue crowded with leucocytes; but this **follicular substance** is everywhere separated from the capsule and trabeculæ by a **lymph sinus**, traversed by a network of fine fibrils with flattened cells lying on them at the points of intersection. The network is coarser than that of the adenoid tissue, and it contains a few leucocytes. It, as well as the trabeculæ, is covered by a layer of sinuous endothelium. The medulla is also occupied by adenoid tissue crowded with leucocytes, but the lymphoid tissue forms

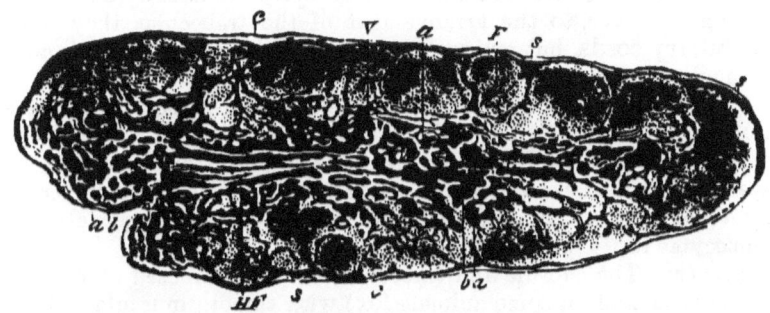

FIG. 222.—L.S. Cervical lymph glands of Dog. *c.* Capsule ; *s.* Lymph sinus ; *F.* Follicle ; *a.* Medullary cord ; *b.* Lymph paths of the medulla ; *V.* Section of a blood-vessel ; *HF.* Fibrous part at the hilum, × 10.

branching or anastomosing cords—**medullary cords**—each one surrounded by its lymph sinus. The medullary cords are continuous with the follicular substance in the cortical alveoli. Near the centre of the lymphoid tissue of the alveoli are clearer areas, the **lymph-knots** or **germ-centres**. In them mitosis goes on rapidly, but it requires a high power to discern the stages of the division of the nuclei. The lymph sinuses are continuous throughout the gland, and they are the channels through which the lymph moves. Some glands are pigmented.

LYMPHATIC GLANDS.

1. Lymphatic Glands.—Harden in 5 per cent. ammonium bichromate or alcohol the lymphatic glands of the mesentery of a cat or calf, or those found in the neck or under the **lower jaw** of a cat. For the germ-centres harden a small gland in Flemming's fluid and stain with safranin. Make sections including both poles of the gland and the hilum. Stain them with logwood and then with eosin. Mount in balsam.

(*a.*) (**L**) Observe the **capsule** (fig. 222, *c*) surrounding the gland, and sending at fairly regular intervals **septa** or **trabeculæ** into the substance of the gland. The trabeculæ and capsule are stained by the eosin. The **trabeculæ** are flattened in the outer part or **cortex**, and divide it into compartments—**follicles** or **alveoli** (F). The trabeculæ are continued into the central part or **medulla**, where they form a network of smaller, branched, more rounded trabeculæ.

(*b.*) The compartments in the cortex are nearly filled by leucocytes lying in a meshwork of adenoid tissue constituting the **follicles** (cortical nodules) of the cortex (fig. 222, F). Between the trabeculæ of the medulla the leucocytes are equally abundant. Owing, however, to the arrangement of the trabeculæ, they form **medullary cords**, but everywhere the adenoid tissue is continuous throughout the gland.

(*c.*) The **lymph channels** (*s*) exist between the capsule and the follicles, and between the trabeculæ and the lymphoid tissue of the cortex and medulla (figs. 222, 223). They form an anastomosing system of paths or channels throughout the gland, and are traversed by a fine network of coarse adenoid tissue with comparatively few leucocytes in its meshes (fig. 223).

(*d.*) (**H**) The capsule and trabeculæ, chiefly composed of connective tissue, and in some animals (ox) with smooth muscular fibres. Continuous with its under surface is a delicate network of adenoid tissue, which stretches across the lymph channel to the follicle, where it becomes continuous with the adenoid tissue supporting the leucocytes (fig. 223). Between the trabeculæ and the medullary cords in the cortex similar lymph paths with an adenoid network.

(*e.*) The follicles and medullary cords, everywhere crowded with leucocytes.

2. **Adenoid Reticulum.**—(i.) As this is largely obscured by the presence of the leucocytes, these must be got rid of. This is readily done by making sections of the lymph gland of an ox, hardened for two or three days in 5 per cent. ammonium bichromate, and then shaking the sections in a test-tube containing water, or the leucocytes may be "pencilled" out by a camel's-hair pencil.

(ii.) A better method, perhaps, is to inject by the puncture method (p. 237) a $\frac{1}{4}$ per cent. solution of silver nitrate into the lymph gland of an ox. The fluid being driven in forcibly, causes an artificial œdema and forces the parts asunder. Harden in alcohol, stain in logwood and eosin, and mount in balsam. The reticulum in the ox has a brownish appearance from the deposition of a brownish pigment. Sinuous outlines of endothelial cells may be seen on the trabeculæ, and branched cells of the lymph sinus (fig. 223), and even the follicular substance. It is continuous with the endothelial lining of the lymphatic vessels.

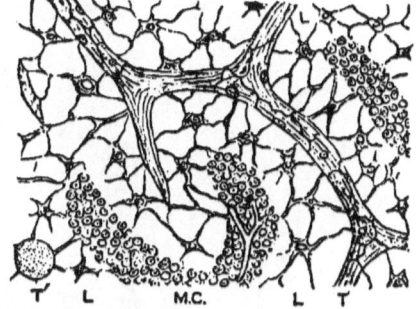

FIG. 223.—Lymph Sinuses from the Medulla of a Lymphatic Gland. *T'* and *T*. Trabeculæ; *L*. Lymph-path; *M.C.* Medullary cord. Silver nitrate and hæmatoxylin, × 300.

3. **Injection of the Lymph Channels.**—Fill a hypodermic syringe with a watery solution of Berlin blue, force the nozzle of the syringe into a small lymph gland of an ox, and inject the blue fluid haphazard into the gland. The blue passes into the lymph channels. Harden in alcohol. Make sections by freezing, stain them with picro-carmine, and mount in Farrant's solution.

(L) The channels are filled with a blue mass, while the leucocytes are red and the septa yellowish-red. The blue mass lies under the capsule, and in the lymph paths around the trabecula. Notice the difference in the distribution of the blue mass in the cortical and medullary parts of the gland. If it be desired to study the endothelium covering the trabeculæ, inject the gland as above, but with $\frac{1}{10}$ per cent. silver nitrate, and harden in alcohol.

4. **Central Tendon of Diaphragm.**—Lave the central tendon of the diaphragm of a newly-killed rabbit in distilled water. Place it for an hour in .2 per cent. silver nitrate in a dark place. Remove it, wash again, and place it for twenty-four hours in water containing a little alcoholic solution of thymol (10 per cent.), or a drop of carbolic acid to prevent the formation of fungi. Maceration in water enables the endothelium on the surface to be readily pencilled off. Mount it in balsam.

(L) Examine the pleural surface with the naked eye or with a lens, and a plexus of lymphatic vessels—clear on a dark-brown ground—will be seen (fig. 224). The vessels anastomose, and lead into narrow vessels, which run more or less parallel to each other, and correspond to inter-tendinous spaces.

(H) Observe the dilations and constrictions in the finer lymphatic vessels, and their sinuous epithelium (fig. 225, L), and the communications between the cell-spaces and the lymphatics. (Lesson XL 12.)

238   PRACTICAL HISTOLOGY.   [XX.

5. Septum Cisternæ Magnæ Lymphaticæ.—Open the abdomen

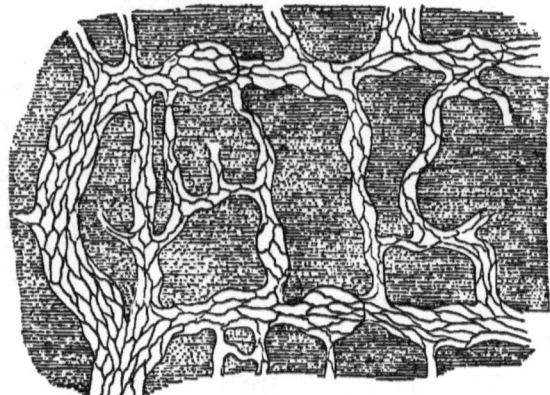

FIG. 224.—Lymphatic Vessels of the Central Tendon of the Diaphragm of a Rabbit, the lining endothelium shown. Silver nitrate.

of a newly-killed frog and remove the intestines. Turn the frog

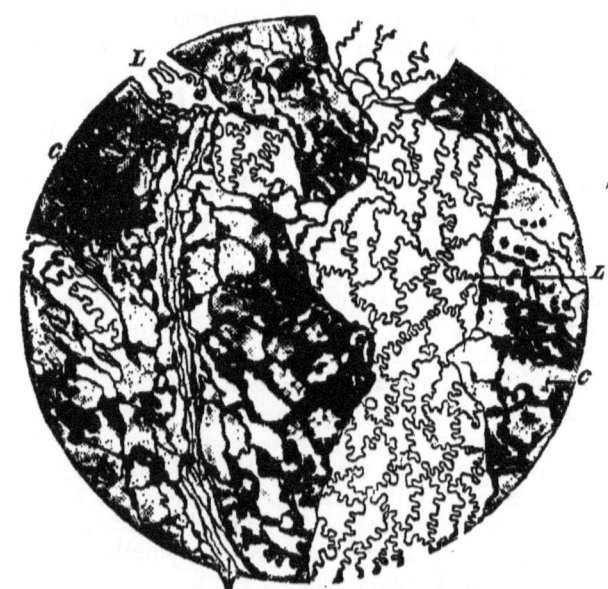

FIG. 225.—Pleural Surface of the Central Tendon of Diaphragm of Rabbit. *L.* Lymphatic vessel lined with sinuous endothelium; *c.* Cell-spaces of connective tissue. Silver nitrate, × 110.

on its belly. Cut through the vertebral column at its lower end, raise it, and cut out a square window—about ¾ inch in diameter, including the whole of the posterior body-wall. This exposes the dorsal surface of a thin membrane, the septum of the great lymph-sac. It is attached to the kidneys on each side. Pin the membrane, by means of hedgehog-spines, to a thin ring of cork with a hole in it. Wash the membrane with distilled water and place it for ten minutes in .5 per cent. silver nitrate solution. Wash it—still on its ring of cork—in distilled water, and expose it to light. After it has become brown, cut it into pieces, and mount them in Farrant's solution, one with the peritoneal surface uppermost, the other with its dorsal surface uppermost.

A. (H) **Peritoneal Surface.**—Observe the slightly sinuous "silver lines," indicating the existence of a single layer of endothelium. Here and there small **stomata** or openings, which lead from the peritoneal cavity to the great lymph-sac (fig. 226). The stomata may be closed or open, and are recognised by their brownish appearance. They are surrounded by a few finely-granular brown-stained cells—**germinating epithelium**. The pointed angles of several of the larger endothelial cells radiate from these apertures.

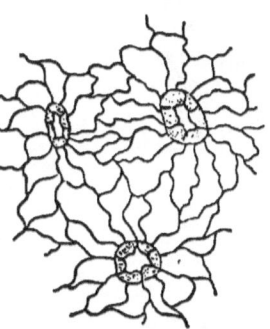

FIG. 226.—Endothelium and Stomata of the Peritoneal Surface of the Septum Cisternæ Lymphaticæ Magnæ of the Frog, × 200.

Focus through the thickness of the membrane, and note its fibrous character. On the deeper surface another layer of endothelial cells comes into view. They are more sinuous and narrower than those on the peritoneal surface.

B. **The Dorsal or Cisternal Surface.**—Observe the more sinuous, polygonal, and broader endothelial cells covering this surface of the membrane; and the other openings of the stomata.

The *serous cavities communicate by means of stomata with the lymphatic system*. Stomata occur on the pleura, under-surface of the diaphragm, and mesentery.

THE TONSILS.

6. **Tonsils.**—Use those of rabbit or cat, as the human tonsil does not give such good results. Fix them in Kleinenberg's fluid (12 hours), and then harden in gradually-increasing strengths of alcohol. Or use mercuric chloride or Flemming's fluid (3–6 hours). Make sections by the paraffin method, and stain them

with logwood or logwood and eosin, and mount in balsam. Sections may be stained first in acid-hæmatoxylin and then counter stain them with eosin. For the lymph-knots or germ-centres stain with safranin and counter stain with picric acid.

For fixing the sections on a slide, Gulland [1] uses a modification of the capillary-attraction method of Gaule. The sections in paraffin are placed in warm water as recommended by Gaskell.[2] Float the sections on to a slide, pour off the surplus water, and expose the slide for several hours to a temperature under that of the melting-point of the embedding paraffin. When dry remove the paraffin by xylol and then stain the sections. They adhere to the slide by capillary attraction.

(L) Observe on the surface the stratified squamous epithelium, and under it numerous round or oval aggregations of adenoid tissue (fig. 227). These form but imperfect nodules. Pit-like recesses are seen, lined by stratified epithelium, and into them mucous glands sometimes open.

(H) Trace some of the leucocytes of the adenoid tissue upwards

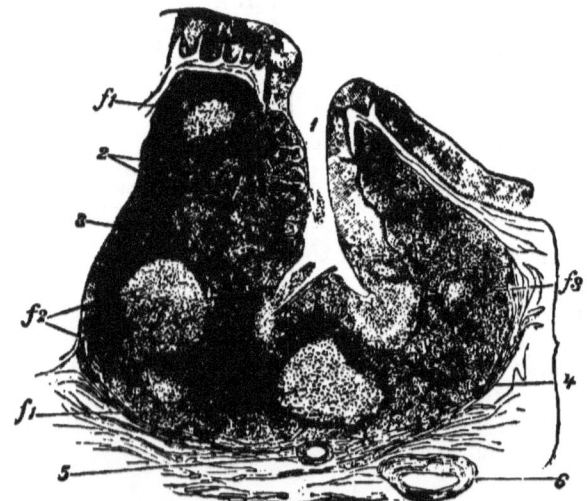

FIG. 227.—Single Follicle of Tonsil, × 20. 1. Cavity of follicle; 2. Epithelium infiltrated with leucocytes; 3. Adenoid tissue; f_1, 2, and 3. Follicles cut in various directions, f_1 with a lymph-knot; 4. Fibrous sheath; 5. Section of duct of mucous gland; 6. Blood-vessel.

between the epithelial cells, so that the epithelial layer is at places infiltrated with leucocytes.

[1] *Jour. of Anat. and Phys.*, xxvi., 1891, p. 56.
[2] *Quart. Jour. Micros. Sci.*, xxxi., 1891, p. 382.

THE THYMUS GLAND.

This gland is very large in the embryo and infant, but it begins a retrograde development about the sixth year, and is eventually replaced by fat and connective tissue. In the rabbit it retains its structure. It is composed of a number of **lobes**, and these again of smaller **lobules**. A **capsule** composed of connective tissue holds all together, and sends in septa—carrying blood-vessels and lymphatics—between the lobes and lobules, and also fine prolongations into the interior of the latter. There are no smooth muscular fibres in the septa. Each lobule consists of a **cortical** and a **medullary** part. Within each lobule is a delicate network of reticular connective tissue, finer and more like adenoid tissue in the medulla. It appears to consist of branching cells, and is coarser in the cortical part. The meshes are crowded with leucocytes, which, however, are most abundant in the cortex. The medullary substance contains the **concentric corpuscles**, like nests of concentrically-arranged flattened epithelial cells (fig. 228). The blood-vessels run along the septa and form a capillary plexus within the lobules.

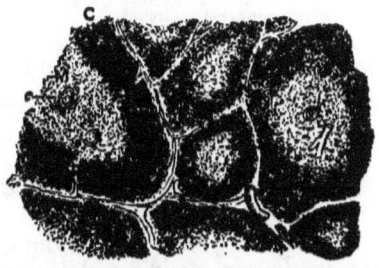

FIG. 228—Section of a Few Lobules of a Child's Thymus. *C.* Cortical, *M.* Medullary part; *c.* Concentric corpuscles, × 20.

7. Thymus Gland.—Harden the thymus of a young animal or child in Müller's fluid (three weeks), and then in gradually-increasing strengths of alcohol. Sections stained with logwood are mounted in balsam.

(*a.*) (**L**) Observe the **capsule** sending septa between the larger **lobules**, and finer septa into the lobules, thus subdividing them into smaller secondary lobules. Each such small lobule is about 1 mm. in diameter, and as one is exactly like the others, it suffices to study one.

(*b.*) A darker, denser peripheral zone, the **cortex**, and a more open light central part or **medulla**, the former surrounding the latter (fig. 228).

(*c.*) (**H**) The **septa** consist of fibrous tissue with some elastic fibres, with numerous blood-vessels and slits; the latter are the lymphatics. The **lobule** consists of adenoid tissue—the mesh-work not visible because it is crowded with **leucocytes**.

(*d.*) A variable number of concentrically-striated bodies, con-

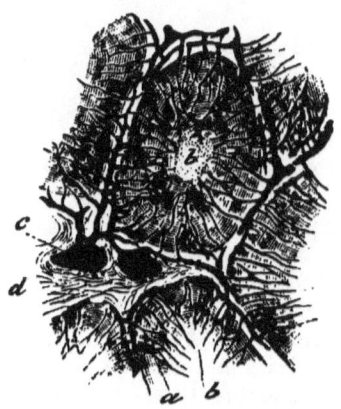

FIG. 229.—Injected Lobules of Thymus of a Cat. *a.* Cortex; *b.* Medulla; *c.* Blood-vessels; *d.* Septum of connective tissue.

FIG. 230.—Elements of the Thymus Gland. *a.* Leucocytes; *b.* Concentric corpuscle, × 300.

centric corpuscles or Hassall's corpuscles. Sections of capillaries amongst the adenoid tissue (fig. 230).

THE SPLEEN.

The **spleen**, like the thymus, thyroid, and some other glands, is a "ductless gland," and is invested by a fibrous **capsule**. The capsule consists of an outer layer of connective tissue covered by endothelium—the serous or peritoneal covering—and a deeper layer of connective tissue with networks of elastic fibres, and in some animals (dog, cat, pig) smooth muscular fibres. From the deeper surface of the latter flattened or rounded **trabeculæ** pass into the organ, and as they do so they branch and anastomose, thus forming a spongy meshwork with a labyrinth of communicating spaces. These spaces are filled with a reddish-purple soft substance, the **splenic pulp**. The blood-vessels are ensheathed by connective tissue, to which the trabeculæ are attached. The red colour of the pulp is due to the large number of blood-corpuscles. The trabecular framework is continuous with the connective tissue entering the organ along with and covering the blood-vessels—"adventitial sheath"—at the hilum of the organ. In the splenic pulp are small spherical whitish bodies—**Malpighian corpuscles** (0.2–0.7 mm.). The Malpighian corpuscles are small groups of leucocytes developed here and there in the adventitia of the splenic artery.

In the guinea-pig the lymphoid mass forms almost a continuous covering.

The Malpighian corpuscles occur chiefly at the bifurcations of the artery, so that the artery perforates them usually at one side, and thus the mass is arranged in a lob-sided manner on the artery. In structure they resemble the follicular substance of lymphatic glands, and in the centre of some of them is a lymph-knot, in which mitosis occurs.

The **splenic pulp** consists of a mesh-work composed of branched cells with membranous expansions, and the processes of neighbouring cells anastomose to form a fine reticulum, which occupies the irregular chambers of the trabecular framework, and with which it is continuous. The meshes of this fine network are occupied by cellular elements in considerable variety. This reticulum is permeated by blood-corpuscles—in fact, the blood stream of the spleen seems to bear the same relation to this reticulum that the lymph stream bears to a lymph gland. Besides red blood-corpuscles there are larger cells called splenic cells; some of these often contain degenerated blood-corpuscles or pigment. There are also leucocytes (fig. 232). The **arteries** enter the spleen at the hilum, and run for a short part of their course in the trabeculæ,

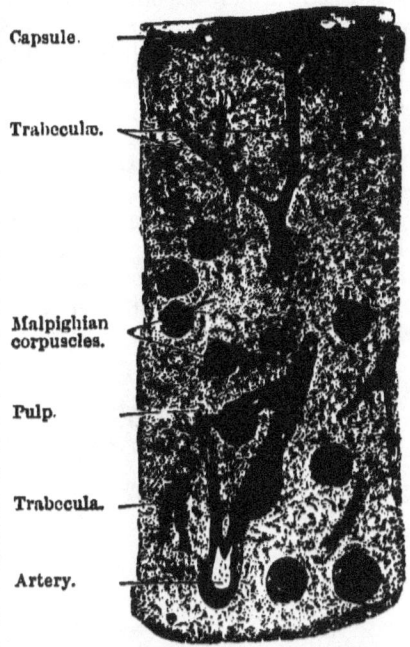

FIG. 231.—T.S. Part of the Human Spleen. Müller's fluid, hæmatoxylin, × 10.

which they soon leave, enter the reticulum, and break up into pencils or groups of small arteries. Some of these open into true capillaries in the Malpighian corpuscles; other fine branches open directly into the reticulum. The endothelial lining, instead of forming a continuous membrane, leaves apertures between the cells, through which the blood escapes. The **veins** arise from the spaces of the reticulum, and rapidly pass into the trabeculæ, in which they are firmly fixed. Near the hilum, and for part of their course, the arteries and veins lie in a "common sheath," with the corresponding nerves. The vein is always much wider than the corresponding artery. The lymphatics are not very numerous. For the nerves see Lesson XVII. 7.

8. Spleen.—Tie the blood-vessels at the hilum of the spleen of a cat, so as to keep the blood in the spleen. Cut it out and place it in a large volume of Müller's fluid (two weeks) or 2 per cent. bichromate of potash. Wash it thoroughly in running water for an hour or two. Cut out small pieces and harden them in alcohol. Make transverse sections, *i.e.*, across the long axis of the organ; stain one in logwood, and mount in balsam. Other sections are to be stained in picro-carmine and mounted in Farrant's solution, and a thin set placed in 1 per cent. osmic acid (twenty-four hours), and mounted in Farrant's solution. This sharpens the outlines of the elements. Stain other sections in eosin-logwood or safranin (forty-eight hours).

9. T.S. Spleen (L) (*Cat; hæmatoxylin* or *eosin* and *hæmatoxylin*). —The Malpighian corpuscles are visible as small blue spots to the naked eye.

(*a.*) Externally the **capsule**, fibrous, thick, firmly adherent and closely applied to the organ, sends **trabeculæ** into the spleen, where they branch and anastomose to form a trabecular framework. Some of them will be cut longitudinally, others obliquely, and some transversely. In the larger trabeculæ, sections of large blood-vessels

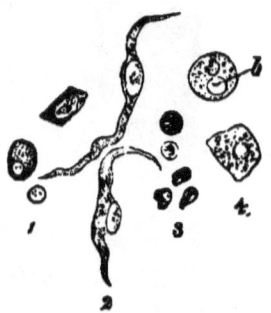

FIG. 232.—Elements of Human Splenic Pulp. 1. Leucocytes; 2. Epithelial cells; 3. Coloured blood-corpuscles; 4. Cells containing pigment-granules.

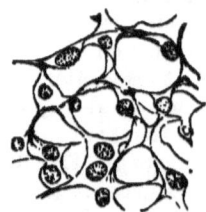

FIG. 233.—Reticulum of the Splenic Pulp.

(fig. 231). Note that there is no lymph space between the capsule and the gland substance, as is the case in lymph glands.

(*b.*) Filling the interstices of this network, the **splenic pulp**, and in it oval or rounded bodies—**Malpighian** or **splenic corpuscles**— as blue-stained bodies contrasting with the yellowish brown pulp in which they lie. In the centre of each is a lighter area, the "**germ-centre**" of Flemming. In each corpuscle a section of a small artery lying excentrically in the mass. The splenic corpuscles are small lob-sided aggregations of lymphoid tissue around branches of the splenic artery. They are relatively more numerous than in the human spleen. The track of the blood in the pulp is mapped out by the yellow blood-corpuscles.

(c.) (H) The capsule and trabeculæ, composed of fibrous tissue, with elastic fibres and smooth muscle. The Malpighian corpuscles, consisting of leucocytes in an adenoid reticulum. The centre is lighter in tint than the circumference, which is more condensed. The lighter centre is due to the larger cells present there. They are undergoing proliferation. The cells formed in the splenic corpuscles pass into the spaces of the pulp and leave the organ by the venous blood stream.

(d.) In the pulp irregular rows of coloured blood-corpuscles—yellow—and between these leucocytes and other cells.

The exact structure of the pulp can only be properly studied in a section which is very thin, and especially at the edges of the section, or best of all in a section of a dog's (or cat's) spleen, whose blood-vessels have been washed out, and cleared of all blood-corpuscles by a warm stream of normal saline solution. The vessels are then injected with a 5 per cent. solution of ammonium bichromate, and the organ hardened in a large quantity of the same fluid, and subsequently in alcohol.

(e.) In a section prepared in this way, or at the edges of a very thin section, the fine reticulum of branching cells may be seen (fig. 233) with the cells of the splenic pulp washed out of it.

10. Human Spleen.—Harden this in the same way as **8**. Note that, as a rule, the Malpighian corpuscles are less numerous. In other respects the general structure is the same.

11. Injected Spleen.—It is very difficult to inject the finer splenic blood-vessels. They should be washed out first with normal saline, and preferably a watery solution of Berlin blue, or Berlin blue with gelatine, should be used as the injection. Note that an artery and capillaries exist in the Malpighian corpuscles, but the splenic pulp seems to be infiltrated with a blue mass. The capillaries open into this system of labyrinthine blood-passages. These intermediate blood-passages are merely the spaces amongst the cells of pulp and are not lined by epithelium. The terminations of the capillaries in some situations are surrounded by thick sheaths or collars of tissue, perhaps derived from the cells of the pulp (*Bannwarth*).[1]

12. The **varieties of Leucocytes** in lymph glands are best studied by fixing the gland in $HgCl_2$ and staining sections with Ehrlich-Biondi fluid (*Hoyer*).[2] There are at least four varieties, not including phagocytes.

---

[1] *Archiv f. mik. Anat.*, xxxviii. p. 345.   [2] *Ibid.*, xxxiv.

LESSON XXI.

TONGUE—TASTE-BUDS—SOFT PALATE.

TONGUE.

PLACE small portions in Müller's fluid or 2 per cent. potassic bichromate for fourteen days, and complete the hardening in alcohol, or harden it in mercuric chloride. Make vertical transverse sections. It is well to have the tongue of a small cat or kitten, and parts of the human tongue also—the former because a complete transverse section can be put on a slide. The structure will vary according as the section is made through the anterior or posterior part of the organ, as the latter contains many lymph follicles and mucous glands. The sections may be stained in logwood and mounted in balsam.

1. **T.S. Tongue of Cat.**—(*a.*) (**L**) Observe the **papillæ**, of various shapes, on the dorsum of the tongue, and covered by stratified epithelium. Under this the connective tissue of the mucous membrane (fig. 234).

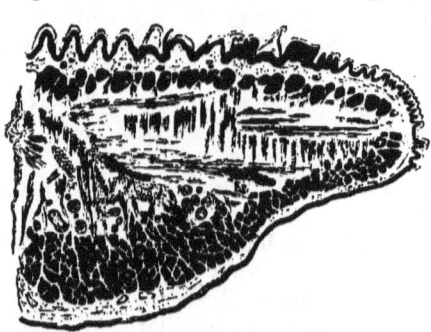

FIG. 234.—T S. of One-half of the Tongue of a Cat.

(*b.*) **Muscular Fibres.**—Many cut transversely and arranged in groups under the dorsal mucous membrane and elsewhere; others which run from the vertical mesial plane or *septum* horizontally outwards, and some which pass vertically. The last may be seen to become conical and end in the connective tissue of the mucous membrane. Some of these fibres branch. (The methods of isolating branched fibres are referred to in Lesson XVI. **3.**)

(*c.*) **Lingual Papillæ.**—The dorsum of the tongue is beset with elevations of the mucous membrane covered by stratified epithelium, and constituting three varieties of papillæ.

(1.) **Filiform** (.7-3 mm long), by far the most numerous, and are placed all over the dorsum. They are conical elevations of the mucous membrane, the upper end of which is beset with five, fifteen, or thirty secondary papillæ. Each papilla is composed of fibrous tissue with elastic fibres, and covered by many layers of stratified epithelium, the superficial cells of which are often corneous (fig. 235).

(2.) **Fungiform** (0.5–1.5 mm. long), are not nearly so numerous as the foregoing, and are scattered over the dorsum. Each papilla is club-shaped or lenticular, with a constricted base. The apex is beset with secondary papillæ, but the epithelium covering them is thinner than in (1) (fig. 236).

(3.) **Circumvallate** (1–1.5 mm. high and 1–3 mm. broad), are confined to the posterior part of the tongue, where they (8–15 in number) are arranged in the form of a V, the apex of the V being directed backwards. Each circular elevation is raised above the level of the tongue and surrounded by a circular trench or fossa. Secondary papillæ occur only on their surface. Taste-bulbs occur in the wall of the papilla directed toward the fossa. They are the organs of taste, and are supplied by the glosso-pharyngeal nerve.

FIG. 235.—Filiform Papillæ, × 30. 1. Primary papilla; 2. Secondary papillæ on its summit; 3. Epithelial process on papilla; 4. Single process, with entangled loose epithelial cells.

It may require several sections to obtain views of all three forms of papillæ.

(*d.*) (H) The *stratified squamous* epithelium covering the papillæ and sides of the tongue. The superficial cells are very thin.

(*e.*) *Glands*, in the back part mucous glands with clear contents, and it may be also serous glands in which the acini are more granular (fig. 237).

(*f.*) *Fat-cells*, like padding between the striped muscular fibres here and there. Near

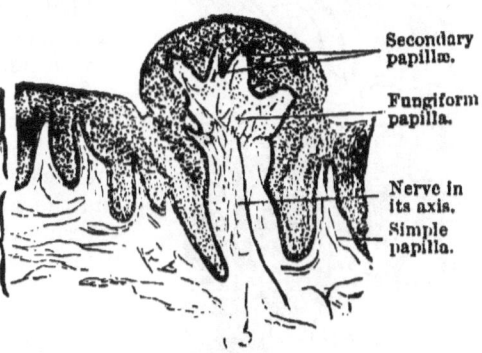

FIG. 236.—Fungiform Papilla, × 30.

the lower surface, sections of the lingual artery and nerves. In the latter *ganglionic cells* may sometimes be seen.

At the back part of the tongue are little depressions of the mucous membrane called **crypts** (fig. 237). In the walls of these are spherical masses of adenoid tissue, and into some crypts open the ducts of small mucous glands (fig. 237).

2. The **mucous glands** occur chiefly at the base of the tongue and along its edges. They have the same structure as the salivary glands of the same name (fig. 237), *i.e.*, their acini are lined by a single layer of mucous cells, but there are no demilunes. They are small compound tubular mucous glands.

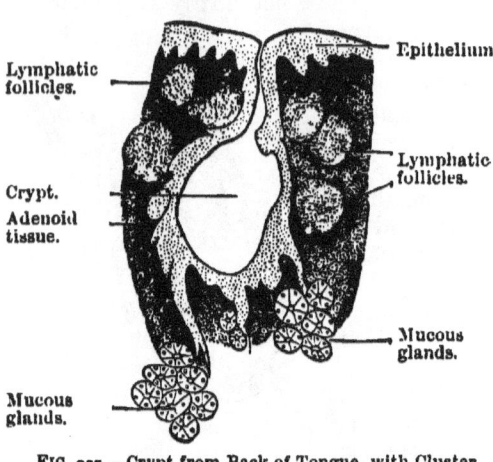

FIG. 237.—Crypt from Back of Tongue, with Cluster of Lymphatic Follicles.

3. The **serous glands** occur only near the circumvallate papillæ and taste-bulbs. Their acini are granular and resemble those of the parotid gland in structure (fig. 240, *d*).

4. **T.S. Tongue** (*Double-Stained*).—Stain a section from the posterior part of the organ, first with methyl-green and then with eosin. Mount in balsam. The connective tissue and papillæ are reddish; the serous glands are reddish also, while the mucous glands have a purplish-green colour.

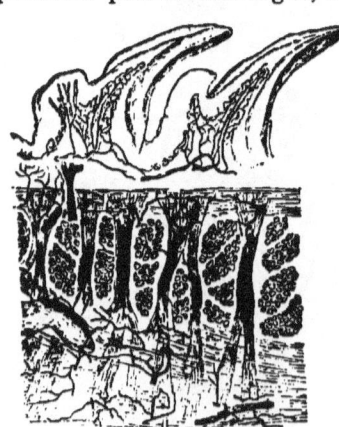

FIG. 238.—T.S. Injected Tongue of Cat.

5. **T.S. Injected Tongue (L).**—This is obtained when the head or whole body is injected. Observe the very vascular muscular portion and the papillæ, each with an artery entering it. If the papillæ be compound, *i.e.*, beset with other smaller secondary papillæ, a small capillary loop passes into each of these secondary papillæ; sections of large blood-vessels in the connective tissue of the mucous coat (fig. 238). If desired, another section may be *faintly* stained with logwood and mounted as above.

TASTE-BUDS.

6. Taste-Buds, or the peripheral organs of taste, occur on the fungiform papillæ and lateral surface of the circumvallate, soft palate, posterior surface of the epiglottis, and a few amongst the epithelial cells on the dorsum and sides of the tongue. It is more convenient, however, to study them in the rabbit. On either side of the posterior part of the rabbit's tongue are two oval patches with transverse ridges and intervening furrows, the **papillæ foliatæ** (fig. 239).

(i.) Cut out these parts and harden them for fourteen days or so in Müller's fluid and then in spirit. Stain in bulk in borax-carmine or hæmatoxylin and cut in paraffin.

(ii.) The excised organ, with as little adherent muscle as possible, is placed for one hour in 1 per cent. osmic acid, or pinned on a cork and exposed for the same time to the vapour of osmic acid. Fine sections are made across the laminæ, and stained with logwood and mounted in balsam.

(a.) (L) Observe the sections of the laminæ, each one with a central papilla or projection of connective tissue (fig. 240, l). This is covered by many layers of stratified epithelium.

(b.) Between the laminæ a furrow, and embedded in the epithelium, on each side of this furrow, the taste-buds (g), which are oval in shape, and composed of epithelial cells, whose bases touch the connective tissue of the mucous membrane, where they receive a branch of the glosso-pharyngeal nerve. The apex has an open mouth—*gustatory pore*—which communicates with the furrow. The cells composing the bud are arranged somewhat like the staves in a barrel.

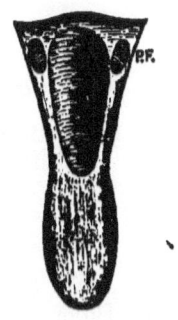

FIG. 239.—Tongue of Rabbit. *P.F.* Papillæ foliatæ.

(c.) Gland-ducts open at the bases of the furrows, and if these ducts be not seen, sections of their acini—serous gland—are sure to be seen deep in the corium (fig. 240, d). The corium has what look like secondary papillæ on it, but they are really septa (l').

(d.) (H) Study a single taste-bud (80 μ long and 40 μ broad) It is composed of two kinds of elongated epithelial cells.

(1.) The **sustentacular cells**, which are most numerous. They are elongated, flattened, and either of uniform breadth or narrowed at their base. They form a protective covering for the true gustatory cells, which lie between and within them.

(2.) The **gustatory cells** consist of narrow fusiform nucleated cells, whose lower pointed end is continuous with a branch of the axial cylinder of a nerve-fibre, while the free end is continued into a fine point or cilium, which projects through the gustatory pore.

If it be desired to study the mode of termination of the nerves in these organs, use the lemon-juice gold chloride method, with subsequent exposure of the tissue to sunlight in water acidulated with acetic acid, Golgi's rapid hardening method, or methylene-blue. For an elaborate research, with beautiful plates showing the terminations of the nerves in the papillæ foliatæ, see Drasch.[1]

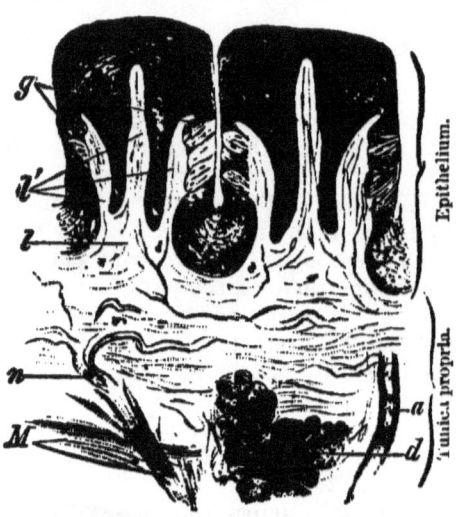

FIG. 240.—Papillæ Foliatæ in the Rabbit. *l*, *l'*. Primary and secondary septa; *g*. Taste-buds; *n*. Medullated nerve; *d*. Serous gland; *a*. Its duct; *M*. Muscular fibres, × 80.

SOFT PALATE.

1. **T.S. Soft Palate.**—Harden the soft palate of a rabbit or dog in Müller's fluid, alcohol, or corrosive sublimate. Make transverse sections by freezing, or stain in bulk in borax-carmine and cut in paraffin.

(*a*.) (**L** and **H**) One of the most beautiful methods of staining is that recommended by List, viz., to stain with aniline-green and eosin, and mount in balsam. Even with the naked eye the thick layer of mucous glands can be seen.

(*b*.) The stratified epithelium and connective tissue are rosy-red, the nuclei blue. The glands are bluish, and are seen to be mucous in character, lined by a single layer of mucous cells without demilunes. In the borax-carmine section, the cells lining the acini of the glands are clear and transparent and show no demilunes, a typical example of a pure mucous gland.

---

[1] "Unters. über d. Pap. fol. et circumvall. d. Kaninchens," *Abhand. d. math.-phys. Classe d. K. Sächs. Gesell. d. Wissensch.*, Bd. xxiv.

ADDITIONAL EXERCISES.

Glands of Tongue or Palate.—Harden the tongue or soft palate of a rabbit in 3 per cent. nitric acid for 1-2 hours. Wash out all the acid and stain the sections with methylene-blue. Wash out the blue with alcohol until only the glands remain blue. Mount in xylol-balsam.

Terminations of Nerves in the Lingual Papillæ and Glands.—These may be studied by staining small pieces of the tongue of mouse or rat by Golgi's silver nitrate method (p. 78), or by the rapid hardening method (bichromate of potash and osmic acid, Lesson XXVI. 14). The nerve in the papillæ contains nerve-cells which are in connection with the nerve-fibres. The nerve-fibres form a plexus of fibres outside the basement membrane of the serous gland acini—epilemmal plexus—and one within this membrane amongst the secretory cells—hypolemmal plexus (*Fusari* and *Panasci*).[1]

LESSON XXII.

TOOTH—ŒSOPHAGUS.

TOOTH.

The chief mass of a **tooth** consists of **dentine**. It is capped by **enamel**, and the root or fang is invested by a layer of bone, the **crusta petrosa**. All three tissues are calcified, and contain calcic phosphate. The enamel, however, is an epithelial structure, and consists of modified and calcified epithelial cells, while the dentine and crusta petrosa belong to the connective tissue group.

Unsoftened Tooth.—This is one of the few preparations which had better be bought.

1. **Longitudinal Section of a Dry Tooth** (fig. 241).

(*a*.) (L) Observe the **crown** and **fang**, and, connecting the two, the **neck**.

(*b*.) The **dentine** surrounding the **pulp cavity**, the **enamel** covering the dentine of the crown, and the **crusta petrosa** or **cement** covering the dentine of the fang. The wavy black lines in the dentine or **dentinal tubules** are really tubules filled with air, hence they appear black. Note their direction from the pulp cavity towards the outer margin of the dentine. Quite at the apex of the crown of the tooth they run vertically; in the fang they run nearly horizontally, and in the part of the dentine intermediate between

[1] *Archiv. ital. de Biol.*, xiv. p. 240, 1891.

both they gradually become more and more oblique from the centre of the crown.

Arched or curved lines—**incremental lines or Schreger's lines** —may sometimes be seen crossing the course of the dentine tubules.

(*c.*) The **enamel** covering the dentine on the crown of the tooth; somewhat brownish-coloured concentric lines may be seen in it.

(*d.*) (**H**) The **enamel** consists of striated prisms, hexagonal when seen in transverse section (fig. 243).

The **dentinal tubules** lie in a homogeneous matrix, and are wavy tubes, which divide dichotomously, and give off lateral branches which anastomose with other lateral branches from adjacent tubules. At the outer part of the dentine are irregular **interglobular spaces**, which appear black when they contain air (fig. 244).

The **crusta petrosa** consists of bone—a thin layer—composed of lamellæ and bone-corpuscles, but no Haversian canals (fig. 244).

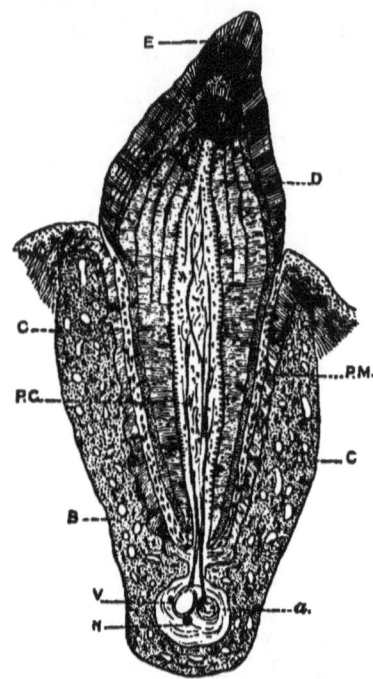

FIG. 241.—V.S. Tooth in Jaw. *E.* Enamel; *D.* Dentine; *P.M.* Periodontal membrane; *P.C.* Pulp cavity; *C.* Cement; *B.* Bone of lower jaw; *V.* Vein; *a.* Artery; *N.* Nerve.

2. Softened Tooth.—Select the jaw of a small mammal, *e.g.*, a cat, and decalcify a short length—¼ inch—of the lower jaw in chromic acid and nitric acid. It will take two or three weeks to remove all the bone-salts, and the decalcifying fluid must be frequently renewed. The tooth is sufficiently soft to be cut when a needle can be pushed into it. Make vertical sections through the whole jaw and a tooth *in situ*.

Stain one section in picro-carmine, another in osmic acid (twenty-four hours), and mount both in Farrant's solution or in glycerine-jelly.

(*a.*) (**L**) Observe the tooth in the **alveolus** or depression of the jaw in which it is fixed. The enamel has disappeared. The bone of the jaw with its periosteum, lining the alveolus and forming there the **periodontal membrane** (fig. 241, P.M.).

(*b.*) Next the latter on the fang the **cement** (C), the dentine, pulp-cavity and its contents. If the section passes directly through the middle of the tooth, the orifice in the fang of the tooth may be seen.

(*c.*) (H) The **dentinal tubules**, not so distinct as in the dry tooth.

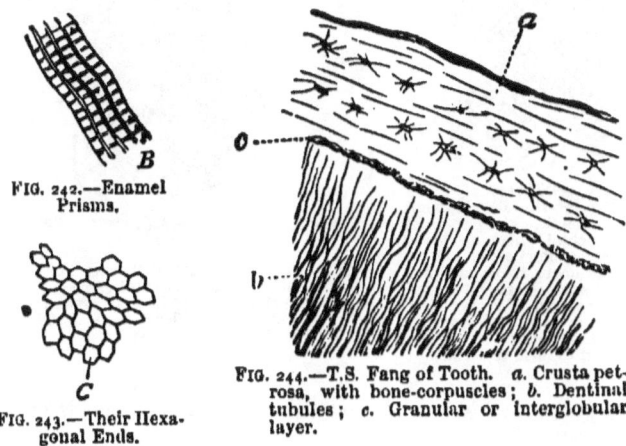

FIG. 242.—Enamel Prisms.

FIG. 243.—Their Hexagonal Ends.

FIG. 244.—T.S. Fang of Tooth. *a.* Crusta petrosa, with bone-corpuscles; *b.* Dentinal tubules; *c.* Granular or interglobular layer.

If they are cut obliquely, they appear merely as tailed dots in a homogeneous matrix.

(*d.*) The **pulp-cavity** contains blood-vessels and fine connective tissue, but next the dentine there is a layer of large cubical cells—**odontoblasts**—which give off fine processes which enter the dentinal tubules—**the fibres of Tomes**. They are best seen, however, in a tooth which has not yet cut the gum.

3. Development of Tooth.—Without entering into all the details of the development of the teeth, the following directions will suffice as to the method of preparing sections so as to show the various stages. What may be called the first stage—that shown in fig. 245—is to be obtained from the lower jaw of a sheep's embryo 7 cm. in length. At this stage only a very little bony matter exists. Harden the whole jaw in corrosive sublimate and decalcify in dilute hydrochloric acid. Stain in bulk in borax-carmine, embed in paraffin, and make T.S. across both rami of the jaw and the tongue. Or harden and decalcify at the same time the jaw of a fœtal kitten by placing small pieces containing embryonic teeth in Flemming's fluid. This yields excellent results,—the tissues are thereby sufficiently differentiated and may be cut in paraffin.

The second stage, fig. 246, is obtainable from the upper jaw of an embryo sheep 15 cm. long. It is treated in the same way.

The third stage, fig. 247, is obtained from the lower jaw of a dog six days old or thereabout.

With a high power it is easy to observe the structure of the

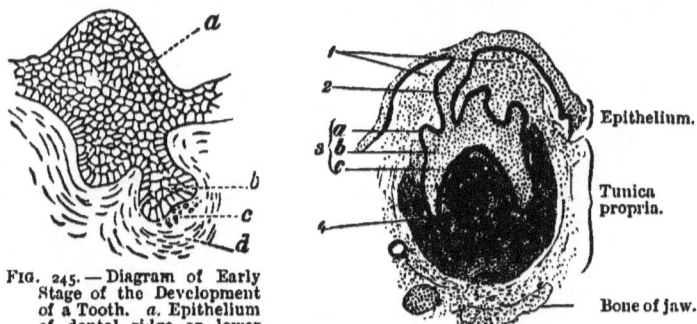

FIG. 245.—Diagram of Early Stage of the Development of a Tooth. *a.* Epithelium of dental ridge on lower jaw; *b.* Portion of epithelium about to be modified into enamel (enamel organ); *c.* Beginning of germ of dentine in the tooth papilla; *d.* Lamination of corium about to form tooth sac.

FIG. 246.—Lower Jaw of Human Fœtus at 4th Month, × 40. 1. Dental ridge; 2. Stalk of enamel germ; 3. Enamel organ; *a.* Peripheral cells; *b.* Germ pulp; *c.* Cylindrical cells of enamel; 4. Papilla.

several parts, and to see the odontoblasts lining what is to be the

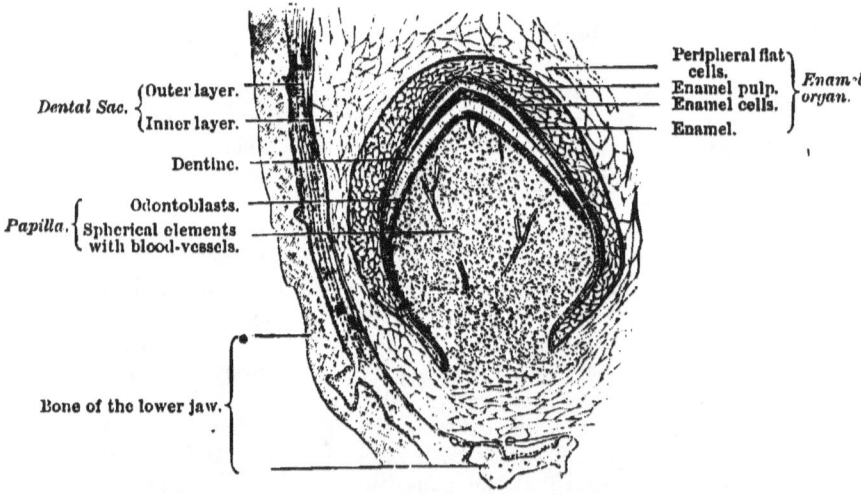

FIG. 247.—T.S. Lower Jaw of New-Born Dog, × 40. The dental sac is shown only in the left side. The tissues originating from connective tissue are shown on the left, and those of epithelial origin on the right.

pulp-cavity. If the latter happen to be partially detached, their processes—fibres of Tomes—may be seen partially withdrawn from the dentinal tubules in which they lay.

ŒSOPHAGUS.

The **Œsophagus** consists from within outwards of—

(1.) **Mucous coat**, composed of stratified squamous epithelium, into which project small simple **papillæ** from the corium or connective-tissue basis of the membrane. At the outer part of the corium is a narrow layer of smooth muscular fibres—**muscularis mucosæ**—arranged for the most part longitudinally.

(2.) **Submucous coat**, consisting of connective tissue and the larger blood-vessels and some nerves. In those animals (*e.g.*, dog) in which glands occur, the acini of the glands lie in the submucous coat, so that their ducts have to perforate the muscularis mucosæ and traverse the mucous membrane before they open on the surface of the epithelium.

(3.) **Muscular coat.**—This will vary with the animal used. In man this coat in the upper third of the œsophagus is composed of striated muscular fibre, the lower two-thirds of smooth muscle. The outer layer of fibres runs longitudinally, the inner circularly.

(4.) **Fibrous coat**, composed of fibrous tissue.

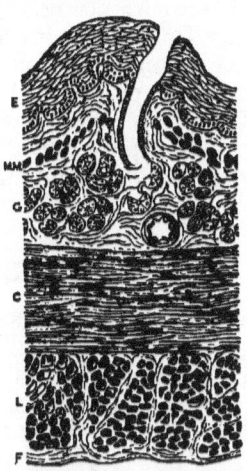

FIG. 248.—T.S. Small Part of Œsophagus of Dog. *E.* Epithelium; *M.M.* Muscularis mucosæ; *G.* Glands; *C.* Circular, and *L.* External or longitudinal muscular coat; *F.* Fibrous layer. Müller's fluid, picro-carmine.

4. **The Œsophagus.**—Cut out a piece of the œsophagus of a dog or cat—2 cm. in length—and harden it in equal parts of chromic acid (½ per cent.) and spirit, or Müller's fluid (fourteen days), or mercuric chloride, and then in alcohol. Make transverse sections; stain one in picro-carmine and mount it in Farrant's solution, and another in logwood and mount it in balsam. Perhaps even more instructive sections are obtained from a small animal, such as a rat, where the whole œsophagus (with trachea) can be stained "in bulk" in borax-carmine. Cut sections of both tubes to show their structure and relations.

(*a.*) (**L** and **H**) The circular tube has several coats, the innermost one being thrown into folds. The **mucous membrane** covered by stratified squamous epithelium, under this the connective tissue with small *simple papillæ*. In the deeper part of the mucous membrane are several layers of non-striped muscle, the *muscularis mucosæ* (248, *M.M*).

(*b*.) Outside this is loose **submucous** connective tissue with a few blood-vessels, and in the dog and some other animals the acini of mucous glands. The ducts of the latter traverse the coats lying internal to the glands, and open on the inner surface by funnel-shaped openings.

(*c*.) Outside this, again, is the **muscular coat**, which varies in the upper and lower parts of the tube, and also with the animal examined. In the upper part there is striped muscle, in the lower part two layers of non-striped muscle, an inner circular and an outer longitudinal. Between the two layers may be seen ganglionic cells of Auerbach's plexus.

(*d*.) Outside all is the fibrous coat or **adventitia**, composed of coarser connective tissue, with elastic fibres and blood-vessels.

For the epithelial cells lining the œsophagus see Lesson IV. 5.

It is to be remembered that there are very great differences as regards the presence of **glands** in the œsophagus. Some animals have a considerable number—*e.g.*, dog—and others very few.

ADDITIONAL EXERCISE.

5. **Other Methods (Œsophagus).**—Very good results are obtained by hardening in absolute alcohol containing methyl-green, the gland-cells being thereby sharply defined (*Rubeli*).[1] Also double stain with borax-carmine (extract with acid-alcohol), then alcohol, and stain again with iodine-green (twenty-four hours), extract with alcohol, embed, and cut in paraffin. The mucous membrane of some animals (*e.g.*, pig) contains **lymph follicles**, which can readily be detected in a part of the tube stained in bulk in borax-carmine.

LESSON XXIII.

THE SALIVARY GLANDS AND PANCREAS.

THE SALIVARY GLANDS.

ALL these glands have not the same structure, hence it is necessary to classify them.

Mucous Salivary Glands.—The sub-maxillary and sub-lingual glands of the dog and sub-lingual of guinea-pig.

[1] *Zeits. f. mik. Anat.*, vii. p. 224, 1890.

Serous Salivary Glands.—The parotid of man and mammals, and the sub-maxillary of the rabbit.

Mixed or Muco-Salivary.—The human sub-maxillary, retro-lingual of the dog, or sub-maxillary of the guinea-pig.

The salivary glands are compound tubular glands, *i.e.*, the duct is branched, while the acini or alveoli—the true secretory parts of the glands—are tubular in form. Each gland consists of **lobes**, held together by connective tissue, which forms a capsule for the whole gland and gives septa to enclose the lobes and lobules. Each lobe in turn is made up of numerous smaller **lobules** also held together by connective tissue, which carries the blood-vessels, nerves, lymphatics, and larger ducts. From mutual pressure the lobes and lobules are usually polygonal in shape. The main **duct** is made up by the convergence of ducts from the lobes—**lobar ducts**—while from each lobule there is a duct—**lobular ducts**, which unite to form **lobar** ducts. Each lobule is made up of a number of **alveoli or acini.** Each alveolus, which has a closed extremity, leads into or discharges its secretion into a fine duct or **ductule,** and these ductules by their union form the intralobular ducts. Practically the arrangement of the ducts is the same in all this set of glands; the differences in structure are in the alveoli. The **alveoli** consist of a **basement membrane,** which by appropriate means can be shown to consist of branched cells forming a reticulated or basket-like membrane. This is lined internally by the secretory epithelium, leaving a larger or smaller lumen in the centre, which leads into a fine duct or ductule by means of a narrow *junctional piece* or *intermediary part* or *ductule,* in which the epithelium is somewhat flattened. Usually several alveoli open into one intermediary tubule or ductule. The ducts with a fibrous wall are lined by a single layer of columnar epithelium, which is striated or "rodded" in its outer part, and granular towards the lumen of the tube (fig. 249); a little inwards from the centre of each cell is a nucleus.

In **mucous glands** the acini (35 μ in diameter) are lined by a layer of polyhedral clear cells, whose broader bases rest on the basement membrane, while their apices abut on the lumen, which is small (fig. 249). Usually in a transverse section of an acinus five or six cells are seen. The appearance of these cells varies according as a gland is at rest or in a state of activity, *i.e.*, whether the gland is "loaded" or "charged" (resting phase), or "unloaded" or "discharged" (active phase). In a **resting gland** the mucous cells are clear, for the most part, while at the outer part of the cell is a flattened nucleus surrounded by a very small quantity of granular protoplasm. The clear part is traversed by a network of fibrils, which includes in its meshes mucigen. The granular matter

R

and nuclei stain readily with the ordinary dyes, while the clear part does not do so.

In some mucous glands, *e.g.* dog, but not in all mucous glands, here and there between the bases of some of these cells, and the basement membrane, are groups of small, granular, nucleated cells, the group having a somewhat crescentic shape; they are called **demilunes** or crescents of Gianuzzi (fig. 250). They stain readily with dyes, and are darkened by osmic acid, and contain two or more nuclei.

In the discharged or **active gland**, the acini are smaller, the lumen wider, the clear part of the cell diminished in volume, while the outer part of the cell is wider, and appears to have encroached on the clear part. The nucleus is usually spherical, and placed nearer the centre of the cell.

In **serous** or **albuminous glands** the chief differences are in the cells lining the alveoli. In serous alveoli there is but one layer of cells, and nothing corresponding to the demilunes. The cells are somewhat smaller than mucous cells; they are more granular, and stain more uniformly with dyes. The nucleus is spherical, and placed nearer the centre of the cell. The differences between active and passive phases are not so marked as in mucous acini.

During activity the cells become smaller, and the "granules" disappear from the outer part of the cell; the cells become more sharply defined, while the nuclei are large and spherical.

N.B.—In all cases examine the acini of the glands in the fresh condition.

A. **Mucous Salivary Glands.**—These must be prepared in several ways.

Methods.—(i.) For the general structure of salivary glands:—

Harden small piece; for 2 or 3 days in the following mixture:— 3 parts 90 per cent. alcohol and 2 parts .5 per cent. chromic acid. To see the finer points—after staining—mount the sections in glycerine (*Langley*).

(ii.) Small pieces of a perfectly fresh dog's sub-maxillary gland are placed for an hour in 75 per cent. alcohol, then for five hours in absolute alcohol, which is then changed, and the hardening is completed in fresh absolute alcohol in twenty-four hours. Sections of the unstained gland are apt to fall to pieces, although the small pieces show the structure sufficiently well. A part of the alcohol-hardened gland should be stained "in bulk" in borax-carmine, and cut in paraffin. In this way the relative position of the parts is retained.

(iii.) Harden very small pieces in 1 per cent. osmic acid (24 hours); wash thoroughly, and complete the hardening in alcohol.

(iv.) Harden other pieces in Flemming's mixture, and stain "in bulk" in borax-carmine as in (ii.).

(v.) **Heidenhain's Method.**[1]—This method is also applicable to the pancreas. Small pieces of glands—*e.g.*, sub-maxillary of guinea-pig, dog, and cat—hardened in alcohol (ii.) are placed in 10 cc. of 0.5–1 per cent. watery solution of hæmatoxylin (6–8 hours); and then for an equal period in 0.5–1 per cent. potassic bichromate, or 1 per cent. watery solution of alum, or stain with .3 per cent. hæmatoxylin (distilled water to be used), and differentiate with 1 per cent. neutral chromate of potash, which forms a steel-grey compound with hæmatoxylin (p. 70). The stain does best with objects hardened in alcohol or picric acid. The pieces are quite black when removed from the second fluid. In this method, the union of the reagents takes place in the tissue itself. The nuclei are bluish-black, the cell-substance a steel-grey, while the demilunes stand out distinctly. This is an excellent method for these glands. Cut in paraffin.

(vi.) Mucous glands harden well in picric acid.

My experience leads me to believe that in studying a mucous salivary gland, it is best to begin with one whose acini contain only mucous cells and no demilunes. Such glands are the sub-lingual of a guinea-pig and sub-maxillary gland of the mole. Then proceed to the sub-maxillary of a dog, which has demilunes at intervals, and lastly, take the acini of a cat's sub-maxillary, where the demilunes form a nearly complete layer outside the true mucous cells.

In order to obtain a general view of the origin of the ducts from the acini of the lobules, and the union of small lobular ducts to form larger ducts, it is well to examine a preparation of the salivary glands of the cockroach, which can be removed *en masse* (p. 265).

1. Sub-Lingual Gland of Guinea-Pig.—This shows acini lined with mucous cells *without demilunes* (L and H). Note the acini, each lined by a single layer of clear transparent mucous cells, resting directly upon a basement membrane without the intervention of any demilunes. In the arrangement of capsule, septa, and ducts, it resembles the salivary glands of the dog. The gland is like fig. 249 without the demilunes.

2. Sub-Maxillary Gland of Dog.—(*a.*) (**L**) Observe the capsule, which sends off thin connective-tissue septa into the gland, mapping it out in polygonal lobes, which are further subdivided by finer septa into lobules. In the larger septa, sections of blood-vessels and gland-ducts (fig. 249).

(*b.*) Within each lobule aggregations of acini or alveoli, which make up the smaller lobules. The shape of the lobule depends on the way it has been cut. Branches of the finer gland-ducts—few—

[1] *Archiv f. mik. Anat.*, 1884, p. 468, and 1886, p. 383.

between the lobules. Some of the acini appear to be crowded with cells inside the basement membrane. The gland-ducts have a distinct lumen.

(c.) (H) Although numerous acini are visible, only those that have been cut across so as to show the lumen are satisfactory; the others appear merely to be filled with cells, and vary much in size, according to the plane of section through the alveolus. Each acinus has a clear, transparent basement membrane, and inside, and on it, is arranged the secretory epithelium.

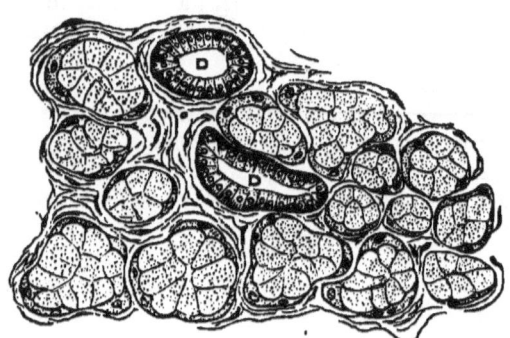

FIG. 249.—Small Lobule of a Sub-Maxillary Gland, Dog. *L.* Lobule; *D.* Duct. Osmic acid.

(d.) The **mucous cells** form a single layer, and are large, clear, cubical cells. Each has a nucleus which is usually flattened and placed near the attached end of the cell. In borax-carmine preparations, the nucleus, surrounded by granular protoplasm, is stained red, while the rest of the cell appears clear, and traversed, it may be, by fine threads or fibrils. In reality, the cell-substance consists of a network containing a clear substance — **mucigen**.

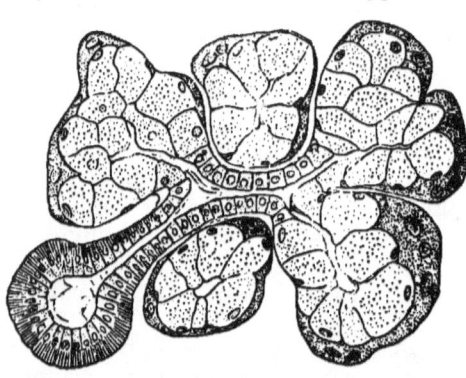

FIG. 250.—Sub-Maxillary Gland, Dog, showing Duct Communicating with an Alveolus by a Narrow Ductule. The alveoli with mucous cells and dense demilunes. Osmic acid and hæmatoxylin, × 300.

(e.) The **demilunes** lie singly next the basement membranes. Two or three may be seen in the section of each alveolus as more granular deeply-stained bodies, sometimes with two nuclei. As their name indicates, they are somewhat half-moon shaped, but they send processes between the mucous cells. As they are deeply stained by pigments and also darkened by osmic acid, their shape and distribution are readily recognised.

(f.) The **lumen** of each acinus is a small more or less regular space in the centre of the acinus, but many acini may be so divided as not to show it. It is difficult to find the connection between the lumen of an acinus and a duct (fig. 250).

(*g.*) **Duct.**—This is best studied in a transverse section of one of the finer ducts lying within the lobules or the larger ones in the septa (fig. 251). The wall consists of circularly disposed connective tissue, and is lined by a single layer of tall, narrow, cylindrical epithelium. The outer part of each cell is distinctly striated or "**rodded**," and the spherical nucleus is placed about the middle of the cell.

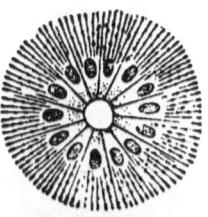

FIG. 251.—T.S. Salivary Duct showing only the "Rodded" Epithelium Lining it, × 300.

N.B.—The student should compare with this a section of a gland hardened in **osmic acid** or Flemming's fluid, or one stained by Heidenhain's method.

3. **Sub-Maxillary Gland of Cat.**—In some of the acini there may be a nearly complete layer of demilune cells between the basement membrane of the acinus and the lining layer of mucous cells.

B. Serous Salivary Glands—Methods.—Use the parotid of any mammal—rabbit, cat, or dog—or the sub-maxillary gland of a rabbit. Harden small parts of the gland in the same way and by the same means as for mucous glands. A saturated watery solution of mercuric chloride is to be preferred to picric acid for serous glands. I find that Flemming's mixture is specially good for the sub-maxillary gland of the rabbit. Sections are made and stained—the hardened gland stained "in bulk" —just as for mucous glands.

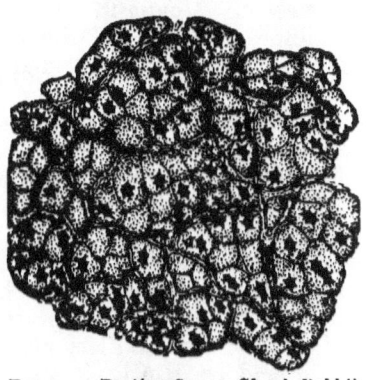

4. **T.S. Parotid Gland.**—(*a.*) (L) Observe the capsule, septa, lobes, and lobules as in the mucous glands, but the alveoli or acini are smaller. More sections of gland-ducts will also be seen.

FIG. 252.—Resting Serous Gland, Rabbit. Alcohol and carmine.

(*b.*) (H) Observe an *acinus*. It is lined by a layer of polyhedral cells, leaving a very small lumen. The cells are very granular, with a spheroidal nucleus placed near the centre of the cell (fig. 252). Numerous sections of ducts, some cut transversely, others longitudinally. They are like those of mucous glands.

5. **Fresh Serous Gland.**—Tease a fragment of a parotid gland in normal saline, and observe how the cells are crowded with granules.

C. Muco-Salivary Glands, *e.g.*, sub-maxillary of man or retro-lingual of the dog, are treated as the other salivary glands.

6. Human Sub-Maxillary Gland.—(L and H) Observe that some of the acini are like those of mucous glands, and others like serous acini, while some of the acini contain both mucous and serous cells. Acini, serous and mucous, may be found lying side by side (fig. 253).

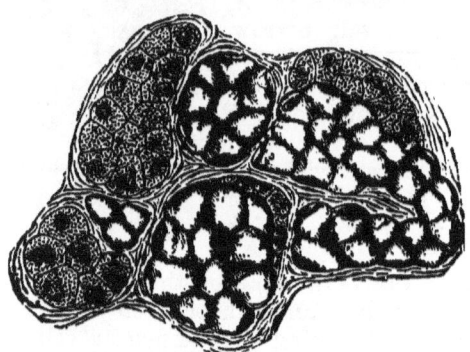

FIG. 253.—Human Sub-Maxillary Gland. On the right groups of mucous, and on the left of serous alveoli, × 300

7. Dog's Sub-Maxillary Gland (Double-Staining).—(i.) Stain alcohol-hardened sections, first in a watery solution of aniline green and subsequently in eosin. An easier plan is to stain first in aniline green, rapidly dehydrate the section in alcohol, taking care that all the green is not washed out of the gland-cells, and clarify with clove-oil in which is dissolved some eosin. Mount in balsam. The mucous cells are green, the demilunes pinkish, and the nuclei generally are green. The cells of the ducts are green, and the interlobular connective tissue pinkish.

(ii.) Stain sections in aniline blue, to which is added a saturated watery solution of picric acid. Mount in balsam. The cells of the acini are blue, while the ducts are yellowish-green.

(iii.) Schiefferdecker's method is also to be recommended. Add a few drops of a 5 per cent. alkaline alcoholic solution of eosin to a watch-glassful of alcohol. Allow the sections to stain in this for half an hour or so, and place them for a few minutes in 1 per cent. watery solution of aniline green, and mount in balsam.

(iv.) The sub-maxillary gland of a dog or guinea-pig hardened in picric acid or $HgCl_2$, if stained with aniline blue and safranin, shows the demilunes red and the mucous cells blue in balsam preparations.

THE PANCREAS.

The **pancreas** is a compound tubular gland, and resembles the serous salivary glands in the arrangement of its capsule, lobes, lobules, and duct, with its branches. The epithelium of its ducts, however, is not so distinctly striated, and in a section, as a rule, not many ducts are visible. Curious groups of cells, "**inter-tubular cell clumps,**" each supplied by a glomerulus-like tuft of capillaries, lie in the interlobular septa or amongst the acini. The alveoli, tubular or flask-shaped, with a very small lumen, consist of a basement membrane lined by a single layer of columnar or pyramidal cells, each showing two zones; an **outer zone** nearly homogeneous, and staining with logwood and some other dyes, and an **inner zone** crowded with "granules." The spherical nucleus lies about the middle of the cell. Sometimes sections of Pacinian corpuscles and groups of nerve-cells are found in the pancreas.

This gland, like other glands, should be examined in different phases of physiological activity. One, the **active state** ("discharged"), when the gland is removed from the body (rabbit) two or three or four hours after a full meal; and the other, the passive, or better, the **resting state** ("charged" or "loaded"), when the pancreas is not secreting actively, which can be secured by allowing an animal to fast for fourteen hours (dog or rabbit). The human pancreas is rarely satisfactory.

Methods.—(i.) One of the best methods of fixing the pancreas is 1 per cent. osmic acid or Flemming's mixture (24 hours). It is then to be thoroughly washed and hardened, first in 75 per cent., and afterwards in absolute alcohol. In such a preparation the "**granules**" are usually well preserved, and they stain deeply with safranin.

(ii.) A piece hardened in absolute alcohol and stained in bulk in borax-carmine or Heidenhain's logwood (p. 70), and afterwards cut in paraffin, shows well the general arrangement.

(iii.) Fix a small piece in corrosive sublimate, and, after the usual precautions, stain the sections with picro-carmine.

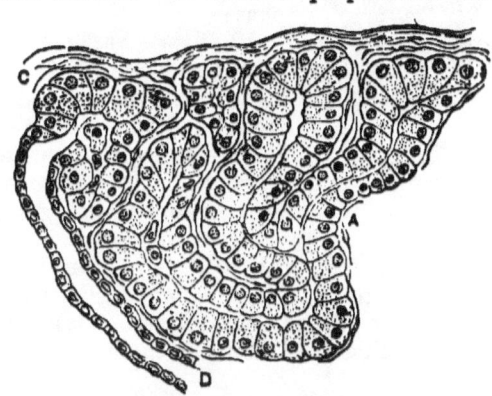

FIG. 254.—T.S. Pancreas, Dog. *A*. Acinus; *C*. Capsule; *D*. Duct. Corrosive sublimate and picro-carmine. × 300.

Mount the osmic acid sections in glycerine, and the stained ones in Farrant's solution or balsam.

8. Resting Pancreas.—(*a.*) (**L**) Observe the capsule (thinner), septa (thinner), lobes, and lobules, as in salivary glands. This arrangement is well seen in the carmine specimen (fig. 254, C).

(*b.*) (**H**) In the osmic acid preparation observe the alveoli (fig.

254, A), with a basement membrane lined by a single layer of columnar cells, tapering somewhat at their central ends, leaving a small irregular central lumen.

(*c.*) In each cell a crowd of dark "**granules**," occupying about the inner two-thirds of each cell, while the outer third or zone is comparatively homogeneous and free from granules (fig. 254). The nuclei of the cells are apt to be obscured by the presence of the granules. In the borax-carmine preparation the nuclei and granules are stained red ; the outer zone, about one-third, is also stained red, the inner two-thirds being either not stained or only faintly so, and granular, but the granules are not so sharply defined as in the osmic acid preparation.

(*d.*) The **ducts** (few), lined by a single layer of columnar epithelium, with very faint longitudinal striation (fig. 254, D).

9. Active Gland (H).—The cells of the alveoli are less granular, so that each cell shows an *outer zone* with no granules, occupying about one-half of the cell, and an *inner granular zone* with granules, which, however, are not nearly so numerous as in the resting alveoli. The nucleus (sometimes with a nucleolus and accessory nucleoli) is distinct, and near the centre of the cell. It is to be noted, however, that all the alveoli are not in the same phase.

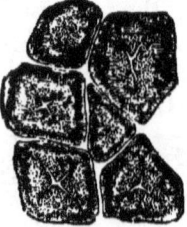

FIG. 255.—T.S. of the Acini of a Fresh Pancreas.

10. Fresh Pancreas.—A very good view of the granular character of the inner zone of the pancreatic cells is obtained by examining a piece of fresh pancreas—*e.g.* ox—teased in normal saline. It is easy to observe the difference between the outer homogeneous zone and the inner granular one. Many of the granules are liberated in the process (fig. 255).

11. Injected Pancreas (L).—Study a section with its blood-vessels injected. It is very vascular, and in the inter-tubular cell clumps are groups of capillaries. It is best injected with a Berlin-blue gelatine mass from the thoracic aorta.

ADDITIONAL EXERCISES.

12. Active Mucous Gland.—The student should examine a section of a sub-maxillary gland (dog), which has been in action for several hours. He cannot without a license prepare such a gland for himself. It is prepared, however, by stimulating at intervals for several hours the chorda tympani of a dog. In this way the gland is kept secreting, or it may be stimulated by injecting pilocarpin. It is then removed from the body and hardened in one of the ways stated on p. 258. It is essential that the active and non-active glands be hardened in the same way, so as to show that such differences as exist are not due to the method of hardening.

(H) Observe that the mucous cells are not so clear as in the passive gland, but are more granular and smaller, while in stained sections part of their cell-substance is stained by the pigments, and thus there is less difference between the demilunes and the mucous cells. The nuclei become more spheroidal. All the acini in the section are not necessarily in the same phase of activity, so that the appearances in any two acini may not be identical.

13. **Isolated Mucous and Demilune Cells.**—(i.) Place fragments of a fresh dog's sub-maxillary gland in 5 per cent. ammonium chromate (4–6 days). Tease a small piece in the same fluid. Note the isolated mucous cells, each with its fibrillar network, spherical nucleus embedded in protoplasm, and what was the attached end of the cell prolonged into a process.

(ii.) The cells, membranes, &c., are readily isolated in 33 per cent. caustic potash.

14. **Mucous Granules.**—The mucous granules are readily seen by "fixing" a small piece of the gland with the vapour of osmic acid (*Langley*).[1]

15. **Salivary Glands of Cockroach.**—It is better to use the species *Periplaneta americana*. Kill the animal with chloroform. Pin it out on its back on a cork plate and make the dissection in normal saline. On cutting open the thorax longitudinally one sees the intestinal tract, and on each side of this, lying on the wall of the latter, are the flattened salivary glands. They can readily be removed. They may be examined fresh, or stained in picro-carmine, or exposed to the vapour of osmic acid, and mounted in glycerine. They show well the general arrangement of ducts and lobules. Each duct is lined by a spiral chitinous fibre like the tracheae.

16. **Terminations of Nerves in Serous Glands.**—The rapid hardening method of Golgi (Lesson XXX.) as directed for the pancreas has been used by Fusari and Panasci[2] for the nerve terminations in the serous glands of the tongue of the rat, rabbit, and cat. The terminal fibrils form an epilemmal plexus, *i.e.*, outside the basement membrane, and other fibrils pass between the gland cells, *i.e.*, are hypolemmal. Fusari's paper is accompanied by a plate.

17. **Gland-Ducts.**—Sometimes on using Golgi's method one gets the lumen of the ducts black. If this happens, then an excellent view is obtained of their course and connections.

18. **Fresh Pancreas.**—In the rat and rabbit the pancreas is spread out in lobules in the mesentery, and if a piece of this containing a thin part of the pancreas be stretched on a ring of cork the granules can be seen in the fresh alveolar cells. Osmic acid does not alter the granules much, but alcohol does.

19. **Changes in Pancreas Cells** may be seen in frogs—one fed say three or four days previously, and the other a few hours before it is required.

20. **Rodded Structure in Cells.**—Macerate a small piece of fresh pancreas in 5 per cent. ammonium chromate for 2–3 days; tease and examine in the same fluid. The outer part of some of the cells will be found to be "rodded."

21. **Outer and Inner Zones of Pancreas Cells.**—(*a.*) Use the pancreas of a starving animal. To stain the outer zone, use ammoniacal or borax-carmine, and for the granules of the inner zone stain a section (fixed on a slide) with methyl-green-fuchsin-S solution made by mixing methyl-green (1 per cent.) 60 cc. with fuchsin-S solution (1 per cent.) 20 cc. Stain for 10 minutes. Wash quickly in water. Mount in balsam. The granules are red, and the outer zone clear. [Fuchsin-S is acid-fuchsin.] For preparations from Flemming's

---

[1] *Journal of Physiology*, vol. x. p. 433.
[2] *Archiv. ital. de Biol.*, xiv. p. 240, 1891.

fluid, stain with safranin and wash with alcohol containing picric acid. The safranin stains the granules.

(*b.*) A pancreas fixed in Flemming's fluid or sublimate is beautifully stained by means of the following modification by Oppel of Biondi's fluid:—

Methyl-green (1 per cent.)	120 cc.
Eosin (1 per cent.)	2 ,,
Acid-fuchsin (1 per cent.)	40 ,,
Absolute alcohol	40 ,,

The granules are red and the nuclei green.

22. **Terminations of Nerves in Pancreas.**—By means of Golgi's rapid hardening method (Lesson XXX.)—*i.e.*, osmico-bichromate fluid and then silver nitrate—Ramón y Cayal and Sala,[1] and more recently Erik Müller,[2] have traced the terminations of nerve-fibres in the pancreas (dog, rabbit). Numerous nerve-fibres enter the pancreas along its ducts, and terminate in a rich plexus of fibrils around the individual acini. The nerve-fibres which surround the acini are derived from two sources, some fibre fibrils are branches of Remak's fibres, and others are processes of very characteristic cells, which R. y Cayal calls "visceral sympathetic ganglion cells." These cells send off processes which are arranged as part of the "peri-acinous" plexus, and some of them penetrate between the gland-cells. Müller finds that the fibrils lie in relation with the secretory cells.

LESSON XXIV.

THE STOMACH.

THE walls of the **stomach**, like the intestine, are composed of four coats, named from within outwards—

(1.) **Mucous coat,** *i.e.*, the glandular coat.
(2.) **Submucous,** composed of loosely-arranged connective tissue, the larger blood-vessels, lymphatics, and nerves.
(3.) **Muscular,** composed of three layers of non-striped muscle —(*a.*) longitudinal, (*b.*) circular, (*c.*) oblique. In some situations only two layers exist.
(4.) **Serous,** from the peritoneum.

The **mucous coat** is lined by a single layer of tall, narrow, cylindrical mucous cells—in reality mucus-secreting goblet-cells—for they have open mouths and contain mucigen. The cardiac portion of the mucous membrane is composed of tubular glands—**fundus glands**—placed side by side. Several gland-tubes may open into

---

[1] *Terminacion de los nervios y tubos glandulares del pancreas de los vertebrados,* Barcelona, 1891.
[2] *Archiv f. mik. Anat.,* xl. p. 405, 1892.

one common duct. The duct is short, and is lined by cells like those covering the surface of the stomach. The secretory part of each gland is lined throughout by a layer of polyhedral or short columnar, granular, nucleated cells, called **chief, principal, inner,** or **adelomorphous cells.** At intervals between these and the basement membrane of the gland are large, ovoid, conspicuous, granular cells—**outer, parietal, delomorphous,** or **oxyntic.** The lumen of the gland is small and ill-defined.

The pyloric mucous membrane is beset with **pyloric glands**, which have a long duct or neck, and are usually branched at their lower ends. The duct is lined by a layer of cells like those lining the stomach, and the secretory part by a single layer of short, finely granular, short columnar cells. The lumen is well-defined. There is more connective tissue between the glands than in the cardiac portion. Masses of **adenoid** tissue are not unfrequently seen between the bases of the pyloric glands.

Methods.—(i.) Select a cat or dog that has hungered for two days. Kill the animal, open the stomach, and place small parts of the cardiac and pyloric ends (all the coats) in equal parts of chromic acid ($\frac{1}{2}$ per cent.) and spirit (7–10 days). Change this fluid within twelve hours. Complete the hardening in alcohol.

(ii.) Place small pieces of the cardiac and pyloric mucous membrane ($\frac{1}{8}$ inch cubes) in 1 per cent. osmic acid (24 hours). Wash well and harden in alcohol.

(iii.) Fix pieces of the pyloric and cardiac mucous membrane in mercuric chloride (2–3 hours). Take care to remove all the salt by prolonged washing in alcohol. Perhaps this is one of the best methods to use.

(iv.) Absolute alcohol is also a good hardening medium.

To bring out all the chief characters of the glands, some sections are to be stained in logwood and mounted in balsam. The sections may be cut by freezing or in paraffin, after staining in bulk. Stain others in picro-carmine, and mount in Farrant's solution; others in dilute carmine (24 hours). Stain others with 1 p.c. watery solution of aniline-blue (Nicholson's No. 1) for twenty minutes. Wash in glycerine and water, and mount in Farrant's solution. For other methods (p. 271).

1. V.S. Cardiac End.—(*a.*) (**L**) Observe the relations, relative thickness, and structure of the several coats. The **mucous coat**, with its gastric or **cardiac glands,** or **glands of the fundus,** set vertically like so many tubes in a rack (fig. 256). The glands are simple tubular glands, and some of them have several secretory parts opening into one duct; at their bases is delicate connective tissue, adenoid tissue, and blood-vessels. Below the closed ends of the glands, in the cat, a clear homogeneous layer of condensed connective tissue, and under this two or three thin layers of non-striped muscle—the **muscularis mucosæ**. The clear layer of condensed tissue is not present in the rabbit, dog, or man.

(b.) The **submucous coat**, composed of loose connective tissue, with large blood-vessels and a few fat-cells. If the mucous membrane be folded and rugæ are present, the connective tissue will be seen to run up into the folds.

(c.) The **muscular coat** consists of two or three layers of smooth muscular fibres. The appearance varies according to the manner in which the plane of the section cuts them, for the stomach has an outer longitudinal and an inner circular, and in some places an oblique muscular coat.

(d.) Outside all a thin layer—**peritoneum or serous coat**—consisting of fibrous tissue covered by a layer of endothelium.

(e.) (**H**) Study specially the **mucous coat**. Observe the **columnar epithelium** lining the stomach, and dipping into the mouths and lining the ducts, which are slightly funnel-shaped. The cells are tall and narrow, and if the section be not too thin, the ends of them may be seen as very small, clear, sharply-defined, polygonal areas. As they secrete mucus, they have been called mucous cells. The upper two-thirds or so of the cell is much clearer than the lowest third, which tapers somewhat, and is more granular. Each cell contains an oval nucleus in its lowest third, *i.e.*, near its attached end.

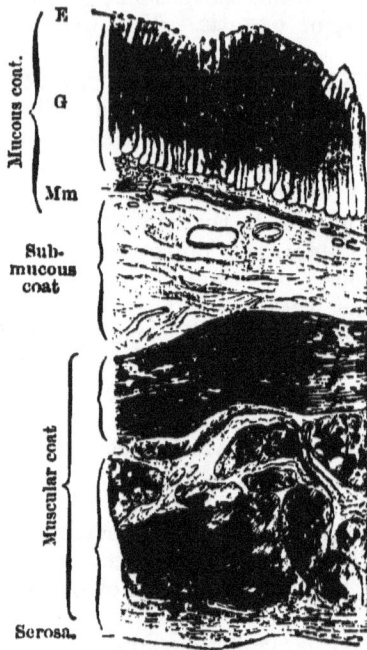

FIG. 256.—V.S. Wall of Human Stomach. *E.* Epithelium; *G.* Glands; *Mm.* Muscularis mucosæ, × 15.

(f.) Select a **fundus gland** (fig. 257). Trace its duct downwards, and note that perhaps it is common to two secretory portions. In the true secretory part, note the large **ovoid, border, parietal,** or **outer cells,** lying next the basement membrane,—ovoid in shape, granular in appearance, and containing an ovoid nucleus. They do not form a continuous layer, but bulge out the basement membrane here and there. At the upper part of the gland-tube the parietal cells are smaller, and placed nearer each other, while towards the fundus or base of the gland they are larger and further apart, the interval between any two being occupied by the inner or central cells. The parietal cells are deeply stained by aniline-blue and carmine, and osmic acid, while in a logwood-stained section the inner cells are usually more deeply stained.

Observe a continuous layer of cells—**inner, central, or chief**—lying internal to the parietal cells. They are smaller, and belong more to the columnar type of cell, but they are not of uniform height throughout; thus they are shorter over a parietal cell, and larger between two parietal cells. Each cell contains a spherical nucleus.

The **lumen** of the gland-tube is very narrow. The secretory part is much longer than the duct.

(*g.*) Arrange the preparation so that the lower ends of the glands come into view. Note their closed extremities, and if the granular epithelium have shrunk somewhat, the basement membrane of each gland-tube may be seen; very probably transverse or oblique sections of the lower part of the gland will be found (fig. 257).

(*h.*) At the bases of the gland-tubes, but outside their basement membrane, note the delicate connective tissue or adenoid tissue, with a number of leucocytes. If the section be from a cat's stomach, a clear homogeneous layer runs outside this.

(*i.*) Outside this the **muscularis mucosæ** composed of at least two layers of non-striped muscle arranged in opposite directions. Especially in a balsam specimen, fine strands of muscular fibres are always to be seen passing from the muscularis mucosæ inwards towards and between the glands. Usually these fine strands pass inwards between groups of glands. In the osmic acid preparations, the parietal cells are much darker than the central cells.

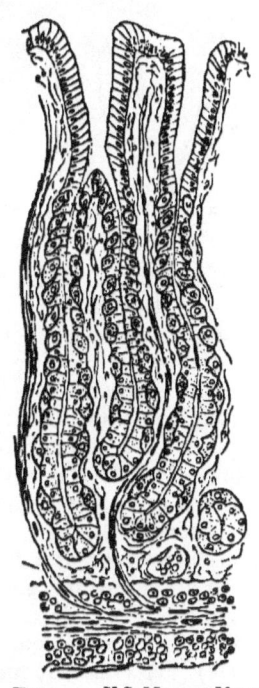

FIG. 257.—V.S. Mucous Membrane of the Stomach, Cat. Osmic acid.

(*j.*) There is nothing particular to note in the structure of the other coats.

2. T.S. Fundus Glands.—Make a section parallel to the surface of the mucous membrane, stain with logwood, or picro-carmine and aldehyde-green, or use an osmic acid preparation.

(H) Observe that all the gland-tubes are not cut at the same level; some are divided through the duct, others through the secretory part of the tube. In the latter observe the large parietal cells—few in number—and inside these a complete layer of inner cells bounding the small lumen of the tube (fig. 258). The glands are arranged in groups as shown by the connective tissue surrounding several tubes. This is very marked in the pig.

3. Fresh Fundus Glands.—From the mucous membrane of the stomach of a newly-killed guinea-pig make a thin vertical section with scissors, and tease it in normal saline to isolate some of the glands. This animal is selected because its gland-tubes are short, but a rabbit does very well. The elevations of the basement membrane due to the bulging of the parietal cells are usually well seen.

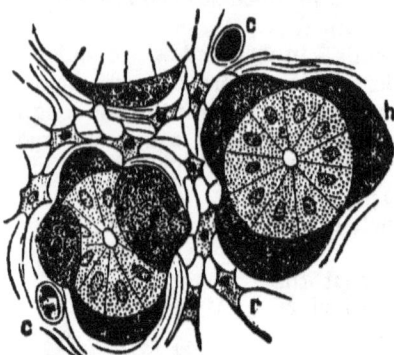

FIG. 258.—T.S. Duct of Gland of Fundus. *a.* Chief, *h.* Parietal cells; *r.* Adenoid tissue; *c.* Capillaries.

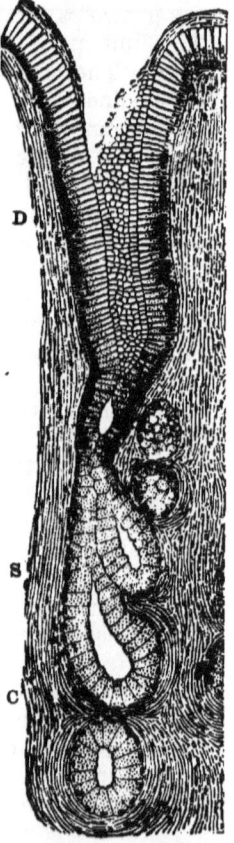

FIG. 259.—V.S. Pyloric Mucous Membrane. *D.* Duct; *S.* Secretory part of gland; *C.* Connective tissue.

4. V.S. Pyloric Mucous Membrane (fig. 259).—(*a.*) (**L**) Observe the same arrangement of coats as in the cardiac end; but the muscular coats are thicker; the mouths of the gland-tubes are wider and longer, the secretory part more branched and shorter than in the cardiac portion. There is also much more connective tissue between the glands.

(*b.*) (**H**) The wide mouth of the glands, lined by narrow columnar epithelium, the secretory part consists of several tubes opening into one gland-duct. The secretory part lined by a single layer of cells, somewhat cubical, but there are no parietal cells.

(*c.*) There is much more connective tissue between the tubes. A mass of adenoid tissue —**solitary follicle**—may be seen in the deeper part of the mucous coat. In some cases its pointed apex may be seen reaching nearly to the surface of the mucous membrane.

5. Osmic Acid Preparations.—Few reagents are so good for fixing the cells of the glands of the mucous membrane (mount in Farrant's solution). In sections of the fundus, the outer cells are more deeply stained, and so are readily distinguished. The columnar cells lining the ducts of the cardiac and pyloric glands may be blackened by the osmic acid where they contain mucigen.

6. **Blood-Vessels of the Stomach.**—Make V.S. of an injected stomach embedded in paraffin. Mount in balsam. They must not be too thin. Note the large vessels in the submucous coat, and from them smaller vessels proceeding vertically upwards, splitting up into capillaries between the tubules, and forming a capillary network round the mouths of the glands. Beautiful plates by Mall.[1]

7. **Pyloro-Duodenal Mucous Membrane.**—It is well to have a section through the pyloric valve to include the mucous membrane on its gastric and duodenal boundaries. To ensure this, the mucous membrane or entire thickness of the stomach must be pinned out on cork before it is hardened by any of the methods, *e.g.*, sublimate (p. 267). It is treated like the sections of the stomach, and does best when stained with eosin-hæmatoxylin. On one side of the thickened pyloric valve—the increased thickness being due chiefly to an increase of the circular muscular fibres—one sees the pyloric structure, and on the other that of the duodenum. The tubular glands of the stomach are confined to the mucous membrane, but the acini of Brunner's glands lie in the submucous coat of the duodenum.

8. **Junction of Œsophagus and Stomach.**—Similar preparations may be made. The transition from the œsophageal mucous membrane with stratified epithelium and few glands in the œsophagus to that of the stomach with its columnar epithelium and mucous glands is sudden and abrupt.

ADDITIONAL EXERCISES.

9. **Double-Staining the Stomach.**—Harden the stomach in Müller's fluid, and stain (24 hours) the sections in indigo-carmine (p. 67); afterwards extract the excess of pigment by steeping them for half an hour in a saturated solution of oxalic acid. Mount in Farrant's solution or balsam. The parietal cells are grey or blue, the central ones coloured, but with red nuclei, and the smooth muscle blue with red nuclei.

10. **Ehrlich-Biondi Fluid.**—Stain in this fluid sections of the mucous membrane fixed in sublimate (saturated in .6 per cent. NaCl). If the fluid be kept for some time, an additional quantity of acid-fuchsin must be added to it. The parietal cells are red and their nuclei blue; the chief cells are scarcely stained at all, but their nuclei are faintly blue. Vacuoles may be seen in some of the outer cells.

11. **Aniline-blue and Safranin.**—Sections of the cardiac end fixed in

---

[1] Vessels and Walls of Dog's Stomach, *Johns Hopkins Hosp. Rep.*, vol. i. 1893.

mercuric chloride are stained with aniline-blue and then with safranin. In balsam preparations the parietal cells are pale blue, the inner cells red.

12. **Isolated Cells of Gastric Glands.**—Macerate fragments of the gastric mucous membrane of a newt in 5 per cent. ammonium chromate (24-48 hours). Stain in picro-carmine, and tease in glycerine. Numerous isolated cells from the ducts and secretory parts of the glands are obtained.

LESSON XXV.

THE SMALL AND THE LARGE INTESTINE.

SMALL INTESTINE.

It has four coats—mucous, submucous, muscular, and serous.

Study specially the mucous coat. In man, in certain parts, there are permanent folds of the mucous membrane—**valvulæ conniventes**—and everywhere the surface is beset with small conical elevations—**villi.** At the bases of the villi is a layer of simple tubular glands—**Lieberkühn's glands**—embedded in an adenoid tissue matrix. Underneath this is the **muscularis mucosæ,** usually consisting of three thin layers of smooth muscle.

A **villus** (.5-3 mm. long) consists of a central core, enclosing a **lacteal,** and covered by a single layer of columnar epithelium with goblet-cells (Lesson V.). The body of a villus consists of a tissue like adenoid tissue with leucocytes and other cells. The central lacteal is really a lymphatic, and begins by a closed extremity. Several strands of smooth muscle pass from the muscularis mucosæ into the villus, and reach its upper extremity. It is very vascular, and the blood-vessels are distributed immediately under the epithelium.

The mucous membrane also contains **solitary follicles,** and in some situations **Peyer's patches,** the latter are most abundant in the ileum.

Methods.—Make transverse sections of the small intestine of a cat or dog hardened in a mixture of potassic bichromate and chromic acid; Klein's fluid; $\frac{1}{8}$ per cent. chromic acid; Kleinenberg's fluid; or mercuric chloride.

(i.) Stain a section in hæmatoxylin and mount it in balsam, or stain another in picro-carmine and mount it in Farrant's solution.

(ii.) All the parts and their relations are best preserved by staining in bulk in borax-carmine and cutting in paraffin, or embed in paraffin, cut, fix on a slide, and then stain.

(iii.) Cut sections by freezing and place some in 1 per cent. osmic acid (24 hours). This sharpens the outlines of many of the structures.

(iv.) It is convenient in teaching to give a complete transverse section of the small intestine of a small animal, *e.g.*, mouse or kitten. In herbivora the wall of the gut is very thin. Stain in bulk and cut in paraffin. Flemming's fluid is an excellent "fixing" reagent both for the small and large intestine. Stain the sections in safranin.

1. T.S. Small Intestine (L).—Observe the serous, muscular, submucous, and mucous coats (fig. 260).

(*a.*) In the **mucous coat**, the surface beset with small conical projections—**villi**—which, if they are contracted, exhibit wrinkles on their surface. At the bases of the villi a single layer of simple test-tube-like glands—**glands of Lieberkühn**—or intestinal glands, embedded in an adenoid tissue matrix. Outside this the **muscularis mucosæ**.

(*b.*) The **submucous coat**, composed of fibrous tissue with nerves and blood-vessels.

(*c.*) The **muscular coat**, composed of two layers, an *outer longitudinal* and an *inner circular layer*. In the cat, the latter is much thicker than the former.

(*d.*) The **serous coat**.

(*e.*) (H) Study a **villus**. Observe the single layer of granular nucleated columnar epithelium covering it, each cell with its free end covered by a clear disc, with vertical striæ (fig. 260). The succession of these free clear borders looks like a clear hem round the circumference of the villus. Occasionally leucocytes may be seen between the cells.

(*f.*) The **goblet cells**, *chalice*, or *caliciform* cells scattered among the former (fig. 260). They may be seen from the side, or their open rounded mouths may be directed toward the observer. When seen from the side, they are ovoid, with a larger and clearer upper part containing mucigen, and a smaller, lower, granular, nucleated part (Lesson V.). Sometimes a plug of mucus may be seen protruding from the mouth of a cell. It is stained blue with logwood and brown with Bismarck brown. If the mucous glands be active and the fresh tissue be fixed in osmic acid, then the plug of mucus is black.

(*g.*) In the centre of the villus a vertical space, the radicle of a **lacteal**, with a thin nucleated wall. The substance of the villus consists of adenoid tissue beset with leucocytes. Close under the epithelium, perhaps, sections of capillaries, and a little farther in one or more strands of non-striped muscle, which can be traced to the apex of the villus, and downwards to the muscularis mucosæ (fig. 260, MM).

(*h.*) The **glands of Lieberkühn** lined by short columnar nucleated cells, the nucleus near the attached end of the cell. There is a gradual transition from these cells to those covering the villi. A clear border may be seen on their free ends. Bizzozero has called these "protoplasm cells" to distinguish them from the goblet-cells which lie amongst them. Goblet-cells, however, are far more abundant in the large intestine. The lumen of each gland is distinct, and, especially if the lining cells be raised slightly, the basement membrane of the gland-tube may be seen. Between the gland-tubes numerous leucocytes and adenoid tissue; in fact, the glands are set in a meshwork of this tissue.

(*i.*) The **muscularis mucosæ** sends delicate processes into the villi. This is best recognised in balsam specimens, by the arrangement of the fusiform nuclei of the smooth muscle cells.

(*j.*) A solitary gland or a Peyer's patch may be cut, but it is better to have special preparations for these.

2. Peyer's Patches or Agminated Lymph Follicles.—(i.) Make V.S. through a hardened Peyer's patch (sublimate or alcohol). Stain with eosin and hæmatoxylin and mount in balsam.

(ii.) Fix a Peyer's patch of a rabbit or guinea-pig in Flemming's fluid. Fix a section on a slide and stain it first in 1 per cent. aniline-blue (watery). Wash out in 1 per cent. ammonia, then in .5 per cent. HCl, and stain in safranin (*Garbini*). Mount in balsam.

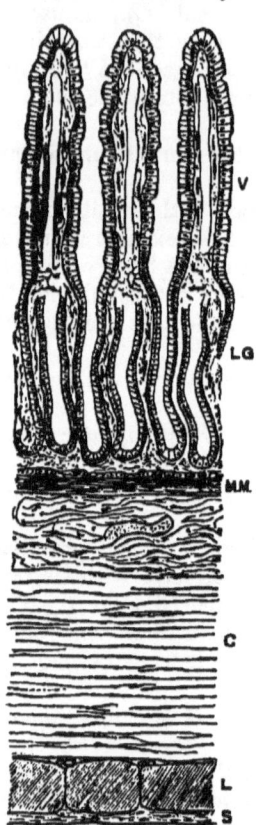

FIG. 260.—T.S. Small Intestine (Cat). *V.* Villi; *LG.* Lieberkühn's glands; *MM.* Muscularis mucosæ; *C* and *L.* Circular and longitudinal fibres of muscular coat; *S.* Serous coat.

(L) Observe a group of oval or roundish masses of adenoid tissue crowded with leucocytes confined to one side of the gut. The conical apices of some of them may be seen projecting upwards quite to the mucous surface, covered only by a layer of columnar epithelium. Between the epithelial cells may be seen colourless corpuscles which have wandered from the adenoid mass. The lower ends of the masses usually pass down into the submucous coat. No villi exist over the apices of these masses of adenoid tissue if they project well into the mucous layer. If, however, they do not, but exist merely as rounded masses of adenoid tissue in the submucous coat, then they present the appearance seen in fig. 261.

A **solitary follicle**—exactly like one of the numerous follicles which compose a Peyer's patch—may be seen (fig. 262).

3. Blood-Vessels of the Small Intestine.—

Make a pretty thick transverse section of a well-injected small intestine (cat). Cut in paraffin and mount in balsam.

(**L**) The mucous coat is the most vascular part; the larger vessels lie in the submucous coat, and few vessels in the muscular coat.

(**H**) An artery runs to the upper part of each villus and gives off a plexus of capillaries; a vein on the opposite side. A rich plexus of capillaries between Lieberkühn's glands. If the section pass through a solitary follicle, note that the capillaries pass into it and form loops (fig. 262).

The general distribution of the blood-vessels is shown in fig. 263, which shows how the blood reaches the various coats.

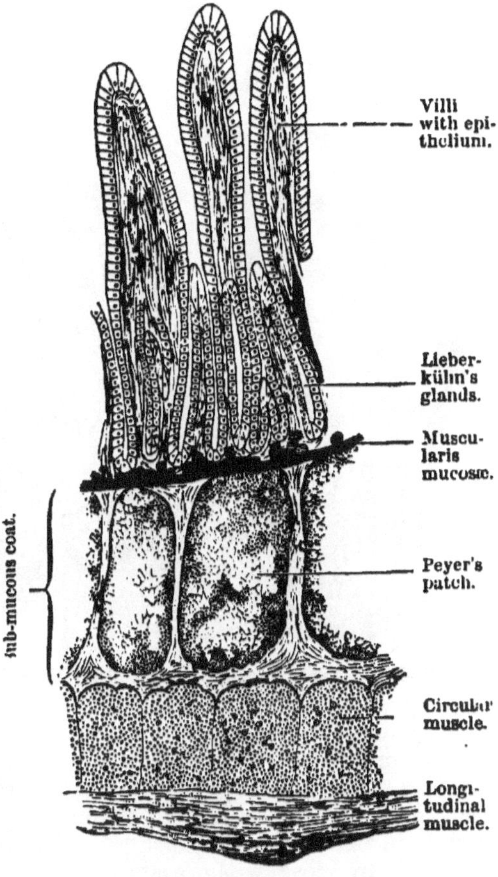

FIG. 261.—Longitudinal Section through a Peyer's Patch of the Small Intestine of a Dog.

4. Surface View of Injected Villi.—

To see the general arrangement of the blood-vessels in the mucous membrane, inject the blood-vessels of a rabbit with a red gelatine mass. Mount in balsam a part of the wall of the small intestine, placing the mucous surface uppermost.

(**L**) Note the villi, and trace the artery to its origin from a larger artery in the submucous coat (fig. 264) The artery runs on one side of the villus quite to the apex of the latter, and the vein—a wider vessel—descends on the opposite side. There is a dense plexus of capillaries (C) placed close under the epithelial covering. The best figures of the blood-vessels of the intestine are to be found in Mall's paper.[1]

5. **Nerve-Plexuses in Intestine.**—(*a.*) *Gold Chloride Method.*—Wash the small intestine of a rabbit with normal saline, fill it with

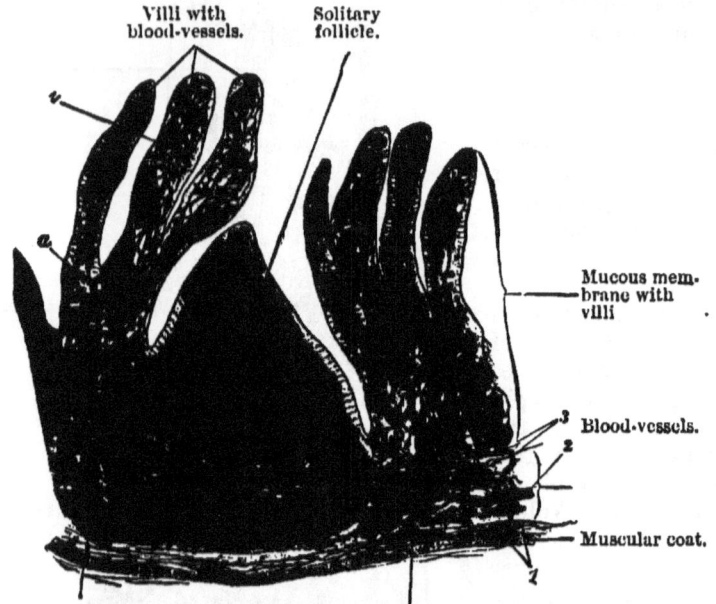

FIG. 262.—Mucous Membrane of Small Intestine (Rabbit), Injected, showing villi and a solitary follicle, × 50.

lemon-juice or 5 per cent. arsenic acid, and leave it in lemon-juice or acid for five minutes. Allow the juice to escape, wash the gut in water, fill it with 1 per cent. gold chloride, and place it in $\frac{1}{2}$ per cent. gold chloride solution (30 mins.). Wash it in water, and keep it in the dark in 25 per cent. formic acid (48 hours). It is now easy to separate the coats of the intestine from each other. Wash in water to remove all the acid, and with a pair of forceps strip off the longitudinal muscular coat (Auerbach's plexus adheres to this); it separates quite easily from the circular coat. Preserve what remains. Mount a small part of the longitudinal muscular coat in Farrant's solution.

(*b.*) *Methylene-blue Method* (p. 284).

[1] "Die Blut- u. Lymph-wege im Dünndarm des Hundes," *Abhand. d. math.-phys. Classe d. K. Sächs. Gesell. d. Wissens.*, Bd. xxiv., 1887.

XXV.] SMALL INTESTINE. 277

6. Auerbach's Plexus lies between the muscular coats, but when they are separated it usually adheres to the longitudinal coat. The general arrangement of the plexus can be seen with the naked eye (fig. 265).

(L) The polygonal meshwork of purplish stained fibres, with slight swellings at the points of intersection (fig. 266).

(H) The meshes are so large that only a part of them comes into the field of view at once. Note the non-medullated nerve-fibres, and at the nodes, groups of nerve-cells. From the plexus numerous fibres are given off to supply the smooth muscle of the intestine.

7. Ganglionic Cells.—In a vertical section of the gut (prepared as in 1) look for groups of ganglionic cells in the submucous coat and others between the two layers of muscular fibres (fig. 267).

8. Meissner's Plexus. — Spread what remains after removal of the longitudinal muscular coat on a slip of glass, mucous surface

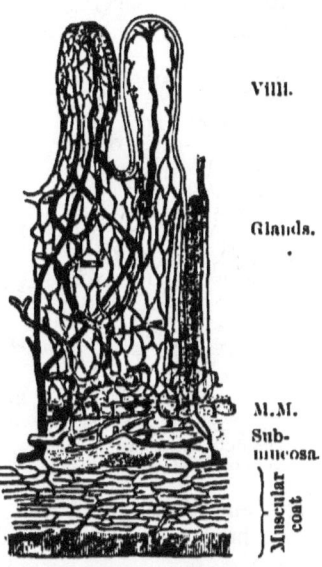

FIG. 263.—Scheme of the Distribution of Blood-Vessels in the Small Intestine of a Dog. *M.M.* Muscularis mucosæ.

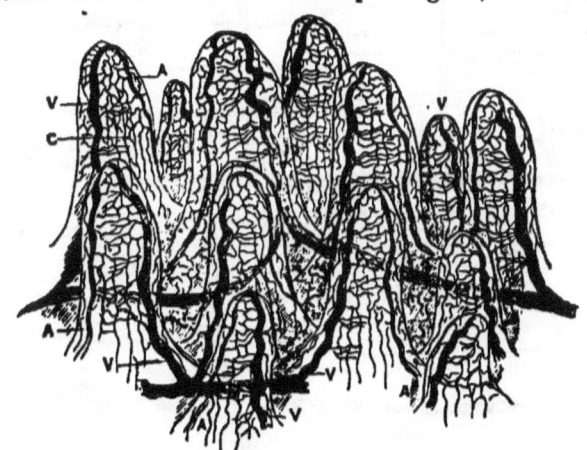

FIG. 264.—Injected Villi of Small Intestine of Rabbit, seen from above and laterally. *A.* Artery; *V.* Vein; *C.* Capillary network.

uppermost. With a knife gradually scrape away the mucous coat.

With care—observing the preparation from time to time under a low power of the microscope—the progress of the process of denudation can be easily observed. By and by the plexiform arrangement of Meissner's submucous plexus will be seen. Mount in Farrant's solution (fig. 268).

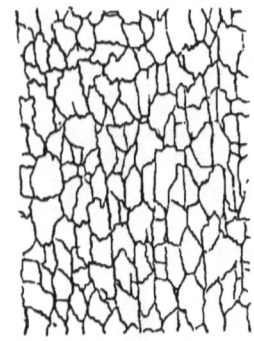

FIG. 265.—Auerbach's Plexus (Rabbit). Lemon-juice and gold chloride, × 8.

(L) Observe the large wide meshes of the plexus. Compare the general arrangement of the plexus with that of Auerbach's plexus. The fibres are finer, and the groups of nerve-cells smaller.

(H) Note the ganglionic nerve-cells at the nodes (fig. 268).

9. Brunner's Glands.—These glands are confined to the duodenum. Proceed exactly as recommended for the small intestine. Stain a section with logwood and use eosin as a counter stain. This gives excellent results.

Perhaps the best method is to slit up the duodenum longitudi-

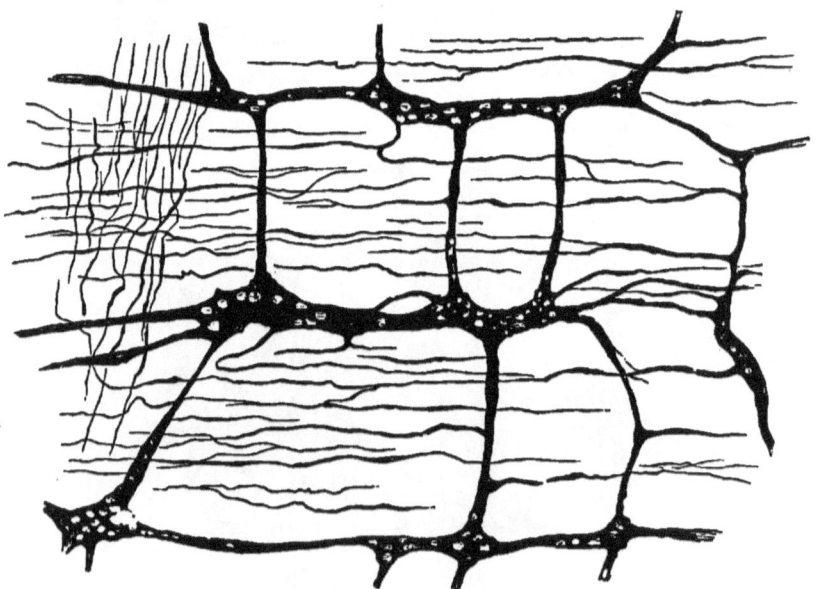

FIG. 266.—Auerbach's Plexus (Dog).

nally, pin it out on a cork plate, and fix it in mercuric chloride (an hour at 40° C. in a thermostat, or 24 hours at ordinary temperature). The surplus $HgCl_2$ is extracted with water at 30° and afterwards with alcohol at 40° C. It is then removed from the cork plate and hardened in alcohol. Embed and cut in paraffin. Kuczyński[1] has investigated these glands in a great number of

FIG. 267.—Auerbach's Plexus in Section. *a.* Ganglionic cells; *b.* Nerve-fibres; *c.* Circular, and *d.* Longitudinal muscular fibres.

animals and man, and tried a great variety of stains. Kleinenberg's fluid is also very good as a fixing reagent.

(L and H) Note villi, Lieberkühn's glands and the usual intestinal coats. In the submucous coat, which is thick, are the acini

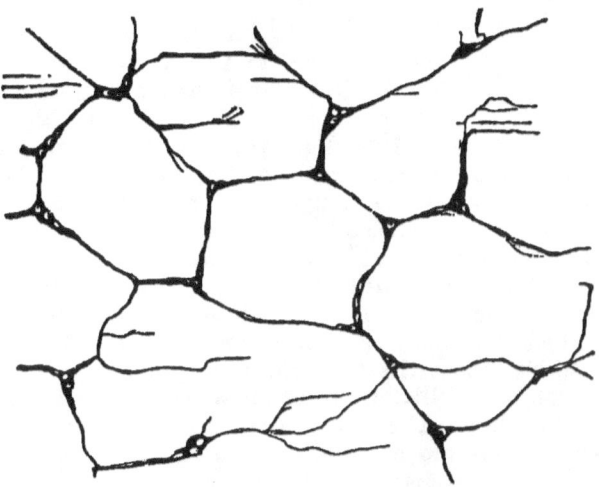

FIG. 268.—Meissner's Plexus Intestine (Rabbit). Lemon-juice and gold chloride.

of Brunner's glands (fig. 269), each composed of a basement membrane lined by cubical or short columnar cells. A duct from each

[1] *Internat. Monats. f. Anat. u. Phys.*, vii., 1890, 419.

gland perforates the muscularis mucosæ, passes up between the intestinal glands, and opens on the free surface at the bases of the villi. It has a distinct lumen, and is lined by low cubical epithelium.

10. Fresh Villus.—Preferably that of a mouse. Examine in normal saline.

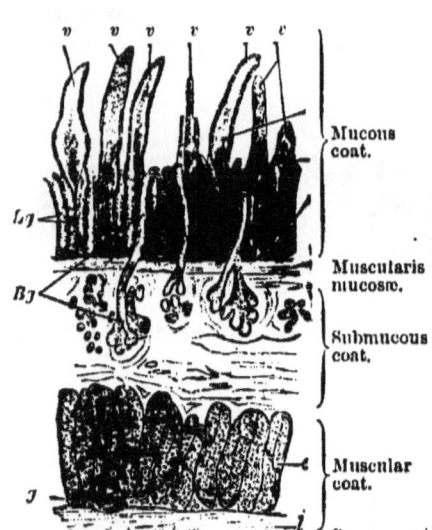

FIG. 269.—V.S. Duodenum of Cat. *c, l.* Circular and longitudinal layers of muscle; *Lg.* Lieberkühn's glands; *Bg.* Brunner's glands; *g.* Ganglion-cells; *v.* Villi.

LARGE INTESTINE.

Methods.—Prepare it in the same way as the small intestine, and make vertical transverse sections. The details of the structure will necessarily vary with the animal used, but perhaps the cat is as convenient an animal as any to employ. I can strongly recommend Flemming's fluid. Sections to be stained with safranin.

11. T.S. Large Intestine (fig. 270).—As the large intestine is wide, it is necessary in making the section to select a portion of the great gut which shows the longitudinal coat. In man, the coat is not a continuous one, the longitudinal fibres being grouped for the most part into three flat bands of fibres.

(L) Beginning from without inwards, observe four coats. In the muscular coat, note the peculiarities of the longitudinal coat and the thick well-marked continuous circular coat. There is nothing special about the submucous coat, unless it be the existence of **solitary follicles**, which, however, are not confined to the large intestine. The mucous coat is characterised by negative characters. There are no villi. It may exhibit folds, into each of which there runs a projection of the submucous coat. Lieberkühn's glands, cut vertically, and some of them obliquely or horizontally.

(H) In the mucous coat the **glands of Lieberkühn**, larger than those of the small intestine, and lined by cells—protoplasm cells—with nuclei near their attached ends. Amongst them are very many goblet-cells (fig. 271).

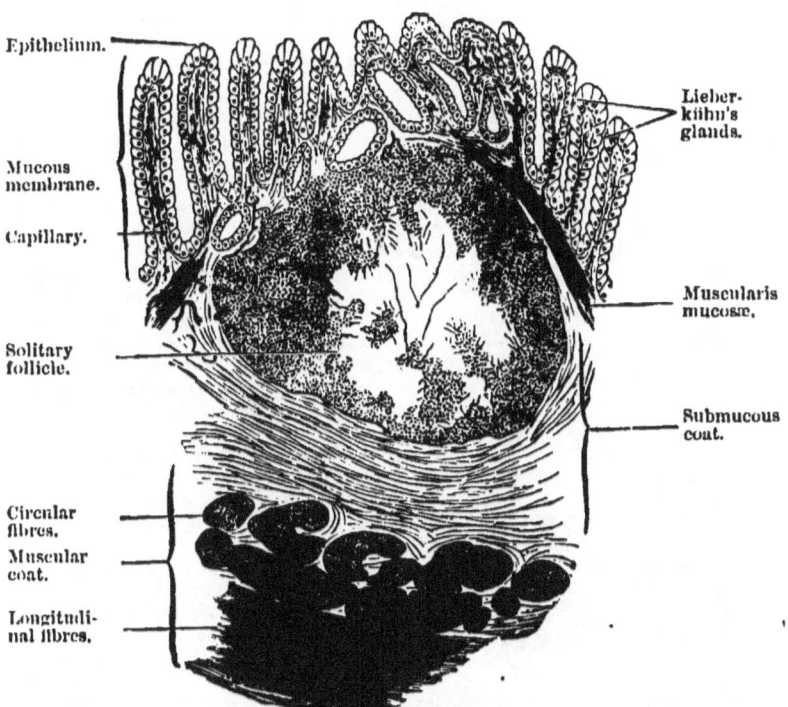

FIG. 270.—L.S. Large Intestine.

12. Vermiform Appendix (Rabbit).—This is instructive on account of the lymphatic follicles it contains and the micro-organisms contained in these follicles (p. 284). Fix it in mercuric chloride and treat it as recommended for the other parts of the intestine. The lymph-sinuses are readily injected by the "puncture" method with watery solution of Berlin-blue. Fix in Müller's fluid and harden in alcohol.

The lymph follicles are arranged in two or three rows, the one inside the other, but the innermost layer does not project beyond the level of the mucous membrane. The apices of the inner row are covered only by cylindrical epithelium.

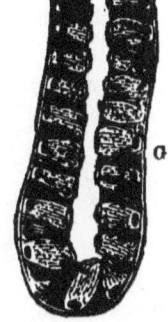

FIG. 271.—Lower End of a Lieberkühn's Gland, Large Intestine of Dog. *G.* Goblet-cells.

ADDITIONAL EXERCISES.

13. T.S. Villus.—Stain a hardened mucous membrane (with villi) of a dog or cat "in bulk" in borax-carmine or Kleinenberg's logwood, embed and cut T.S. in paraffin, fix on a slide, remove the paraffin with turpentine, clarify with clove-oil, and mount in balsam. Many of the villi will be cut obliquely.

(H) Observe the **lacteal** (L) in the centre, and round it the structure of the stroma of the villus, with several groups of non-striped muscle-cells (*m*) close to and surrounding the lacteal. In some animals (dog) there is a double row of these smooth muscles (fig. 272). The capillaries immediately under the epithelium (*c*), the smooth muscular fibres parallel to the lacteal, and the stroma—composed of anastomosing fine trabeculæ with parenchymatous cells and leucocytes—make up the elements present in the core of a villus.

Great differences exist in the relative size of the stroma and lacteal in villi. In the dog and cat the lacteal is relatively small, and the stroma abundant; in the rabbit the lacteal is very large, and the stroma scanty.

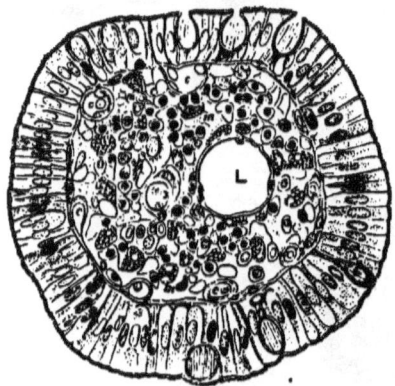

FIG. 272.—T.S. Villus (Dog). *L.* Lacteal; *m.* Muscle; *c.* Capillaries, × 400.

14. Non-Striped Muscle in Villi.—The method of Kultschitzky enables the course of the fibres to be more clearly traced. Harden (for twenty-four hours) a piece of dog's small intestine in the following fluid:—A saturated solution of potassic bichromate and copper sulphate in 50 per cent. alcohol (in the dark), to which, immediately before using, is added 5-6 drops (to 100 cc.) of acetic acid. The preparation and fluid must be kept in the dark. · Complete the hardening in absolute alcohol. Make sections, and stain them in *acid chloral-hydrate carmine*, which is made as follows:—Chloral hydrate, 10 grams; hydrochloric acid (2 per cent.), 100 cc. Add to this dry carmine (.75 to 1.5 gram), according to the strength of stain desired. Boil for one and a half hours, preventing evaporation by means of a cooling apparatus. Allow it to cool and filter. If the preparations stained with this dye be washed in 2 per cent. alum, the nuclei and other tissues become violet.

Sections stained thus show the course of the smooth muscle ·from the muscularis mucosæ obliquely between Lieberkühn's glands into the villi, where they are arranged in several bundles near the lacteal. They curve as they ascend in the villus, the concavity looking outwards, and are fixed or inserted close under the epithelium. They pass quite to the apex of the villus, becoming thinner as they go, where they each split up into a pencil of fibres, the fibres being inserted close under the epithelium.

15. Heidenhain's Method.—Harden small pieces of the small intestine, *e.g.*, dog or cat (24 hours), in a .5 solution of common salt saturated with mercuric chloride. Place it for twenty-four hours in alcohol 80, 90, and 95 per cent., and finally in absolute alcohol. Saturate with xylol, embed in paraffin, and cut thin sections, which are fixed on a slide with a "fixative." After the paraffin has been got rid of by turpentine or xylol, and the turpentine displaced by alcohol, the sections are stained on the slide with Ehrlich-

Biondi's fluid (p. 81), diluted with 40 or 50 volumes of water. It requires 10-12 hours to stain the sections, which are then mounted in balsam. This preparation is particularly valuable for studying the various forms of cells that occur in the stroma of a villus.[1]

16. **Absorption of Fat.**—(i.) Feed a frog on fat bacon; after two days kill it, and tease a portion of the mucous membrane of the intestine in normal saline, or dissociate it in dilute alcohol. Observe the isolated columnar cells crowded with fine granules of oil, which are blackened on the addition of osmic acid.

(ii.) A better plan is to stain the mucous membrane in osmic acid (24 hours), and embed it in paraffin. It is to be remembered, however, that steeping the

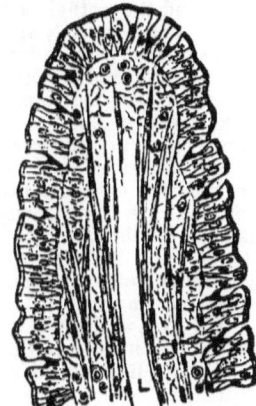

FIG. 273.—Villus of Dog's Intestine, with Lacteal and Non-Striped Muscle. *L.* Lacteal, × 250.

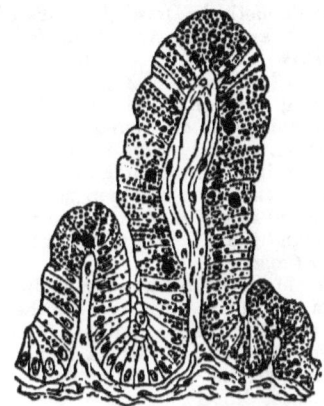

FIG. 274.—Section of Intestine of Frog. Absorption of fat. Osmic acid.

mass or sections of it in paraffin discharges in part the black colour of the fatty granules (p. 33).

(iii.) Feed a rat on bread and fat of bacon; kill it four hours afterwards by means of curare. Harden the small intestine in OsO_4.

(H) Observe the projections like folds of the mucous membrane. The columnar cells covering them are crowded with blackened particles—fatty granules blackened by osmic acid (fig. 274).

17. **Tubular Glands of the Intestine.**—In studying these, different fixing fluids are used according to the animal selected (Bizzozero).[2] Amongst the best fixing reagents are saturated watery solution of picric acid or picro-sulphuric acid (2 days), and wash in water (1 day), subsequently hardening in alcohol. This is suitable for the rectum of the mouse and dog. For the duodenum, either alcohol or Flemming's fluid, or Hermann's fluid (Lesson XXXV.). They are stained in safranin (1-2 hours), washed (10-15 secs.) in absolute alcohol, and then stained in hæmatoxylin. Cleared in bergamot oil and mounted in balsam. Safranin stains the nuclei of the cells, but not the mucin. The latter is stained by the logwood, and one can trace it from the goblet-cells passing into the lumen of the tubule. **Mitosis** may be observed both in the protoplasmic and goblet cells.

[1] *Pflüger's Archiv*, Supp. Bd., 1888.
[2] *Archiv f. mik. Anat.*, p. 325, 1892.

The mucus in the fresh goblet-cells occurs in the form of granules. This is best seen in a fresh preparation teased without the addition of any fluid, or at most in Müller's fluid. In sections fixed in Flemming's fluid and stained with safranin, the goblet-cells are reddish. The network in these cells may be stained with Bismarck-brown.

According to Bizzozero, the epithelial investment of the villi is not renewed by the proliferation of the cells covering it; but by the proliferation and upward growth of the cells lining Lieberkühn's follicles.

18. Terminations of Nerves in Stomach and Intestine.—Besides the gold-chloride method (Drasch,[1] nerves of the duodenum), and the methylene-blue methods as used by Aronstein,[2] Golgi's silver method has been recently applied by Erik Müller[3] for this purpose. The method used was Golgi's "rapid method" (Lesson XXX.), i.e., osmico-bichromate solution and subsequent staining with silver nitrate. The nerve-fibres become black, and numerous communications are found to exist between the plexuses of Auerbach and Meissner. A very large number of nerve-fibrils enter the villi, and are distributed in them, reaching to the cylindrical epithelium covering them. They end free and have not been seen to end in or between the cylindrical cells. Some end in the smooth muscular fibres of the villi. Other fibrils surround the gland tubes. It is said that branched cells, like nerve-cells, lie in the villi (*R. y Cayal*).

19. Methylene-Blue.—Immediately after death inject into the thoracic aorta of a guinea-pig or rabbit the following fluid:—1 gram methylene-blue BX in 300 cc. normal saline. Open the abdomen and expose the gut to the air for 2-3 hours. Place a thin piece of the wall of the small intestine in picrin-glycerine of S. Mayer (p. 192) and search for the reddish-stained nerve plexus (*S. Mayer*).

If a piece of the gut be teased in picrin-glycerine, it is easy to isolate blood-vessels with their nerves. It is to be noted that methylene-blue not only stains the axis-cylinders of nerves, but also the cement of epithelial cells, so that in some respects it acts like silver nitrate.

20. Vermiform Appendix (*Rabbit*).—Fix in Flemming's fluid, Fol's solution, or absolute alcohol.

A. *For Mitosis.*—(1.) Stain the sections (5-10 mins.) in

Gentian-violet	1 gram.
Absolute alcohol	15 cc.
Aniline oil	3 ,,
Water	80 ,,

(2.) Wash in absolute alcohol. (3.) Immerse (30-40 secs.) in 1 per cent. chromic acid. (4.) Again absolute alcohol (30-40 secs.); and (5.) in chromic acid. Then in absolute alcohol to remove all surplus dye. Balsam. In this way the chromatin of the nuclei is stained.

B. *For Micro-Organisms.*—Stain as above in (1). (2.) Absolute alcohol (5 secs.). (3.) Weak iodine solution, *i.e.*, Gram's method (p. 105) (2 mins.). (4.) Then alternately chromic acid and alcohol as above. Balsam (Bizzozero). The secret of getting good preparations is to wash them well in absolute alcohol until all surplus dye is removed.

21. Peyer's Patches and Phagocytosis.—(*a*.) Harden a Peyer's patch (rabbit) in absolute alcohol, cut sections in paraffin, and stain them in alum-carmine, or in addition with gentian-violet by Gram's method (p. 284)

---

[1] *Sitzb. d. k. Akad. d. Wissensch.*, Bd. 81, iii. Abth., Wien, 1880.
[2] *Anat. Anzeiger*, Bd. ii., 1887.
[3] *Archiv f. mik. Anat.*, Bd. xl., 1892.

(*Ruffer*[1]). (H). In the sections numerous small leucocytes will be found to have wandered from the lymph follicles between the epithelial cells. There are always to be seen several much larger cells—mono-nucleated—in the lymph follicles. The smaller forms—which may be mono-nucleated or poly-nucleated—have been called **microphages**, and the largest **macrophages** because both are capable of taking up bacteria—dead and alive—into their protoplasm and changing them by a process of intracellular digestion. Transition forms between microphages and macrophages of leucocytes may be found. In a preparation stained by Gram's method the bacilli—which are found in the lymph-cells, but not in the epithelial-cells—are stained deep blue-violet.

(*b.*) Sections of glands fixed in sublimate may be stained with Ehrlich-Biondi fluid and the same kinds of lymph-cells as occur in lymph-glands are found.[2] The most numerous are (1) small lymph-cells with a nucleus (stained green) surrounded by a small quantity of rose-coloured protoplasm. (2.) Larger cells with rose-red protoplasm. (3.) Granular cells which seem to correspond to eosinophilous cells. (4.) Cells undergoing degeneration; and (5.) Phagocytes.

LESSON XXVI.

LIVER.

It is composed of a large number of **lobules** (1 mm. ($\frac{1}{20}$ inch) in diameter), held together by a greater or less amount of connective tissue. Each lobule practically resembles its neighbour, and is composed of a mass of polyhedral or cubical liver-cells which have in relation with them blood-vessels and bile-ducts. The liver is covered by a **capsule**, which sends processes into the organ at the portal fissure, forming **Glisson's capsule**, which lies between the lobules in the portal canals, and surrounds the portal vein, bile-duct, and hepatic artery. The liver is supplied with blood by the **vena portæ** and hepatic artery; they enter at the portal fissure, and the former divides into branches which ramify between the lobules, constituting the **interlobular veins**. From these veins capillaries pass into and traverse the substance of the lobules, and converge to a veinlet in the centre of each lobule—the **central** or **intralobular veins** or rootlets of the **hepatic vein**, which form sublobular veins, and these in their turn form the hepatic vein which carries the blood from the liver to the inferior vena cava. The substance of each lobule between the capillaries is composed of liver-cells ($20\ \mu$, $\frac{1}{1000}$ inch in diameter), which form anastomosing columns, being more radiate next the centre of the lobule. Between the liver-cells,

---

[1] *Quart. Jour. Micr. Sci.*, xxx. p. 481.
[2] Hoyer, *Archiv f. mik. Anat.*, xxxiv., 1889.

but always separated from the blood-capillaries by hepatic cells or part of a cell, is a fine polygonal plexus of channels—the **bile capillaries**, which become continuous with interlobular bile-ducts at the margin of the lobule. The smaller interlobular bile-ducts unite to form larger bile-ducts, which are lined by columnar epithelium, and in the walls of the largest of them are mucous glands. The **hepatic artery** supplies chiefly the connective tissue between the lobules, and accompanies the branches of the bile-duct and portal veins, so that these three structures lie together in **portal canals**.

Methods.—Harden portions of the liver of a, *e.g.*, pig, rabbit, cat, and man in Müller's fluid or 2 per cent. potassic bichromate (10-14 days). If it be desired to retain the chromatin fibrils in the liver-cells avoid the chromic acid salts. Use sublimate, and then gradually increasing strengths of alcohol. Cut sections parallel to the surface of the organ, and others at right angles to it, the latter to include the capsule. The sections can be stained in hæmatoxylin and mounted in balsam; or picro-carmine, or picro-lithium carmine, and Farrant's solution may be used. Staining in bulk and cutting in paraffin is also good.

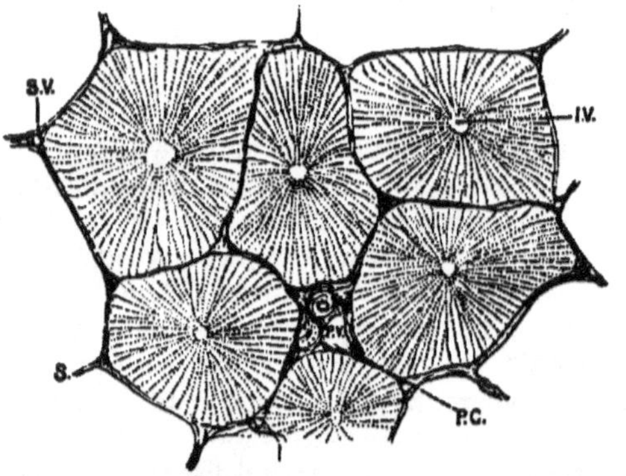

If unstained sections are mounted, the outlines of the tissues will be much better defined if they be soaked in 1 per cent. osmic acid (24 hours) previous to being mounted in Farrant's solution.

FIG. 275.—Liver of Pig, showing Lobules. *P.C.* Portal canal, containing bile-duct, hepatic artery, and portal vein (*P.V.*); *S.* Septa; *S.V.* Sublobular vein; *I.V.* Intralobular vein.

1. **Liver of Pig** (*Hæmatoxylin and Balsam*, or *Picrolithium Carmine*, or *Methylene-Blue*).

(*a.*) (**L**) Observe the polygonal lobules (fig. 275) mapped out from each other by a network of septa of connective tissue, or **Glisson's capsule** (fig. 275, S). In the centre of each lobule a small thin-walled vein, a rootlet of the **intralobular** or **hepatic vein** (I.V.).

(*b.*) At the periphery of each lobule sections of branches of the **portal** or **interlobular vein**; if possible, find a transverse section of the latter, and it will be found to be accompanied by similar sections of the bile-duct (one or more) and hepatic artery (fig. 277,

P.V.). Branches of all three structures always run together in the portal canals, and are surrounded by Glisson's capsule.

(c.) **Capillaries**—for the most part empty—running from without inwards in each lobule. Radiating from the hepatic veinlet columns of liver-cells, forming a network of secretory cells intertwining with the capillary plexus within the lobule.

(d.) (H) The columns of **hepatic cells** radiating from the hepatic veinlet, composed of polygonal or cubical cells, with granular contents and an excentrically-placed spherical nucleus. At the outer part of the lobule the network of cells is more polygonal, corresponding to the arrangement of the capillaries. Between the columns of cells, capillaries, sometimes with a few blood-corpuscles. The vessels, however, are not as distended as normal. Transverse

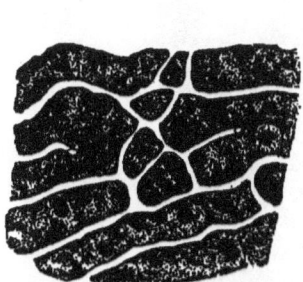

FIG. 276.—Columns of Liver-Cells from a Starving Dog.

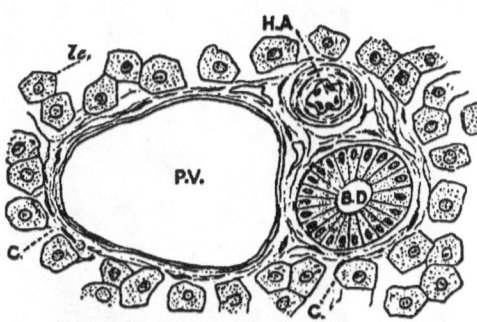

FIG. 277.—T.S. Portal Canal with Glisson's Capsule, enclosing Portal Vein (*P.V.*), Bile-duct (*B.D.*), and Hepatic Artery (*H.A.*). The slit is a lymphatic. *C.* Capillaries. *l.c.* Liver-cells.

branches running between adjoining capillaries, and the whole interwoven with the cellular network.

In a hæmatoxylin-stained section, note that the nuclei of the liver-cells are stained to a less degree than the other nuclei. The nuclei of the cells of the connective-tissue, bile-ducts, and capillaries are more deeply stained. In the picro-lithium preparations, the cells are yellowish with nuclei red, but of different shades of the same.

With the low power find a portal canal with its contents; fix them under the microscope and examine with (H).

(e.) One or more sections of interlobular **bile-ducts**, lined by a single layer of shorter or taller columnar cells; outside them connective tissue, disposed circularly and continuous with that of Glisson's capsule (fig. 277). A section of a branch of the hepatic artery and—the largest opening of all—of the portal vein. Some slits may be seen in Glisson's capsule; they are lymphatics.

2. Liver of Rabbit.—It is convenient to study this liver, because in some respects it shows a transition between that of the pig and man, as it occupies an intermediate place with regard to the demarcation of its lobules. In it the lobules are not nearly so well defined by connective tissue as in the pig or camel, while their separation from each other is more definite than in man.

3. Human Liver.—Hæmatoxylin, balsam.

(L) Observe the greater fusion of the lobules. Practically, the arrangements of blood-vessels and cells in other respects is the same as in the rabbit's liver (fig. 278). If the liver is anæmic, the intra-

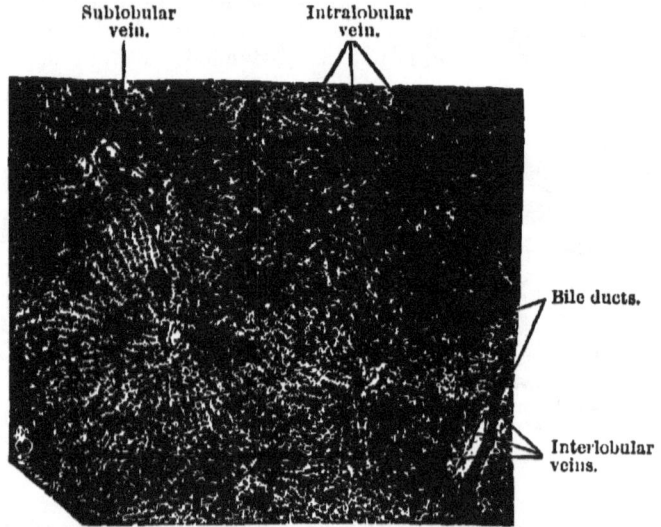

FIG. 278.—Section of Human Liver, showing Liver Lobules and the Radiate Arrangement of the Hepatic Cells from the Centre of each Lobule, × 20.

lobular blood-capillaries are narrow, and the liver-cells appear to compress and narrow them.

4. Liver of Frog or Newt.—Harden a small piece in absolute alcohol or ½ per cent. osmic acid (24 hours). Stain the mass "in bulk" in Kleinenberg's logwood or borax-carmine, and mount in balsam. Mount the osmic acid sections in Farrant's solution without staining, or stain them in safranin (48 hours).

(*a.*) (L) Observe the anastomosing system of gland-tubes made up of hepatic cells. Between them the narrower blood-capillaries, many of them filled with blood-corpuscles. Here and there black patches of pigment—melanin—especially in winter frogs (fig. 279).

(*b.*) (H) Each cell is polygonal, with a large spherical nucleus; the contents may be more or less granular, according to the phase of secretory activity of the cells. In some conditions the granules within the cells may be arranged next the capillary; in others they are more regularly scattered throughout the cell substance. The cells are arranged round the **bile-capillaries**. This is best seen in transverse sections of the latter, which appear as very small circular apertures bounded by four or five cells (fig. 279). When the tubes are cut longitudinally, the bile capillaries are seen to pursue a zigzag course between the cells, but this will be better seen in injected specimens.

(*c.*) The blood-capillaries with their nucleated blood-corpuscles, and note that there is always a cell or part of a cell between the blood-stream with its wide lumen, and the bile-channels with very narrow lumina. Any fatty granules present in the osmic acid section are black.

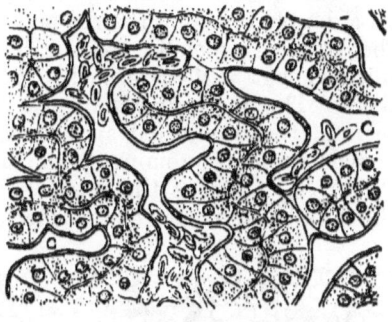

FIG. 279.—Liver of Frog. *C.* Capillaries. Osmic acid, × 300.

5. **Blood-Vessels of the Liver—Opaque Injections.**—Mount in balsam a section of a pig's liver, with the P.V. injected with a red opaque mass, and the H.V. with a similar yellow mass. The light from the reflector must be turned off, and light focussed on the preparation by a condenser (fig. 16).

(*a.*) (L) Observe the polygonal lobules, the branches of the portal veins around the periphery of the lobules, and sending fine branches into the latter. In the centre of the lobules the hepatic veinlets yellow, with capillaries converging to them. The capillaries within the lobules partly filled with red and partly with yellow mass, connecting the portal and hepatic venous systems.

(*b.*) Sometimes a longitudinal section of a lobule may be seen. Trace the veinlet to a larger branch under the lobule, *i.e.*, to a **sublobular vein**. The sublobular veinlets lie between the lobules, but they differ from branches of the portal vein in not being accompanied by a branch of the hepatic artery and bile-duct.

6. **Transparent Injections—Liver of Pig.**—Mount in the same way a transparent injection of the liver of a pig. P.V. blue, and H.V. red.

(*a.*) (L) The hepatic veinlet in the centre of the lobule. If cut transversely, it is circular; if obliquely, oval; and if the lobule be cut longitudinally, it appears as a central channel joining a sublobular vein. Trace outwards from this a radially-arranged capillary network, with its cross-branches, right out to the outer part of each lobule. Notice that the shape of the meshwork of capillaries is different at the centre and periphery. Between the lobules and outside each lobule branches of the portal vein.

7. Liver of Rabbit, Injected (fig. 280), *e.g.*, a section with its blood-vessels injected with different colours, *e.g.*, P.V. red (carmine gelatine) and H.V. blue (Berlin-blue gelatine). Also a section where all the blood-vessels are injected with a mass of one colour. In a double injection the red and blue masses do not always occupy the area corresponding to the blood-vessel into which they were injected, but with care such a piece can be found in an injected liver. The arrangement of the blood-vessels corresponds to that seen in the uninjected specimens, only in the former the blood-vessels are more prominent than the cells.

(*a.*) (**L**) Observe the **interlobular veins** round the periphery of the lobule, the **central or intralobular vein**, and the plexus of

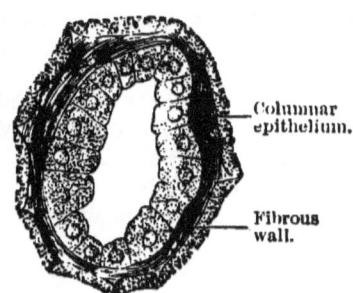

FIG. 280.—Injected Blood-Vessels of a Lobule of a Rabbit's Liver.

FIG 281.—Interlobular Bile-Duct.

capillaries connecting the two. The capillaries converge towards the centre of the lobule, where the meshes are more elongated. At the periphery of the lobule the capillary meshwork is more polygonal.

Very instructive also are those with only one set of vessels injected, *e.g.*, portal or hepatic. They serve to show the capillaries belonging to each area within a lobule.

8. Injected and Stained Section.—A beautiful preparation is obtained by staining a section (blue injection) with picro-lithium carmine. In it the vessels are blue, the cells yellow, and the nuclei of the latter red.

9. Bile-Ducts.—Mount in balsam a section of a rabbit's, or, better still, a guinea-pig's liver in which the bile-ducts have been injected with a concentrated watery solution of Berlin-blue. It is very difficult to get a perfect injection, but it is easy in the guinea-pig to inject the large bile-ducts. They are injected from the common bile-duct after ligature of the cystic duct. Sometimes it is better not to clamp the cystic duct, as after the gall-bladder is full it acts as a reservoir, and prevents too great pressure, causing extravasations of the injection fluid.

(*a.*) **(L) Large Bile-Ducts.**—Numerous sections of these lying in portal canals between the lobules. Each is provided with a fibrous coat containing circular, smooth, muscular fibres, lined by columnar epithelium. In the walls of the largest ducts are mucous glands, which open into the bile-duct.

(*b.*) **(H)** The bile-capillaries within the lobules appear as a fine hexagonal network of blue lines between the surfaces of the hepatic cells (fig. 282). If cut transversely, they appear as mere blue specks between adjacent cells. They are not to be confounded with the blood-capillaries, which are much wider, and are arranged in a different way.

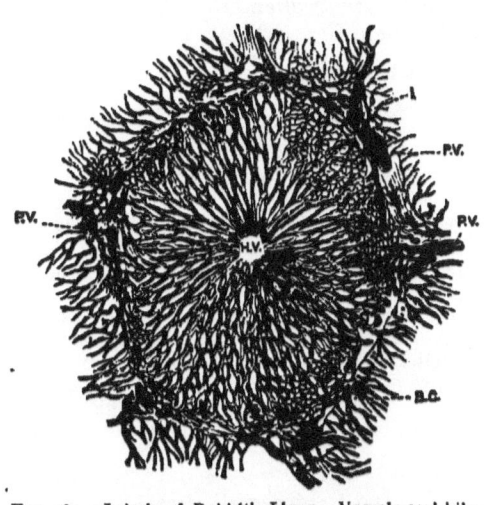

Fig. 282.—Lobule of Rabbit's Liver. Vessels and bile-ducts injected. *P.V.* Portal, and *H.V.* Hepatic vein; *B.D.* Bile-duct, and *B.C.* Bile-capillaries.

(*c.*) The columnar epithelium lining the interlobular **bile-ducts** and their fibrous walls (fig. 281).

10. Auto-Injection of Bile-Ducts.—Place a piece of indigo-carmine, about the size of a split pea, under the skin of the forearm of a pithed (brain-destroyed) frog. Tie the slit to prevent its escape. After twenty-four hours the whole frog will appear quite blue. Kill it, rapidly remove the liver, cut it in small pieces, and place it at once in absolute alcohol, which fixes the blue colour. After it is hardened, cut sections and mount in balsam.

(*a.*) **(H)** The blood-vessels are yellow, with gland-tubes between, but the bile-capillaries are blue. They can be seen as blue zigzag fine streaks between the cells when the tubes are cut longitudinally, and as very small dots when cut transversely. A thin section gives a clear view of the relation of blood-capillaries, cells, and bile-capillaries.

A good plan is to clarify an unstained section with clove-oil containing eosin. The cells are stained red, and form a sharp contrast to the blue. A watery solution such as picro-carmine cannot be used to stain the cells, as it rapidly extracts the blue from the bile-channels.

11. Fresh Liver-Cells.—Scrape the surface of a fresh liver, and observe in normal saline.

(H) Observe the pale, nucleated, polygonal, or cubical faintly-granular cells, often containing small refractive oil globules. Often the cells are broken up. There are always many blood-corpuscles in the field (fig. 94).

ADDITIONAL EXERCISES.

12. Connective-Tissue Stroma.—With a camel's-hair brush pencil away as many as possible of the hepatic cells from a thin section of any properly hardened liver. The connective-tissue network may be afterwards stained with picro-lithium carmine or eosin.

13. Intralobular Connective Tissue.—Various methods have been adopted to show the existence of connective tissue within the liver lobules and between its cells. The following methods reveal a network of black fibrils continuous with the interlobular tissue, and traversing as a network the substance of the lobule. This method is also useful for showing the trabeculæ of the spleen, and the fibrillar structure in the splenic corpuscles and lymph-glands generally.

(*a.*) *Böhm's Method.*

(1.) Harden small pieces (.5 cm. cube) in 0.5 per cent. chromic acid (2 days).
(2.) Silver nitrate .75 per cent. (3 days).
(3.) Wash in distilled water, harden, cut, and mount in balsam.

(*b.*) *Oppel's Method* is available for liver hardened in alcohol.

(1.) Yellow chromate of potash 10 per cent. (1 day).
(2.) Silver nitrate, large volume, .75 per cent. (2 days). Much silver chromate is formed, so that the silver must be frequently renewed.
(3.) Wash in water, harden in alcohol, sections, balsam.

By either method the liver cuts readily on freezing, but hand sections, I think, are best. Paraffin embedding is not to be recommended.

14. Bile-Capillaries (*Böhm*). **Golgi's method.**

(1.) Harden (3–4 days) small pieces of the liver (.5 cm. cube) in

Bichromate potash solution (3 per cent.) . 4 vols.
Osmic acid (1 per cent.) . . . 1 vol.

(2.) Place in 0.75 per cent. silver nitrate (2 days).
(3.) Wash in distilled water, harden in alcohol.
(4.) Sections. Balsam.

The bile-capillaries appear as a black polygonal network on a yellow ground. It is rare, I find, to have the whole thickness of the tissue equally well stained. Only a thin layer on the surface shows the capillaries well, but the result is excellent. I find that the sections keep for a long time if they are mounted under a cover-glass.

15. Glycogen in Liver-Cells.—(*a.*) The animal must be well fed, *e.g.*, rabbit, with carrots, and six hours or so thereafter it is killed. As glycogen is soluble in, or is at least extracted from, the liver by water, small pieces of the liver must be hardened in alcohol. In a section placed in a weak iodine solution or Lugol's solution (p. 93), the glycogen granules in the cell protoplasm are stained of a port-wine colour. In sections of liver sometimes one sees vacuoles from which the glycogen granules have been washed out.

(*b.*) Harden the liver of a well-fed frog in osmic acid, make thin sections, and irrigate with iodine. The granules of glycogen in the hepatic protoplasm are stained brownish.

16. Reaction for Iron (*Tizzoni's*).—(*a.*) Select the liver of a young animal, and harden it in alcohol. Place sections in the following fluid, which should be freshly prepared :—

Water	90 cc.
Hydrochloric acid (25 per cent.)	1.5 ,,
Ferricyanide of potash (1 : 12)	3 ,,

Mount in balsam. Particles of free iron are coloured blue. Particles of free iron are seen in the spleen, liver, and kidney by this reaction.

(*b.*) *Zaleski*.[1]—Harden the liver in 65 per cent. alcohol, then in 96 per cent. alcohol to which a few drops of sulphuretted hydrogen are added. After 24 hours the iron granules assume a green colour.

17. Injected Human Liver.—(*a.*) For a double injection, and in order to save injection-mass, the plan recommended by Orth is excellent, viz., to pass an elastic catheter as far as it will go into one of the branches of the portal vein, and through it to make the injection. The hepatic vein is treated in the same way. Sections of a well-injected part may be afterwards stained with picro-lithium-carmine.

(*b.*) In a way a natural injection of the liver is obtained by hardening a human liver which is congested, and contains a large amount of blood. Such livers are apt to show pigmentation of the liver-cells and other changes due to disease.

18. Pigment in Liver.—The presence of pigment in the liver-cells or capillaries is a matter of considerable importance in regard to the question of the destruction of blood-pigment in this organ, or whether the liver acts as a trap for pigment already altered by its passage through the spleen or gastro-mesenteric capillaries.[2]

In winter frogs the liver contains a large amount of black pigment, which lies in the blood-capillaries.

19. Granular Cells of Ehrlich.—Harden the liver (ox, pig) in absolute alcohol. Place sections for 24 hours in Westphal's fluid (p. 67), wash in absolute alcohol (4–6 hours) until all is clear, only the granular cells remaining stained. Balsam. The granular cells ("Mastzellen") are reddish violet, and lie in the interlobular connective tissue.

---

[1] *Zeitsch. f. Phys. Chem.*, xiv. p. 274, 1890.
[2] W. Hunter, *Brit. Med. Jour.*, Nov., Dec. 1892.

LESSON XXVII.

TRACHEA—LUNGS—THYROID GLAND.

The **trachea**, a fibro-muscular tube, the wall of which contains 16–20 C-shaped pieces of hyaline cartilage held together by a fibrous membrane. Behind, the rings are deficient, and the trachea is membranous, and there it is strengthened by smooth muscle—*trachealis muscle*—which stretches between the ends of the cartilages. It is lined by a mucous membrane, which is united to the outer fibrous coat by a submucous coat. The **mucous coat** consists from within outwards of—(1.) Stratified, columnar, ciliated epithelium; (2.) Basement membrane; (3.) A basis of connective tissue with capillaries, and infiltrated with adenoid tissue; (4.) A layer of elastic fibres arranged longitudinally. Outside this is a loose submucous coat, in which lie the acini of the glands.

The **intra-pulmonary bronchi** are lined by stratified ciliated epithelium resting on a basement membrane. Outside this is a basis of fibrous tissue, with numerous longitudinally-arranged elastic fibres, and some adenoid tissue. Outside this, again, is a completely circular layer of smooth muscle—*bronchial muscle*. Then follows the *submucous coat* with its vessels, glands, and in some animals (cat) masses of adenoid tissue. Most externally is the fibrous coat, in which are embedded *several* pieces of hyaline cartilage of irregular shape. As the bronchi pass into the lung, they divide and form smaller and smaller tubes, until they end in terminal bronchi or **bronchioles**. Each bronchiole, with thin walls, no glands or cartilage, and the epithelium cubical and non-ciliated, opens into several wider expanded parts—**infundibula** or **alveolar passages**—which are beset with **air-cells or alveoli**. The alveoli are spherical or polygonal vesicles, which open by a wide opening into the infundibula; the air-vesicles, however, do not open into each other. The air-cells are lined by a layer of squames—large, flattened, irregular plates—with small granular cells here and there between them.

Blood-Vessels.—The branches of the *pulmonary artery* accompany the bronchial tubes, and finally terminate in a rich capillary plexus over and outside the basement membrane of the air-vesicles. The blood is returned by the *pulmonary veins*. The *bronchial vessels* seem to supply chiefly the connective tissue along the bronchi and in the septa. Numerous ganglia exist in the intra-pulmonary *nerves*. Lymphatics are numerous.

THE RESPIRATORY ORGANS.

Methods.—(i.) Place small portions of the trachea of a cat and the human trachea in .2 per cent. chromic acid (10–14 days) or sublimate, and harden afterwards in gradually increasing strengths of alcohol.

(ii.) Fill the lungs of a cat with .2 per cent. chromic acid, and after closing the trachea suspend them in a large quantity of the same fluid. After two days cut them into small portions and keep them (14–20 days) in fresh chromic acid solution and complete the hardening in alcohol.

(iii.) Harden small parts of the human lung—as fresh as possible—in the same fluids.

(iv.) Fix the trachea in Flemming's fluid, harden in alcohol, and stain with safranin. Admirable for epithelium.

Trachea.—Make transverse sections across the trachea of a cat, and also longitudinal vertical sections, so as to include the cartilage of two or three rings. This is best done by staining in bulk and cutting in paraffin. If done by freezing, stain T.S. and L.S. sections in hæmatoxylin, and mount in balsam; others in picro-carmine, and mount in Farrant's solution.

1. T.S. Trachea (fig. 283).

(*a*.) (**L**) Observe internally the **mucous coat**, outside it the **submucous coat**, and external to this an incomplete ring of hyaline cartilage embedded in the **outer or fibrous coat**.

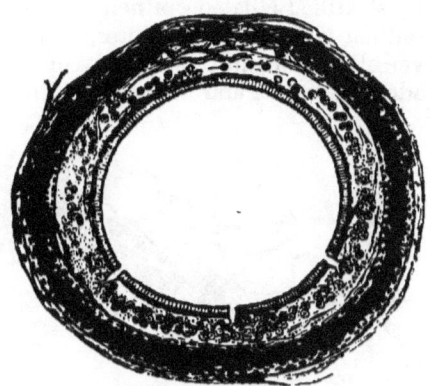

FIG. 283.—T.S. Trachea of Kitten, × 15.

The fibrous coat becomes continuous with the submucous coat.

(*b*.) (**H**) The **fibrous coat** of connective tissue, and embedded in it an incomplete ring of hyaline cartilage. Notice the arrangement of the cartilage cells in the latter. The ring is deficient posteriorly, but bridging over the gap and extending between the ends of the cartilage there is a transverse band of smooth muscle, the *trachealis muscle*. Outside it are muscular fibres cut transversely. The coats just inside the muscle are apt to be thrown into folds in a trachea detached from all its surroundings. Sections of nerves—perhaps with ganglionic cells—may be seen near the muscle.

(*c.*) The **submucous coat**, composed of more open connective tissue, and continuous with the former. In it are the acini of **mucous glands**; but as the glands are more abundant in the spaces between the cartilages, their distribution is better seen in a longitudinal section. It is rare to find a duct, as they pierce the mucous coat obliquely, and open on its inner surface.

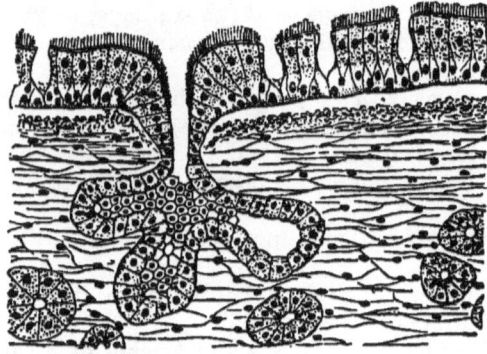

FIG. 284.—L.S. Mucous Membrane of Trachea of Cat.

(*d.*) The **mucous coat**, composed of fibrous tissue covered by stratified ciliated epithelium. Under the epithelium is a longitudinal layer of elastic fibres, which are therefore cut across transversely. Within this, a basis of connective tissue infiltrated with adenoid tissue, and internal to this again a structureless basement

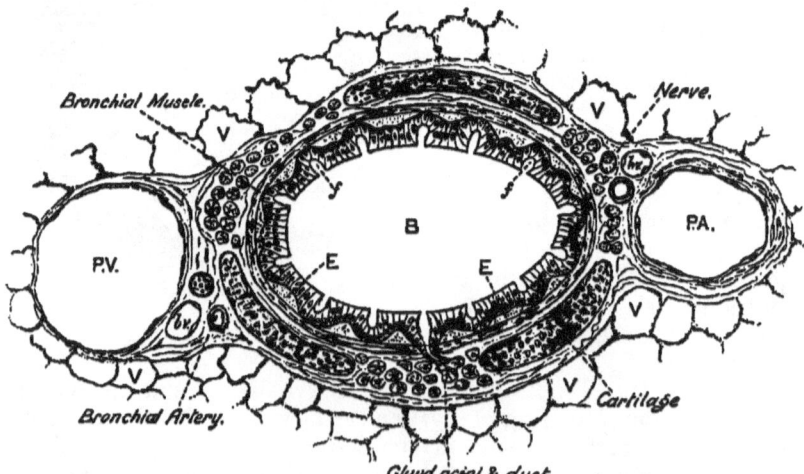

FIG. 285.—T.S. Intra-Pulmonary Bronchus of Cat. *PA* and *PV*. Pulmonary artery and vein; *bv*. Bronchial vein; *V*. Air-vesicles.

membrane, best seen in very thin sections, and best of all in a human trachea. Resting on the basement membrane is the *ciliated epithelium*. It occurs in several layers, but only the superficial layer of cells is ciliated. In the lower layers of cells—three or four layers—some are pear-shaped and the lowest more oval. Very frequently a thin layer of mucus is adherent to the cilia. The arrangement of the cells of the mucous layer is the same as in the bronchus (fig. 286).

2. **L.S. Trachea** (fig. 284) (**L** and **H**).—Compare these with the previous section. The coats are the same, but several oval pieces of hyaline cartilage are seen one after the other embedded in the fibrous coat. The elastic fibres are now seen arranged longitudinally. The acini of the mucous glands are most numerous in the interspaces between the cartilages, and amongst the acini may be seen leucocytes and some adenoid tissue. If a gland-duct be found, it opens by a funnel-shaped expansion on the free surface of the mucous membrane. If it be desired to study the glands, use hæmatoxylin and eosin as stains.

3. **T.S. Human Trachea** (**L** and **H**).—Observe the general similarity to the previous preparation. Here, however, the glands are well-developed; the basement membrane is well marked, and in picro-carmine specimens is stained red. Some of the superficial epithelial cells are apt to be detached.

4. **T.S. Intra-Pulmonary Bronchus.**—(*a.*) (**L**) Observe the *fibrous coat*, and outside it the vesicular tissue of the lung. In the fibrous coat—two or three—pieces of hyaline cartilage. The *submucous coat*, with its glands (fig. 285).

(*b.*) Inside this a complete ring of smooth muscle—*bronchial muscle*—perforated here and there by the ducts of the glands. Immediately inside this, in the *mucous coat*, several layers of longitudinal elastic fibres cut transversely. Most internal, **ciliated epithelium**, like that lining the trachea, and resting on a basement membrane. The mucous membrane is frequently thrown into ridges or folds.

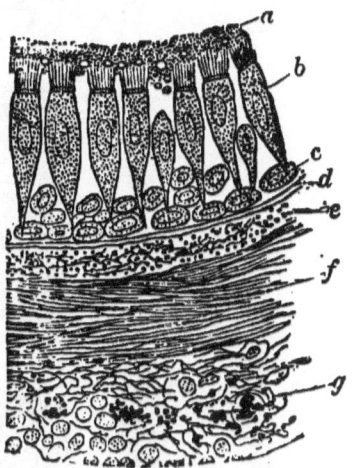

FIG. 286.—T.S. Mucous Membrane of Human Bronchus. *a.* Mucus; *b.* Ciliated cells; *c.* Deep cells; *d.* Basement membrane; *e.* Longitudinal elastic fibres; *f.* Bronchial muscle; *g.* Connective tissue, with leucocytes and pigment.

(*c.*) In the fibrous coat, external to the cartilages, search for sections of two large vessels—the pulmonary artery and vein—and of small branches—the bronchial vessels. Also several sections of nerves; in the course of some of them may be found ganglionic cells.

298   PRACTICAL HISTOLOGY.   [XXVII.

(*d.*) (H) Study specially the mucous membrane. Note the epithelium, basement membrane, and the longitudinal layer of elastic fibres under it (fig. 286).

5. Vesicular Structure of the Lungs.—Make sections of a lung hardened by the freezing method. Let the sections be cut so as to include the pleura and subjacent lung. Stain in hæmatoxylin, and mount in balsam.

(*a.*) (L) Observe the **pleura**, with its two layers, the deeper

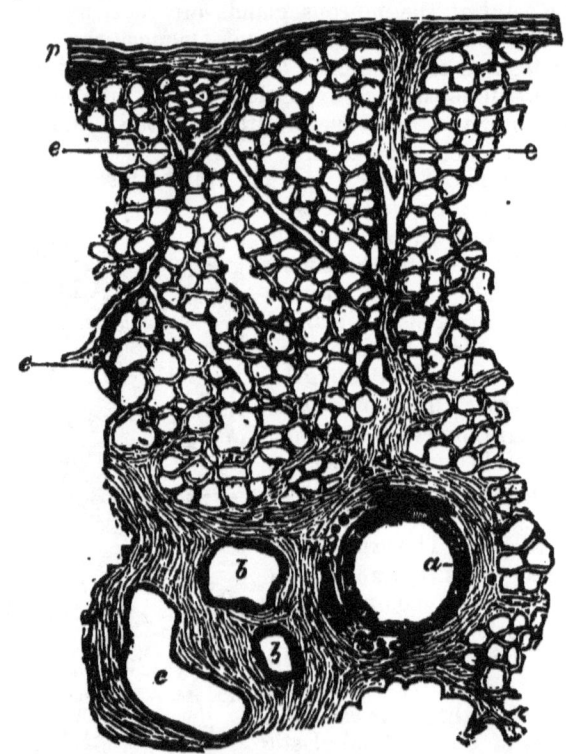

FIG. 287.—V.S. Human Lung. *p.* Pleura; *a.* Epithelium of a bronchus; *b.* Blood-vessel; *c.* Pulmonary vein; *s.* Interlobular septum, continuous with deep layer of pleura; *v.* Air-vesicles (× 50 and reduced ½).

layer sending fine septa—**interlobular septa**—into the lung between its alveoli (fig. 287).

(*b.*) The **alveoli** or *air-cells* cut in every direction; some appear as an open network, and others with a base. The outline of the alveoli may be somewhat irregular, according to the extent of distension of the lung. Here and there, between groups of alveoli, may be seen a wide passage—the *infundibulum*.

(c.) (H) In each alveolus stained oval nuclei, belonging to the squamous epithelium lining it, but the outline of the squames themselves cannot be seen. Sections of the capillaries on the wall of the alveolus and their nuclei stained. In the thin walls, separating adjoining alveoli, fine elastic fibres.

6. **Respiratory Epithelium of the Alveoli.**—(i.) Remove the lungs from a young kitten, and fill them with a ½ per cent. solution of silver nitrate. Run in the fluid from a pipette provided with a bulb on its stem. Pump the lungs from time to time to remove as much air as possible. After half an hour replace as much as possible of the silver solution by means of alcohol. Tie the trachea, and suspend the lungs in alcohol till they are hardened. Cut sections by freezing, mount in Farrant's solution, and expose the sections to light. They become brown. Sections can be stained in hæmatoxylin or picro-carmine.

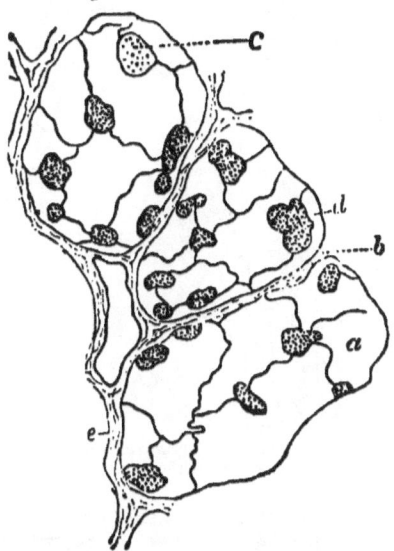

FIG. 288.—Alveoli of Lung of Kitten, Silvered. *a, b.* Squames; *d.* Granular cells; *c.* Young epithelium-cell; *e.* Alveolar wall.

(ii.) Fill the lungs with .25 per cent. $AgNO_3$, and suspend them in .5 per cent. $AgNO_3$ (1–2 hours). Diffusion takes place. Cut into pieces and place in alcohol (80 per cent.).

(H) Select an alveolus which is so divided as to have a base, and observe the silver lines showing the boundary-lines of the squamous epithelium which lines it. At the junction-points of some of these are groups of two or three small, polyhedral, granular cells stained deep brown (fig. 288).

7. **Blood-Vessels of a Mammalian Lung.**—Mount in balsam sections of an injected lung.

(L and H) Observe the alveoli, each beset with a dense plexus of capillaries (fig. 289), and especially where the edge of an alveolus is seen, the wavy course of the capillaries passing from one side to the other of the inter-vesicular septa. Search for the termination of a branch of the pulmonary artery and the commencement of the pulmonary vein.

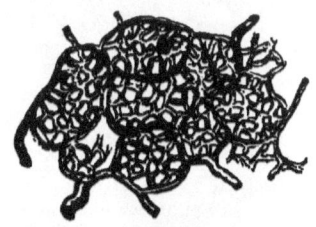

FIG. 289.—Capillaries of Human Lung, Injected, × 90.

8. Fresh Lung.—Tease a small piece of lung in normal saline. It is difficult to get rid of all the air, but this may be done by beating the tissue with a needle.

(H) Observe the large number of fine **elastic fibres**, which branch and anastomose. They can be rendered more evident by running in dilute caustic potash (2 per cent.) under the cover. This destroys to a large extent the other elements.

9. Dried Lung.—(i.) With a dry razor make thin sections of a dried and distended lung. Examine the section in water, taking care that it does not curl up, which it readily does. Get rid of the air-bubbles by pressing on the section with a needle.

(ii.) Make rather thick sections parallel to the pleura, and examine them by reflected light. A very good idea of the air-vesicles is thus obtained.

FIG. 290.—T.S. Dried Lung. *a.* Vesicles; *I.* Infundibula.

(*a.*) (L) Observe the **air-vesicles**, and the thin partitions between them. Also sections of the **infundibula** or **alveolar passages**. Connective-tissue septa may be seen. Of course the finer details of structure cannot be made out (fig. 290).

10. Fœtal Lung.—This serves very well for showing that a lung

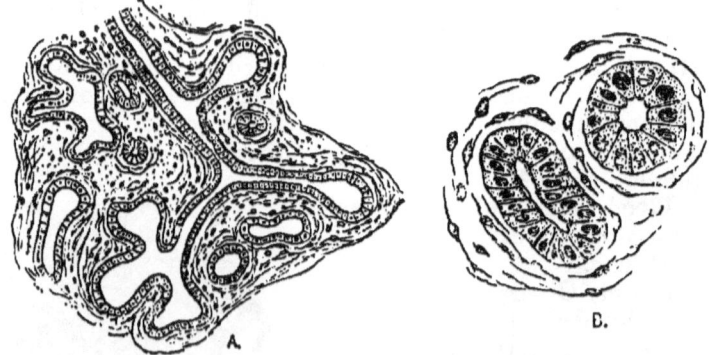

FIG. 291.—T.S. Fœtal Lung, showing a Bronchus Terminating in Air-Vesicles, × 75. *B.* Epithelium lining an alveolus, × 300. Müller's fluid and hæmatoxylin.

is made up of lobules. Harden in Müller's fluid (14 days) the lungs from a human fœtus (12 cm. in length). Cut sections and stain them either in hæmatoxylin or picro-carmine. Stain in bulk and cut in paraffin.

(*a.*) (L) Observe the **pleura** sending well-marked septa into the lung, thus defining the **lobules**, which are polygonal, about 1 mm. in diameter, and separated from each other by connective tissue.

(*b.*) In each lobule sections of bronchi, which can be seen occasionally to terminate in several vesicles, thus presenting a very gland-like arrangement (fig. 291). Many of the alveoli are cut across, and appear like sections of tubes lined by columnar epithelium. There is much embryonic connective tissue between the alveoli.

(H) Select an alveolus, and note that it is lined by a layer of low, columnar, granular epithelium, while individual alveoli are separated from each other by much embryonic connective tissue, with numerous cells, and as yet few or no elastic fibres.

THE THYROID GLAND.

Methods.—(i.) Harden pieces of the human thyroid, or the complete thyroid of a cat or dog, in Müller's fluid (3 weeks) and then in alcohol. Sections are stained with hæmatoxylin and mounted in balsam. Or stain in bulk and cut in paraffin.

(ii.) Fix in Flemming's fluid (1–3 hours), and stain—wash thoroughly, alcohol—with Ehrlich-Biondi fluid, or stain with Heidenhain's logwood.

11. **T.S. Thyroid Gland.** — (*a.*) (L) Composed of small polygonal **lobules** united to each other by loose connective tissue. In each lobule a large number of completely closed **acini** held together by loose connective tissue.

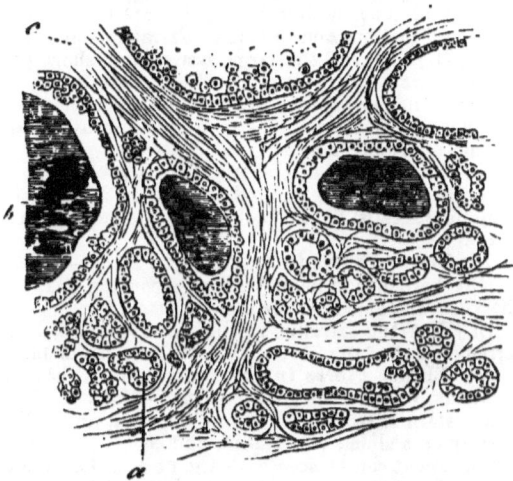

FIG. 292.—T.S. Thyroid Gland. *a.* Closed vesicle.

(*b.*) (H) The spherical **acini** (fig. 292), bounded by a basement membrane, are lined by a single layer of low cubical epithelium, and contain a homogeneous fluid. This fluid, however, is often of a colloid nature, and is then a pathological product. It is stained by logwood. Numerous sections of blood- and lymph-vessels outside the basement membrane of the acini.

12. Injected Thyroid (*thickish section in balsam*).—(L) Numerous large vessels in the connective tissue, with a plexus of capillaries over the acini, but outside their basement membrane.

ADDITIONAL EXERCISES.

13. Lung of Newt.—(i.) This is an elongated sac, and is of comparatively simple structure. Fill a lung with gold chloride (.5 per cent.), and suspend it in a few cc. of the same fluid for twenty minutes. Reduce the gold by exposure to sunlight in water feebly acidulated with acetic acid. Mount a portion of the thin wall in glycerine.

(*a.*) (L) Observe islands or small groups of epithelial cells. They lie in the **intercapillary spaces**. The capillaries are wide, and form an anastomosing network. Outside this capillary layer is a layer of smooth muscle, and one of fibrous tissue.

(ii.) If the gold chloride be reduced by formic acid (25 per cent. in the dark), the epithelium lining the lung is shed, and then the **nerves to the lung**—with many ganglia in their course—can be seen.

14. Lung of Frog.—(i.) Fill a lung with dilute alcohol, suspend it in the same fluid (twenty-four hours), lay open the lung and pencil away the inner lining epithelium and mount in Farrant's solution.

(*a.*) (L) Observe the large, coarse, but short primary septa, which project inwards from the wall of the lung towards the large central cavity. From them secondary septa pass to form a trabecular arrangement, thus giving the interior of the lung a honeycomb-looking appearance. The trabeculæ consist chiefly of smooth muscle.

(ii.) The **nerves** of the frog's lung are readily demonstrated by the gold chloride formic acid method and the methylene-blue method. The numerous ganglionic cells in the course of the nerves have a straight and a spiral process.

15. Elastic Fibres in Trachea and Lung.—(i.) Stain a longitudinal section of the hardened human trachea according to the method described in Lesson X. 10. The elastic fibres become black.

(ii.) Use the safranin method (Lesson X. 9). This shows beautifully the arrangement of the fibres, now of a purplish or black tint.

16. Elastic Fibres in the Lung.—(i.) I have devised the following two methods, which give good results:—Make sections of a dried and distended lung, stain a section in dilute magenta, and allow the section to dry completely on a slide; add balsam and cover. The elastic fibres are red, and their arrangement can be seen with the utmost distinctness.

(ii.) Or, stain a section in methyl-violet, and clarify it with the aniline-oil and xylol mixture (p. 123).

LESSON XXVIII.

KIDNEY—URETER—BLADDER.

KIDNEY.

IF **a kidney** be divided longitudinally, one distinguishes **a cortical and a medullary part**, the latter consisting in different animals of one or more pyramidal portions—the **pyramids of Malpighi**—whose apices project into the pelvis of the kidney, while their bases are surrounded by cortical substances. The medullary portion is subdivided into the **boundary or intermediate zone** and the **papillary portion** (fig. 293). The kidney is covered by a loosely adherent **capsule**. It is a compound tubular gland, consisting of numerous uriniferous tubules closely packed together with very little connective tissue between them, the connective tissue carrying the blood-vessels, lymphatics, and nerves. The uriniferous tubules pursue a straight course in the medulla, but they exhibit a contorted or convoluted arrangement in the cortex, although bundles of straight tubules—the **medullary rays or pyramids of Ferrein**—pass into the cortex from the medulla (fig. 293). Each uriniferous

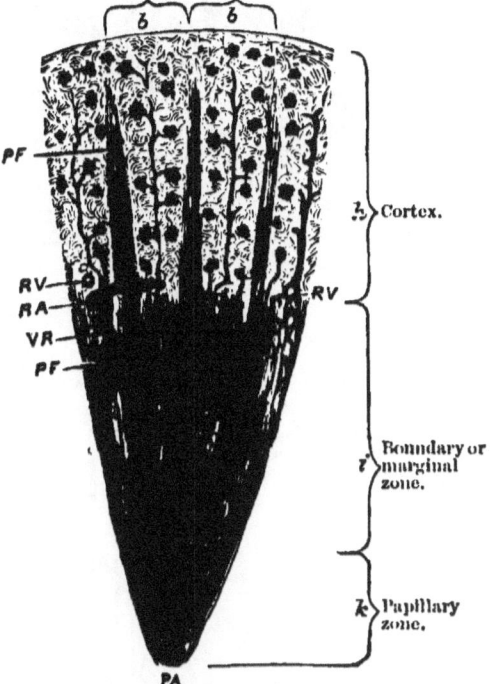

FIG. 293.—L.S. of a Pyramid of Malpighi. *PF.* Pyramids of Ferrein; *RA.* Branch of renal artery with an interlobular artery; *RV.* Lumen of a renal vein receiving an interlobular vein; *VR.* Vasa recta; *PA.* Apex of a renal papilla; *bb.* Embrace the bases of the lobules.

tubule consists of a basement membrane lined by a single layer of epithelium. The tubules alter their character and course in different parts of the organ. The tubules begin in the cortex in a

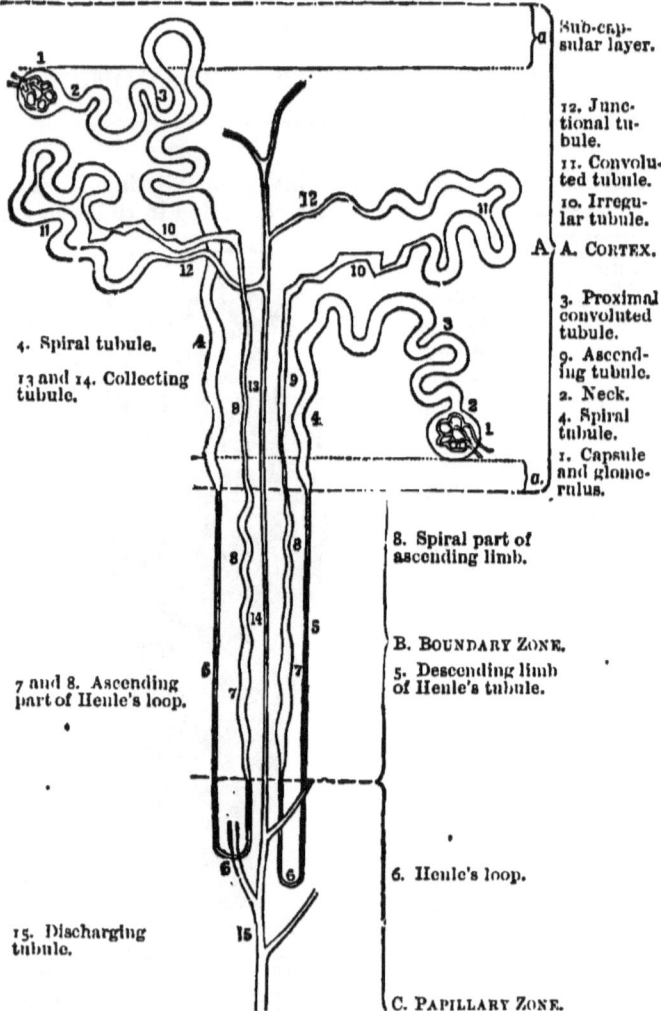

Fig. 294.—Diagram of the Course of Two Uriniferous Tubules.

globular dilatation—the **Malpighian capsule**—which encloses a tuft of vessels—the **glomerulus** (fig. 294). The capsule leads into a narrow **neck**, which passes into the **first** or **proximal convoluted** tubule, which in turn is continued into the **spiral tubule**. After this it narrows suddenly and runs into the medulla as the **descending tubule of Henle**, where it forms a narrow loop—**loop of Henle**—and passes in the reverse direction towards the cortex as the **ascending tubule of Henle**. In the cortex it has a zigzag course—**irregular tubule**—and again becomes wider and convoluted—**second or distal convoluted tubule**—which leads into a straight tube—**junctional tubule**—which joins a straight or **collecting tubule**. This passes

Tabular View (after Schäfer) of the Parts, Situation and Epithelium of a Uriniferous Tubule.

Portion of Tubule.	Kind of Epithelium (always in a single layer).	Situation of Tubule.
1. Capsule . . .	Flattened, reflected over glomerulus	Labyrinth of cortex.
2. First convoluted tube	Cubical, rodded cells, which interlock	,, ,,
3. Spiral tube . .	Like last, but "rods" very distinct	Medullary ray of cortex.
4. Descending limb of Henle's tube	Clear flattened cells	Boundary zone and partly in papillary zone.
5. Loop of Henle .	,, ,,	Papillary zone of medulla.
6. Ascending tube of Henle	Cubical, rodded, sometimes imbricated	Medulla and medullary ray of cortex.
7. Irregular zigzag tube	Cells cubical, very "rodded," lumen small	Labyrinth of cortex.
8. Second convoluted tube	Like those of (2), but longer and more refractive	,, ,,
9. Junctional tube .	Clear, flattened, cubical cells	Labyrinth passing to medullary ray.
10. Straight or collecting tube	Clear cubical and columnar cells	Medullary ray and medulla.
11. Duct of Bellini .	Clear columnar cells	Opens at apex of papilla.

straight through the cortex and medulla, receiving other similar tubes as it goes; and becoming wider, it opens as a **discharging tubule** (duct of Bellini) on the apex of a Malpighian pyramid. The tubes are lined throughout by a single layer of epithelium, which, however, changes its characters in the different parts of the tube. The *Malpighian capsule* is a globular expansion ($200\ \mu$), composed of a basement membrane lined by a single layer of squamous epithelium. In the *convoluted tubules*, proximal and distal, the epithelial cells are somewhat cubical, but their outlines are not well defined. They each contain a spherical nucleus, and their protoplasm is "rodded," especially at the outer part. In the *spiral tubule* the cells are not unlike those of the convoluted tubule, but they are not so tall, and therefore leave a more distinct lumen, and they are not so markedly "rodded." In the *descending part of the loop* and the *loop* itself—very narrow (10–15 μ)—the cells are clear and flattened, with a bulging opposite the nucleus, and these projections alternate with those on the opposite side of the tubule. The *ascending limb* (30 μ wide) has somewhat cubical cells, which leave a regular lumen. They are striated, and often present an imbricate arrangement. The *irregular* or *zigzag tubule* bends on itself with sharp angles, and is wide, with an irregular lumen. Its cells stain deeply with staining reagents, and are conspicuously striated in their outer part. The *second* or *distal convoluted tubule* is like the proximal. The *junctional* and *collecting tubules* are lined by low, columnar, clear, transparent cells with small nuclei. The cells do not stain readily. In the *discharging tubules* the cells have the same general character, but they are taller and more columnar. Fig. 294 shows the general arrangement of the tubules, and from it it is easy to see in what part of the kidney each kind of tubule is placed.

Blood-Vessels.—The renal artery enters the kidney, splits into branches which run towards the cortex, and at the junction of the cortex and medulla form incomplete arterial arches (fig. 293). From these arches arise the **radiate** or **interlobular arteries**, running in the cortex between two medullary rays in a radial direction towards the surface. They give off at intervals on all sides short, slightly-curved vessels—**vasa afferentia**—which run without branching to end within the Malpighian capsules, and there form the **glomeruli**. The **vas efferens** comes out of the Malpighian capsule close to where the afferent vessel enters it and at the pole opposite to the origin of the uriniferous tubule. The efferent vessel splits up into capillaries, which ramify amongst the tubules of the cortex. The blood is returned from the cortex by interlobular veins (fig. 293), which run alongside of the corresponding arteries. The medulla is supplied by leashes of vessels—**vasa recta**—which for the most part proceed from the arterial arches. The vasa recta are pencils of arterioles (10–15), splitting up into capillaries which ramify between the tubules of the medulla; the medulla, however, is not so vascular as the cortex. The blood is returned by corresponding veins. The **connective tissue** is very scanty in the cortex, but abundant in and near the apices of the Malpighian pyramids.

Methods.—(i.) Harden small pieces in a 2 per cent. solution of potassic bichromate (18–20 days) or Müller's fluid. Corrosive sublimate is also very good, and so is alcohol. The pieces should not be large, and should be cut according as a longitudinal or transverse section is desired. Stain and cut in bulk in paraffin. In all cases the pieces should include both the cortical and medullary portions, and should not be more than half-an-inch in thickness.

(ii.) Place small pieces in boiling water and then complete the hardening in Müller's fluid or alcohol.

(iii.) To fix the epithelium use Flemming's fluid, *e.g.*, for mouse's kidney.

Make radial sections from the cortex to the apex of a Malpighian pyramid. For a general view stain a section in hæmatoxylin and mount it in balsam. Another one should be stained in picrocarmine or picro-lithio-carmine and mounted in Farrant's solution. For unstained sections, steeping them for twenty-four hours in 1 per cent. osmic acid before mounting them in Farrant's solution is excellent. The best way is to begin with a section of the entire kidney of a small animal, such as a mouse, rat, or guinea-pig. In this way a good view is got of the entire organ in section.

1. Radial Section of Kidney of a small mammal.

(*a.*) (**L**) Observe the **capsule**; it is thin and apt to fall off; the cortical and medullary portions of the parenchyma. The **medulla**, composed of straight tubules of different sizes, and running radially from the pelvis of the kidney outwards.

(*b.*) Trace some of the straight tubules outwards through the intermediate layer, in bundles—the **pyramids of Ferrein** or **medullary rays**—into the cortical layer (fig. 293, *PF*). Many medullary rays pass from a Malpighian pyramid, and they run radially outwards in the cortex nearly to its outer part—although they do not reach the surface—becoming narrower as they are traced outwards.

(*c.*) In the **cortex**, between every two medullary rays, are convoluted tubes, twisted and cut in every direction, and two rows of glomeruli, enclosed in their capsules—the **labyrinth**. Here and there a glomerulus may have fallen out, and the space it occupied be left as a round aperture. The regular arrangement of the glomeruli will only be seen provided the section runs parallel to the course of the medullary rays. The glomeruli are confined to the cortex, but none of them reach quite to the capsule.

(*d.*) (**H**) **In the cortex**, the **Malpighian capsules**, each enclosing a tuft of capillaries or **glomerulus**. They consist of a structureless membrane lined by a single layer of squames. The oval, flattened nuclei of the latter are seen lying on the inner surface of the capsule. Within each capsule a cluster of capillaries—**glomerulus**—arranged in several groups. Although the squamous lining of the capsule is reflected over the capillaries, it is not easy to distinguish the nuclei of these cells from the very numerous nuclei of the capillaries (fig. 295). The basement membrane of the capsule is continuous by a narrow neck with the basement membrane of a convoluted tubule, but it is only in the rare case where the section cuts the capsule at this level that this connection is seen. This connection must usually be made out on isolated tubules, although sometimes it is seen in sections of a mouse's kidney.

(e.) The **convoluted tubules**, with a sinuous or twisted course, cut some longitudinally and others transversely. Each tube is lined by a single layer of cells, the individual cells not sharply mapped off from each other, and leaving a small lumen in the centre. The outer part of each cell is striated or "rodded," and near its centre it contains a spheroidal nucleus (fig. 296).

FIG. 295.—Glomerulus and Sections of Convoluted Tubules of a Kidney, × 300.

(f.) **Irregular Tubules.**—Here and there in the cortex may be seen zigzag portions of tubules of unequal width, and usually more deeply stained than the rest, and the epithelium distinctly rodded (fig. 297).

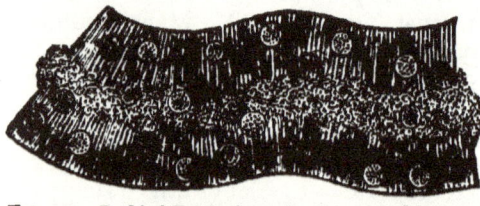

FIG. 296.—Rodded Epithelium of a Convoluted Tubule. Ammonium chromate.

(g.) Note the small amount of **connective tissue** between the tubules of the cortex. It is best seen just outside the Malpighian capsules. It is far more abundant in the medulla.

(h.) In the medulla, the **straight tubes**, which are largest and widest where they are about to open on the apex of a Malpighian pyramid—discharging tubules—but it is not so easy to get a section showing this, as from their radial arrangement they are apt to be cut obliquely. The collecting and discharging tubes have a wide lumen, and are lined by a layer of clear columnar cells with ovoid nuclei. Trace the straight tubules outwards towards the surface; they become smaller, still their lumen

FIG. 297.—Irregular Tubule, Kidney of Dog. Müller's fluid.

remains distinct, and they are lined by clear nucleated short columnar or cubical epithelium.

(*i.*) In the medullary rays in the **intermediate layer** very narrow tubes—the descending portion of the **looped tubule of Henle** (fig. 300), not unlike fine capillaries—may be seen, and also the wider, more deeply stained ascending portion of the same tubule.

2. T.S. of the Apex of a Malpighian Pyramid (*Hæmatoxylin and Balsam*).

(*a.*) (**L** and **H**) Observe the large amount of connective tissue between the tubules. The large collecting tubules are cut across, so that their large lumina and the clear columnar epithelial cells lining them are distinctly seen (fig. 298). Not infrequently the epithelium falls out, and then the connective tissue appears as a network with round or oval holes.

3. T.S. Medullary Ray.—Sections should be cut in various directions and at different levels in the cortex. One of the most instructive is to cut a section across the direction of the medullary rays. In it will be seen groups of transverse sections of the various tubes—collecting, ascending, and descending portions of the looped tubule of Henle—which make up a medullary ray. Between the rays there are sections of glomeruli and convoluted tubules.

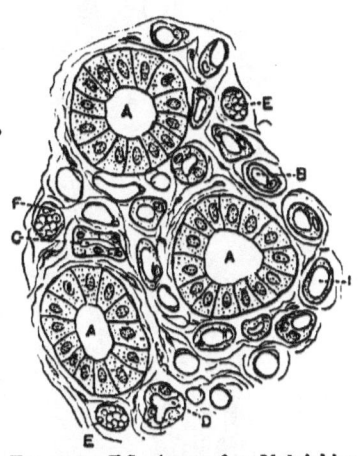

FIG. 298.—T.S. Apex of a Malpighian Pyramid. *A.* Large collecting tubules; *B, C, D.* Wide and narrow parts of Henle's tubule; *E, F.* Blood-vessels.

Blood-Vessels of the Kidney.—The sections should be cut from a kidney injected with carmine gelatine or Berlin-blue gelatine; they should be radial, not too thin; best from the kidney of a small mammal, and mounted in balsam. In injecting a kidney do not use too great pressure, as otherwise the glomerular capillaries are apt to burst, and the injection-mass passes into the tubules. When the stellate veins on the surface are seen to be well injected, one may infer that the mass has traversed the glomerular capillaries. In a small animal inject from the aorta, in a large one from the renal artery.

4. T.S. Injected Kidney.—(*a.*) (**L**) Between the cortex and the medulla, *i.e.*, in the boundary or intermediate layer, sections of the larger branches of the renal artery and vein will be seen (fig. 293, RA), *i.e.*, along the "line of vascular supply." From

310 PRACTICAL HISTOLOGY. [XXVIII.

FIG. 299.—Blood-Vessels of the Kidney. *A.* Capillaries of cortex; *B.* Of medulla; *a.* Interlobular artery; 1. Vas afferens; 2. Vas efferens; *r, e.* Vasa recta; *VV.* Interlobular vein; *S.* Origin of a stellate vein; *i, i.* Bowman's capsule and glomerulus; *P.* Apex of papilla; *C.* Capsule of kidney; *e.* Vasa recta from lowest vas efferens.

these the **interlobular arteries** and **veins** (fig. 299, *a*), running outwards in the middle between every two medullary rays (fig. 293). From these are given off on all sides, at short intervals along the course of the vessels, short arteries —the vasa afferentia. Each **vas afferens**, after a very short course, runs to a Malpighian capsule, and splits up into capillaries to form the glomerulus.

(*b.*) If the capsule is cut at the level where the artery enters, the short and narrower efferent vein or **vas efferens** will be seen coming out at the same pole, and after a similar short course breaking into a network of capillaries, which surrounds the tubules of the cortex. Around the convoluted tubules the meshes of the network are somewhat polygonal, but in the medullary rays in the cortex they are more elongated.

(*c.*) From this capillary network arise short veins, which join the interlobular veins (fig. 299). Quite at the upper end of the interlobular artery a few branches are given off, which end directly in this capillary network without the intervention of glomeruli. Under the capsule the small veins are arranged in a stellate manner, constituting the *venæ stellatæ* (fig. 299, S).

(*d.*) Passing again to the "line of vascular supply," the short vessels which break up into leashes or bundles of small blood-vessels (fig. 299, *r*), the **vasa recta** (arteriæ rectæ, venæ rectæ), which run between the medullary rays into the medulla, where they form an elongated capillary meshwork between the straight tubules of the medulla. The medulla is not so vascular as the cortex, and it has no glomeruli.

5. Injected and Stained T.S. Kidney.—A not too thin injected section—say injected with a blue mass—may be stained with picro-carmine, which makes the tubular structures more distinct by staining their nuclei.

6. Fresh Kidney—Glomeruli and Basement Membranes of Tubules.—Expose a fresh kidney to the air for a day or two according to the temperature. Cover it to prevent evaporation. Tease part of the medulla and cortex in normal saline.

(*a.*) (**H**) Observe the long, partly empty, structureless, **basement membranes** of the tubules, often exhibiting folds; also isolated cells of the tubules.

(*b.*) Isolated **glomeruli**, consisting of several tufts of capillaries. The nuclei of their cells are revealed by dilute acetic acid.

(*c.*) Narrow tubules, not unlike blood-capillaries, but they possess a wall lined by a layer of squames, the nucleated part of the squames alternating on opposite sides of the tubule. These are the descending part of the **looped** tubule (fig. 300).

FIG. 300.—Part of Descending Looped Tubule of Henle.

7. Isolated Cells of the Different Tubules.—Place very small pieces (size of a split-pea) 24-48 hours in a 5 per cent. solution of neutral ammonium chromate. Wash in water, and tease small fragments in a 50 per cent. solution of potassic acetate, or, without washing, tease a fragment in the chromate solution.

(*a.*) (**H**) Note specially the cells of the convoluted tubules and those of the ascending limb of Henle's loop. They show the "rodded" character of the outer part of the protoplasm. Adjoining cells tend to interlock with each other (fig. 301).

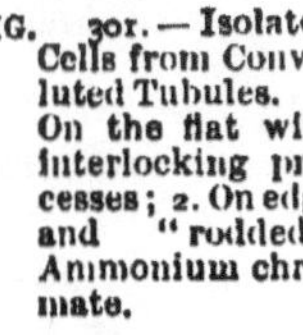

FIG. 301.—Isolated Cells from Convoluted Tubules. 1. On the flat with interlocking processes; 2. On edge and "rodded." Ammonium chromate.

8. Isolated Tubules.—Place small pieces (size of a pea) of the kidney of a mouse, tortoise, or guinea-pig (3-4 hours) in pure hydrochloric acid or 40 per cent. nitric acid (2-4 hours), wash in water and leave them in water for 18-24 hours. They swell up, and their constituents readily fall asunder. Place a fragment in water slightly tinged with iodine and gently tap the glass slide, or stain with dilute acid-fuchsin. This is sufficient to cause the tubules to fall asunder.

(**L** and **H**) In a part from the cortex search for a convoluted tubule still connected with its capsule, the twists on the tube itself,

312　PRACTICAL HISTOLOGY.　[XXVIII.

and the transition to the narrow part of the looped tubule of Henle. This is perhaps easiest to obtain from a mouse's kidney. Isolated straight tubules from the medullary part. To preserve this preparation, suck away the fluid and gently replace it with glycerine.

Ureter and Bladder.—Harden small pieces in Müller's fluid (14 days) or in corrosive sublimate (5-6 hours), and then in gradually increasing strengths of alcohol. Make transverse sections of the one and vertical sections of the other. Flemming's fluid for epithelium. Stain in hæmatoxylin and mount in balsam; and others in picro-carmine and mount in Farrant's solution. Or stain in bulk in borax-carmine and cut in paraffin. The ureter of a cat or monkey does very well, and it is well to use the contracted bladder of a small mammal.

9. T.S. Ureter (fig. 302).—(*a.*) (L) Externally is a thin **fibrous**

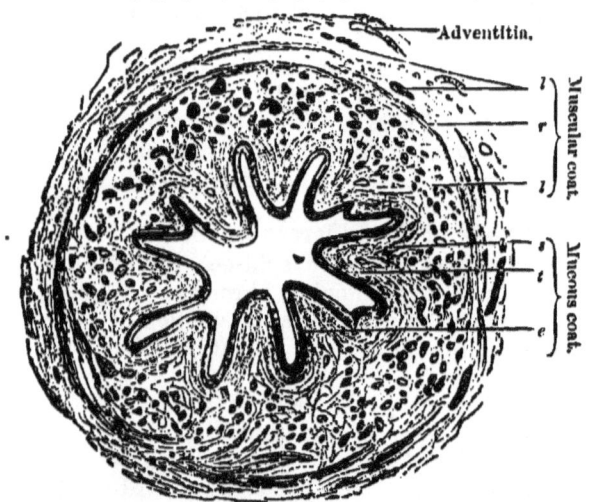

FIG. 302.—T.S. Lower Part of Human Ureter. *e.* Transitional epithelium; *s.* Submucosa; *l* and *r.* Longitudinal and circular smooth muscular fibres; *t.* Tunica propria. Müller's fluid, × 15.

coat or **adventitia**, consisting of connective tissue with the large vessels and nerves.

(*b.*) The **muscular coat** consists of an outer layer of smooth muscle arranged *circularly*, and inside this a *longitudinal* coat arranged in bundles, the latter, of course, divided transversely. In the lower part of the ureter there is an incomplete longitudinal muscular coat outside the circular coat.

(*c.*) The **submucous coat** is thin, and passes gradually into

(*d.*) The **mucous coat**, which is thrown into ridges or folds, and is lined by transitional epithelium.

(*e.*) (H) The transitional epithelium, and the variation in the shape of the cells arranged in several layers, from the free mucous surface outwards.

10. V.S. Bladder (fig. 303).—(*a.*) (L) Most externally a thin fibrous coat, in some places covered by a serous coat.

(*b.*) The **muscular coat**, composed of longitudinal and circular smooth fibres. Usually there is an outer and an inner longitudinal layer with a circular layer between. The appearance of these layers will depend on the plane of the section.

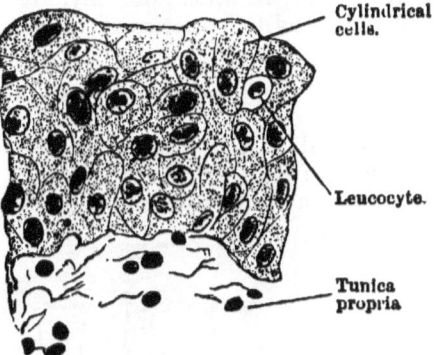

FIG. 303.—V.S. Epithelium of the Mucous Membrane of a Human Bladder. Müller's fluid, × 560.

The **submucous** and **mucous** (H) coats are like those of the ureter.

(*c.*) (H) The transitional epithelium and the great variety in the shape of the cells from below upwards. Occasionally amongst the epithelial cells are leucocytes (fig. 303). It is important to observe that the thickness and shape of the lining transitional epithelium will necessarily vary with the state of distension or contraction of the bladder.

11. T.S. Penis, *e.g.*, of a monkey or other small mammal. Harden it in alcohol or fix in Flemming's fluid. Make T.S. and stain with safranin.

(*a.*) (L) It consists of the two *corpora cavernosa*, placed dorsally, one on each side of the middle line, and inferiorly the *corpus spongiosum*. In the centre of the latter is the urethra as a transverse slit.

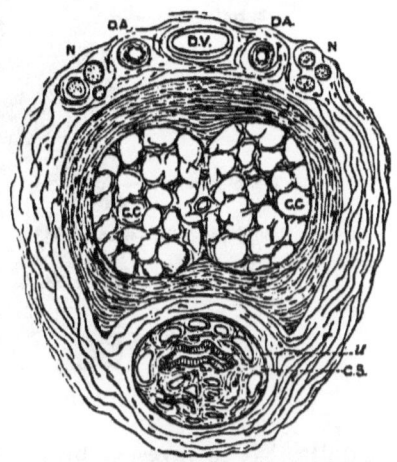

FIG. 304.—T.S. Penis of Monkey. *CC.* Corpus cavernosum; *CS.* Corpus spongiosum; *s.* Septum; *u.* Urethra; *DV.* Dorsal vein; *DA.* Dorsal arteries; *N.* Nerves

In the prostatic part the mucous membrane is lined by transitional epithelium, but in the body of the penis it is lined by the columnar variety, except at the meatus, where it is stratified.

(b.) Note the trabeculæ of connective tissue—bounding wide spaces—in the cavernous part. The whole is surrounded by a tough capsule, in which, dorsally, are sections of blood-vessels—one vein (DV) and two arteries (DA)—and nerves (fig. 304, N).

SUPRARENAL CAPSULE.

It is well to remember that there are great variations in the structure of this gland in different animals. The **suprarenal capsule** is a ductless gland, consisting of a **cortical zone** and a

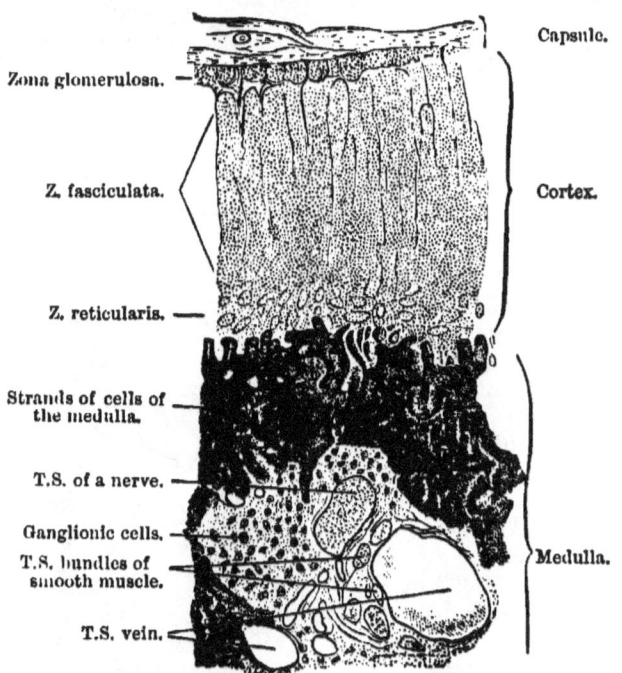

FIG. 305.—T.S. Human Suprarenal Capsule, × 50.

medulla. It is invested by a fibrous **capsule** which sends septa into the gland. Especially in the cortex, these septa run so as to give a columnar arrangement to the cells which lie between them. The parenchyma of the organ consists of cells which vary in their characters in different regions. The cells of the cortex (15 μ) are polyhedral, nucleated, granular, yellowish-coloured cells, arranged under the capsule in rounded groups—**zona glomerulosa**. Next this is the widest zone—the **zona fasciculata**. Next the medulla is the **zona reticularis**. In the **medulla** the cells are often irregular or polygonal with a clearer protoplasm, which is often tinged of a yellowish or brownish colour. There are numerous **vessels** and **nerves**, the latter with ganglionic cells.

Harden the suprarenal capsules of a guinea-pig in Klein's fluid (5-7 days), and then in alcohol, or fix in Flemming's fluid. Harden a human suprarenal in Kleinenberg's fluid (24 hours), and then in alcohol. Make radial sections, and stain some in hæmatoxylin, and others in picro-carmine. Or stain and cut in paraffin.

12. **V.S. Suprarenal Capsule.**—(L) Observe the arrangement already described. It is to be noted, however, that there are great variations in the structure of these organs in different species of animals.

(H) Examine the cells in the various zones (fig. 305).

ADDITIONAL EXERCISE.

Termination of Nerves in Suprarenal Capsules.—The literature and most recent results will be found in Fusari's paper[1] (with a plate). He used capsules of the mouse, rabbit, pig, cat, and new-born infant. The method employed was the quick method of Golgi (Lesson XXX.), *i.e.*, small fresh pieces are placed in the osmico-bichromate fluid (3-10 days), and afterwards in 1 per cent. silver nitrate solution (1-2 days).

LESSON XXIX.

SKIN AND EPIDERMAL APPENDAGES.

THE SKIN.

The **skin** consists of the **epidermis** and **cutis vera, dermis,** or **corium.** The epidermis consists of many layers of stratified squamous epithelium (p. 317). The corium is composed of a basis of fibrous connective tissue—white and yellow fibres—and its surface is thrown into a number of **papillæ**, which differ in size, number, arrangement, and form in different parts of the body. Undivided conical elevations are called *simple papillæ*, but when these are beset with smaller papillæ, they are called *compound papillæ*. The epidermis completely covers in the apices of the

[1] *Archiv ital. de Biol.*, xvi. p. 262, 1891.

papillæ, and also dips down into the furrows between adjoining rows of papillæ, so that the surface of the skin is smooth, although the arrangement of the papillæ is readily detected by the lines on the palmar aspect of the hand and foot. The fibrous tissue of the cutis, next the epidermis, forms a very thin modified layer with scarcely any fibrils and no corpuscles. This layer acts the part of a basement membrane, and is continuous with the basement membrane of a sweat-gland. In the dermis, the bundles of white fibres interweave with each other, and form a dense tissue; at the lower part of the skin it becomes more open in texture, and gradually passes into the subcutaneous tissue. Elastic fibres in the form of networks exist in large numbers in the cutis; they are finer in the papillæ, and coarser lower down.

The **subcutaneous tissue** consists of a complex system of trabeculæ of fibrous tissue, and in some of the meshes are lobules of fatty tissue forming a fatty layer, constituting the *stratum adiposum*.

The arrangement of the *blood-vessels* is stated at p. 325. There are also numerous *lymphatics* and *nerves*—some of the latter with peculiar terminations—*glands* (sweat and sebaceous), and, in some situations, *hairs* with their *hair-follicles*.

It is important to distinguish between the hairy skin and the parts of the skin without hairs. The *non-hairy parts* are the volar surfaces of the hands, feet, fingers, and toes, nails, lips, mammary papillæ, certain parts of the external genitals, and the inner part of the external auditory meatus. The *hairy parts* are the remainder of the skin. The non-hairy parts are concerned with direct tactile sensations, the hairy parts with indirect tactile sensations, the hairs themselves being the chief tactile organs (*Blaschko*[1]). This observer has shown that the epidermis projects into the cutis vera in the form of septa, varying in form and distribution in different parts of the skin (p. 325).

Methods.—The skin must be prepared in various ways according to the particular part which it is desired to study. For a general view proceed as follow:—(a.) Procure a fresh portion of human skin from the palm of the hand or sole of the foot, cut it into pieces about 1 cm. square, and remove most of the subcutaneous fat; pin it, epithelial surface downwards, on a piece of cork, and harden it in absolute alcohol (12 hours). Renew the alcohol for another twenty-four hours. Sections may be cut by freezing, and then stained with hæmatoxylin or picro-carmine (the latter to be mounted in Farrant's solution). Better still, stain the whole "in bulk" in borax-carmine or hæmatein, and embed and cut it in paraffin.

---

[1] *Archiv f. mik. Anat.*, xxx. p. 495.

Mount in balsam. Or skin so hardened may be double stained in bulk, first in borax-carmine and then in hæmatoïn.

(*b.*) Harden the skin in Müller's fluid.

(*c.*) For the layers of the epidermis fix say small pieces of the skin in 1 per cent. osmic acid or Flemming's fluid, and harden in alcohol. Stain sections in safranin.

1. V.S. Skin, Palm of Hand.

(*a.*) (**L**) The **epidermis**, consisting of many layers of stratified squamous epithelium, resting on the **cutis vera, dermis, corium**, or **true skin**. The latter consists of connective tissue, and is provided with finger-shaped elevations or **papillæ**, which project into the deeper layers of the epidermis, the latter in the form of septa, dipping in between the papillæ (fig. 306).

(*b.*) The **epidermis**, composed entirely of stratified epithelial cells. Proceeding from the outside (fig. 307), observe—

(i.) The **stratum corneum**, of variable thickness, consisting of many layers of flattened or slightly fusiform, clear, non-nucleated cells united to each other. As the cells are seen on edge, they are very thin. Those on the surface are about to be shed, and consist of keratin.

(ii.) The **stratum lucidum**, a thin, narrow, clear, homogeneous layer, composed of two or more layers of flattened cells, containing sometimes a rod-shaped nucleus. The cells do not stain well with dyes. The eleidin granules seem to become fused together and form the basis for cornification, as the cells are changed and become corneous.

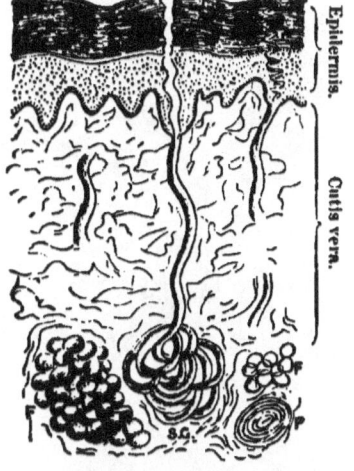

FIG. 306.—V.S. Skin of Palmar Surface of Finger. *F.* Fat; *P.* T.S. Pacinian corpuscle.

(iii.) The **stratum granulosum**, a somewhat thicker layer, composed of ovoid cells two or three rows deep. Each cell is distinctly granular, and usually this layer stands out deeply stained, because its granules of **eleidin or keratohyalin** are stained with the carmine. The cells, like the foregoing, are devoid of "prickles."

(iv.) The **stratum Malpighii**, several layers of more plastic cells. At the lowest part, where they rest on the papillæ of the true skin, the cells—prickle-cells—are smaller and columnar in shape (with oval, vertically-placed nuclei), but above this they become more spheroidal or polygonal, and each one is distinctly nucleated.

(c.) The **cutis vera**. The **papillæ**, conical elevations projecting into the Malpighian layer. They consist of compact fibrous tissue. The rest of the skin consists of bundles of white fibrous tissue interwoven with networks of elastic fibres, and at its lower part masses of fat-cells. The connective tissue and fat-cells below become continuous with the **subcutaneous tissue,** which is of a more open texture; but there is a gradual transition from the one to the other. The nuclei of the connective tissue corpuscles appear as red oval dots.

In sections of the **sudoriferous glands,** their coils (in the deeper layers of the corium), their ducts running vertically through the skin, and a corkscrew passage in the epidermis may be seen. In some of the papillæ observe a **touch-corpuscle** (p. 320), and in the subcutaneous tissue occasionally sections of **Pacinian bodies** (p. 320).

(*d.*) (**H**) Observe in the **epidermis** the shape and characters of the successive layers of epithelium. In the Malpighian layers, "**prickle-cells,**" *i.e.*, cells connected with each other by fine "inter-cellular bridges," are better studied in an osmic acid section (fig. 97). (See also Lesson IV.)

(i.) In the *stratum Malpighii* the lowest cells are arranged in a single layer of elongated, somewhat columnar cells (6–12 μ), with large oval nuclei surrounded by granular protoplasm. The lower ends of the cells frequently exhibit processes which fit into the dermis. The remainder of the cells of this layer are irregularly cubical, and exhibit prickles (Lesson IV., and p. 127). In the dark races the particles of melanin, which give the dark colour to the skin, are present in the cells of this layer, especially in the deepest layer of cells. Nuclei are sometimes seen in process of division.

(ii.) The cells of the *stratum granulosum* are arranged in two or more layers, and are flattened horizontally, so that they are lozenge-

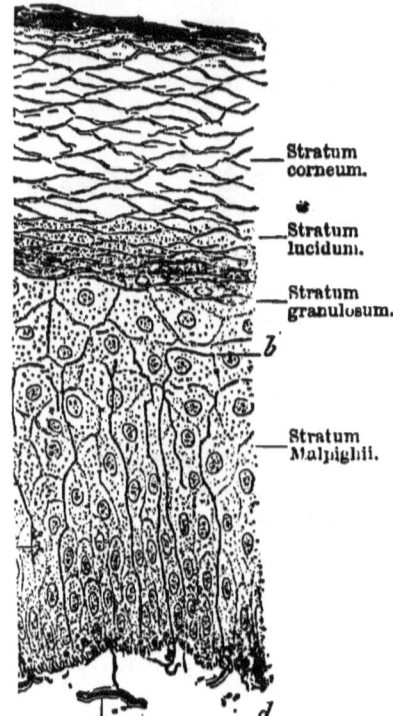

FIG. 307.—V.S. Human Epidermis with Terminations of Nerve-Fibrils. *n.* Nerve; *d.* Dermis; *b.* Branches of nerve-fibrils.

shaped, devoid of prickles, and are crowded with granules of *eleïdin*, a substance, apparently one of the stages on the way to the body keratin. These cells stain deeply with picro-carmine, hæmatoxylin, and osmic acid. They are soluble in caustic potash, and in this respect differ from keratin.

(iii.) The *stratum lucidum* is composed of two or more layers of flattened transparent cells, with no prickles, and free from granules, but with a horizontally-placed, rod-shaped nucleus. They do not stain readily.

(iv.) The *stratum corneum* or horny layer consists of horny squames composed of keratin (Lesson IV.).

(e.) The **sweat-glands** are most numerous in the palm of the hand and sole of the foot. Each gland is a simple tube coiled up at its lower extremity into a coil $\frac{1}{50}$ inch in diameter. To see their whole course—coil and duct—the sections must not be too thin, and should be parallel to the course of the gland. The coil of the gland lies in the subcutaneous tissue. The **secretory** part of the tube consists of a basement membrane lined by a single layer of nucleated transparent cubical or columnar cells surrounding a small but distinct lumen. Between the epithelium and the basement membrane is a longitudinally-disposed layer of smooth muscular fibres. The coil also contains a part of the sudoriferous canal or duct (fig. 308). The latter is narrower than the secretory part, and consists of a basement membrane lined by several layers of polyhedral cells. There is no muscular layer, but internal to the epithelial lining there is a delicate membrane or cuticle.

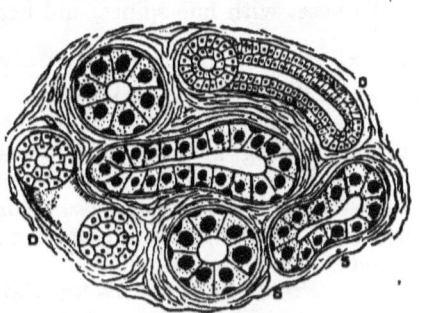

FIG. 308.—Section of Part of Coil of a Sweat-Gland. *D.* Duct; *S.* Secretory part, × 300.

Trace the coil into its duct, which runs vertically through the cutis vera with a slightly wavy course. It has a basement membrane lined by two or three layers of short cubical cells, which, if traced upwards, become continuous with those of the Malpighian layer of the epidermis, looking like a funnel-shaped expansion. The lumen of the duct is distinct. The *basement membrane* becomes continuous with the altered superficial layer of the corium just under the epidermis. The *lumen* of the duct is continued upwards in a corkscrew spiral through the epidermis. A complete view of its course is only obtained in a thickish section. In a thin section the twistings are of course divided.

2. **V.S. Skin of Negro** (H).—Harden in alcohol, and stain the sections slightly in eosin. Mount in balsam. Observe the granules of melanin in the deepest layers of the epidermis.

3. **V.S. Skin of Finger** (*Double Stained*).—Stain a section first in methyl-green iodide and clarify it with clove-oil coloured with eosin. Wash out the clove-oil with xylol and mount in balsam. The stratum corneum is green, and so are the nuclei of the other epidermic and connective-tissue cells.

4. **Prickle-Cells and Touch-Corpuscles.**—Place in 1 per cent. osmic acid or Flemming's fluid (24 hours) a very small piece of fresh skin from the palmar surface of a finger. Wash it well in water and complete the hardening in alcohol. Make V.S. and mount in Farrant's solution. Or cut in paraffin, stain in safranin, and mount in balsam.

(*a.*) (H) Observe the various layers of the epidermis, but in the Malpighian layer note the *prickle-cells*. The cells appear to be joined by their edges by fine striæ, the striæ leaving small spaces between them (fig. 97). The striæ are fine "intercellular bridges," stretching from one cell to another, and it is only when the epidermis is dissociated and the bridges broken that these cells appear as cells beset with fine spines, and hence they were called "prickle-cells."

(*b.*) In a papilla search for a Wagner's **touch-corpuscle** (fig. 351). It is an oval body, with its long axis in the long axis of the papilla, but they are not present in all papillæ (fig. 350). They consist of a fibrous-looking material, with flattened nuclei arranged transversely. To their lower end passes a medullated nerve-fibre, which usually twists round the corpuscle before it enters it. The ultimate distribution of the nerve is best seen in a gold chloride preparation (Lesson XXXIV.).

5. **V.S. Fœtal Skin for Sweat-Glands and Pacinian Corpuscles.** —Harden the skin of the tips of the fingers of an infant in alcohol and make V.S. Stain with hæmatoxylin and eosin.

(*a.*) (L) Observe the general arrangement already described, but the sweat-glands are much more closely placed than in adult skin, and there is less intervening connective tissue. A child at birth has its full complement of sweat-glands, and hence they must be more crowded together than in the adult.

(*b.*) In the subcutaneous tissue are lobules of fat and sections of Pacinian corpuscles (Lesson XXXIV. **5**). The latter appear to consist of concentric laminæ surrounding a central core.

(*c.*) (H) Observe two or three layers of more or less cubical cells lining the duct of the gland, while the true secretory portion is lined by a single layer only of low, clear, columnar cells.

Hair Follicles.—To see their structure, harden the scalp in alcohol. V.S. must be made parallel to the course of the hair-follicles, which requires some care. Others are made across the follicles at different levels; but in this case care must be taken not to make the section parallel to the surface of the skin, but at right angles to the course of the hair-follicle. If an oblique section be cut, the hair-follicles are cut at different levels.

Harden a small piece of the human scalp (2 cm. square) in Müller's fluid and afterwards in alcohol. Stain a section in logwood, or first with logwood and then with picro-carmine. Mount in balsam.

6. V.S. Hair-Follicle.—(*a.*) (**L**) and (**H**) Observe the very thin epidermis, the thick cutis vera, and deep down the subcutaneous masses of fat; the **hair-follicles**, running obliquely through the skin, each one with a *hair* in it. At the lower part the hair has a bulbous end implanted on a papilla; the various coats of the hair-follicle, some continuous with the corium, and others with the epidermis. The following scheme shows the layers of the hair-follicle:—

Coverings of a Hair-Follicle from Without Inwards.

1. *Fibrous layers* { (*a.*) Longitudinally-arranged fibrous tissue.
   (*b.*) Circularly-arranged spindle-cells.
2. *Glass-like* or *hyaline membrane.*
3. *Epithelial layers* { (*a.*) Outer root-sheath. { Henle's layer.
   (*b.*) Inner root-sheath. { Huxley's layer.
   (*c.*) Cuticle of the hair. { Cuticle of root-sheath.
4. *The hair.*

(*b.*) **Dermic Coverings.**—(1.) (*a.*) The **outer fibrous sheath** denser than and continuous with the corium. The fibres run for the most part longitudinally. (*b.*) The *inner fibrous sheath* of fibrous tissue has a more circular arrangement and is seen as fibres cut across transversely, with a few nuclei interspersed.

(2.) The *hyaline* or *basement membrane*, clear, structureless, and well marked. It separates the dermic from the epidermic coverings of the hair.

(3.) **The Epidermic Coverings.**—(*a.*) The *outer root-sheath*—the most obvious part of the covering—consisting of several layers of nucleated cubical cells, continuous with and resembling those of the Malpighian layer.

(*b.*) The *inner root-sheath*, much narrower and paler, consisting of three layers of cells of different characters, is present only in the lower part of the follicle, *i.e.*, below the sebaceous gland.

(4.) The **hair** with its *cuticle.*

(*a.*) At its lowest part the bulbous enlargement of the hair, with

322 PRACTICAL HISTOLOGY. [XXIX.

the *papilla of the hair follicle* (fig. 309) projecting into it, and continuous with the corium.

(*b.*) The **sebaceous gland** or glands. In a balsam preparation

FIG. 309.—V.S. Hair-Follicle of Human Scalp. 1 and 2. Outer and inner fibrous sheaths of hair-follicle; 3. Hyaline layer; 4. Outer, and 5 and 6. Inner root-sheaths; *p*. Root of hair, with its papilla; *A*. Arrector pili muscle; *C*. Cutis vera; *a*. Subcutaneous tissue, with fat lobules; *b*. Epidermis (horny layer); *d*. Rete Malpighii; *g*. Blood-vessels; *v*. Lymphatics of papilla; *h*. Fibrous part, *i*. Medulla, *k*. Cuticle of hair; *K*. Sweat-gland and its coil.

its acini are yellowish and clear, opening by a duct into the hair-follicle at about its upper third (fig. 309, T).

(*c.*) The **arrector pili muscle** (smooth muscle) stretching obliquely from the deeper part of the hair-follicle to the upper part of the corium (fig. 309).

7. T.S. of Hair-Follicles (Scalp).

(*a.*) (L) In the human scalp the hair-follicles are arranged in groups of three or four, with interweaving strands of connective tissue between them. The various coverings—dermic and epidermic—can now be distinctly seen, especially if the section be through the lower half of the hair-follicle.

(*b.*) (H) Observe both V.S. and T.S. to see the structural elements forming the outer coverings of a hair-follicle.

(i.) The *inner root-sheath* consists of an outer layer of cells, clear and non-nucleated—*Henle's* layer—and an inner nucleated layer—*Huxley's layer*. Both are best seen in T.S.

(*c.*) The hair has a *cuticle*, while the hair itself may or may not have a medulla.

Carefully compare the structures of the hair-follicle in the T. and V. sections.

(*d.*) The *sebaceous gland*, its acini lined by cubical cells, containing fat, and rendered clear by the balsam.

The epithelium of the duct continuous with that of the outer root-sheath.

8. Sebaceous Glands.

—(*a.*) Harden the alæ of the nose of a new-born child in corrosive sublimate. (*b.*) Or the ala of nose or adult scrotum in picro-sulphuric acid (2–3 hours). Stain all with hæmatoxylin. Large sebaceous glands opening free on the surface without any hair-follicle are found. In other situations they open into the neck of a hair follicle. They are saccular glands with oval alveoli, which lead into a short duct. The alveoli are lined by a layer of polyhedral cells, and internal to this are larger cells containing fatty matter. The sebaceous secretion is formed by these cells undergoing disintegration, and liberating the fatty matter they have formed. They are developed from the outer root-sheath. In balsam they are clear, but in water they appear dark and granular.

9. Human Hair (H).

—Place it in water, cover, and examine. A rod-shaped body covered by a single layer of thin, non-nucleated transparent imbricate scales arranged transversely—*cuticle*. In some hairs it is seen merely as fine, more or less transverse, irregular, or wavy lines joining each other. These indicate where the one cell overlaps the other. The substance or *cortex* of the hair, composed of horny, fibrous substance—*hair fibres*—finely striated longitudinally, with, in some hairs, fine pigment-granules scattered along the course of the hair between the hair-fibres. In some hairs a darker central core or *medulla* composed of polyhedral cells.

10. Elements of a Hair (H).—Place a small piece of a hair in a drop of strong sulphuric acid. Cover and press lightly with a needle. Be careful to avoid letting a drop of the acid fall on the brasswork of the microscope. The hair splits up longitudinally into what look like fibres, but by gentle tapping on the cover they split into cells, so that a hair is composed of epithelial cells joined together, having previously undergone conversion into keratin.

11. Rabbit's Hair (H).—Mount in balsam. This hair contains one or more rows of cubical cavities containing air. The cavities appear black, and are surrounded by a small quantity of cortex.

12. Wool (Lesson I. 10).

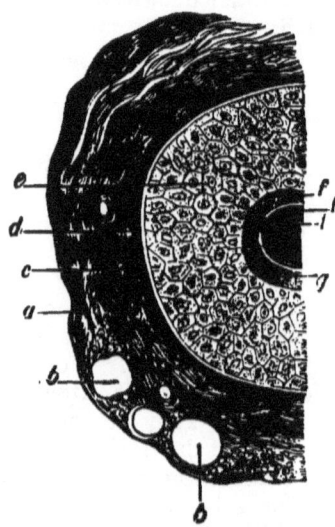

FIG. 310.—T.S. One-half of a Hair in its Follicle; *a.* Outer, *c.* Inner fibrous sheath; *b.* Blood-vessels; *d.* Hyaline layer; *e.* Outer, *f*, *g.* Inner root-sheath (*f.* Henle's layer, *g.* Huxley's layer); *h.* Cuticle; *i.* Hair.

FIG. 311.—V.S. Injected Skin, Palmar Surface of Finger.

Blood-Vessels of the Skin.—V.S. of a piece of skin cut from a limb injected with a gelatine mass. They must not be too thin. After injection the skin is hardened in Müller's fluid and afterwards in alcohol (three weeks). Mount (balsam) a section of the palmar surface of a finger, and one from the general surface of the body.

A good injection-mass is a watery solution of china-ink. It is rubbed down on a hone until a moderately thick black solution is obtained, so that when a drop is placed on blotting-paper it holds together, and no grey ring is formed round the drop. It has this advantage, that it is not changed by exposure to light, but the tissue must be hardened before sections are cut.

13. **V.S. Injected Skin**, *e.g.*, **Palm**.—The section should be cut in paraffin, and include the subcutaneous tissue (fig. 311).

(*a*.) (L) The arteries of the skin are branches of the larger arteries in the subcutaneous tissue. A branch may be seen running towards the surface. In its course it gives off three independent sets of branches, which end in capillaries:—

(i.) The lowest to the groups of fat-cells, where it forms a network of capillaries around and between the fat-cells.

(ii.) The short branch to the coil of a sweat-gland, forming a rich network of capillaries between the coils of the tube.

(iii.) The highest is from the terminal branches of the artery, and splits up into capillaries, which form a network chiefly in the upper part of the corium, and from this branches pass which form capillary loops in the papillæ of the skin. From it also proceed branches to the hair-follicle and its sebaceous gland.

(*b*.) The **vein** arises from the capillaries of the papillæ and the branches of the arteries to the upper part of the cutis, and in its course—running near the corresponding artery—it collects the veinlets from the coil and masses of fat. For blood-vessels of skin see W. Spatleholz.[1]

14. **Under-Surface of Epidermis**.—Separate by one of the following methods the epidermis from the cutis, most easily done in the fœtus:—(*a*.) Use a fœtus that has died and been macerated in utero.

(*b*.) Place pieces of skin in $\frac{1}{3}-\frac{1}{4}$ per cent. acetic acid (1–3 days), adding a drop or two of chloroform to prevent putrefaction (*Philippson*).

(*c*.) Macerate at 40° C.—preferably fœtal or young—skin pinned out on cork in 6 per cent. (or weaker) wood vinegar for 1–2 days. The epidermis separates rapidly (*Loewy*).[2]

After the epidermis peels off, in all cases turn its deeper surface upwards, and stain it for 3–4 minutes with a watery solution of Boehmer's logwood (p. 68). Wash and mount in balsam, preferably without a cover-glass.

(L) Observe a system of septa crossing each other, and forming longitudinal and transverse ridges, which project into the cutis. They form, as it were, the negative picture; the papillæ of the cutis vera represent the positive image. Part of the cells lining the sweat-glands and hair-follicles may also be pulled out, and are turned toward the observer. Figures of the arrangement of these septa are given in the papers of Blaschko and Loewy (pp. 316, 325).

---

[1] *Archiv f. Anat. v. Phys.*, Anat. Abth., 1893.
[2] *Archiv f. mik. Anat.*, xxxvii. p. 159, 1891.

THE NAILS.

The body of the nail rests on the *nail-bed*, the root of the nail on the matrix, and the part at the root and sides from which the nail springs is the *nail-groove*. The body of the nail is made up of numerous clear horny cells, each containing a rod-shaped nucleus. The nails are really the stratum lucidum, the stratum corneum being absent, and this rests on the Malpighian layer like that of the epidermis. The *corium* or nail-bed, on which the nail rests, is beset with very vascular longitudinal *ridges*, papillæ being absent.

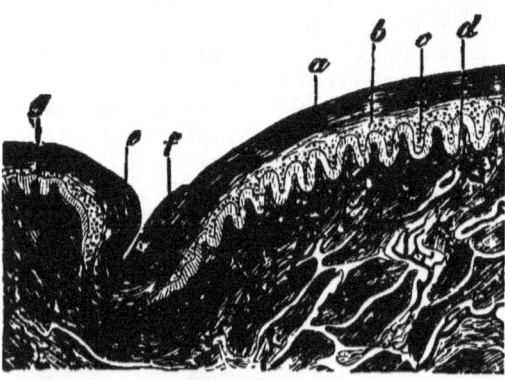

FIG. 312.—T.S. Through Half the Nail, Injected. *a.* Nail substance; *b.* More open layer of cells; *c.* Stratum Malpighii; *d.* Transverse sections of ridges; *e.* Nail-groove; *f.* Horny layer of *e* projecting over the nail; *g.* Papillæ of skin.

Harden the nail of a child and its subjacent bed in alcohol. Make T. and L. sections. Stain in hæmatoxylin (balsam) or picro-carmine (Farrant's solution).

15. **T.S. Nail (L and H).**—Observe the substance of the nail (fig. 312), and under it a series of transverse sections of the ridges of the corium of the nail-bed projecting into the epidermis. Under this the dense fibrous matrix.

16. **L.S. Nail.**—Observe the same general arrangement, but note that no papilla-like sections of ridges are to be seen.

ADDITIONAL EXERCISES.

17. **Elastic Fibres in the Skin.**—(i.) These resist gastric digestion; hence, add some pepsin to .2 per cent. of hydrochloric acid. If small pieces of skin be partially digested in artificial gastric juice at 40°C., part of the connective tissue is dissolved and the elastic networks remain.

(ii.) **Herxheimer's Method.**

(1.) Harden in Müller's fluid.
(2.) Stain 3–5 minutes in

Hæmatoxylin . . .	1 gram.
Absolute alcohol . . .	20 cc.
Water . . .	20 ,,
Sat. sol. lithic carb. . . .	1 ,,

(3.) Extract (5-15 secs.) with official perchloride of iron solution.
(4.) Wash in water.
(5.) Alcohol, oil, balsam.

The elastic fibres are bluish-black or black, and the surrounding tissue light blue. What Herxheimer described as "spirals" in the lower layer of the epidermis are spiral fibrils proceeding from the lowest layer of the cells of the epidermis, as shown by Kronmayer (p. 327).

(iii.) **Unna's Method.**—Place sections of the skin for 12-24 hours in the following mixture:—

Orcein	0.5 gram.
Absolute alcohol	40 cc.
Water	20 ,,
Hydrochloric acid	20 drops.

The sections are afterwards decolorised in very dilute HCl containing some alcohol. A full account of the arrangement of the elastic fibres is given by Lenthoefer.[1]

18. **Sweat-Glands of Axilla.**—Make V.S. of the skin of the axilla hardened in alcohol. Stain in hæmatoxylin, or stain in bulk and cut in paraffin. The sweat-glands, and particularly the coils, are very large. In these glands it is easy to see the smooth muscular fibres outside the lining epithelium of the secretory part of the coil (fig. 313).

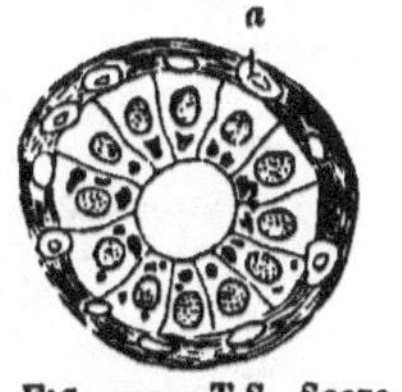

19. **Development of Hairs.**—Make V.S. of the skin of a fœtus at the fourth to the fifth month, after being hardened in Müller's fluid (12-15 days) and then in alcohol. Stain a small piece "in bulk" in borax-carmine, and cut in paraffin. The sections may also be stained subsequently with methyl-green. Mount in balsam.

20. **Double-Staining of Hair-Follicles.**—(a.) Make V. and T. sections, and stain some with eosin and hæmatoxylin, and others first with picro-carmine (12 hours) and then with methyl-green iodide. The latter preparation is specially beautiful, and both T.S. and L.S. should be stained by this method. The scalp is best hardened in potassic bichromate. Harden other pieces in alcohol. Methyl-green stains the inner root-sheath green.

FIG. 313.—T.S. Secretory Part of Sweat-Gland of Axilla. *a.* Nuclei of smooth muscle.

(b.) Stain with aniline-blue and safranin. Henle's sheath is rosy, Huxley's blue.

(c.) Or stain in safranin in a weak alcoholic solution of picric acid.

21. **Tactile Hairs.**—Harden the skin containing the large tactile hairs of a rabbit or cat in alcohol or Müller's fluid. Make T.S. and L.S. Outside the sheaths of the hair-follicle, already described, there is a large blood-space traversed by trabeculæ, and thus presenting the characters of cavernous tissue.

22. **Nail** (*double-stained*).—Picro-carmine and methyl-green.

23. **Blood-Pigment in Hair (H).**—Examine in normal saline one of the large "feelers" from the lip of an albino rabbit (*S. Mayer*). At some part of the hair in its centre a red pigment—hæmoglobin—may be seen.

24. **Fibrillation of Protoplasm of Epithelial Cells** (*Kronmayer*).[2] By means of Weigert's fibrin-staining method (Lesson III. 20) Kronmayer has demon-

---

[1] *Topographie d. Elastischen Gewebe*, Leipzig, 1892.
[2] "Die Protoplasmafaserung d. Epithelzelle," *Archiv f. mik. Anat.*, xxxix. p. 141, 1892.

strated the passage of the fibrils of one epithelial cell into adjacent epithelial cells, thus giving rise in the intercellular spaces to the appearances known as "intercellular bridges" (p. 128). The fresh skin of the sole of the foot or palm of the hand is hardened in absolute alcohol. Failing this, use part of an epithelioma. It may be stained with alum-carmine or borax-carmine, and then embedded and cut in paraffin. The sections must be *very thin* (at least 0.005 mm.), and in order to obtain these the knife must be placed obliquely and the sections cut from the epithelial surface towards the cutis vera.

The sections are placed in a watch-glass and xylol added to dissolve the paraffin. Add fresh xylol, pour it off, and then slowly add absolute alcohol. Gradually add water until the sections float flat on the surface. They are then carefully transferred to a slide, and by the aid of a pad of filter-paper gently fixed on the same. The sections so fixed are stained on the slide.

The best staining reagent is methyl-violet-6B, prepared as follows:—Mix equal parts of aniline-water and a saturated watery solution of methyl-violet. Pour a few drops of this fluid on the section fixed on the slide, and in five minutes lave the section in water.

Decolorise (few seconds) in iodine solution (p. 93). Wash in water. Remove the water by pressing blotting-paper on the section, and then flood it with aniline-xylol (aniline oil, 1 : xylol, 2). This mixture extracts the surplus dye very rapidly, so that the preparations must remain but a few seconds therein.

The march of events, supposing one wishes to stain the cell nuclei before staining the protoplasm fibrils, is as follows—the cells being previously hardened:—

(1.) Alum-carmine.
(2.) Wash in water (HCl-alcohol, absolute alcohol).
(3.) Methyl-violet-aniline-water.
(4.) Wash in water.

(5.) Iodo-potassic-iodide fluid.
(6.) Wash in water.
(7.) Aniline-xylol.
(8.) Xylol-balsam.

LESSON XXX.

SPINAL CORD.

The **spinal cord**, like the brain, is invested by three **membranes**, named, from without inwards, **dura, arachnoid,** and **pia mater**. The pia mater closely invests the cord, and sends processes into its substance as well as into its fissures. The cord itself is composed of white matter externally and grey matter internally. Running along the cord anteriorly and posteriorly are the **anterior** and **posterior median fissures**; the former is the wider, the latter rather a groove than a fissure. The two fissures do not meet in the middle line, but they serve to divide the cord incompletely into two lateral halves, which are united across the middle line by a **com-**

missure, composed anteriorly of white fibres crossing from one side of the cord to the other—**white commissure**; and posteriorly of grey matter—**posterior commissure**. In the middle of the latter is the minute **central canal**, lined by columnar ciliated epithelium. In each half of the cord is a crescent-shaped mass of **grey matter** as seen in transverse section, the two masses connected across the middle line, and presenting more or less the form of an **H**, with the extremities of its vertical limbs turned outwards. Its extremities are the anterior and posterior cornua. The anterior cornua are generally wider and shorter than the posterior, which are narrow, and come nearer the surface of the cord. The **nerve-roots** arise from the cornua, the anterior root by several bundles from the anterior cornu, and the posterior root by a single bundle from the posterior one. In this way, and by the existence of the fissures, each half of the white matter of the cord may be conveniently described as divided into an **anterior, lateral,** and **posterior column,** or more correctly into an antero-lateral and a posterior. The anterior cornu contains numerous large multipolar nerve-cells arranged in groups. Each cell is continuous, through its axis-cylinder process, with a nerve-fibre. The arrangement and number of cells, however, vary in different parts of the cord. There are no large nerve-cells—only small fusiform ones and some small isolated or "solitary" cells—in the posterior cornu, which is capped by a peculiar greyish matter—the **substantia gelatinosa** of Rolando. The white matter is composed of medullated nerve-fibres—small and large—arranged for the most part longitudinally, so that in a transverse section, when stained with carmine, they appear like clear rings with a central, red-stained spot—the axis-cylinder. In Pal's method the myelin is stained, and so they appear as blackish circles with a clear centre. The nerve-fibres, and grey matter as well, are supported by a peculiar sustentacular tissue—**neuroglia**—composed of **glia-cells** (p. 343), and both are supplied by blood-vessels, the grey matter, however, being far more vascular than the white. The grey matter, besides blood-vessels, lymphatics, glia-cells, nerve-cells, and their numerous processes, also contains nerve-fibres.

The **nerve-cells** vary in size, shape, and other characters in the different parts of the cord. Golgi speaks of two types of nerve-cells in the spinal cord.

Type 1, or *Motor Type*, corresponding to the multipolar cells of the anterior cornu. In these one process retains its individuality, and passes directly to become an axis-cylinder of a medullated motor nerve-fibre in the anterior root. This axis-cylinder process gives off a few secondary lateral processes or collaterals, which divide and enter into the formation of the nerve-complex in the grey matter. The protoplasmic processes subdivide, and also form part of the same grey complex, but they do not unite with fibres from adjoining cells or with nerve-fibres.

Type 2, or *Sensory Type*. Also multipolar, smaller cells, with no axis-cylinder process. All the processes subdivide into finer processes, lose their individuality, and pass *in toto* into the nervous complex, or diffuse nervous network in all strata of the grey matter of the cord. These cells are therefore connected only indirectly with nerve-fibres, *i.e.*, through the intervention of the grey nervous network; and it is only in this indirect way that they come into relation with—contact, not actual union with—the branches of axis-cylinders of the nerve-fibres of the posterior root.

Arrangement of Nerve-Cells.—In the grey matter of the anterior cornu, the large multipolar nerve-cells—*cells of the anterior cornu*—are arranged in groups varying in different parts of the cord. Many of their axis-cylinder processes pass out as axis-cylinders into the nerves of the anterior root on the same side of the cord. These cells are in relation with the fibrillar nerve-endings of the fibres of the pyramidal tracts, and also with the collateral fibres of the posterior root-fibres. In the upper dorsal and lower cervical regions is a group of nerve-cells—*intermedio-lateral tract*—lying well forward in a projecting part of the grey matter known as the lateral cornu (fig. 317, *Til*). In the middle of the crescent is the *middle cell group*. At the base of the posterior cornu, on its inner aspect, is a group of large cells—best marked in the thoracic region—*Clarke's column* (figs. 318, 319, *CC'*). The axis-cylinder processes of the cells of Clarke's column pass into and become the nerve-fibres of the direct cerebellar tract (p. 331). The cells of the posterior cornu are small, and for the most part isolated.

Diffuse Nervous Grey Network.—The network or complex of nervous fibrils in the grey matter of the cord is formed by—(1.) the fibres and their branches of the 2nd type of nerve-cells (p. 330); (2.) the fibrils and prolongations of the protoplasmic fibres of the 1st type; (3.) the lateral branches of the axis-cylinder processes of the 1st type; (4.) the fibrils produced by the terminal arborisations of the axis-cylinders entering the cord by the posterior roots.

Thus it is evident that some nerve-fibres spring from the cord directly from nerve-cells, and others indirectly from the nervous complex in the grey matter. Thus they are different, both morphologically and physiologically.

By several lines of research—including the facts of development, experimental and pathological evidence—the anatomical columns of the cord can be shown to be further subdivided. By these methods it is found that the **posterior column** consists of a narrower internal part, the *postero-internal* or *postero-mesial column*, or *funiculus gracilis* or *column of Goll*, and an outer, *postero-lateral tract* or *funiculus cuneatus*, or *column of Burdach*. In the upper part of the cord these two tracts are mapped off from each other by a septum of connective tissue (fig. 317).

The *postero-internal tract* is composed chiefly of rather small fibres, derived from the fibres of the posterior roots and fibre of the postero-lateral column. They end above in grey matter in the nucleus gracilis of the bulb.

The *postero-lateral tract* is chiefly composed of nerve-fibres of the posterior roots which run in it a certain distance before they pass into the grey matter and Goll's column. They end in grey matter in the cord and in the nucleus cuneatus of the bulb.

In the postero-lateral column is a small zone which undergoes descending degeneration, but only for a short distance, after section of the cord—the so-called *comma tract*.

Lissauer's Tract, or "*marginal bundle*," lies near the entrance of the posterior roots, either in the lateral column or postero-external column. It is derived directly from the fibres of the posterior roots. It undergoes ascending degeneration.

The **antero-lateral column** contains a large tract—the *crossed pyramidal tract*—which lies external to the posterior half of the grey matter, and in the greater part of its course is separated from the surface of the cord by the direct cerebellar tract. It consists of fibres descending from the central areas (motor) of the cerebral cortex, which have crossed at the decussation of the pyramids in the bulb. It consists of moderately large and some small fibres. Its fibres end by breaking up into fibrils, which form arborisations around the nerve-cells in the anterior cornu. It gradually diminishes in size as it passes down the cord, and can be traced as far as the origin of the 3rd or 4th sacral nerve, where it reaches the surface.

The *direct pyramidal tract* or *column of Türck* bounds the anterior median fissure, and consists of fibres coming from the motor areas of the cerebral cortex, which do not cross at the bulb. This tract gets smaller, and gradually disappears about the mid-dorsal region. Its fibres perhaps cross in the cord. These two are descending tracts.

Beginning at the lower dorsal region, and increasing in size as it passes upwards, is a tract lying external to the crossed pyramidal tract—the *direct cerebellar* or *dorso-lateral tract*—which consists of large fibres derived from the cells of Clarke's column, which pass up to the cerebellum, and enter it on the same side by its inferior peduncle or restiform body. Near the surface of the cord, and lying more anteriorly, is a tract which extends ventrally into the anterior column—the *antero-lateral ascending tract of Gowers*. Sometimes the term ventro-lateral is applied instead of antero-lateral. It enters the cerebellum by the superior cerebellar peduncle.

In the dog, as a result of excision of one-half of the cerebellum, a circumferential tract, occupying three-fourths of the surface of the antero-lateral tract—the *antero-lateral descending cerebellar tract*—has been mapped out. The remainder of the antero-lateral column is spoken of as the *antero-lateral ground bundle*.

As regards the results of **degeneration** following section or lesion of the cord, those parts that undergo degeneration above the lesion are called "ascending tracts"; and "descending tracts," are those nerve-fibres that degenerate below the lesion or seat of section. The parts which may undergo these respective degenerations are :—

Descending degeneration.
- Direct pyramidal tract.
- Crossed pyramidal tract.
- Antero-lateral descending cerebellar tract (limited).
- Comma tract.

Ascending degeneration.
- Goll's column.
- Direct cerebellar tract.
- Tract of Gowers.
- Tract of Lissauer.

It will thus be seen that in the antero-lateral columns there are some nerve-fibres which do not undergo degeneration.

Course of Fibres of Nerve-Roots.—The fibres of the *anterior roots* arise in several bundles from the axis-cylinder processes of the multipolar nerve-cells in the anterior cornu.

The fibres of the *posterior roots* enter the cord by a single bundle, but each one is an axial cylinder process of a nerve-cell in the corresponding spinal ganglion. They pass into the postero-lateral column, some pass into the posterior cornu, and a few small fibres form the marginal bundle (p. 331). Many fibres pass up in Goll's column, and the postero-external tract to end in terminal arborisations of fibrils around nerve-cells in the bulb, the fibres of Goll's column in arborisations around the cells of the nucleus gracilis, and those of the postero-external tract in the nucleus cuneatus. After entering the cord, the fibres in the posterior columns bifurcate, one branch passes upwards and one downwards (fig. 322). Collateral fibres are given off from the original fibre, and also from its branches, which enter the grey matter and end by terminal arborisations of fine fibrils which come into relation—but not direct union—with the nerve-cells of the grey matter, notably with the cells of Clarke's column.

THE SPINAL CORD.

It is convenient to begin with the cord of a small animal, *e.g.*, a cat or dog, but the student must also be provided with sections of the human cord. The same methods are applicable to both. Remove the whole length of the spinal cord from a cat, taking care not to squeeze or crush it in the process. Make transverse cuts into it about $\frac{3}{4}$ of an inch apart, and suspend it in a tall vessel in a large quantity of Müller's fluid, or 2 per cent. ammonium or potassium bichromate. Keep it in a cool place in the dark. Bichromate of ammonium hardens the cord very slowly indeed. In fact, to get a properly hardened cord months are required. To test if the cord is properly hardened, cut a section of the cord taken from the ammonium bichromate fluid, place it in water, and if it curls up, it is not properly hardened. It ought to remain flat. The process may be expedited by hardening first for 4 or 5 weeks in the bichromate, and completing the hardening for 2–3 weeks in $\frac{1}{8}$ per cent. chromic acid. Change the hardening fluid on the second day, and repeatedly thereafter. After 4–5 weeks, when it becomes tough, wash it, and harden in the various strengths of alcohol, beginning with 50 per cent. If, however, the spinal cord is to be used for Weigert's hæmatoxylin stain, it must not be washed in water, but placed in alcohol direct from the Müller's fluid or potassic bichromate.

T.S. are made from the cervical, dorsal, and lumbar regions. They may be made by means of the freezing microtome, the cord having been previously saturated in the sugar and gum mixture; or small pieces of the cord may be stained in borax-carmine (1 week), and then cut in paraffin. Or they may be embedded and cut in celloidin, and afterwards stained (p. 60). The paraffin sections are fixed to a slide by a "fixative," the paraffin removed by turpentine or xylol, the sections clarified by clove-oil, and mounted in balsam. Sections, if made by freezing, may be stained with carmine, hæmatoxylin, aniline-blue-black, safranin, methylene-blue, or by other methods.

1. T.S. Spinal Cord of Cat (L) and (H).—Speaking broadly, the same general arrangement obtains as in fig. 314. Suppose it to be the *dorsal* region, and the cord to be stained in carmine and mounted in balsam, observe:—

(*a*.) (L) The nearly circular outline of the section, covered externally by the **pia mater**, composed of two layers. From its under surface septa run into the white matter of the cord, and some of them carry blood-vessels, and processes pass into the fissures of the cord.

(*b*.) The **anterior** and **posterior median fissures**. The anterior fissure is wider and better marked. In it run both layers of the pia mater. The posterior fissure appears as a septum due to the prolongation of the inner layer of the pia mater into it, the outer layer of the pia runs over the fissure. The cord is thus almost completely divided into two symmetrical halves (fig. 314).

(c.) If the section is made at the level of the **origin of the nerve-roots**, these may be seen. The mode of origin of some of the fibres of these roots can always be seen. The *anterior root* passes

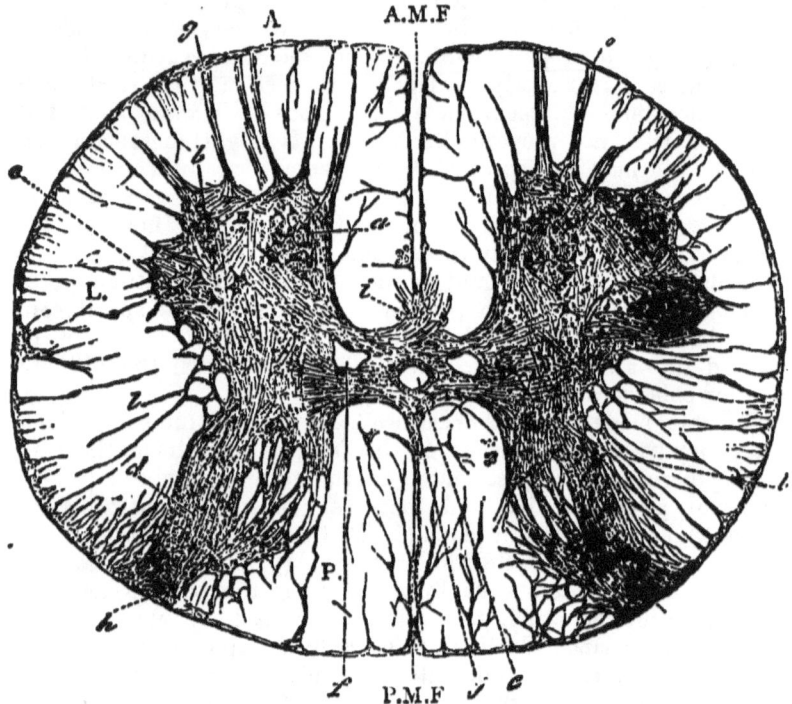

FIG. 314.—T.S. Lower Dorsal Cord, Human. *A, L, P.* Anterior, lateral, and posterior columns; *A.M.F*, *P.M.F*. Anterior and posterior median fissures; *a, b, c.* Nerve-cells of the anterior horn; *d.* Posterior cornu and substantia gelatinosa; *e.* Central canal; *f.* Veins, *g.* Origin of anterior nerve-root; *h.* Posterior nerve-root; *i.* White, and *j.* Grey commissures; *l.* Reticular formation.

out of the cord in several bundles, while the *posterior root* enters the cord in one compact bundle.

(d.) The **white matter** externally, and the grey matter internally, the latter deeply stained. The origin of the nerve-roots dividing the white matter into the **anterior, lateral,** and **posterior columns,** but owing to the anterior root leaving the cord in several bundles, there is not an exact anatomical limitation of the anterior from the lateral column.

(e.) The **grey matter,** deeply stained—a crescentic mass in each half of the cord, with a broader *anterior cornu*, with numerous large multipolar nerve-cells, arranged in groups—which does not reach the surface of the cord, and a *posterior cornu*—with a few small nerve-cells—which comes to the surface, and is prolonged into the posterior root. At the hinder part of the posterior cornu, an oval, deeply stained part, the *substantia gelatinosa*. A neck connecting the two cornua. There is also a group of nerve-cells placed laterally in the grey matter, the *vesicular column of Clarke* (fig. 318). In the anterior horn, note especially the entrance of the fibres of the anterior nerve-root and the large multipolar nerve-cells (fig. 315).

(*f.*) Connecting the two halves of the cord, the **anterior** and **posterior commissures**, running between the grey matter of opposite sides, and in the middle between them the **central canal**, which is surrounded externally by a deeper stained layer of neuroglia.

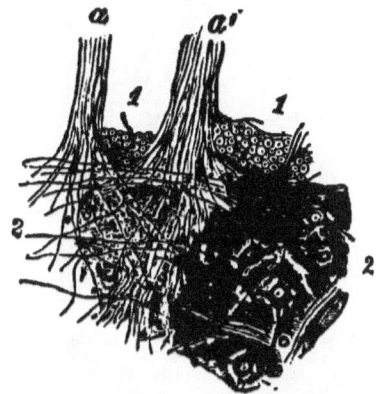

FIG. 315.—Entrance of Anterior Root into the Anterior Cornu of Lumbar Region. 1. Part of anterior white column; 2. Anterior grey matter with four multipolar cells; *a, a'*. Two rootlets, × 30.

(*g.*) In front of the anterior grey commissure is the **anterior white commissure**, with large medullated fibres crossing each other at an angle.

(*h.*) (H) The **pia mater** surrounding the **white matter**. The cut ends of the medullated nerve-fibres of various sizes. In the centre of each circular area the stained axis-cylinder, surrounded by a concentric clear area—the white substance of Schwann (fig. 316). In a preparation hardened in chromic acid not infrequently a number of concentric lines are seen in the myelin. Between the fibres here and there the **neuroglia**, composed of **glia cells**. On the surface of the cord there is a thin layer, in which the fibrils

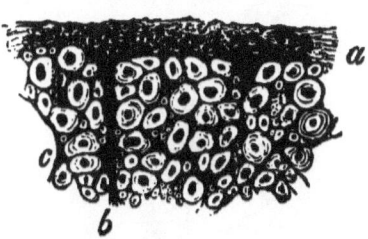

FIG. 316.—T.S. White Matter of Cord. *a.* Peripheral layer; *b.* Septum; *c.* Branched glia-cell; the remainder nerve-fibres, small and large, × 150.

of the neuroglia can readily be seen. Note also the prolongations, as fine septa, of the deeper part of the pia into the substance of the cord. The larger septa carry blood-vessels. The nerve-fibres of the antero-lateral columns are generally larger, *i.e.*, broader, than those of the posterior column, while those of the column of Goll are remarkable for their small size. Measure the sizes.

(*i.*) The **anterior cornu**, with large multipolar nerve-cells arranged

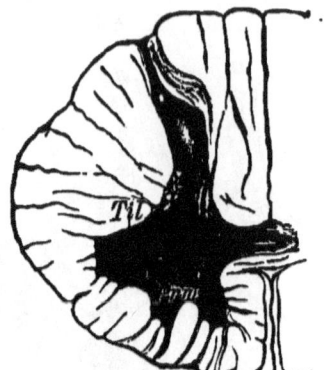

FIG. 317.—T.S. of Human Spinal Cord, Level of Sixth Cervical Nerve. *Prm.* Middle cervical process of the anterior cornu; *Til.* Lateral horn.

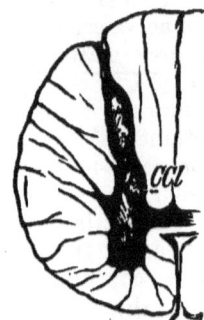

FIG. 318.—At the Level of the Third Dorsal Nerve.

in groups, chiefly in the anterior and lateral parts of the cornu. Each cell has numerous processes, a spherical well-defined nucleus with a membrane and a nucleolus (fig. 202). Owing to the method

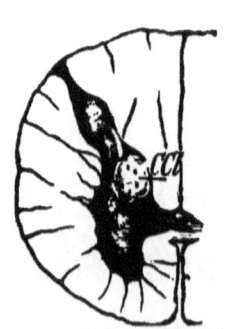

FIG. 319.—At the Twelfth Dorsal Nerve. *CCl.* Clarke's column.

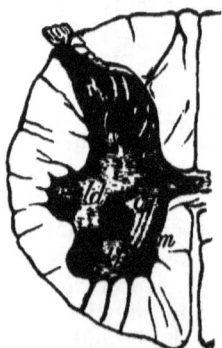

FIG. 320.—Level of Fifth Lumbar Nerve. *m.* Median group of nerve-cells of anterior-cornu; *lv, ld,* and *c.* Latero-ventral, latero-dorsal, and central groups of cells, × 5.

of hardening, all the cells are somewhat retracted, each cell apparently lying in a cavity.

(*j.*) Trace medullated nerve-fibres across the white commissure from the grey matter of one side to the white matter of the opposite side.

(*k.*) The **posterior cornu.** Note the absence of large cells. Only a few small fusiform nerve-cells are seen. Some of the fibres of the posterior root pass into the posterior cornu; some pass directly through the substantia gelatinosa, and others sweep with a curve through the posterior column before they enter the grey matter. The grey matter generally is traversed by fine fibrils, and has a somewhat finely granular appearance.

(*l.*) The **central canal,** lined by a single layer of columnar ciliated epithelial cells. The cilia may be wanting. Near it sections of blood-vessels. In the grey matter observe the plexus of fine fibrils and numerous axis-cylinders. (Multipolar nerve-cells, Lesson XVIII. 9.)

2. T.S. Human Spinal Cord in several Regions.—Stain sections with carmine or lithium-carmine, hæmatoxylin, or aniline blue-black, and mount in balsam. Mount sections from the cervical enlargement (fifth nerve) and lumbar enlargement, and compare them with the dorsal section. With the naked eye (and **L**) observe:—

(*a.*) The cervical and lumbar sections (figs. 317, 320) are not only larger, but the amount of grey matter is greater than in the dorsal region. Note the large expanded anterior cornu with numerous nerve-cells. In the lumbar region the grey matter is large in amount, the white matter smallest. The white matter is more abundant in the dorsal region and most abundant in the cervical region. This can, to a certain extent, be determined by the amount of white matter lying between the grey matter and the pia mater. There is a gradual transition from the one region of the cord to the other.

(*b.*) In the *cervical region* note a thin septum, which dips into the white matter and divides its posterior column into a smaller internal part—the *postero-internal column* or *column* or *fasciculus of Goll*, and a larger external or lateral part—the *postero-external column*, or *fasciculus cuneatus*. The grey matter, containing some nerve-cells, also projects into the white matter about midway between the anterior and posterior cornua, forming the *intermediolateral tract*.

(*c.*) In the dorsal region, observe a group of nerve-cells at the inner part of the neck of the grey matter. It lies behind the plane of the central canal—*column of Clarke*, or *posterior vesicular column* (fig. 319, *CCl*).

In sections of the cord at different levels it is important to note the sectional area of the different columns. Thus, Goll's column is largest in the upper cervical region, and diminishes from above downwards. The increase or decrease in the sectional area of some other columns can only be made out by a study of degenerated or developing cords.

3. **L.S. of Cord (Antero-posterior direction).**—Stain as for T.S. (**L** and **H.**)

(*a.*) Observe the longitudinally-arranged nerve-fibres in the white matter of the anterior and posterior columns. In the anterior cornu the rows of multipolar nerve-cells.

(*b.*) The anterior roots passing obliquely through the anterior column.

(*c.*) The posterior cornu, with its gelatinous substance and nerve-fibres passing into and through it.

4. **Weigert's Hæmatoxylin Method.**[1]—The myelin of medullated fibres is stained of a deep blue-black tint, while degenerated parts are not so stained; they are lighter just in proportion to the disappearance of medullated fibres. The spinal cord, or the brain, is hardened for four to six weeks in Müller's fluid, and the hardened pieces are placed direct into 70 per cent. alcohol, and must not be washed with water. Keep in the dark. Next day add 90 per cent. alcohol, and harden completely in 95 per cent. or absolute alcohol. Instead of Müller's fluid, Erlicki's fluid may be used. This is best done by keeping the fluid at the temperature of the body in a warm chamber, when the hardening process is completed in a few days; but the results are not so satisfactory as in the case of a cord hardened in the ordinary way. Small pieces of the hardened tissue are placed in a half-saturated solution of *neutral* acetate of copper (*i.e.*, a saturated solution is mixed with its own volume of distilled water), where they remain for three to five days. They are then transferred direct to 90 per cent. alcohol, and can then be cut in alcohol, or the pieces can be embedded in celloidin and then cut (p. 47).

(**A.**) The sections are placed for two or three hours in a few cc. of Weigert's hæmatoxylin, in which they become black.

Weigert's Hæmatoxylin.

Hæmatoxylin	1 gram.
Alcohol absolute	10 cc.
Cold saturated solution of lithium carbonate	1 ,,
Distilled water	90 ,,

Dissolve the hæmatoxylin in the absolute alcohol, add the water and boil. After it is cool, add the lithium carbonate. The time which the sections remain in this fluid depends on what it is

[1] *Fortschrit. d. Med.*, 1884 and 1885; *Zeits. f. wissensch. Mik.*, 1884 and 1885.

desired to show. Two hours or so are enough for the cord, but if the fine plexuses of medullated fibres—described by Exner—in the cerebral cortex are to be well seen, let them stain for twenty-four hours. After they are sufficiently stained, throw the watch-glass and the stained sections into a large basin of distilled water. Remove the sections from the water at once, and place them, for about half an hour or more, in the following mixture:—

Potassic ferricyanide	2.5 grams.
Borax	2 ,,
Water	100 cc.

This fluid decolorises and differentiates the black sections. The grey matter becomes of a brown or light-yellow tint, and should remain so while the white matter becomes violet. The sections must remain in the decolorising fluid until the deep blue or violet-coloured medullated nerve-fibres are seen. This can readily be determined after a little practice. The sections are then washed in water—in which they can be kept for a considerable time—transferred to 90 per cent., and finally to absolute alcohol, clarified by xylol or origanum oil, and mounted in xylol-balsam. This method stains the medulla or myelin of the nerves—especially of the central nervous system—of a deep blue or violet tint, while the nerve-cells, neuroglia, and axis-cylinders are not stained. The method, however, may be combined with staining methods. After the organ is hardened in the chromium salt and alcohol, small pieces are embedded in celloidin and cut in a microtome. The sections are placed in 80 per cent. spirit—not water. The celloidin enables the sections to be readily handled and transferred to a slide. They are then stained in the special hæmatoxylin fluid already described. If a series of sections is to be mounted on the same slide, see the method described at p. 60. The sections are clarified in origanum oil or a mixture of xylol and carbolic acid (p. 83). Mounted in balsam. This method is also applicable to the spinal ganglia.

(B.) **Freezing Method.**—The piece of cord or brain, after hardening in Müller's fluid, is placed in spirit and transferred to a mixture of equal parts of absolute alcohol and ether, and then embedded in celloidin. The embedded tissue is placed for forty-eight hours in Erlicki's fluid, to get rid of the spirit, and they are then placed in the following mixture, and kept in stoppered bottles in a warm chamber at 38° C. for two or three days:—

Cupric sulphate	0.5 gram.
Potassic bichromate	2.5 grams.
Mucilage of syrup and gum	100 cc.

Sections are cut in a freezing microtome and received into Erlicki's fluid, washed in methylated spirit, stained in the hæmatoxylin fluid, decolorised by the ferricyanide mixture, clarified, and mounted in balsam (*Hamilton*).

Stages of Weigert's Method.

(1.) Harden nervous system in Müller's fluid.
(2.) Harden in alcohol without previous washing in water.
(3.) Embed in celloidin.
(4.) Place celloidin block in a half-saturated solution of copper acetate (24-48 hours).
(5.) Alcohol, 70 per cent. (24 hours).
(6.) Cut sections and stain them in Weigert's hæmatoxylin (24 hours).
(7.) Wash in water.
(8.) Partially decolorise in ferricyanide fluid until grey matter becomes yellow.
(9.) Wash in water, dehydrate in absolute alcohol, clarify in xylol, and mount in xylol-balsam.

Pal's method is referred to on p. 343, but the following combination of the Pal and Weigert method gives good results:—

5. Modified Weigert-Pal Method.

(i.) Harden in Müller's fluid and then in alcohol, without washing pieces in water.
(ii.) Embed and cut sections in celloidin.
(iii.) Wash in water and transfer to Marchi's fluid (5-10 hours).
(iv.) Wash and transfer for 10-16 hours to

Kultschitzky's Hæmatoxylin.

Hæmatoxylin	1 gram.
Absolute alcohol	2-5 cc.

Dissolve, and to the fluid add

Acetic acid (2 per cent.)	100 cc.

The sections become black.

(v.) Bleaching process—(*a.*) Wash in water and bleach for 5 mins. in .2 per cent. permanganate of potash. Wash again, and transfer to
(*b.*) Pal's solution (p. 344), in which they are rapidly (5-10 mins.) bleached, the medullated fibres alone remaining black.
(vi.) Wash in water, dehydrate in alcohol, balsam, oil (xylol), balsam (*Schäfer*).

6. Weigert-Stained Cord.—(*a.*) (L and H)

Observe the medulla of the medullated nerves, stained purplish. There are so many fine medullated nerve-fibres revealed in the grey matter that it is impossible to reproduce their complexity in a woodcut. The nerve-cells are not specially in evidence, although they can be stained with picro-carmine.

Formerly Weigert used acid-fuchsin for staining the medullated nerve-fibres of the central nervous system, but this method is now given up in favour of the hæmatoxylin copper method.

7. Neuroglia (H).—Stain in safranin (24 hours) thin sections of the white matter of the cord hardened in Müller's fluid, partially decolorise in absolute alcohol, and mount in balsam.

(*a.*) Observe the branching neuroglia-cells between the white nerve-fibres (fig. 316, *c*). The connective-tissue elements have a tint more towards the violet, and are thus differentiated from the nervous elements. The network of neuroglia-fibres is readily seen near the surface of the cord.

8. Blood-Vessels of the Spinal Cord.—Mount a transverse section of the cord, with its blood-vessels injected. Cut the cord in paraffin. The sections must be rather thick (**L** and **H**). Inject the animal, *e.g.*, cat or rabbit, from the aorta with a blue or red mass. Harden in alcohol.

(*a.*) Observe the greater vascularity of the grey matter as compared with the white. A blood-vessel may be seen running into the anterior median fissure, and at the bottom of it dividing, and giving a branch to each mass of grey matter.

(*b.*) The dense plexus of capillaries in the grey matter. Branches of blood-vessels passing into the cord along the roots of the nerves and along the larger septa, which pass from the pia mater into the cord.

9. Nerve-Fibres of the Spinal Cord (H).—Crush a piece of the white matter of the cord, either fresh, or after maceration in $\frac{1}{8}$ per cent. bichromate of potash, between a cover-glass and a slide.

(*a.*) Observe the nerve-fibres, many of them with lateral bulgings, or presenting a beaded or moniliform appearance. This is due to the fact that these nerve-fibres are devoid of a primitive sheath.

(*b.*) Droplets of "myelin" with concentric markings are seen in the field.

10. Nerve-Fibres of the Spinal Cord (H).—By means of a hypodermic syringe make an interstitial injection of 1 per cent. osmic acid into the white matter of the antero-lateral column of the spinal cord of an ox. Tease a piece in glycerine.

(*a.*) Observe tubes of different sizes, many varicose, with incisures and cylinder-cones, but no primitive sheath.

11. Staining the Cord.—To stain sections the following dyes may be used:—

(1.) Ordinary carmine, picro-carmine, or acid-carmine. In using the last, use a very dilute (scarcely coloured) solution, and let the sections remain in the solution for 1–2 weeks. (2.) These stains may be combined with hæmatoxylin. (3.) Benzo-azurin. (4.) Aniline-blue soluble in water but insoluble in spirit. (5.) The same blue with eosin or Magdala-red. If (5) be used the axis-cylinders are blue, the myelin rose, and there is also a sharp distinction in colour between the nerve-cells and glia-cells. (6.) Watery solution of Congo-red. The sections are dipped for a moment into very dilute sulphuric or hydrochloric acid (1 drop to 10 cc. water). They become blue. (7.) Weigert's hæmatoxylin-copper. (8.) Golgi's silver and mercuric-chloride method. The Müller's fluid should contain 3–3.5 grams of potassic bichromate to 1 gram of sodic sulphate. (9.) Golgi's silver method and sections stained by Magdala-red (*Lavdowsky*).

12. Tracts in the Spinal Cord.—These are made out histologically by studying (*a.*) embryonic cords in mammals from the 5th to the 9th month. These sections are stained either by Weigert's method (p. 338) or by the modified Pal method. A T.S. of the cord of a human fœtus just before birth shows the nerve-fibres in the pyramidal tracts still devoid of myelin, and thus they are easily mapped out from the other parts of the cord which are already medullated.

(*b.*) The degeneration changes resulting after hemi-section or section of the cord. Section of the cord is practised say 6–10 days before the cord is required. Parts of the cord above and below the seat of injury are hardened for 10 days or so in Müller's fluid. Thin pieces are then placed for several days in Marchi's fluid (p. 347).

Sections are mounted in balsam. All the nerve-fibres which have undergone degeneration are stained black; healthy nerve-fibres are yellowish (p. 347). Or instead, Weigert's process may be used, as then the degenerated tracts remain unstained owing to the absence of myelin.

ADDITIONAL EXERCISES.

13. Dry Preparation.—Stain a T.S. with methylene-blue (1 per cent.). Wash it, and allow it to dry on a slide. Add a drop of balsam. This shows very well the general characters of the cord; the multipolar nerve-cells are somewhat shrunken, but still they and their processes are well stained.

14. Sections in Ehrlich-Biondi's Fluid.—Place sections in this fluid (p. 81) well diluted (1 : 40), and heat them in a watch-glass until vapour is just given off. Mount them in balsam, clarifying either with xylol or aniline-oil and xylol (Lesson III. 16). The glia-fibrils are violet in tint, and so is the connective tissue generally; the nuclei of the glia-cells bluish, the myelin orange, the axis-cylinder somewhat violet. The cells in the grey matter have a pleasant violet tint. Benzo-azurin can be used in the same way.

15. T.S. Cord. Eosin and Logwood.—The neuroglia-cells, connective-tissue, and epithelium of the central canal have a logwood tint, the other parts are rosy.

SPINAL CORD.

16. Staining the Cord in Bulk in Aniline Blue-Black.—Small pieces of the cord are placed in the following fluid for a day or two:—

    Aniline blue-black . . . 2 grams.
    Water . . . . . 60 cc.
    Alcohol . . . . 40 ,,

Mount the sections in balsam.

17. Transverse Markings on Axis-Cylinders and Nerve-Cells.—Place small pieces of a perfectly fresh cord in 1 per cent. silver nitrate solution, and keep them in the dark for forty-eight hours, renewing the fluid several times. Wash the pieces and place them, exposed to light, in the following mixture:—Formic acid (1 part), amylic alcohol (1 part), and water (100 parts), for 5-7 days. Tease a fragment in glycerine and observe the alternate brown and clear markings on the axis-cylinders (Frommann's lines) and on the nerve-cells (*Jakimovitch*).[1]

18. Isolated Neuroglia-Cells.—(*a.*) These are obtained by the interstitial injection of osmic acid into the white matter of the cord (Lesson XXX. 10.) A small piece is teased and stained with picro-carmine.

(H) Observe the branched cell, with a granular body and long processes (fig. 321). It requires considerable care to dissociate such a cell, and it must usually be looked for and isolated with the aid of a dissecting microscope (p. 22).

(*b.*) In sections of the cord prepared by Golgi's method, neuroglia-cells may be stained along with the nerve-cells, and on other occasions they may be the only elements on which the silver or mercury takes effect. Note that each cell gives off many very fine processes. Some of the latter may be seen to become attached to the walls of a capillary or other blood-vessel.

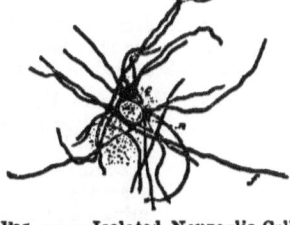

FIG. 321.—Isolated Neuroglia-Cell of Spinal Cord of Ox. *n.* Nucleus ; *c.* Granular protoplasm ; *f.* Fibres of neuroglia.

(*c.*) Isolated glia-cells may be found after maceration in Landois' fluid (p. 26), and subsequent staining with Magdala-red.

19. Isolated Nerve-Cells of the Cord.—The best dissociating reagents are dilute alcohol or Landois' fluid (3-4 days, p. 26). A staining fluid may be added to the dilute alcohol, and thus dissociation and staining go on simultaneously. With cells—either nerve or glia—isolated by Landois' fluid, it is better to stain after maceration. The best stains are Magdala-red, methyl-blue II. (0.5-1 per cent., 5-10, drops added to 10 cc. of the macerating fluid). Lavdowsky[2] uses a "semidesiccation method" like that used for connective tissue ; the isolated cells are allowed to become nearly dry, and then alcohol is slowly added to remove the remainder of the water (see also Lesson XVIII.). Or after maceration shake the tissue in a very small quantity of water in a test-tube, and place it in a watch-glass, add 6-10 drops of glycerine and a little picro-carmine, and dry the whole over a sulphuric acid desiccator. This removes all the water, and one has then the cells stained in glycerine ready to be mounted.

20. Pal's Method of Staining Nerve-Fibres.—This method requires considerable care. Harden the cord or brain in Müller's fluid, make T.S., and place them in alcohol. Stain in .75 per cent. watery solution of hæmatoxylin con-

---

[1] *Journ. de l'Anat. et de la Phys.*, xxiv. p. 142, 1888.
[2] *Archiv f. mik. Anat.*, xxxviii. p. 264.

taining some alcohol and a few drops of a saturated solution of lithium carbonate (6-10 hours). They are then "differentiated" in ¼ per cent. solution of permanganate of potash (10-15 seconds), and subsequently in the following mixture:—

>   Oxalic acid . . . 1 gram.
>   Potassic sulphite . . 1 ,,
>   Distilled water . . . 200 cc.

When the grey and white matter are differentiated, the sections may be stained in carmine, safranin, or alum-carmine, and mounted in balsam.

Stages of Pal's Method.

(1.) Harden in Müller's fluid.
(2.) Cut sections and stain in Weigert's hæmatoxylin (24-48 hours).
(3.) Wash in water to which 1-2 per cent. lithium carbonate is added. The sections must be deep blue.
(4.) Differentiate the sections in 0.25 per cent. potassic permanganate solution (20-30 secs.) till grey matter becomes yellow.
(5.) Transfer to oxalic acid solution (few secs.).
(6.) Wash in water, and dehydrate in alcohol, xylol, balsam.

This method can be done rapidly. Only the nerve-fibres are stained; the intervening parts may be stained subsequently with picro-carmine.

21. Vessale's Modification of Weigert's Method.[1]

(1.) Harden in Müller's fluid and afterwards in alcohol.
(2.) Stain celloidin sections (3-5 mins.) in 1 per cent. hæmatoxylin dissolved in warm water. Allow it to cool. The sections become black.
(3.) Place in saturated filtered solution of neutral copper acetate (3-5 mins.) and then lave in water.
(4.) Differentiate in a solution—

>   Borax . . . 2 grams.
>   Potassic ferridcyanide . 2.5 ,,
>   Water . . . 300 cc.

Ganglion-cells, glia-cells, and degenerated parts soon become decolorised, only the medullated fibres remain dark-violet.

(5.) Wash thoroughly in water. They may be stained with alum-carmine as a contrast stain.
(6.) Absolute alcohol. Xylol-carbolic acid (3 : 1), remove surplus with bibulous paper. Balsam.

22. Kulschitzky's Method for Medullated Fibres.[2]

(1.) Harden in Müller's or Erlicki's fluid (1-2 mins.), then wash in water (1-2 days). Alcohol. Make sections in celloidin.
(2.) Stain (1-3 hours) in his acid hæmatoxylin (p. 342).
(3.) Differentiate (2-3 hours) in

>   Sat. sol. lithium carb. . . 100 cc.
>   Ferridcyanide of potash (1 per cent.) 10 ,,

(4.) Wash thoroughly in water. Balsam.

23. Golgi's Silver Methods[3]—(a.) *Slow Method.*—Small parts of the cord or

---

[1] *Zeit. wiss. Mikros.*, vii. p. 517, and *Archiv. ital. de Biol.*, xv. p. 158, 1891.
[2] *Anat. Anzeig.*, iv. p. 519, 1890.
[3] *Sulla fina anatomia degli organi centrali del sistema nervoso*, Milano, 1886.

brain, after being hardened in 2 per cent. potassium bichromate for 3-4 days, are then transferred to a stronger solution, say 3 per cent., for 4 days. Increase successively the quantity of the bichromate until 4-6 per cent. is reached. It takes 30-50 days, according to temperature or other circumstances, to get the tissues properly hardened. The hardened parts are then transferred for 24-48 hours to .5-.75 per cent. silver nitrate. It is better to place them first of all in a small quantity of silver nitrate, and to wash them in this fluid. The fluid becomes of a dark orange tint from silver chromate. Place the pieces in fresh silver nitrate. Harden in alcohol. Make hand sections, clarify them in the usual way, and mount in balsam, but *do not apply a cover-glass.*

(b.) *Rapid Method.*—Make the following mixture :—

Osmico-Bichromate Mixture.

Potassic bichromate (3 per cent.) . . 20 cc.
Osmic acid (1 per cent.) . . . 5 „

Place small pieces of the fresh organ in 15 cc. of this fluid for 2-3 days, and then place them in the silver solution (.5-1 per cent.) for 1-2 days. I have found the cells of the cerebral cortex stained black two or three hours after immersion of the organ in silver nitrate.

This is an excellent method, specially useful for young or embryonic nervous system, as the fluid penetrates more readily where the myelin is scanty. It is very capricious in its action. Sometimes only small parts are stained. The nerve-cells and their processes, and nerve-fibres without myelin, and axis cylinders are stained black.

(c.) *Medium Method.*—Small pieces of the fresh tissue are placed for 3-5 days in 3 per cent. potassic bichromate; then 3-4 days in the osmico-bichromate mixture as above, and then in silver nitrate.

It is not a matter of indifference which method is used. Golgi's method only stains certain elements—nerve-cells, neuroglia-cells, and sometimes blood-vessels. Thus, by the rapid method, in the case of the cerebellum, the parallel fibres and the granular layer are the chief parts stained; Purkinje's cells stain best by the slow method. As a general rule, the axis cylinders and their processes stain best by means of the rapid and medium methods, the slow method, as a rule, staining best the protoplasmic processes.

The rapid silver method has been extensively used by Golgi and his pupils, by Kölliker, but above all by Ramón y Cayal. The black deposit Golgi calls a "black reaction."

24. Golgi and Mondino's Sublimate Method.—Small pieces of the central nervous system, after hardening in bichromate of potash, are placed in a watery solution (.25 per cent.) of corrosive sublimate. The volume of fluid must be large, and renewed frequently, *i.e.*, as often as it is yellow. After 10-15 days the reaction has occurred in small pieces; but it is better to expose the pieces longer than this to the action of the salt; in fact, prolonged immersion rather improves it. The tissues may remain for months in the fluid without disadvantage. The sections must be very thoroughly washed, else, after being mounted, needle-shaped crystals of the sublimate are apt to form. The sections are mounted in balsam or glycerin; only they are mounted without a cover-glass.

In Golgi's silver method the cells, and in some preparations the blood-vessels as well, are opaque and black. Sometimes the body of the cell and its finest ramified processes can be seen with the utmost sharpness. The silver is deposited only on the cells and their processes, not on the nerve-fibres. The sections, however, are often dotted over with a black metallic deposit. In the mercury preparations, if the cells do not appear to be very black, they may be darkened by washing them in sodic sulphite.

The sublimate may act on (a.) the ganglionic cells, (b.) the neuroglia cells, and (c.) the blood-vessels. The best results with these methods are obtained with the cerebral cortex. The sections can be stained afterwards by the usual methods, and particularly by Magdala-red.

The parts acted on by the sublimate are white by reflected light, and appear black by transmitted light, because they are opaque. Golgi[1] finds that for the study of the diffuse network in the central nervous system, by this method the best results are obtained when the "metallic white" is converted into a metallic black. For this purpose he uses the "fixer" employed for fixing positive photographs on aristotypic paper.

(A.) Water . . . . . 1 litre.
Hyposulphite of soda . . . 175 grams.
Alum . . . . . 20 ,,
Ammonium sulphocyanide . . 10 ,,
Chloride of sodium . . . 40 ,,

Leave the mixture for 8 days and then filter.

(B.) Water, ... 100 cc.
Gold chloride. . . . 1 gram.

For the toning and fixing fluid, mix of

A . . . . . 60 cc.
B . . . . . 7 ,,

Procedure.

(1.) Wash in distilled water.
(2.) Immerse sections (1-2 mins.) in the above mixture. The sections become black.
(3.) Prolonged washing in water.
(4.) (Optional.) Faint coloration of the sections with acid carmine.
(5.) Wash again, then mount in balsam.

25. Double Impregnation Method of Ramón y Cajal.[2]—Morsels of tissue are placed in the dark in the osmico-bichromate mixture as for Golgi's rapid method (2-3 days), and then they are gently washed and placed in .5-.75 per cent. silver nitrate (1-2 days). They are retransferred for 3-4 days to the osmico-bichromate fluid (3-4-5 days), and then again to silver. They are then hardened in alcohol and cut. This method certainly gives good results in some cases, and is very useful for the brain and embryo preparations.

26. Embryo Cords.—(a.) Harden the spinal cord of an embryo chick at the 9th day as above. Remove as much as possible of the surrounding vertebral column before placing it in the osmico-bichromate mixture.

(L) Observe the axis-cylinders of the fibres of the anterior roots springing from the axis-cylinder processes of the multipolar branch cells of the anterior cornu, themselves black. Note that the fibres of the posterior root enter the cord and split up into fibrils.

(b.) **Collateral Fibres.**—Harden the spinal cord of an embryo mammal (e.g., sheep, 20-25 cm. long) or embryo of chick in the above fluid, and make L.S. in the line of entrance of the posterior roots. The sections need not be very thin, but one may have to make several preparations before one gets a satisfactory result.

(H) Note that when a fibre of the posterior root enters the white matter of the cord it divides, sending one branch upwards and one downwards (fig. 322), which run for a distance more or less longitudinally, but many enter the

---

[1] *Archiv. ital. de Biologie*, xv. p. 462, 1891.
[2] *Internat. Monatsch. f. Anat. u. Phys.*, vi. p. 170.

grey matter, and end free in fine branches without forming connections with nerve-cells. The fibres give off at right angles to their course fine fibrils —collaterals—which enter the grey matter, divide into fibrils, and end free.

27. Degeneration of the Cord—
(*a.*) **Method of Marchi** for degeneration, in central nervous system or in nerves (Lesson XVII.) :—

(1.) Harden very small pieces of a nerve or spinal cord—either of which is undergoing degeneration—in Müller's fluid for at least 8 days.
(2.) Place for 6 days in the following :—

Müller's fluid . . 2 parts.
Osmic acid (1 per cent.) 1 part.

(3.) Wash in water, harden in alcohol.
(4.) Embed and cut in celloidin.

All degenerated parts appear black, all the others light grey or yellowish (see p. 212).

(*b.*) **Weigert's hæmatoxylin method** may be used for the same purpose. The degenerated fibres are unstained, *i.e.*, on differentiating the section the degenerated parts rapidly give up their stain, and thus appear unstained.

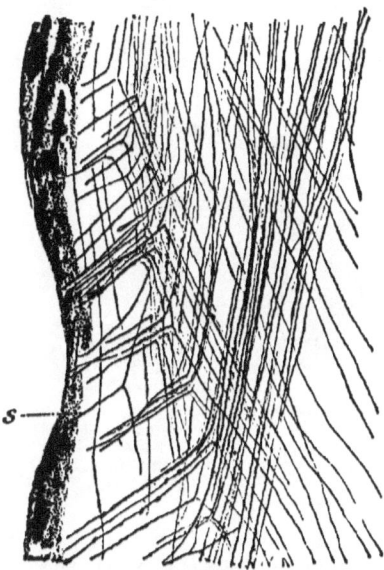

FIG. 322.—L.S. of the Cord of the Cervical Region of an Embryo Sheep (22 cm. long), to show division of posterior roots after entering the spinal cord.

LESSON XXXI.

MEDULLA OBLONGATA—CEREBELLUM— CEREBRUM.

MEDULLA OBLONGATA OR BULB.

THE medulla oblongata is hardened in the same way as the cord. T.S. are made at different levels, and stained in the same way as the cord (**L** and **H**).

1. T.S. Decussation of Pyramids.

(a.) Observe the shape of the cord, the decussation of the anterior pyramids, *i.e.*, bundles of fibres are seen coming from the lateral column of one side, and running inwards and towards the middle line, thus separating somewhat the anterior from the posterior cornua (fig. 323). They actually pass to the anterior pyramid of the opposite side of the medulla.

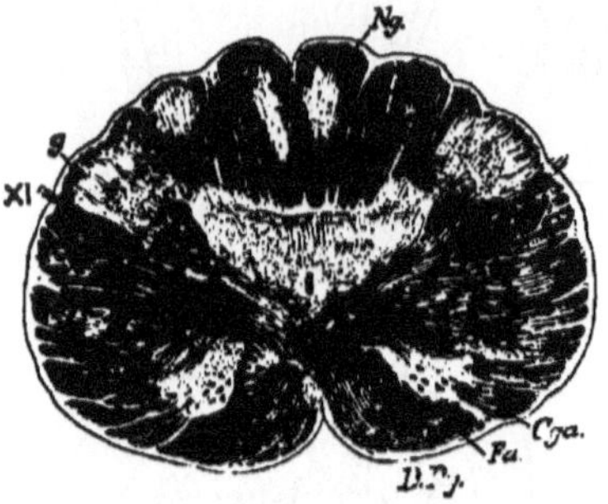

(b.) At the anterior part, on each side of the anterior median fissure, sections of the anterior pyramid.

(c.) At the posterior part, a small part of the posterior cornu extending backwards as the nucleus of the clava in the funiculus gracilis. Another mass of grey matter in the funiculus cuneatus.

FIG. 323.—T.S. Medulla Oblongata through Decussation of Pyramids. *D.Py.* Anterior pyramid; *Fa.* Anterior Cornu; *Ng.* Nucleus of the funiculus gracilis; *g.* Substantia gelatinosa; *XI.* Spinal accessory nerve.

2. T.S. of Olivary Bodies (L and H).—Observe the folded mass of grey matter, with many multipolar nerve-cells, constituting the **olivary nucleus** (fig. 324); the complex of fibres, horizontal, vertical,

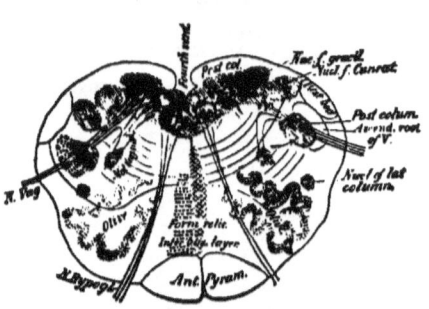

FIG. 324.—T.S. Medulla Oblongata at the Level of the Olivary Body; partly Diagrammatic.

and those cut longitudinally, constituting the **formatio reticularis**; the much altered arrangement of the grey matter, which appears in the floor of the fourth ventricle; and, according to the level at which the section is made, there may be met with the origin of certain of the cranial nerves (fig. 324).

Of course sections should be made from higher levels in the medulla, and also through the pons. If a student has the requisite time, it is best to make a series of sections from below upwards, fixing them in order on slides.

The two following tables show, the one the fissures, areas, and mouldings to be noted on the medulla oblongata, and the other the grey matter of the cord and medulla oblongata.

MEDULLA OBLONGATA.

Objects seen on the Surface of the Medulla Oblongata.

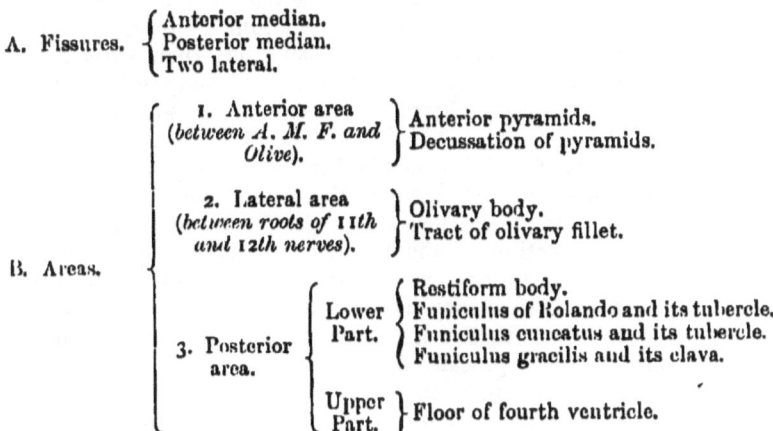

A. Fissures.
- Anterior median.
- Posterior median.
- Two lateral.

B. Areas.
- 1. Anterior area (*between A. M. F. and Olive*).
  - Anterior pyramids.
  - Decussation of pyramids.
- 2. Lateral area (*between roots of 11th and 12th nerves*).
  - Olivary body.
  - Tract of olivary fillet.
- 3. Posterior area.
  - Lower Part.
    - Restiform body.
    - Funiculus of Rolando and its tubercle.
    - Funiculus cuneatus and its tubercle.
    - Funiculus gracilis and its clava.
  - Upper Part. — Floor of fourth ventricle.

C. External Arciform Fibres.

Table of Grey Matter of the Medulla Oblongata.

	CORD.			MEDULLA.
Grey Matter of the Spinal Cord.	Anterior Cornu.	Head	...	Nucleus lateralis.
		Neck	...	Anterior part of formatio reticularis.
		Base	...	Nucleus of the fasciculus teres.
	Posterior Cornu.	Head	...	Nucleus of Rolando.
		Neck	...	Posterior part of formatio reticularis.
		Base	...	Nucleus of the funiculus gracilis (clava). Nucleus cuneatus. Nucleus on floor of fourth ventricle.

Isolated Grey Nuclei in the Medulla.
- Nucleus of olivary body.
- Accessory olivary nuclei.
- Nucleus of external arciform fibres.

CEREBELLUM.

Methods.—The cerebellum is hardened in the same way as the cord. It is advantageous, however, to wash out the blood-vessels of the whole brain with Müller's fluid, and afterwards to distend them with the same fluid.

If ammonium bichromate be used (2 per cent.), the pieces harden quicker than in the case of the cord.

For the application of Golgi's method, see p. 345.

3. V.S. Cerebellum, *i.e.*, including the grey and white matter cut at right angles to the direction of the folds. Stain it with carmine, aniline blue-black, or, better still, first with logwood and then with eosin. Mount in balsam. The best way for the student is to stain a small piece of the cerebellum in bulk in borax-carmine for two or three days, or longer, and then embed and cut in paraffin. In this way there is no fear of the sections breaking up, and the relative positions of the several parts are accurately maintained.

In making sections of the cerebellum, it is important, if one wishes to see the wide expanse of the protoplasmic processes of Purkinje's cells, to make sections across the direction of the laminæ. If made in the direction of the laminæ the leash of protoplasmic processes appear quite narrow (fig. 330, B). The protoplasmic processes, or *dendrites*, as they have been called by His, spread out in planes transverse to the direction of the lamellæ, hence the necessity for the above precaution.

Suppose a hæmatoxylin-eosin specimen to be prepared.

(*a.*) (L) Observe the primary and secondary convolutions (fig. 325). In each leaflet from within outwards, the white matter,

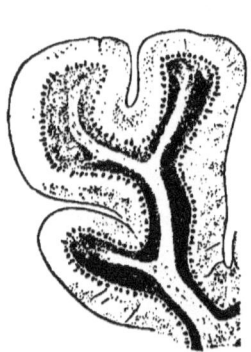

FIG. 325.—Leaflet of the Human Cerebellum, × 10.

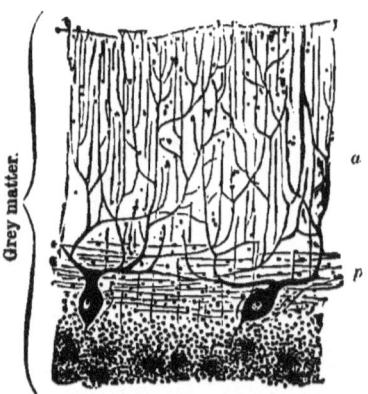

FIG. 326.—Cortex of the Cerebellum, × 90. *a.* Outer; *b.* Inner or granular layer; *p.* Cells of Purkinje.

composed of medullated nerve-fibres, and outside this the **grey matter**, composed of two layers, viz.:—(i.) The **nuclear layer**, composed of many layers of small nuclei—stained blue—each surrounded by a very small quantity of protoplasm. (ii.) The **outer layer** of the cortex, thicker than (i.), with a somewhat granular appearance, and containing branches of Purkinje's cells and some small branched angular nucleated cells (fig. 326).

Numerous blood-vessels enter the surface of the cerebellum from the pia mater covering it.

(*b*.) At the boundary-line between (i.) and (ii.) a row of large cells—**Purkinje's cells**—each with a somewhat oval or globular body, with a single central process, which becomes continuous with a nerve-fibre, although this is not seen in this preparation. Each cell gives off a peripheral process, which immediately branches, the larger branches running laterally for a short distance, and each branch divides again and again, the fine branches—protoplasmic processes or dendrites—running vertically through the outer layer of the cortex nearly to its free surface. The branched arrangement of these fibres has been compared to the antlers of a stag.

By means of the rapid Golgi method, a basket-shaped complex of fibrils can be seen round the basis of Purkinje's cells.

(*c*.) (H) The **granular** or **nuclear layer**, with two kinds of nuclei or rather cells. The most numerous are small and granular, and arranged in groups; they are stained violet by the hæmatoxylin. The others (larger and spherical, with a nucleolus) are the nuclei of small ganglionic nerve-cells. They are stained reddish. Amongst the granules may be seen medullated fibres stained reddish.

The more recent methods of Golgi and Ramón y Cayal show that these two kinds of cells correspond to nerve-cells. The small cells have an axis-cylinder and several protoplasmic processes, while the large cells in some respects resemble small Purkinje's cells.

4. **Blood-Vessels of the Cerebellum.**—These are injected when the blood-vessels of an animal are injected from the aorta with a Berlin-blue or carmine-gelatine mass. Mount an unstained sectin in balsam (**L** and **H**).

(*a*.) Observe the vascular pia mater sending at intervals **long** or **medullary** branches through the cortex to the medulla, and **short** or **cortical** branches which break up into a rich plexus of capillaries in the cortex, so that the latter is far more vascular than the white matter. Each vessel is surrounded by a perivascular lymph-space.

CEREBRUM.

Methods.—Harden it in the same way as the cord and cerebellum. For Golgi's methods (p. 345).

If Weigert's method is used, the sections must remain for twenty-four hours in the hæmatoxylin solution in order to see the plexus of medullated fibres in the superficial layers of the cortex, and three hours for the fibres which ascend between the pyramidal cells of the cortex cerebri.

5. **V.S. Cerebrum.**—Make V.S. through the central part of the

brain of a small animal. In the case of a human brain, select the ascending parietal or ascending frontal convolution. Make V.S. either by freezing or in paraffin after staining in bulk in borax-carmine. Other sections may be stained in aniline blue-black, and all are mounted with balsam.

With the naked eye observe the shape of the convolution, the grey matter outside, of a certain thickness ($\frac{1}{12}$–$\frac{1}{8}$ in. thick), and deeply stained, and inside it the white matter less deeply stained.

(L and H) (*a.*) Observe the white matter, composed of medullated fibres, with leucocytes here and there between them.

(*b.*) Outside this the grey matter, composed of several layers, recognised by the arrangement and shape of the cells present in it.

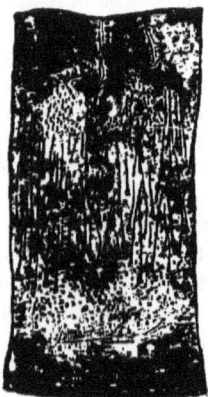

FIG. 327.—V.S. of Cortex of the Ascending Frontal Convolution, *i.e.*, a Motor Area. Carmine, × 20.

FIG. 328.—V.S. Middle Frontal Convolution. Carmine, × 20.

It will depend very greatly upon the plane of the section whether the student sees all the layers in any single section. They are usually in a five-layer type, but the relative thickness of the layers varies in different parts of the cerebrum.

A. **Arrangement of Nerve-Cells.**

(*c.*) The **layers** from the surface inwards are :—

(i.) The narrow *outer* or *first layer* (or finely-granular or molecular layer) consists of a network of fibrils with a very few small cells. Chiefly neuroglia cells, mostly vertical to the surface.

By Golgi's method, it can be shown to contain a layer of medullated fibres just under and parallel to the pia, and also some branched non-medullated fibres. It also contains a few small nerve-cells with two or more axis-cylinder processes. The latter details can only be detected in a preparation made by Golgi's method. In it may be seen sections of blood-vessels passing from the pia.

(ii.) The *second layer*, or *layer of small pyramidal cells*, is usually narrow, with several rows of small pyramidal cells, the peripheral processes of the latter pointing towards the surface of the convolution. It passes gradually into

(iii.) The *third layer*, or *layer of large pyramidal cells*, which is much thicker than the others, and contains large *pyramidal cells*, each with a peripheral process, which can sometimes be traced outwards for a considerable distance. The axis-cylinder process gives off several collaterals, and can be traced into a medullated fibre of the white matter. The cells, however, may be cut obliquely, and then they appear triangular. Usually the cells are larger in the deeper layers, and become smaller in the outer layers.

(iv.) The *fourth layer*, or *layer of irregular* or *polymorphous cells*. This is a narrow layer of small, usually angular but irregularly-shaped cells. They lie between the nerve-fibres which pass into the cortex. In the motor areas of the frontal and parietal convolutions large pyramids arranged in groups or nests may be found between these cells.

(v.) The *fifth layer*, or *layer of fusiform cells*, also a narrow layer, with a few fusiform cells with nerve-fibres between them. This layer abuts on the white fibres of the medulla, and may in some regions be fused with the previous layer. In the grey matter superficial to the island of Reil, it exists as a separate layer constituting the *claustrum*.

Not unfrequently in bichromate and chromic acid preparations, the place of the pyramidal cells is partly represented by clear spaces, produced by vacuolation of the cells. It is by no means easy to distinguish a nerve-cell from a neuroglia-cell in the cerebrum. Usually the glia-cell nuclei are smaller. Between the rows of cells may be seen fine longitudinal striation, indicating the existence of nerve-fibres in the grey matter.

The structure of the cortex is not identical, although similar, throughout. In the motor areas especially, *i.e.*, in the ascending frontal, ascending parietal, and part of the marginal convolutions, the pyramidal cells are usually larger than in those areas which are described as sensory—*e.g.*, the occipital, temporo-sphenoidal, or even in the anterior part of the frontal.

It is advisable, therefore, to provide the student with sections from these different areas. There is no abrupt transition between one type of cortical structure and another.

B. Course of White Fibres in the Cortex Cerebri.—This cannot be made out in sections prepared as above. All that can be seen is that fine strands—*radii*—of medullated fibres pass from the white centre at intervals into the grey matter, and run outwards as *medullary rays* between the nerve-cells. By other methods it can be shown that some of them become continuous with axis-cylinder processes of the pyramidal and polymorphous cells, while, as shown in fig. 331, others end in free arborisations. Tracing the fibres the other way, *i.e.*, as axial-cylinder processes of pyramidal cells, it has been shown by other methods that some of them—chiefly from the large pyramids of the motor areas—pass into the white matter, enter the corona radiata, and pass through the inner and anterior two-thirds of the posterior division of the internal capsule to enter and form the pyramidal tracts. They constitute the **projection fibres**, and end in free arborisations in relation with the multipolar nerve-cells of the anterior cornu of the grey matter of the cord.

The so-called **commissural** or **callosal fibres**, which join opposite halves of the brain, pass into the corpus callosum either directly or by collateral fibres (fig. 331) and pass to the opposite side to end as

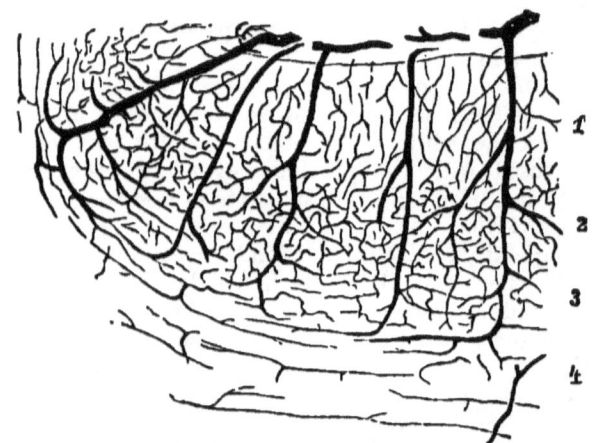

FIG. 329.—Injected Cerebral Cortex of Dog. 1. Layer with few vessels; 2. Layer of large pyramidal cells; 3. Deepest layers of cortex; 4. Medulla.

free arborisations in the grey matter there (fig. 331, *f*). Others are said to join the antero-posterior fibres (fig. 331, B) or **association fibres**, which run between grey matter in different parts of the same hemisphere.

6. **V.S. Middle Frontal Convolution** (fig. 328).—Compare this with that from a motor area, and note the absence of the very large pyramidal cells. In the **gyrus hippocampi** or uncinate gyrus there are other peculiarities, its grey matter consisting of numerous conical cells with very long processes. It is hardened in the same

way. The peculiar shape of the cells in this region is best shown by the "double-impregnation" method of Ramón y Cayal (p. 222).

7. **Blood-Vessels of the Cerebrum** (L and H).—Make rather thick sections of an injected brain, and mount them in balsam. They are best embedded and cut in paraffin. The whole head should be injected from the aorta.

(*a.*) (L) The larger vessels in the pia mater send into the cerebral cortex two sets of arteries: those that perforate the grey matter and proceed to supply the medulla—the long or **medullary arteries**; and a more numerous, shorter set, that ramify chiefly in the grey matter—the short or **cortical arteries**. The grey matter is much more vascular than the white, and the vessels are surrounded by perivascular sheaths (fig. 329).

(*b.*) Study the arrangement and relative vascularity of the capillary plexus in the cortex. At the surface it is less dense, and it is most dense in the region of the large pyramidal cells.

The best *resumé* of the researches of Golgi, R. y Cayal, and Kölliker is given by Waldeyer,[1] whose papers contain numerous woodcuts showing not only the histological results, but their bearing on physiological problems.

ADDITIONAL EXERCISES.

8. **Cerebellum in Osmic Acid.**—Place very small pieces of the grey matter (1 mm. cubes) in 1 per cent. osmic acid (24-48 hours) as recommended for the cerebrum (Lesson XXXI. 13). Wash in water and harden in alcohol. In the sections note the medullated fibres in the granular layer, but none of them pass into the outer layer of the cortex.

9. **Cerebellum in Ehrlich-Biondi's Fluid.**—Stain sections in this fluid as directed under Spinal Cord (Lesson XXX. 13) and mount in balsam. The outer grey layer and the medulla are red, the granular layer violet. In the latter can be seen the two kinds of cells, one stained red, the other violet.

10. **Purkinje's Cells by Golgi's Method** (p. 344).—(*a.*) The easiest method is the slow silver nitrate one (p. 344). If the preparation is successful, one is rewarded by the beauty of the preparation. It is almost impossible to reproduce the complexity of the processes of Purkinje's cells (fig. 330). They are stained black, the nerve-fibres may or may not be stained. The sections are mounted in balsam *without a cover-glass*. The much-branched protoplasmic processes, or *dendrites*, are shown in fig. 330, A, as they appear when a section is made across the laminæ, and in B when the section is in the direction of the laminæ. The dendrites do not anastomose with each other, and all lie in one plane in the transverse direction of the leaflet.

(*b.*) **Fibres in Cerebellar Grey Matter.**—By using the rapid hardening method of Golgi and Cayal many other details may be made out, but not all necessarily in one section. Round the body of each cell of Purkinje is a plexus of fibrils—basket-work of fibres—produced by the division or arbori-

---

[1] *Deutsche med. Wochensch.*, Nos 44, 45, 46, 47, 1891.

sation of the axis-cylinder processes of nerve-cells lying in the molecular layer of the grey matter. The dendrites are in part also invested by a basket-work due to the arborisation of fibres proceeding from the medulla—*i.e.*, fibres which do not terminate in the cells of Purkinje. Some of the cells of the granular layer are nerve-cells with protoplasmic and axis-cylinder processes. The small nerve-cells, by far the most numerous, give axis-cylinder processes which run out into the molecular layer, where they divide and become continuous with a fibre running at right angles to it. In the molecular layer there is an arrangement of nerve-fibres running more or less parallel to the surface of the leaflets, and joined here and there by the axis-cylinder processes of the nerve-cells of the granular

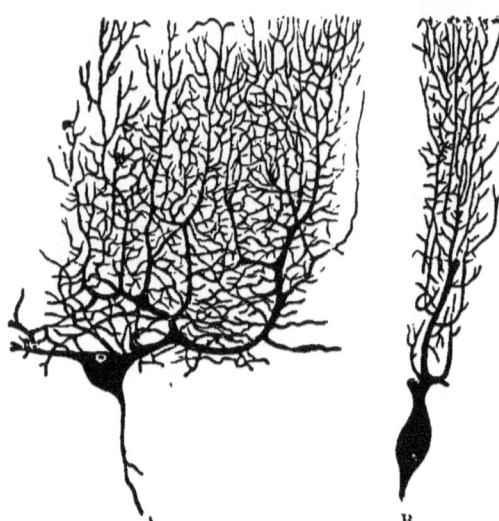

FIG. 330.—Cell of Purkinje. Corrosive sublimate. A. Seen on the flat, and B, from the side, × 120.

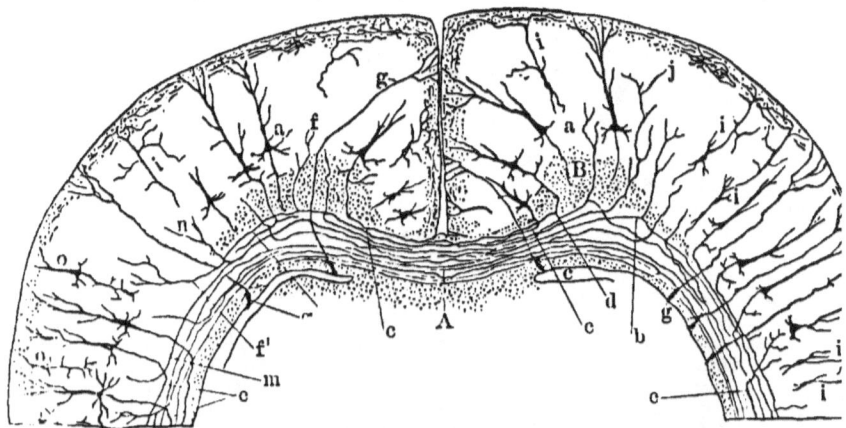

FIG. 331.—Scheme of T.S. of Cerebrum of New-born Rat, prepared by Cayal's Double-impregnation Method. *A.* Corpus callosum ; *c.* Lateral ventricle ; *B.* antero-posterior, or association fibres arising from large pyramidal cells ; *a.* Large pyramidal cells whose axis-cylinder processes pass into the antero-posterior layer ; *b.* Fibre of corpus callosum bifurcating ; *c.* Callosal fibre ; *d.* Callosal fibre arising from a pyramidal cell ; *e.* Axis-cylinder process descending obliquely to enter the corpus callosum ; *f.* Final ramification of a callosal fibre in the grey matter of the cortex ; *h.* Collateral fibre from a large pyramidal cell ; *i.* Fusiform cells with axis-cylinder processes passing into outer layer of cortex ; *j.* Final ramification of a callosal fibre arising in opposite side of cortex.

layer already described. If the cerebellar lamina be cut in a direction transverse to the course of these fibres, then they merely appear as black dots. My experience leads me to believe that this system is best demonstrated by the rapid Golgi method.

11. **Cerebrum by Golgi's Method** (see Lesson XXX. 19).—Do the same as for the cord and cerebellum. Either the $AgNO_3$ or $HgCl_2$ methods may be adopted. In the latter case the best results are obtained by keeping the portions of the brain for months in the mercuric-chloride solution. I have usually got good results with the brain of the rat or rabbit. In all cases mount the sections in balsam without a cover-glass. In new-born animals better results are obtained—especially with the rapid method (osmico-bichromate method, p. 345)—than with adult brains. Fig. 331 shows some of the results which may be obtained in a T.S. of the brain of a newborn rat. The results have been pieced together from a study of many sections.

By the rapid Golgi method it is easy to obtain beautiful preparations showing the pyramidal cells of the cortex with their apical, lateral, and axis-cylinder processes; in fact, I have often obtained such in the superficial sections after a few hours' immersion in silver nitrate. For the cornu ammonis cells the double-impregnation method of R. y Cayal is excellent. Sometimes, however, one gets the blood-vessels stained.

12. **Cerebrum by Weigert or Weigert-Pal's Method** (see Lesson XXX. 4).—This method reveals the existence of medullated fibres in the cerebral cortex arranged according to the scheme shown in fig. 332.

At the lower part of the grey matter a large number of fine medullated nerve-fibres enter it—*radii*—which run outwards in the cortex as *medullary rays*. Between the latter is an *inter-radial plexus*, and above them is a *supra-radial* plexus, and quite under the pia are tangential fibres. On the boundary-line between the supraradial plexus and the inter-radial plexus is a white stripe, visible to the naked eye, especially well marked in the cuneus. It has been called the stripe of Gennari, Baillarger, and also of Vicq d'Azyr.

FIG. 332.—V.S. Cortex Cerebri. The right side drawn from a Weigert's hæmatoxylin preparation, and the left from a Golgi's sublimate one. On the right the medullated fibres, and on the left only the nerve-cells are shown. There are really more cells than shown in the drawing.

13. **Medullated Nerve-Fibres in the Cortex Cerebri.** —Place very small pieces of the outer part of the cortex in 1 per cent. osmic acid (24 hours). The pieces must be black throughout. Make thin sections, place one on a slide, add a drop of ammonia; the section swells up and the medullated fibres become distinct. Expose the section to the vapour of osmic acid, *i.e.*, place the section on a slide over a glass thimble filled with osmic acid and cover the whole with a bell-jar. After half an hour or less the "fixation" is complete. This is Exner's method as modified by Ranvier. Preserve in glycerine. Observe the narrow medullated nerve-fibres between the nerve-cells; the nuclei only of the latter are distinctly seen.

Hypophysis Cerebri or Pituitary Body.

14. **Hypophysis Cerebri.**—This consists of two parts, the one derived from the brain, a continuation of the infundibulum; the lower part is derived from

an inflexion of the embryonic mucous membrane of the mouth. Harden in Müller's fluid, stain sections in picro-carmine, and mount in balsam. Observe sections of closed gland-tubes or alveoli almost completely filled with cubical, somewhat granular, or clear cells. A lumen is rarely visible. The communication between the mouth and the lower part of the gland is cut off in the process of development.

LESSON XXXII.

THE EYE.

Methods.—(i.) Enucleate the eyeball of a rabbit or cat; remove any adhering fat and muscles. With a sharp razor make a single cut at the equator of the eyeball, and suspend it in 200 cc. of .25 per cent. chromic acid. After twenty-four hours cut the eye into an anterior and a posterior half, and place them for several days in new fluid. Wash and harden in the dark in gradually increasing strengths of alcohol. The eye hardened in this way may be used for preparing sections of some of its parts.

(ii.) Harden the other eyeball for two or three weeks in Müller's fluid, after cutting into it in the same way as above. Complete the hardening in alcohol. It is well to have sections of the cornea of several animals, *e.g.*, the pig and ox.

THE CORNEA.

Make V.S. from a cornea hardened in Müller's fluid or Flemming's fluid. Stain with picro-carmine and mount in Farrant's solution, or in hæmatoxylin and mount in balsam. Stain the one fixed in Flemming's fluid with safranin. A good method is to stain it in bulk in borax-carmine. Eosin-hæmatoxylin is a good stain.

1. V.S. Cornea (L. H.).

(*a*.) Observe the **anterior** or **conjunctival epithelium**, consisting of several layers of stratified epithelium. The most external cells are flattened; those of the middle layers are more oval or rounded, many of them with finger-shaped processes dipping down between the deeper rows of cells. The cells in the lowest layer are columnar, and placed perpendicularly upon the cornea, resting on (*b*) (fig. 333).

(*b*.) The narrow, clear, transparent layer—**anterior elastic lamina**; this is best marked in the human eyeball.

(*c*.) The **substantia propria**, or body of the cornea, composed of

XXXII.] THE EYE. 359

layers of transparent fibrous tissue arranged in laminæ. The laminæ, while arranged in the main parallel to each other, have connecting processes joining adjacent laminæ. Between the laminæ oval spaces, and in each space a nucleated **cornea corpuscle** (fig. 333, c), which, as seen on edge, is thin and flattened. The processes of these corpuscles are best seen in gold specimens.

(d.) The **posterior elastic lamina**, or **membrane of Descemet**, a well-marked, thin, transparent, hyaline lamina, with sharply defined outlines, and covered posteriorly by

(e.) The **posterior epithelium**, consisting of a single layer of large flattened nucleated cells, seen in profile. They are apt to be displaced if the section be roughly handled.

If the posterior elastic lamina be broken across, it is apt to curl up. It stains readily with carmine.

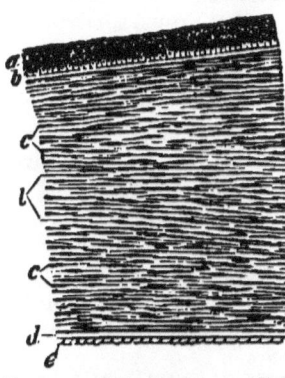

FIG. 333.—V.S. Cornea. a. Epithelium; b. Anterior elastic lamina; c. Corneal corpuscles; l. Lamellæ of cornea; b–d, Substantia propria; d. Descemet's membrane; and e. Its epithelium.

FIG. 334.—Corneal Corpuscles, and a few Nerve-Fibrils of Frog. Gold chloride.

2. **Nerves of the Frog's Cornea.**—In a pithed frog, squeeze the head to make the eyeball project, and with a sharp razor cut off the cornea. Wash it in normal saline, treat it with gold chloride by the lemon-juice method (p. 79).

After impregnation with the gold solution, with scissors make three snips into its margin, dividing it into three sections, so that it may lie flat on a slide. Mount in Farrant's solution.

(a.) (L and H) Observe the branching **cornea corpuscles**, arranged in many layers, one under the other, readily seen by raising or depressing the lens (fig. 334). The numerous processes given off from the cells anastomose with similar processes from adjoining cells, not only in the same plane, but with cells lying in planes above and below them.

(*b.*) **The Nerves.**—If a part of the sclerotic is adherent to the cornea, the fine medullated nerve-fibres, deeply stained, are seen in the sclerotic and passing into the cornea, where they lose their myelin and become non-medullated. The larger non-medullated fibres have nuclei in them, and unite with other fibres to form a coarse plexus — the **ground** or **primary nerve plexus.**

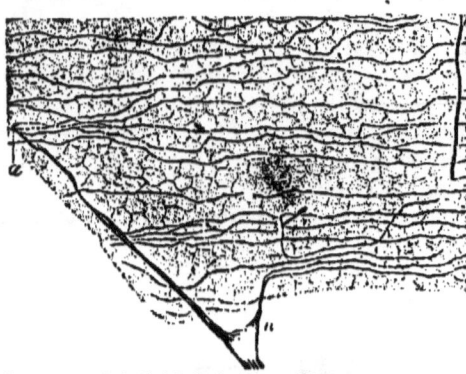

FIG. 335.—Sub-Epithelial Nerve-Plexus, Cornea of Frog. *n.* Non-medullated nerve with nerve-fibrils; *a.* Nerve-fibrils, × 300.

(*c.*) From this plexus finer bundles of fibrils proceed to form a finer plexus, and from the latter numerous fibrils—often with varicose swellings—running chiefly in the planes of the laminæ are seen.

(*d.*) (**H**) Fine fibrils are seen in the larger branches of the non-medullated nerves (fig. 335).

If the cornea of a rabbit be used, then tangential thin sections must be made and mounted in Farrant's solution or balsam.

3. **Inter-Epithelial Termination of the Nerve-Fibres.**—The cornea of a rabbit is stained in 1 per cent. gold chloride (25–30 minutes) and reduced in slightly-acidulated water (acetic acid) by exposure to light (p. 79). The lemon-juice and formic-acid method must not be used, as the formic acid removes the epithelium. Make V.S. either by freezing or free hand and mount them in glycerine (**L** and **H**).

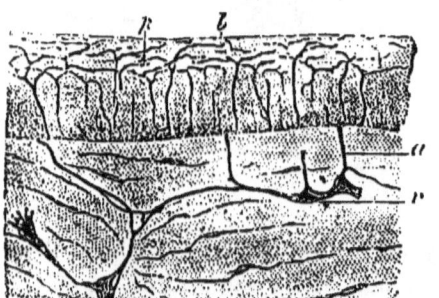

FIG. 336.—V.S. Cornea of Frog. *n.* Nerve-fibres; *a.* Perforating fibrils; *r.* Nucleus; *pb.* Inter-epithelial plexus of fibrils. Gold chloride.

(*a.*) Observe sections of the lamellæ and corneal corpuscles, and between these, parts of the primary nerve-plexus cut across, more or less obliquely, the branches being finer towards the anterior elastic lamina. Bundles of fine fibrils—the **rami perforantes**—perforate this membrane and spread out under the anterior epithelium to form the **sub-epithelial plexus**, from which fibrils proceed to form a plexus of fine varicose fibrils—the **epithelial plexus**—between the epithelial cells, where they terminate in fine points, forming no connection with special end-organs (fig. 336).

4. Cell-Spaces in the Cornea (Silver Method) (L and H).—Pith a frog, scrape off the corneal epithelium, and rub the surface of the cornea with solid silver nitrate. After 20–30 minutes, excise the cornea, mount it in glycerine, and expose to light.

FIG. 337.—Cornea of Frog with Cell-Spaces. *s.* Cell-spaces. Solid silver nitrate, × 300.

(*a.*) The general substance of the section is stained brown, and in it are clear branching spaces, each one corresponding in shape to that of

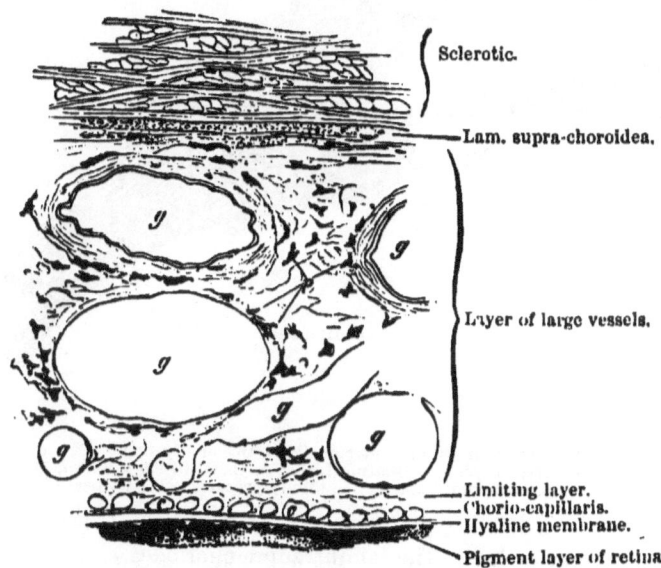

FIG. 338.—V.S. Part of Sclerotic and the Choroid. *g.* Large vessels; *p.* Pigment-cells; *c.* T.S. of capillaries, × 100.

a cornea corpuscle (fig. 337). The branches of adjoining spaces anastomose, so as to form a system of **juice-canals**, not only with

362 PRACTICAL HISTOLOGY. [XXXII.

spaces in the same plane, but also with the spaces in lower and deeper planes.

5. Iron Sulphate Method (Lesson XI. 13).—The matrix is blue and the spaces are clear and not stained.

THE SCLEROTIC, CHOROID, AND CILIARY REGION.

6. Sclerotic and Choroid (L.H.).—From an eye hardened in Müller's fluid and stained in bulk in borax-carmine cut sections in paraffin to include the sclerotic and choroid. Observe—

(1.) The sclerotic (fig. 338). (*a.*) Composed mainly of bundles of white fibrous tissue crossing each other in several directions.

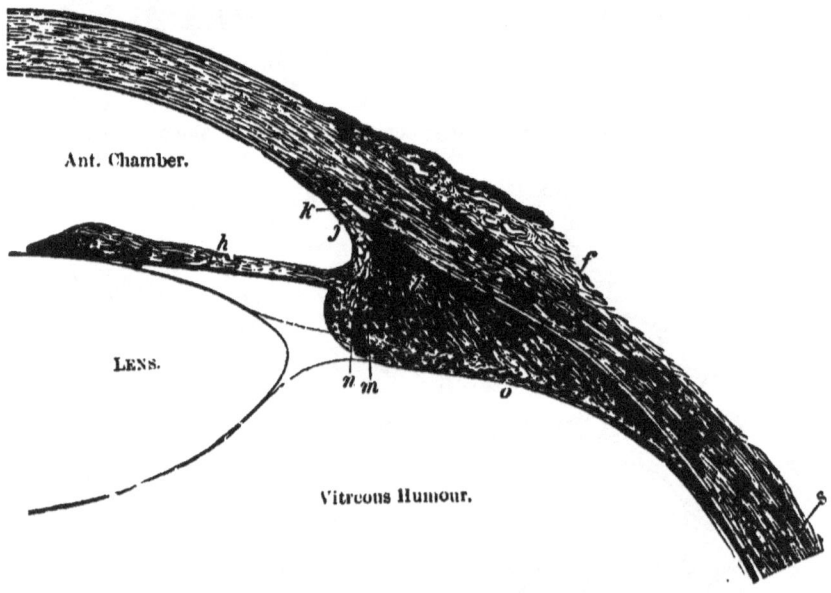

FIG. 339.—Horizontal Section of Anterior Quadrant of Eyeball. *f.* Conjunctiva; *s.* Sclerotic; *h.* Iris; *j.* Lig. pectinatum iridis; *k.* Canal of Schlemm; *l.* Longitudinal; *m.* Circular fibres of ciliary muscle; *n.* Ciliary process; *o.* Ciliary part of retina.

(*b.*) A narrow layer, the lamina supra-choroidea.

(2.) Inside this a section of the choroid composed of the following layers from without inwards—

(*a.*) Layer with large vessels and numerous branched pigment-cells. This forms by far the thickest layer.

(*b.*) Limiting layer.

(*c.*) The chorio-capillaris, containing numerous sections of capillaries.

(*d.*) A basement or hyaline membrane.

(*e.*) Pigment-cells containing melanin.

7. Pigment-Cells of the Choroid (L.H.).—From the inner choroidal surface of an eye hardened in Müller's fluid or spirit, scrape off a little of the black pigment layer and mount it in balsam.

(*a.*) Observe the branched pigmented cells, filled with brown or black granules of melanin. The nucleus contains no pigment granules. If the granules be discharged from the cells and float in a watery fluid, they exhibit Brownian movement. The nucleus stains readily with hæmatoxylin.

8. Ciliary Region, Ciliary Muscle, and Iris.—Make a meridional section through the corneo-scleral junction of a hardened eye (Müller's fluid), *e.g.*, of an ox, cat, or rabbit, so as to include the ciliary muscle and the iris. Stain the section in hæmatoxylin or eosin-hæmatoxylin, and mount in balsam; or it may be stained in picro-carmine and mounted in Farrant's solution. To preserve all the parts exactly in position cut it in celloidin, or stain in bulk and cut in paraffin. Fix the sections on a slide by a fixative.

(*a.*) Observe the cornea passing into the sclerotic, and near it an opening, the **canal of Schlemm** (fig. 339).

(*b.*) From the angle of the iris and cornea, running in a fan-shaped expansion, the radiating fibres of the **ciliary muscle**, consisting of non-striped muscle. The circular fibres cut transversely inside the radiating fibres are not so well marked as in the human eye.

(*c.*) The membrane of Descemet splits up into fibres, some of which curve into the iris, others spread into the ciliary processes. This region constitutes the **ligamentum pectinatum iridis.**

(*d.*) The pigmented folds or **ciliary processes.** A layer of black pigment-cells continuous with those of the retina continued over these processes, and inside the pigment a clear unpigmented layer of columnar epithelium.

(*e.*) The **iris,** composed of five layers from before backwards.

(1.) A layer of *endothelium* continuous with that covering the posterior surface of the cornea (fig. 339, *e*).

(2.) The *anterior boundary layer*, chiefly consisting of branched connective tissue-cells.

(3.) The *vascular layer*, or body of the iris, composed of a stroma of connective tissue with numerous sections of blood-vessels and branched pigment-cells, and at its deeper part transverse sections of the circular smooth muscular fibres, which constitute the circular muscle or *sphincter pupillæ* of this organ. It is doubtful if a dilator pupillæ exists in man.

(4.) The *posterior hyaline layer*, of an elastic nature.

(5.) Two layers of pigment-cells—the *uvea*—the outlines of the cells difficult to define, and continuous with the epithelium of the pars ciliaris retinæ.

(6.) **Blood-Vessels.**—A large number of blood-vessels and nerves pass into the iris, but their distribution must be studied by the methods already described for such purposes.

THE LENS.

The lens is developed by the inflection of part of the epiblast, so that it consists of modified epiblastic epithelial cells invested by a transparent capsule. The **capsule of the lens** is a clear transparent elastic membrane, not unlike a basement membrane, but is less resistant to acids than the latter. Immediately on the inner surface of its anterior part is a layer of clear flattened **epithelial cells**—hexagonal and nucleated—the remains of the original cells. Tracing these cells towards the margin of the lens they become narrower and more cylindrical, until further on they gradually pass into lens fibres. Posteriorly there are no epithelial cells between the lens and its capsule, the lens fibres resting directly on the capsule itself. The **lens fibres** are modified epithelial cells, the fibre constituting the principal part of the cell. The cells are long, soft, hexagonal prisms united to each other by more or less toothed edges. The superficial fibres are longer and larger than the deeper ones, and contain, or have on their surface, nuclei lying on the part that corresponds to the equatorial (nuclear) zone of the fibre. No nucleus is visible in the central fibres.

9. **Methods.**—Harden in its capsule, the lens of a cat or rabbit, in Müller's fluid (2–3 weeks); do not place it in spirit, but cut it in a freezing microtome, or cut in celloidin. Paraffin is not suitable. Stain the sections in eosin and mount in Farrant's solution. Make meridional and equatorial sections.

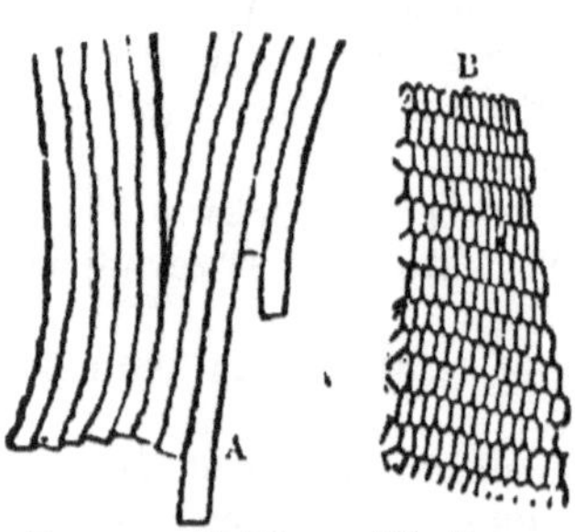

T.S. Lens—(*a.*) (**L.H.**).—Observe the *hyaline capsule*—if present—and just under its front part a layer of epithelial cells, which, when traced outwards, are seen to rapidly elongate into lens-fibres (p. 364). Nuclei may be seen on some of the fibres.

(*b.*) **T.S.** of the fibres are hexagonal (fig. 340, B).

FIG. 340.—*A*. Fibres of the lens; *B*. T.S. of the same.

10. **Isolated Lens Fibres—Methods.**—(*a.*) Place the lens of a frog or fish in a 10 per cent. solution of sulphocyanide of potas-

sium (24–36 hours). Tease out a fragment of the softened lens in glycerine. (*b.*) Or use dilute sulphuric acid (4–5 drops to 5 cc. of water for 24–36 hours). Wash in water to remove the acid. Stain in carmine or hæmatoxylin. (*c.*) Methyl-green feebly acid, does well for fresh-teased fibres, but the preparations do not keep. The nuclei on the fibres may be seen in this preparation.

(H) (*a.*) Observe the flattened bands with serrated edges; the teeth dove-tail into each other. The teeth may be seen from the side or directed to the observer.

RETINA.

Methods.—(i.) With a sharp razor cut the eye of a cat, rabbit, pig, or other animal into an anterior and posterior half. Place the posterior half in 2 per cent. potassic bichromate, or in a mixture of Müller's fluid and spirit (two weeks), and complete the hardening in spirit.

(ii.) Stain "in bulk" in borax-carmine for several days the whole posterior segment of the hardened eyeball, then place it for a day in acid alcohol, and embed in paraffin and cut sections. Perhaps celloidin is to be preferred if the sclerotic be thick. Fix the sections on a slide and mount in balsam. In this way the whole structure of the retina is preserved in its relations to other structures.

(iii.) The same process may be adopted with the posterior segment of the eyeball hardened in 1 per cent. osmic acid.

(iv.) Harden the retina in a 3 per cent. solution of nitric acid (15–20 mins.). Wash out the acid, stain in bulk, and make sections. Stain with eosin and methylene-blue.

(v.) For small eyes, such as those of a lizard, it is sufficient to expose the whole unopened eyeball to the vapour of osmic acid for several hours, and then to stain the eyeball in bulk in borax-carmine. Embed in paraffin and cut sections of the whole eyeball. Or, expose the eye of a frog or triton to osmic acid vapour (10 mins.). Place in 40 per cent. alcohol (4–6 hours); divide eyeball by incision parallel to edge of cornea; bichromate of ammonia (3 per cent.) for 5–10 hours; wash in water; alcohol; borax-carmine; embed and cut in paraffin.

(vi.) Make sections of an eyeball hardened in Müller's fluid and stain them in eosin-hæmatoxylin. The latter stains the granular layers of the retina, while the other parts are rosy in tint.

(vii.) **Methylene-Blue Method.**—This is done either by injecting methylene-blue (1 in 300 normal saline) into the blood-vessels of an animal just killed, and waiting two or three hours thereafter, or by placing the fresh retina in the same fluid and observing with the microscope the coloration of the nervous elements, especially the cells in the retina. Mount in picrate of ammonia and glycerine (p. 192).

11. V.S. Retina.—(L) Beginning from within, *i.e.*, next the vitreous, observe (fig. 341)—

(*a.*) The **internal limiting membrane**, and springing from it by wing-shaped expansions the **fibres of Müller**, which run vertically outwards through the layers of the retina to the outer limiting membrane.

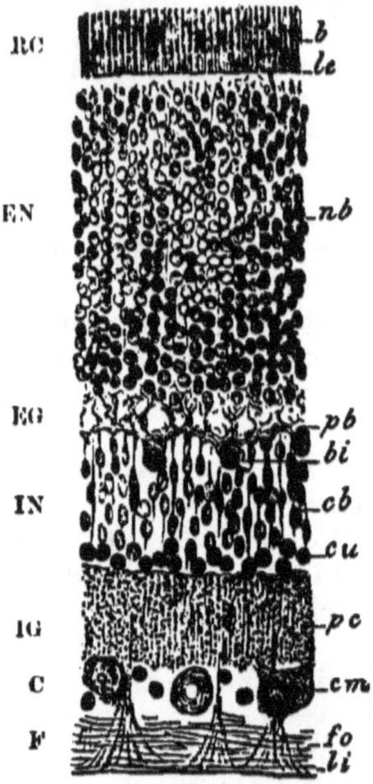

(*b.*) The **fibrous layer**, composed of the non-medullated fibres of the optic nerve. The fibres are medullated in the optic nerve, but they lose their medulla as they enter the retina. The layer is thinner in the anterior part of the retina. Some of the nerve-fibres end in the ganglionic cells of the next layer, but others pass outwards to the inner molecular layer to end by terminal arborisations there (shown by Golgi's method).

(*c.*) The **ganglionic** or **nerve-cell layer**, consisting of a single row of large multipolar nerve-cells, each cell with a large conspicuous nucleus. Each cell gives off (i.) one thick process, which may divide into several processes which pass towards and ramify by terminal arborisations in the layer external to this, and (ii.) one fine axis-cylinder process towards the optic nerve-fibre layer, where it becomes continuous with a nerve-fibre. Between the nerve-cells, or in close relation to them, sections of the larger blood-vessels.

FIG. 341.—V.S. Retina of Cat. *b.* Rods and cones; *le, li.* External and internal limiting membranes; *nb.* Nuclei of rods; *pb.* Basal plexus; *cb.* Bipolar cells; *cu.* Unipolar cells; *bi.* Internal basal cells; *pc.* Cerebral plexus; *cm.* Multipolar nerve-cells; *fo.* Fibres of optic nerve. On the left the initials of the usual names for these layers; *RC.* Rods and cones; *EN., EG.,* and *IN., IG.* External and internal nuclear and granular layers; *C.* Cellular layer; *F.* Fibrous layer.

(*d.*) The **internal molecular layer** appears to consist of fine fibrils or granules. It is not unlike the grey matrix of the cerebrum, and is traversed by the fibres of Müller and also by fibres of the optic nerve, and processes of the cells of the ganglionic and inner granular layers.

(*e.*) The **internal nuclear layer**, deeply stained, and consisting of several rows of large spherical oval nuclei.

Some of the granules, however, by other methods can be shown to be **bipolar cells** with large nuclei. The processes extend into the inner molecular layer and end in arborisation, and by Golgi's method the opposite processes have been traced outwards as far as the external limiting layer. Other branched cells supposed to resemble neuroglia cells lie amongst these cells.

(*f.*) The **outer molecular layer** is thin, and is composed of the terminal arborisations of processes from the rods and cones and inner granular layer.

(*g.*) The **external nuclear layer** consists of many more rows of nuclei than in (*e*). They are the nuclei of the rod and cone fibres.

(*h.*) The **external limiting membrane**, cut across, and therefore appearing as a thin clean-cut line.

(*i.*) The **layer of rods and cones.** The rods are more numerous than the cones, and the latter are shorter than the former. Each rod and cone consists of an outer and an inner segment.

(*j.*) A **layer of hexagonal pigment-cells**, somewhat flattened, which sends pigmented processes or filaments between the rods and cones. The length and condition of these processes depends upon the influences to which the eye has been exposed before death (p. 370). The pigment granules, some of which are crystalline, exist chiefly in the inner part of the cells, and may extend into the cell processes which pass between the rods and cones.

(H) Study the successive layers. Although the layer of rods and cones and external nuclear layer have been described as separate layers, in reality they should be studied together. Each rod is continued inwards by a tapering fibre—the **rod-fibre** ; in the course of the latter is an oval nucleus, the fibre itself ending in an arborisation in the outer molecular layer. Each cone is similarly prolonged into a **cone-fibre**—thicker than the rod-fibre—which also has an enlargement in it containing a nucleus. The cone-fibre ends like the rod-fibre in the outer molecular layer. These two kinds of nuclei intercalated in the course of the rod and cone fibres make up the external nuclear layer, which as the rods and cones are merely modified epithelial cells—must be regarded as epithelial in their origin. In a freshly-teased retina the nuclei of the rod-fibres are marked by alternate transverse light and dim stripes.

Each **rod** and **cone** consists of two segments. The *outer segment* of the rods is clear hyaline and transparent, and striated transversely, and readily breaks up into transverse discs. During life the external segments of the rods contain the visual purple. The *inner segment* is wider, granular and striated longitudinally. The *cones* also consist of a shorter tapering *outer segment* transversely striated, and an *inner* bulging larger, longitudinally striated, segment which is continued through the external limiting membrane into the cone-fibre.

The outer segments are stained black by osmic acid, and the inner segments red by carmine.

Study successively the other layers to make out some of the details already described on p. 366.

12. Retina of Frog.—(i.) Place the posterior segment of the eyeball, with its vitreous humour removed, in 1 per cent. osmic acid (24 hours); wash well, and tease a fragment of the now blackened retina in glycerine. Observe

(a.) The rods, each with an outer and inner segment; the outer segment, however, is blackened by the osmic acid, and usually shows a tendency to split transversely.

(b.) In the cones there is a small refractive oil globule between the outer and inner segments. It first becomes brown, and then black, in osmic acid. In some animals, *e.g.*, some birds and reptiles, the oil globule is pigmented, the pigment being held in solution by the fat of the globule.

(c.) Numerous *pigment cells*, each consisting of a rather thick body, part of which contains the nucleus and is non-pigmented, while the other half is pigmented, and sends numerous fine processes between the outer segments of the rods.

(d.) Other parts of the retina are also isolated, *e.g.*, the nuclei of the nuclear layer and the limiting membranes.

(ii.) Fix the whole eyeball in osmic acid, and make V.S. of all the coats after embedding in paraffin or celloidin.

13. Cones.—To see large cones, place the fresh eye of a cod in 2 per cent. potassic bichromate (2–3 days) or osmic acid (24 hours), or isolate them by dilute alcohol.

14. T.S. Optic Nerve (L and H).—Place the optic nerve in 2 per cent. ammonium bichromate (2–3 weeks) or in Flemming's mixture (10–20 hours). Make T.S. Stain in hæmatoxylin, picro-carmine, or safranin, or by Weigert's method. Mount in balsam. Double-stain some in picro-carmine and hæmatoxylin.

(a.) Observe the sheaths, thick and well-marked, sending septa into the nerve, and thus breaking it up into small bundles of nerve-fibres, each surrounded by a fibrous sheath. In the larger septa are blood-vessels. The anastomosing septa form a kind of alveolar system, so that a section of this nerve is readily distinguished.

(b.) The nerve-fibres are all medullated, but they possess certain peculiarities. They are without sheath of Schwann, and the fibrous tissue runs parallel with the nerve-fibres.

(c.) As to the *sheaths*, one is continuous with the dura mater; under it is a space, the subdural space; inside is a prolongation of the arachnoid with subarachnoid space, and a thin continuation of the pia mater.

If desired, make L.S. This enables one to note the structure of the nerve-fibres and the arrangement of the connective tissue. An interstitial injection of osmic acid is a good method for isolating the structural elements.

15. Blood-Vessels of Eyeball.—It is not difficult to inject from the aorta with a carmine-gelatine mass the blood-vessels of the eyeball of an animal. Select an albino rabbit. Sections are mounted in balsam. One-half of the injected eyeball of an albino rat mounted in balsam shows the connections, distribution, and arrangement of the blood-vessels.

16. The eyelids.—Harden the eyelids of an infant in corrosive sublimate or .25 per cent. chromic acid (3 or 4 days) and then in alcohol, or in alcohol alone. Make sections, vertical to its surface and transverse to its long axis, which must be rather thick, and mounted in balsam to show the general arrangement of the parts. Finer sections are stained in hæmatoxylin or picro-carmine. Besides the sections of hair-follicles, connective tissue, and orbicularis muscle, note the Meibomian glands.

ADDITIONAL EXERCISES.

17. Fibrils of Cornea.—(*a.*) Macerate the cornea in lime-water, baryta-water, or dilute potassic permanganate. (*b.*) A better method is to dry the cornea, make T.S., allow them to swell up in water, and stain with picro-carmine.

18. Epithelium of Lens.—Use a small lens (frog or rat), and place it for a few minutes in $AgNO_3$ (1 : 300). Expose to light in glycerine, and examine. Note on the anterior surface (*a.*) silver lines bounding polygonal areas corresponding to the inter-epithelial cement. (*b.*) Black granular converging lines corresponding to the cement between the crystalline fibres (best seen on posterior surface).

19. Other Methods for the Retina.—By the usual methods of hardening the retina, it is impossible to make out the connections between its several elements. The two methods which have quite recently yielded the best results—apart from the direct use of osmic acid—are the methods of Golgi and Ehrlich's methylene-blue method.

(*a.*) **Golgi's Methods.**—Ramón y Cajal has made a large number of preparations both by the silver method (p. 344) and the rapid osmico-bichromate method (p. 345), especially on the retina of birds. In this way he has shown that the protoplasmic processes of the nerve-cells of the ganglionic layer ramify in the inner molecular layer, and end in terminal arborisations in several planes in this layer. Several varieties of nerve-cells with extensive terminal arborisations in the other layers of the retina are also described by him.

The following medium method gives good results :—

    (1.) Place the retina and sclerotic in weak osmico-bichromate fluid (osmic acid, 1 per cent.), 2 cc., bichromate of potash (3 per cent.), 20 cc., for 24 hours.
    (2.) Transfer to 1 per cent. silver nitrate (24 hours).
    (3.) Again for 24 hours, as in (1).
    (4.) Again for 24 hours in silver.

(b.) **Methylene-Blue Method.**—Dogiel[1] by means of this method finds that in man there are more nervous elements present than one has been in the habit of supposing. He arranges the nervous elements (in three layers) and other layers as follows:—

Membrana limitans externa. ⟶	Pigment epithelium (I). Neuro-epithelial layer (II).		
External reticular layer.	{ Sub-epithelial nerve-cells (a). Stellate nerve-cells (b). Bipolar nerve-cells (c).	}	A. Outer ganglionic layer.
Inner reticular layer.	{ Middle ganglionic layer.	}	B. Layer of spongioblasts of W. Müller.
	{ Inner ganglionic layer. Layer of nerve-fibres.		C. D.
Membrana limitans interna. ⟶			

The neuro-epithelial layer embraces the layer of rods and cones and subjacent external nuclear layer. What he calls the outer ganglionic layer corresponds to the inner nuclear layer, and contains three varieties of nerve-cells. The middle ganglionic layer corresponds to a layer of cells—of which there are several varieties lying on the confines between the internal granular (reticular) layer and the internal nuclear layer. These cells were called spongioblasts by W. Müller, but according to Dogiel, they are nerve-cells. The inner ganglionic layer corresponds to the cellular layer.

20. Epithelium of the Retina.—Harden the eye of a frog in picro-sulphuric acid (24 hours), stain in bulk in borax-carmine, and make vertical sections after embedding in paraffin.

21. Isolated Elements of the Retina.—(a.) Macerate the retina in Schiefferdecker's fluid as described for the isolation of the cells of the cord (Lesson XXX.), using the sulphuric acid method to remove the water.

(b.) For Müller's fibres 10 per cent. chloral hydrate is good.

22. Retina of Frog in Light and Darkness.—(a.) Keep one frog in absolute darkness for thirty-six hours. Kill it in the dark, and harden the eye in alcohol. (b.) Place another frog in direct sunlight for a few hours; kill it, and harden the retina in alcohol. Sections are made and stained with picro-carmine. The pigment-cells covering the rods of the retina in (a) are retracted, while those in (b) are pushed out between the segments of the rods.

23. V.S. Macula Lutea.—Secure a human eye as fresh as possible. Harden it, and make sections through the macula lutea, and observe the peculiarities of its structure. There are no rods in the fovea centralis, while the cones are long and narrow. The other layers are reduced to a minimum at the fovea centralis, but at the margins of the fovea they are thicker. There are a large number of ganglionic bipolar cells.

24. V.S. Entrance of Optic Nerve.—Make from an eyeball stained in bulk in borax-carmine a section longitudinally through the optic nerve to include its entrance into the eyeball. Observe the *lamina cribrosa, i.e.,* the felt-work of fibrous tissue of the sclerotic, perforated by the fibres of the optic nerve.

25. Eye of Triton Cristatus.—Excise one eyeball, pin it to the under surface of a cork, and fix the cork in a glass thimble containing a little osmic acid. In ten minutes or so the retina is "fixed," owing to the tenuity of the sclerotic. Divide the eyeball into two by a cut at the equator, place the posterior half in dilute alcohol (6 to 10 hours), and then in picro-carmine for

---

[1] *Archiv f. mik. Anat.,* xxxviii. p. 317, 1891.

the same period. Fix it finally in osmic acid, embed and cut it in soft paraffin. The retina of this animal is particularly serviceable, because its structural elements are so large.

26. **Lachrymal gland** is treated precisely in the same way as the serous salivary glands (Lesson XXIII.).

LESSON XXXIII.

EAR—NOSE.

EAR.

1. **Membrana Tympani (H).**—Fix it in osmic acid and mount in Farrant's solution. Observe the radiate yellow-looking fibres, and also fibres disposed circularly, the latter best developed near the periphery of the membrane. The thin epithelial coverings on the two surfaces of the membrane can be seen, and so can a few fine blood-vessels.

2. **Ceruminous Glands of the Meatus.**—Harden in absolute alcohol a portion of skin from the external auditory meatus, preferably from a new-born infant. Make rather thick sections, stain in hæmatoxylin and mount in balsam. Observe the **ceruminous glands**, which in some respects are like the sweat-glands. They have a narrow duct and a coil, the latter lined by a single layer of cubical cells, and the former by several layers of cells, as in a sweat-gland. In the secretory part smooth muscular fibres lie between the epithelium and the basement membrane. The lumen of the secretory part is very wide, and the lining cells have a clear cuticular disc, and contain a yellowish pigment and fatty granules.

THE COCHLEA.

Methods.—(i.) In a freshly-killed rabbit or guinea-pig, inject, by means of a hypodermic syringe, osmic acid (1 p. c.) through the membrana tympani into the middle ear. Cut away the lower jaw, so as to expose the large spherical osseous bulla. Open the bulla, cut away its walls, and expose the middle ear. On the inner wall of the latter are seen the turns of the cochlea. Cut away the surrounding parts from the osseous cochlea, and place the latter in Müller's fluid (a week), after opening one of the turns, to allow the hardening fluid to penetrate. It is better to do this under fluid, to prevent air entering, as the perilymph escapes. Decalcify in chromic and nitric acid fluid, and harden in the dark in alcohol. Stain the softened cochlea in bulk in borax-carmine. Embed in paraffin, and cut sections from base to apex of the coiled tube. Fix on a slide, and mount in balsam. Use the other decalcified cochlea, stained as above, and cut it in celloidin.

(ii.) The following is a better method (Ranvier), and preserves the finer structures *in situ*. Open the bulla of a guinea-pig, and cut out the inner ear. Make an opening or two in the turns of the cochlear tube, and place it in 1–2 cc. of 1 per cent. gold chloride. Add to the gold solution from time to time a few cc. of the following mixture :—1 per cent. gold chloride and a fourth part of formic acid, boil, and allow the mixture to cool. Leave the cochlea in the gold (20–30 minutes), remove it, wash, and expose it to light in

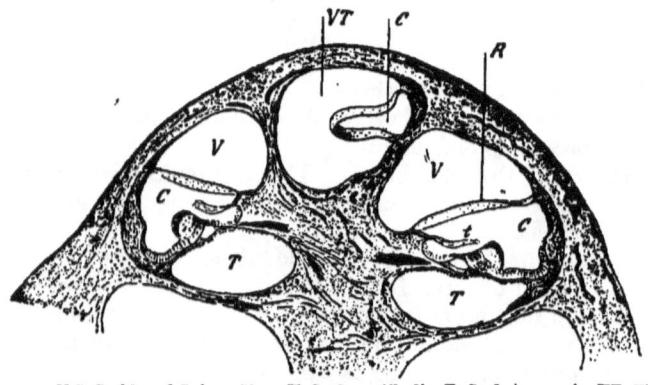

FIG. 342.—V.S. Cochlea of Guinea-Pig. *V.* Scala vestibuli; *T.* Scala tympani; *VT.* There the two communicate; *C.* Cochlear canal or scala media; *R.* Reissner's membrane; *t.* Membrana tectoria.

water slightly acidulated with acetic acid (1–2 drops to 20 cc. water). After 2–3 days harden it in alcohol and decalcify it in picric acid. Embed in gum and harden in alcohol, or embed it in celloidin and make sections parallel to the axis of the cochlea.

(iii.) After removing the inner ear, open its osseous canal under osmic acid .2 per cent. and leave it in this fluid for 10–12 hours. Wash in water and decalcify in chromic acid, 2 per 1000.

(iv.) After treating the cochlea with osmic acid and hardening it in alcohol, the following decalcifying fluid may be used :—1 cc. of a 1 per cent. solution of chloride of palladium, 10 cc. of hydrochloric acid, and 1000 cc. of water.

3. V.S. Cochlea, parallel to its axis.

(*a.*) (**L**) The turns of the cochlear tube cut across (fig. 342), each tube divided by a transverse partition—the **lamina spiralis**—into

XXXIII.]   THE COCHLEA.   373

an upper—**scala vestibuli**—and a lower compartment—**scala tympani**; the central column or **modiolus**; the last is osseous and contains channels for nerves and vessels. The lamina spiralis is partly osseous, viz., that part next the modiolus, and partly membranous, viz., that part next the wall of the cochlea; the latter part is the **membrana basilaris**. The osseous portion terminates in a crest—the **crista spiralis**—and its free end is scooped out into a groove—**sulcus spiralis**.

(*b*.) From the upper surface of the lamina spiralis ossea there runs to the wall of the cochlea a thin membrane—**membrane of**

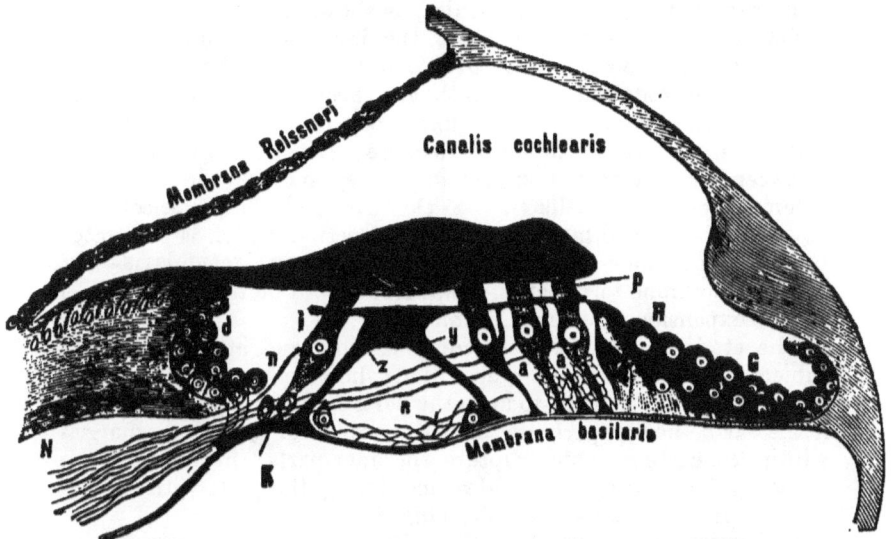

FIG. 343.—Scheme of the Canalis Cochlearis and the Organ of Corti. *N*. Cochlear nerve; *i*. Inner row, and *p*. Outer rows of hair-cells; *n*. Nerve-fibrils terminating in *p*; *a*. Supporting cells for *p*; *d*. Cells in the sulcus spiralis; *G* and *H*. Epithelial cells; *o*. Membrana reticularis; *z*. Inner, and *y*. Outer Rod of Corti.

Reissner—which completely shuts off a small three-sided cavity from the scala vestibuli. This is the **ductus cochlearis, canalis cochlearis,** or **scala media**. It is bounded by the wall of the cochlea, the membrane of Reissner—and its base is formed by the outer part of the osseous and the whole of the membranous spiral lamina. The latter is called the **basilar membrane**. This cavity contains a fluid, the endolymph, and into it pass the cochlear branches of the auditory nerve to end in a complex structure—**organ of Corti**—which rests on the basilar membrane.

(*c*.) Note in the scala media a membrane—**membrana tectoria**

—which arises from the lamina spiralis ossea, and spreads over so as partially to cover the organ of Corti. The **organ of Corti** consists of two rows of pillars—inner and outer—the **rods of Corti**—which meet above, forming arches, and leave a three-sided tunnel between them. Internal and external to these are rows of cells, some of them provided with fine bristles—**hair-cells**—constituting the cells of Corti and Deiters. Some of these cells are the actual terminal end-organs of the cochlear nerve, and others are sustentacular in function. The scala vestibuli and scala tympani communicate at the apex of the cochlea, as shown in fig. 342, V.T.

(**H**) Observe the rods of Corti, the inner more numerous than the outer, and how the head of the one lies over the head of the other. The rods vary in length and span, in different parts of the cochlea. Internal to the inner rods is a single row of hair-cells, and external to the outer rods are several rows of hair-cells. Between the latter are supporting cells, and beyond them are columnar epithelial cells (fig. 343). The rods and hair-cells are covered by a special membrane, seen in section, which is perforated by the upper ends of the hair-cells—**membrana reticularis**—the basilar membrane terminating towards the wall of the cochlea in a fibrous expansion, the spiral ligament.

The student must give considerable time and attention to the subject if he wishes to get preparations showing the structure of all the parts. It is difficult to keep the parts from falling asunder, and if the ear be not properly decalcified, bubbles of gas are discharged within the cochlea, which rupture the finer parts; hence arises the necessity for opening the cochlea and fixing the parts with a "fixing" medium previous to decalcifying it.

4. Semicircular Canals.—Fix the membranous semicircular canals, and their ampullæ, of a skate in Flemming's or Fol's fluid. Harden in alcohol. In the ampullæ are the terminations of the auditory nerve in the **crista acustica**. Make T.S. of the canals and V.S. of the crista acustica; in the latter case take care to include the entrance of the nerve-fibres. The preparation of suitable specimens to stain the termination of the nerves in the crista acustica presents very considerable difficulties. The student may have to repeat the process several times if he wishes to get typical specimens.

THE NOSE.

5. Olfactory Mucous Membrane.—Divide longitudinally the head of a freshly-killed rabbit. Place small pieces of the olfactory mucous membrane—readily recognised by its brownish colour—in dilute alcohol (2 hours), and then in 1 per cent. osmic acid

(24 hours). Harden in alcohol, make V.S., stain with hæmatoxylin, and mount in balsam.

(*a.*) (**L**) Observe on the surface the row of columnar epithelial cells with oval nuclei, and under these numerous rows of cells with spherical nuclei. At the base of the latter a row of more granular-looking basal cells (fig. 344). The basis of the mucous membrane composed of connective tissue with sections of numerous glands—**Bowman's Glands.** Sections of blood-vessels and branches of the non-medullated olfactory nerve may also be seen.

6. **Isolated Olfactory Cells.**—Place small pieces of the olfactory

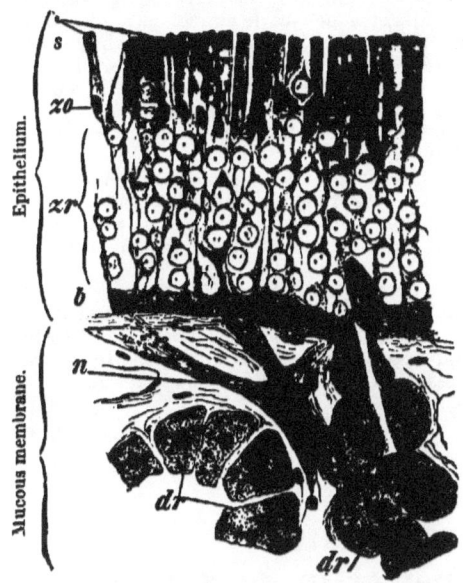
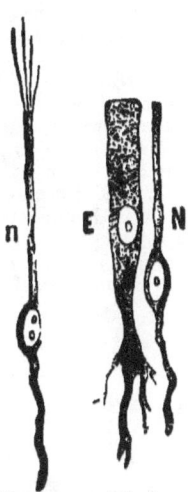

FIG. 344.—V.S. Olfactory Region (Rabbit). *s.* Disc on cells; *zo* and *zr.* Zones of oval and round nuclei; *b.* Basal cells; *dr.* Part of Bowman's glands; *n.* Branch of olfactory nerve.

FIG. 345.—Olfactory Cells. *N.* Human; *n.* Frog; *E.* Supporting cell.

mucous membrane in dilute alcohol (24 hours). Fix the membrane in 1 per cent. osmic acid (5 minutes). Stain the pieces in bulk in picro-carmine (24 hours). Scrape off a little of the stained and softened epithelial covering and mount it in glycerine.

(**H**) (*a.*) Observe the **olfactory cells** as very narrow, cylindrical, elongated cells with a large spherical nucleus, much broader than the body of the cell. The free surface carries fine cilia, but they are apt to be displaced in the process of teasing the tissue (fig. 345).

(*b.*) **Supporting cells,** not unlike columnar epithelium, but they have a large oval nucleus (fig. 345, E). They have a long central process. The free surface of the cell is covered by a clear hem, but its exact constitution and significance are unknown.

(*c.*) There may also be found large granular polygonal cells derived from Bowman's glands—their acini and ducts.

ADDITIONAL EXERCISES.

Olfactory Bulb.

1. This is a very complicated structure, and has been investigated recently by Golgi and R. y Cayal. The bulb of a new-born animal is hardened by Golgi's rapid method and sections made. It contains a layer of white nerve-fibres, and under this large nerve-cells—mitral cells—and one remarkable arrangement known as olfactory glomeruli, which are nests of fibrils, formed partly by the terminal arborisations of the processes of mitral cells and by non-medullated nerve-fibres of the olfactory nerve, which are passing onwards to terminate in the olfactory epithelium.

2. **Olfactory Region by Golgi's Method.**—The mucous membrane of the olfactory region is placed for 7 days in Golgi's osmico-bichromate fluid, then in silver nitrate, and hardened in alcohol. One often fails, but if a successful preparation be obtained the ends of the olfactory nerve are seen passing into the olfactory cells.

3. **T.S. Nose.**—Harden the whole nose of a mouse in Müller's fluid, stain in bulk in borax-carmine, and make T.S. across the whole organ to show its walls, septum, turbinated bones, respiratory and olfactory regions. In such animals as possess a well-developed **organ of Jacobson** study it. In it also, if a young animal, will be found beautiful sections of developing tooth.

LESSON XXXIV.

TERMINATION OF NERVES IN SKIN AND SOME MUCOUS MEMBRANES.

Sensory nerve-fibres terminate in the skin and mucous membranes in three ways :—

(I.) **Free nerve-endings**, *i.e.*, by intra-epidermic fibrils.
(II.) In special **terminal corpuscles**.
(III.) In **neuro-epithelial cells**, *i.e.*, specially modified epithelial cells, as the rods and cones of the retina, the auditory hairs in the ear, the inner cells of taste-buds and olfactory cells.

I. **Free nerve-endings** occur especially in stratified epithelium, *e.g.*, in the skin, mouth, and cornea (Lesson XXXII.). The nerve-fibres lose their myelin and primitive sheath, and the axis-cylinder splits up into fine fibrils, which penetrate into the epithelial layer and run between the epithelial cells, and sometimes anastomose with each other, to end in free points which do not form connections with any structure in the epidermis. In the skin these fibrils are confined to the rete Malpighii, and do not reach the stratum granulosum.

1. **V.S. Skin** (*gold*).—Use boiled gold chloride and formic acid (p. 372). Take very small pieces (2 mm. cubes) of the palmar skin of the fingers or toes, preferably from a new-born child or a young infant; cut away all the adipose tissue, and place the pieces in the gold mixture for at least one hour. Then place the tissue in slightly acidulated water, and expose to sunlight until the gold is reduced. Harden in alcohol, and mount the sections in formic glycerine.

(H) Observe the blackened fine nerve-fibrils. Some of them beaded, running between the epithelial cells (fig. 336).

II. **Terminal Corpuscles.**—They are very varied in their form, and include the following:—

    A. *Simple tactile cells.*
    B. *Compound tactile cells.*
    C. *End-bulbs with many modifications.*
    D. *Touch-corpuscles.*

A. **Simple Tactile Cells—V.S. Skin** (*gold*) (H).—Treat very small pieces of the human skin (volar surface of finger or toe) or the snout of a pig by the boiled gold formic acid method.

(*a.*) Observe a nerve-fibre (fig. 346, *n*) passing towards and entering the epidermis, where it is non-medullated and splits into fibrils, which terminate in oval, nucleated, *tactile discs* or menisci (*m*), each of which lies under a *tactile cell* (*a*). The cells are 6–12 μ in length, and lie in the deeper part of the epidermis.

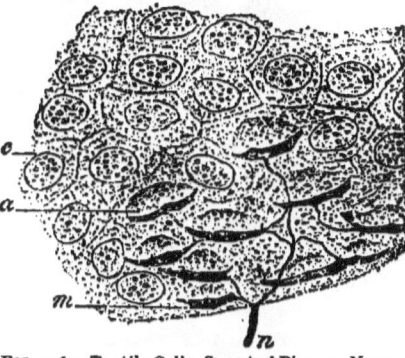

Fig. 346.—Tactile Cells, Snout of Pig. *n*. Nerve-fibre; *a*. Tactile cell; *m*. Tactile disc.

B. **Compound Tactile Cells**, called also **Grandry's** and **Merkel's Corpuscles**, consist of two or more cells (15 μ × 50 μ) piled one on the other; between each two cells is a disc or plate—**tactile disc**—which is connected with the axis-cylinder of a nerve. The cells have a large pale nucleus, and the whole is surrounded by a fibrous capsule. The nerve loses its myelin after it passes into the organ, and its sheath becomes continuous with the fibrous capsule. So far these structures have been found only in the bill and tongue of birds, *e.g.*, duck. They lie in the corium, close under the epidermis (fig. 347).

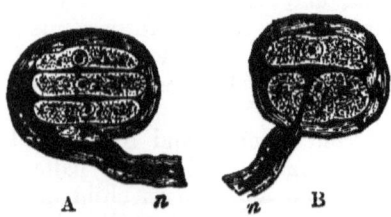

FIG. 347.—Tactile Corpuscles, Bill of Duck. *A.* With three, and *B.* With two tactile cells. The axis-cylinder terminates in the tactile disc (black).

3. **T.S. Tongue or Bill of Duck (Grandry's Corpuscles) (L and H).**—Harden in 1 per cent. osmic acid small pieces of the marginal part of the tongue or the sieve-like structure on the edges of the mandibles of a duck. Treat other small pieces by the boiled gold chloride method.

(*a.*) Observe the tactile cells and discs as in fig. 347.

C. End-bulbs are small, oval, or cylindrical bodies of various shapes. A nerve-fibre enters one pole, and as it does so it loses its myelin and terminates in a softer granular inner core. The bulb—which may be cylindrical in shape—consists of a tough fibrous capsule continuous with the sheath of the nerve; and a softer inner core in which the axis-cylinder terminates. Such bodies are found in the connective tissue of the mucous membrane of the mouth and conjunctiva. It is not difficult to isolate them from the conjunctiva of a calf (fig. 348). They lie in the sub-epithelial connective tissue. Certain special forms of them occur in the genital organs, *e.g.*, clitoris, and in connection with joints. A special form is known as **Pacinian corpuscles.**

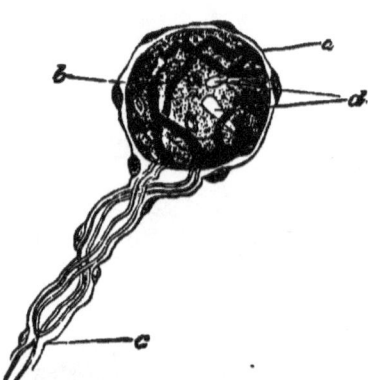

FIG. 348.—End-Bulb from Human Conjunctiva. *a.* Nucleated capsule; *b.* Core; *c* and *d.* Nerve terminating in *d.*

4. **Pacinian Corpuscles** (*Cat*).—These are elliptical transparent bodies (2–3 mm. long and 1 mm. thick), readily found in the meso-colon and mesentery of a cat. They consist of many concentric fibrous laminæ, arranged like the coats of an onion, surrounding a central core, which is continuous with the axis-cylinder of a nerve. The lamellæ are lined on both surfaces by endothelium, which can be stained with silver nitrate (fig. 350). The capsules are closer together near the centre than they are at the periphery. Examine one fresh in normal saline, and "fix" another with osmic acid and mount it in Farrant's

solution. A medullated nerve-fibre surrounded by perineurium passes to each corpuscle. The lamellæ of the perineurium become

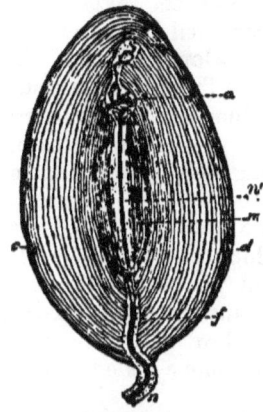

FIG. 349.—Pacini's Corpuscle, from Mesentery of Cat. *c.* Capsules; *d.* Endothelial lining separating the latter; *n.* Nerve; *f.* Funicular sheath of nerve; *m.* Central mass; *n'.* Terminal fibre; and *a.* Where it splits up into finer fibrils. Examined in fresh normal saline.

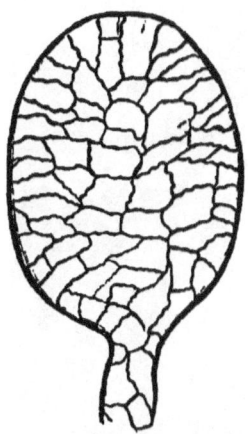

FIG. 350.—Endothelium of Lamellæ of a Pacinian Corpuscle. Silver nitrate.

continuous with the lamellæ of the corpuscle, and the nerve—still provided with its myelin and surrounded by its endoneurium — pierces the lamellæ and reaches the central part of the corpuscle. The endoneurium seems to form a soft core round the axis-cylinder, which usually splits up at the further end into a tuft of fibrils (fig. 349).

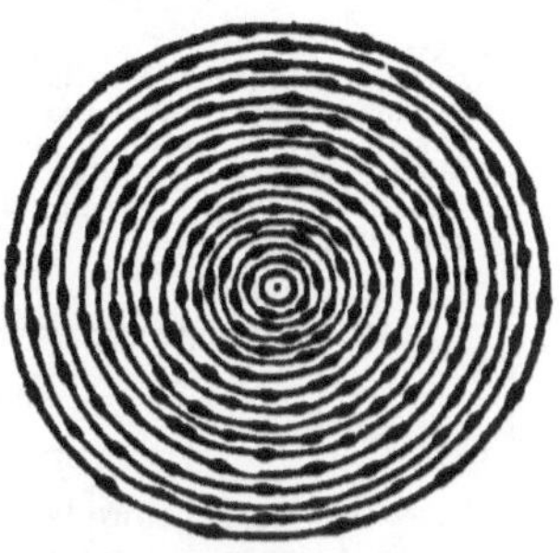

5. **T.S. Pacinian Corpuscles.**—They lie in the subcutaneous tissue. In a V.S. of fœtal skin from the palmar surface of a digit (Lesson XXIX. 5) it is easy to find these bodies cut obliquely

FIG. 351.—T.S. Pacinian Corpuscle, Fœtal Skin.

or transversely. Fig. 351 shows their appearance when cut transversely. Harden the skin in alcohol or osmic acid. Stain in picro-carmine or hæmatoxylin.

6. **Herbst's Corpuscles (H)** are modified Pacinian corpuscles, occurring in the mucous membrane of the tongue of the duck. They lie in the corium, just under the epidermis. Harden the tongue or bill as for tactile corpuscles (p. 378), or use absolute alcohol, and stain the tissue in bulk with borax-carmine. Mount in balsam.

(*a.*) Observe the same general arrangement as in Pacinian corpuscles. These bodies, however, are more elongated, and the axis-cylinder in the centre is bordered by a row of nuclei.

D. Touch-Corpuscles of Wagner or Meissner. — These are elliptical bodies (40–150 μ long and 30–60 μ broad), which occur in the papillæ of the skin, especially in the volar side of the fingers and toes. A nerve-fibre passes to each corpuscle, which has a fibrous sheath. Owing to the way in which the divisions of the nerve twist round the corpuscle, they have an irregular transversely striated appearance. The nerve-fibre enters the corpuscle, and its branches take a coiled course round the corpuscle.

7. V.S. Skin (Touch-Corpuscles). — (i.) Make vertical sections of skin (palmar surface of digit), harden in alcohol, stain them in hæmatoxylin or safranin, and mount in balsam (fig. 352).

(ii.) Treat skin by the boiled gold chloride method to see the terminal branches and distribution of the nerve within the corpuscle (fig. 353). In some situations the corpuscles are compound.

The following Table shows the modes of termination of nerve-fibres in sensory surfaces :—

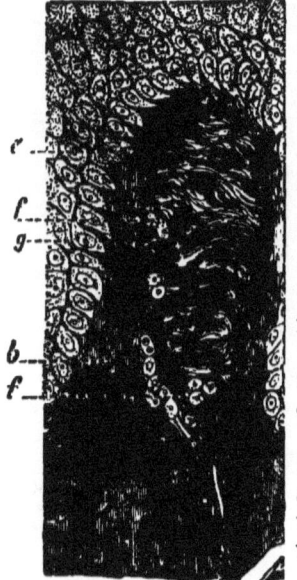

FIG. 352.—V.S. Skin, Palm of Hand. *b.* Papilla of cutis; *d.* Nerve-fibre in touch-corpuscle; *e, f.* Nerve-fibre in touch-corpuscle; *g.* Cells of Malpighian layer. Alcohol and safranin.

I.	II.	III.	IV.
Free nerve-endings (cornea, skin).	Simple tactile-cells (human skin). Compound tactile-cells (birds, *e.g.*, duck's bill and tongue).	Cylindrical end-bulbs (conjunctiva). Herbst's corpuscles (tongue of duck). Pacinian or Vater's corpuscles (deep part of skin, mesentery of cat, and near joints). Genital corpuscles (clitoris). Wagner's touch-corpuscles (papillæ of skin).	Neuro-epithelium (retina, internal ear, olfactory region of nose, and organ of taste).

ADDITIONAL EXERCISES.

8. Tactile Hairs and Nerves to Hair-Follicles.—These are to be studied by the boiled gold chloride method.[1]

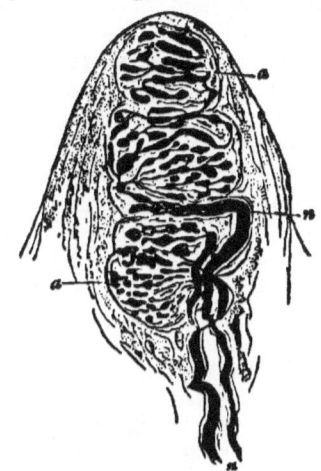

FIG. 353.—Wagner's Touch-Corpuscle, Skin of Hand. *n.* Nerve; *a.* Terminations of *n.* Gold chloride.

FIG. 354.—Organ of Eimer, Nose of Mole. *n.* Nerve; *e.* Epithelium. Gold chloride.

9. Organ of Eimer.—Prepare the nose of the mole by the gold chloride method. In sections study the termination of the nerve-fibrils in the structure known as the organ of Eimer (fig. 354).

LESSON XXXV.

THE TESTIS.

THE *framework* of the **testis** (fig. 355) consists of a strong, tough, fibrous capsule—the **tunica albuginea**—which is covered externally by endothelium reflected from the serous membrane—the **tunica vaginalis**. At the back part of the organ, the tunica albuginea is prolonged for some distance (8 mm.) into the gland to form the **corpus Highmori** or **mediastinum testis**. From the under surface of the tunica albuginea **septa** or **trabeculæ** pass towards the corpus Highmori, and thus subdivide the gland into compartments or **lobules** (about 120 in number), with their bases directed outwards and

[1] Hoggan, *Jour. of Anat. and Phys.*, 1892.

their apices towards the corpus Highmori. The tunica albuginea consists of dense fibrous tissue, but it is of looser texture internally, where it has numerous vessels and lymphatics. This inner layer is sometimes called the **tunica vasculosa**, and is continuous with the septa. Each lobule contains 2-8 **seminal tubules.** These convoluted tubules (800) begin by a blind extremity, and are of considerable length (30–50 cm.) and of nearly uniform diameter (.3 mm. or 140 μ). The tubules of any one compartment unite at acute angles to form a small number of much narrower straight tubules (20–25 μ)— **tubuli recti**—which pass into the mediastinum and there form a network of inter-communicating tubules of irregular diameter (24–180 μ)—**rete testis** (fig. 356).

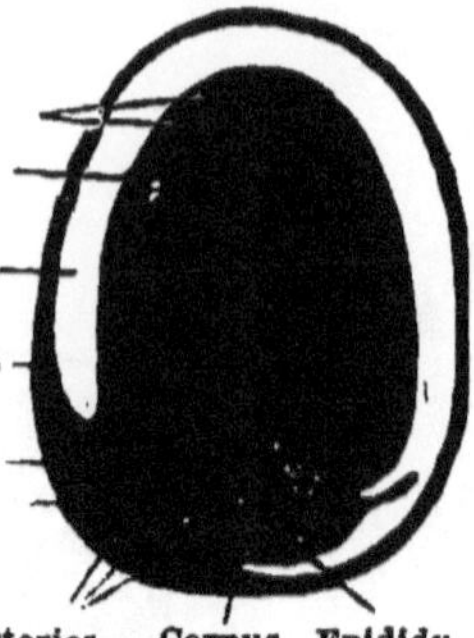

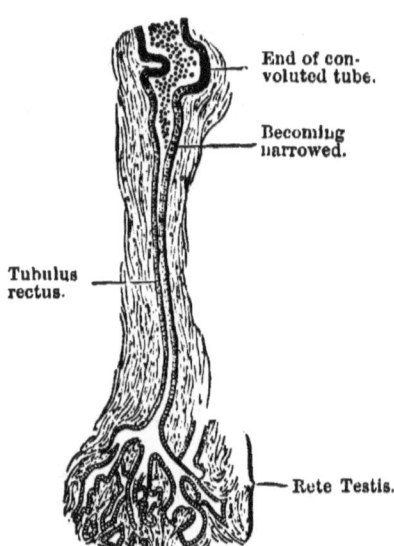

FIG. 355.—T.S. of Testis.

FIG. 356.—Convoluted and Straight Tube of Testis.

From this proceed 12–15 wider ducts—**vasa efferentia**—which are straight at first, but some become convoluted and form a series of conical eminences—**coni vasculosi**—which together form the head of the **epididymis**. The epididymis, consisting of a single convoluted tube (600 cm., or 20 feet long), is continued into a thick-walled muscular tube—**vas deferens** (60 cm., or 2 feet long)—which conducts the secretion to the urethra, and is, in fact, the excretory duct of the testis.

The **seminiferous tubules** consist of a thick membrane composed of flattened nucleated cells, arranged like membranes and lined internally by a basement membrane. The latter is lined by several rows of cubical cells, which differ much in appearance, according to the functional activity of the organ. In the resting condition, the tubes are lined by several layers of more or less cubical cells, whose nuclei stain more or less deeply with staining reagents. In an active gland in some tubules some of the clear cubical cells of the outer row of cells—**lining epithelium**—contain nuclei undergoing mitotic division. The rows of cubical cells internal to these have a radiate arrangement, and are separated into groups by larger structures, sometimes called spermatoblasts, and by other observers **sustentacular cells**, which grow up between the smaller cells, and at their upper ends are connected with the spermatozoa. Inside the layer of lining epithelial cells are several rows of larger cells—**spermatogenic cells**—with nuclei showing different stages of mitosis. Next the lumen, which is always well defined, in some of the tubules are spermatozoa in different stages of development. In others, however, not so far advanced, the inner row of cells consists of small protoplasmic cells—the true *spermatoblasts*—as from these cells the spermatozoa are developed. The developing spermatozoa rest by means of their heads on the sustentacular cells. The spermatozoa are arranged in tufts or groups, with their tails towards the lumen. The spermatozoa gradually develop from the true spermatoblasts and pass towards the lumen, and as they are set free new spermatoblasts are formed by the mitotic division of the spermatogenic cells.

The straight tubules are lined by a single layer of flattened epithelium; the rete testis has no basement membrane, but it also is lined by a single layer of flattened epithelium. The vasa efferentia and the tube of the epididymis have smooth muscular fibres in their walls, and are lined by columnar ciliated epithelium, the cilia being very long.

The **interstitial tissue** of the testis between the tubules is very loose in texture and laminated, and has numerous lymphatic slits. In some animals (rat, boar) are numerous polyhedral, nucleated, sometimes pigmented, cells, the remains of the cells of the Wolffian bodies.

The **vas deferens** consists of a fibrous coat investing an outer thick layer of smooth muscle arranged longitudinally; inside this is a thick layer of smooth muscle arranged circularly; inside this again is a submucous coat of connective tissue. In some parts of the tube there is a layer of smooth muscle arranged longitudinally just internal to the circular coat. Then follows the mucosa lined by columnar non-ciliated epithelium.

Methods.—(i.) Harden the fresh testis in Müller's fluid (2 weeks). If it be large, cut it into small pieces. The capsule exerts considerable pressure on the gland substance, so that when the testis is cut into, the latter projects somewhat. Complete the hardening in alcohol. Corrosive sublimate is a good hardening reagent. Stain sections in hæmatoxylin or picro-carmine, and mount in balsam. The hardened testis of a small animal, e.g., rat, should also be stained in bulk in borax-carmine or Kleinenberg's logwood, and cut in paraffin, so as to include the body of the testis and the epididymis. Mount in balsam.

(ii.) Harden very small pieces of the testis of a freshly-killed rat or mouse in Flemming's mixture and stain the sections in safranin. A very small testis is taken, as Flemming's fluid only penetrates a few mm. into the tissue. This is for the study of spermatogenesis.

(iii.) Harden the testis of a frog in absolute alcohol and stain it in bulk in Kleinenberg's logwood.

(iv.) Harden the testis of the dogfish in Flemming's fluid and stain the sections in hæmatoxylin or safranin. This, as shown by Swaen, is an excellent object for the study of spermatogenesis.

(v.) Harden the testis of a mouse, salamander, or frog in the following fluid [1] in which platinic chloride is substituted for the chromic acid in Flemming's chromo-aceto-osmic acid mixture:—

Hermann's Fluid.

Platinic chloride (1 per cent.)	15 parts.
Osmic acid (2 per cent.)	4 ,,
Glacial acetic acid	1 part.

The above is used for the tissues of mammals; but if a salamander testis is used, then take only 2 parts of the osmic acid.

After 2–3 days the details of the cells are better brought out than with Flemming's solution. Complete the hardening in alcohol. Make sections in paraffin, fix them on a slide with albumin fixative, and stain for 24–48 hours in the following safranin fluid:—

Safranin	1 gram.
Alcohol	10 cc.
Water	90 ,,

After staining, wash in water, acid-alcohol and alcohol, but do not remove all the surplus safranin. Then stain for 3–5 minutes in a gentian-violet solution. Decolorise the sections by means of Gram's method (p. 105), i.e., after washing in alcohol, place them for 1–3 hours in the following fluid—

Iodine	1 gram.
Potassic iodide	2 grams.
Water	300 cc.

until they are quite black, and then differentiate them in alcohol. Mount in balsam.

1. T.S. Testis (*Rat*).—(*a.*) Observe the thick, tough, fibrous **tunica albuginea** with sections of large blood-vessels and lymphatics

[1] *Archiv f. mik. Anat.*, xxxiv. p. 58.

in its deeper part. From its under surface *septa* pass into the gland at fairly regular intervals, thus dividing it into a series of compartments or **lobules**. At the upper and back part is the fibrous septum—**corpus Highmorianum** or **mediastinum testis**. Many of the septa are connected with it.

(*b.*) In the lobules lie twisted or convoluted tubules—the **seminiferous tubules**—which converge towards the mediastinum and form near it a number of straight tubes—the **tubuli recti**—which in their turn unite and form the **rete testis** in the mediastinum, and from this proceed the **vasa efferentia**, which run to join the canal of the **epididymis** (fig. 356).

(*c.*) The tubules in a state of activity are distinguished from the resting ones by the intensely stained heads of the young spermatozoa. An active tubule is lined by several rows of polygonal cells, some of which are larger than the others. Embedded amongst the cells, near the lumen of the tube, are bunches or tufts of spermatozoa, best seen in longitudinal sections of the tubes. The heads are directed towards the wall, and the tails towards the lumen of the tube.

(*d.*) In some animals the interstitial tissue between the tubules is chiefly formed by thin flattened membranes of connective tissue; but in others, *e.g.*, boar, the matrix consists of numerous pigmented, polygonal, very granular cells—**interstitial cells**.

2. **Spermatogenesis** (*Rat*).—In order to see the structure of the seminiferous tubules to the best advantage, harden small pieces of the testis of a rat, or a guinea-pig, as directed under v. and ii. (p. 384). To keep the parts together, cut in paraffin. The best stain is safranin. Fix a thin T S. of a tubule under a high power.

(*a.*) Observe the rather thick wall—*membrana propria*—of the tubule, composed of flattened cells, perhaps of a connective tissue nature. This is lined by three or more layers of glandular cells, which vary in appearance according to their condition of physiological activity. In a *state of rest* each tube is lined by several layers of large polygonal cells placed one inside the other.

(*b.*) In an *active gland*, known by the evidences of division of cells and by the development of spermatozoa (fig. 357), there is an outer layer of cubical-looking cells, and internal to it several layers of round or ovoid cells, which are called **spermatogenic cells**. In the latter may be seen nuclei undergoing mitotic division. The lining epithelial cells seem to divide; one part of each cell passes into the second layer of cells, and becomes a spermatogen or spermatogenic cell, while the remainder of the original cell enlarges and grows up as a **sustentacular cell**. The spermatogenic cells divide and redivide by mitosis, and yield the small daughter-cells or

spermatoblasts of the innermost layer. The latter, arranged in groups, gradually elongate, and from them the groups of spermatozoa are formed. Each group rests on and is connected with a sustentacular cell; hence arose the old view that these sustentacular cells produced the spermatozoa, and for this reason they were formerly called spermatoblasts. The nucleus of a daughter-cell forms the head and body of the spermatozoon, while the tail is formed within the protoplasm, but it is connected with the nucleus, and in its development grows out and forms a cilium.

3. Isolated Cells from the Tubules.—Macerate small fragments of the testis in dilute alcohol 10–12 hours. Tease a small fragment in glycerine. Observe the various forms of cells isolated.

4. Spermatozoa (H).—Make a cut into the epididymis of a testis removed from a rabbit as soon as possible after death. A milky

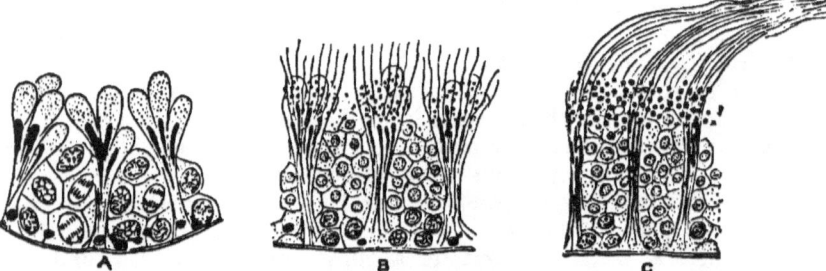

FIG. 357.—Tubules of Testis of Rat, showing Spermatogenesis. *A.* Less advanced stage: *B* and *C.* More advanced stages. *A* and *B*=T.S.; *C*=L.S. Flemming's fluid and safranin, × 300.

juice exudes. Place a little of this on a slide and dilute it with normal saline.

(*a.*) The spermatozoa, each one with a head and a long vibratile tail or cilium. By the side-to-side movement of the tail the whole spermatozoon is moved onwards in a zigzag course.

(*b.*) Add a drop of distilled water. The movements are arrested. The spermatozoa are slowly killed.

They may be readily preserved by smearing a little of the milky juice from the epididymis upon a slide and allowing it to dry. It is then covered and sealed.

5. Spermatozoa of Newt and Frog (H).—The testis of a newt is teased in a mixture of alcohol and glycerine.

(H) Observe the pointed head, body, and long tail of each spermatozoon. There is an intermediate part, which is best seen in a stained preparation. From it springs a filament which appears to be prolonged as a spiral filament around the cilium or tail. The spiral filament is in reality the optical expression of a thin membrane attached in a spiral direction to the cilium. Mount the spermatozoa of a frog. The male frog is easily known by the wart on his thumb. Open the abdomen, and low down in the hollow on each side of the vertebral column is an oval, kidney-shaped, whitish body—the testis. It is to be treated like the testis of other animals. At certain seasons of the year no spermatozoa are to be found. The cells are in a resting phase.

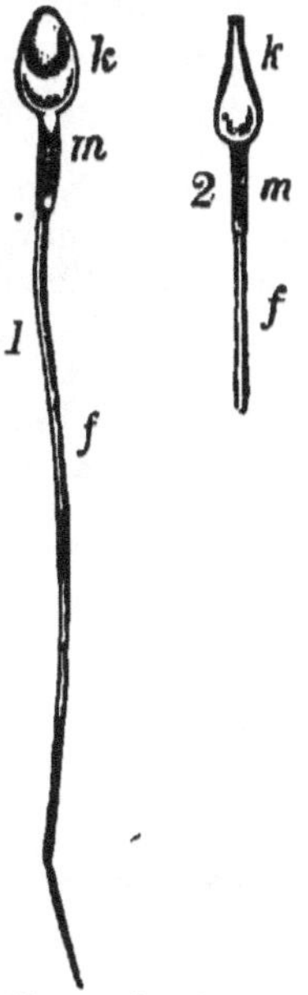

6. **Human Spermatozoa.**—Mount in glycerine some spermatozoa obtained from a spermatocele (fig. 358). Observe the head (k), long tail (f), and middle piece (m).

7. **Cover-Glass Preparations (H).**—Compress a small piece of the testis of a newt or frog between two cover-glasses. Separate them, and allow the film on them to dry. Stain one in methylene-blue, and the other for twenty-four hours in Ehrlich-Biondi's fluid. Dry and mount in balsam.

(*a.*) In the methylene-blue specimen only the heads of the spermatozoa are stained. In the other, the head of the spermatozoon is stained one colour, while the tail and the intermediary or junction piece is of a different hue.

FIG. 358.—Spermatozoa. 1 2. Human.

8. **Testis of Salamander.**—Fix the testis (in September or October) in Hermann's fluid, and stain as directed at p. 384. The bundles of ripe spermatozoa are very marked. The body of each spermatozoon is reddish, its head bluish-violet, the middle piece also of the same colour, while the tail is brownish-violet. In Hermann's paper figures will be found showing the remarkable stages of development of the bodies of the spermatozoon from the nucleus of the spermatids. Numerous mitotic figures may be seen.

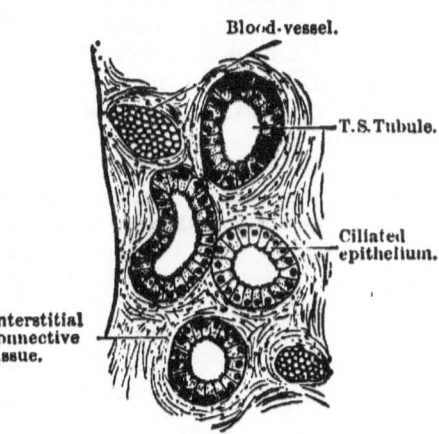

FIG. 359.—T.S. Tubules of Epididymis.

9. **Epididymis (L and H).**—Prepare like the testis. Make T.S., stain in hæmatoxylin and mount in balsam.

(*a.*) Observe numerous sections of the twisted tube of the epididymis (fig. 359); each tubule has a fibrous wall, and is lined by a single layer of tall, slender, columnar ciliated epithelial cells. Each cell is provided with a fringe of long cilia. The lumen is wide and distinct, and may contain a confused mass of spermatozoa. At the bases of the cells are sometimes seen small cells, and in pigmented animals pigment granules are not unfrequently seen in the connective-tissue stroma supporting the tubule. There is a fair amount of connective-tissue stroma between the tubules, and in it are blood-vessels, lymphatics, and nerves.

10. T.S. Vas Deferens.—Harden it like the epididymis, make T.S., stain, and observe the arrangement of its parts as described at p. 383.

LESSON XXXVI.

THE OVARY—FALLOPIAN TUBE—UTERUS.

THE OVARY.

The **ovary** consists of a **stroma**, covered on the surface with a single layer of short columnar or **germinal epithelium**. Embedded in the stroma are numerous **Graafian follicles** in all stages of ripeness. Each Graafian follicle contains one ovum. The unripe follicles, with **primitive ova**, form a superficial layer close under the surface, while the more mature ova and follicles lie deeper. The coverings of the follicle and the structure of a ripe ovum are given at p. 389.

Methods.—Select the ovaries of small animals, *e.g.*, mouse, rat, as they can be better "fixed" than large ones.

(i.) Harden the **ovary** of small animals *in toto* in Müller's fluid (2 weeks), and then in alcohol. Make two or three transverse cuts in the human ovary before hardening it in the same fluid.

(ii.) Picro-sulphuric acid, for a day, is also a good hardening reagent. Complete the hardening in alcohol. The pieces had better be stained in bulk in borax-carmine, or Kleinenberg's logwood, and embedded and cut in paraffin, or embedded and cut in celloidin.

(iii.) For small ovaries use Flemming's fluid, embed and cut in paraffin. Stain with safranin.

For obtaining a general view, the sections must be pretty thick, otherwise the follicles are cut into, and the ova are apt to fall out. Thin sections must also be made.

1. **T.S. Ovary** (fig. 360).—(*a.*) (**L**) The body of the ovary, covered on its surface by a single layer of short, columnar, nucleated cells — the **germinal epithelium**.

(*b.*) The substance or **stroma**, composed of connective tissue with numerous blood-vessels, and in some places smooth muscular fibres. The connective tissue is denser, and is arranged in several layers, near the surface.

(*c.*) Lying in the stroma — the **Graafian follicles**. Near the surface is a layer of smaller unripe ova— **primitive ova** (40 μ)—

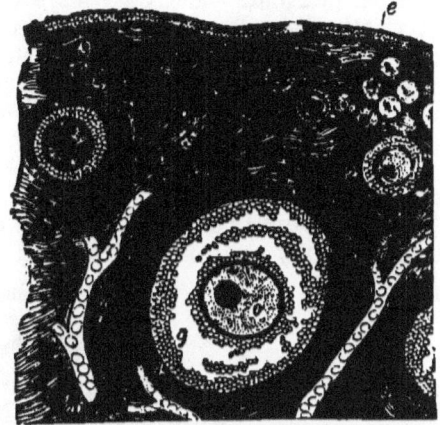

FIG. 360.—T.S. Ovary. *e.* Germ epithelium; 1. Large Graafian follicle; 2. Middle-sized, and 3. Small ova; *o.* Ovum; *g.* Membrana granulosa; *s.* Stroma.

while the riper and larger follicles are situated deeper in the stroma. Each follicle contains an ovum, but if the section be thin the ovum may drop out and leave only the follicle and its coverings.

(*d.*) If the ovary be from an adult animal, a **corpus luteum** may be found. Its appearance will depend upon whether it has been recently formed or not, and whether the animal was recently pregnant. If it is of some standing, it is large and more or less spherical in outline, occupying a considerable part of the stroma, with an umbilicus-like centre and radiating lines of connective tissue. Between these are numerous large granular cells. In a recent one there may be a yellow pigment—**lutein**—staining the scar.

(*e.*) (**H**) The germinal epithelium, and under it the **tunica albuginea**, composed of two or more layers of connective-tissue lamellæ—with fusiform cells—crossing each other. It gradually passes into the stroma.

(*f.*) **The Coverings of the Follicle.**—Select a well-developed or nearly ripe ovum (.5–5 mm. in diameter). Outside is the theca folliculi, composed of two layers, an outer one fibrous—the **tunica fibrosa**—and inside it a layer with blood-vessels—the **tunica propria**. Inside this several layers of cubical cells, constituting the **membrana granulosa**. From this there projects a mass of cells at one part, forming the **cumulus** or **discus proligerus**. The cells are continued as the **tunica granulosa** around the ovum.

Between the tunica and membrana granulosa is a space which encloses a fluid in the large follicles—the **liquor folliculi**.

(*g.*) The **ovum**, composed externally of a hyaline membrane—the **zona pellucida**—enclosing more or less granular cell-contents; the **vitellus or yelk**; and placed excentrically in this the **germinal vesicle** (corresponding to a nucleus), with its small excentric **germinal spot** (corresponding to a nucleolus).

(*h.*) **Small Unripe Ova.**—Besides the difference in size, the follicles show no separation between the tunica and membrana granulosa. They form a layer of cells in which the small ovum appears to be embedded.

2. **The Ovum** (fig. 361).—A fresh ovum may be obtained thus. Take the ovary of a cow or sheep, which on its surface shows clear

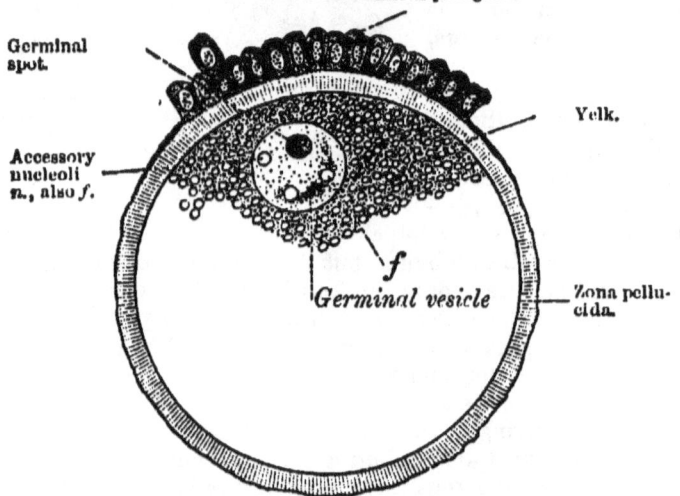

FIG. 361.—Ripe Ovum of Rabbit highly magnified.

elevations about the size of peas, filled with a watery-looking fluid; these are ripe follicles. Prick the follicle and receive its contents upon a slide. The liquor folliculi escapes, and with it the ovum surrounded by the cells of the tunica granulosa. With (**L**) search for the ovum in the fluid on the slide. Do not cover the preparation while doing so. If a cover-glass be applied, two strips of paper must be placed under the cover-glass to avoid pressure upon the delicate ovum. Observe the ovum, and in it will be seen the parts already described and shown in fig. 361.

3. **T.S. Fallopian Tube** (**L** and **H**).—This is hardened in Müller's fluid or Flemming's fluid. Treat it in the same way as

XXXVI.]   THE OVARY.   391

the intestine. Make transverse sections, and stain them either with picro-carmine or logwood.

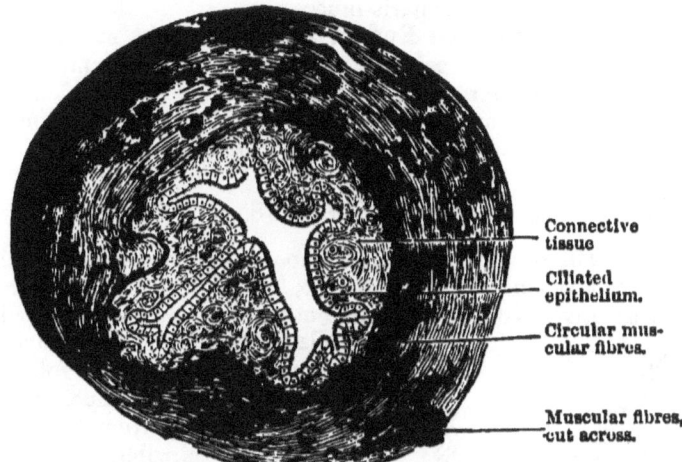

Fig. 362.—T.S. of Fallopian Tube.

The Fallopian tubes in animals often pursue a curved course. The peritoneum should be cut off as close to the tube as possible, and the latter stretched if complete T.S. are required.

(*a.*) Observe, most external, the thin **serous coat**.

(*b.*) Inside this the **muscular coat**, composed of non-striped muscle, a very thin longitudinal layer of fibres, and a much stronger circular layer.

(*c.*) A thin **submucous coat**.

(*d.*) The **mucous coat**, thrown into numerous ridges or folds, *i.e.*, sections of longitudinal folds, so that the lumen of the tube appears somewhat star-shaped or branched (fig. 362). Sometimes the processes of the mucous membrane are very complex in their arrangement, and give rise to arborescent-looking folds in transverse sections (fig. 363).

Fig. 363.—T.S. of the Fimbriated Extremity of the Fallopian Tube of Pig.

(*e.*) (**H**) The thick mucous coat is covered by a single layer of low cylindrical ciliated cells, but there are no glands, although the depressions act the part of glands. Under this is connective tissue with a very thin muscularis mucosæ.

4. T.S. **Fimbriated End of Fallopian Tube** stained in borax-carmine and cut in paraffin. Observe the very complex folds of the mucous membrane, each with secondary folds and covered by ciliated epithelium. The folds are much higher and far more complex than they are in the Fallopian tube (fig. 363).

UTERUS.

The wall of the uterus is composed of smooth muscle lined by a mucous membrane. As a rule, two layers of muscular fibre—imperfectly separated from each other—can be distinguished in the **muscular coat**; the external layer having an irregular course; then follow blood-vessels and connective tissue, which separate it incompletely from the inner one arranged more circularly. This thick inner layer of smooth muscle is said to represent a much hypertrophied muscularis mucosæ. The **mucous membrane** (1 mm. thick) consists of connective tissue containing a very large number of cells and nuclei, and embedded in it are tubular glands—**utricular glands**—some of them simple, and others branched (fig. 364), especially at their lower ends. These glands seem to be devoid of a membrana propria, but the surrounding connective tissue is arranged around them in a concentric manner. The body of the uterus, upper half of the cervix, and the uterine glands are lined by a single layer of short columnar ciliated epithelium. The vaginal portion of the uterus is lined by stratified squamous epithelium, and carries small vascular papillæ covered in by the epithelium. The mucous membrane rests directly on the muscular coat, there being no proper submucosa. Small processes of smooth muscle from the muscular coat are prolonged inwards between the bases of the glands. The mucous coat is very vascular and contains many lymphatics.

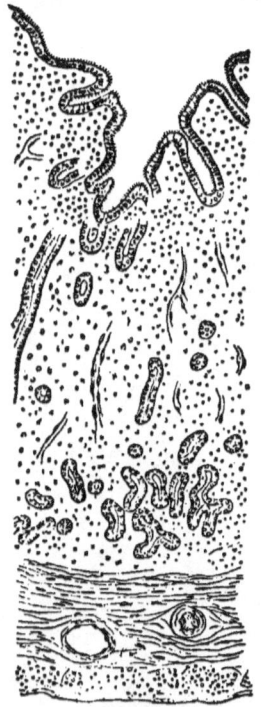

FIG. 364.—V.S. Uterus of a Bitch.

Methods.—It is rare to obtain the human uterus sufficiently fresh for microscopical preparations. Harden the uterus of a bitch, cat, or rabbit, in chromic and spirit fluid, alcohol, or Müller's fluid. Make T.S., and treat them as the Fallopian tube.

T.S. Uterus.—(*a.*) Observe the serous, muscular, and mucous coats as in the tube, but here the muscular coat is very thick, and is composed of numerous fibres arranged in bundles and running in all directions. The arrangement of these bundles is much simpler in animals than in the human uterus.

(*b.*) The mucosa is very thick, and is covered by a single layer of cylindrical nucleated cells, and has numerous glands, which are lined by similar cells. Not unfrequently the gland-tubes branch, especially near their lower extremities. It is difficult to retain the cilia on the epithelium lining the cavity of the uterus and its glands. Between the tubules is a relatively large amount of connective tissue containing many nucleated corpuscles and blood-vessels. As the glands pursue a curved course, and do not always run at right angles to the mucous surface, it is difficult to obtain a section through the entire length of a gland (fig. 364).

LESSON XXXVII.

MAMMARY GLAND, UMBILICAL CORD, AND PLACENTA.

THE **mammary gland** is a compound racemose gland, but it has about twenty galactoferous ducts which open on the nipple, each duct being dilated into a small reservoir just before it ends on the surface. The ducts, when traced backwards, branch and end in acini or saccular alveoli. The alveoli—as in all glands—vary in appearance according as the gland is or is not active. The walls of the ducts and acini consist of a basement membrane, said to be composed of branched cells, which in the acini is lined by a single layer of somewhat flattened polyhedral secretory cells. A cluster of acini gives origin to one of the larger ducts, and a considerable amount of connective tissue lies between groups of acini. In fact, the connective tissue greatly preponderates. During lactation the secretory cells are taller and larger, and in their interior—probably formed from and by the protoplasm of the cells themselves—are formed the fatty granules which are discharged to form the milk globules. The active gland presents some resemblance to a salivary gland, so closely are the alveoli pressed together, with only a small quantity of interstitial connective tissue between them. The acini are in groups and separated from each other by fibrous imperfect septa. Numerous corpuscles, including granular cells, occur in the alveolar connective tissue. The ducts are lined by columnar epithelium, and in a section of a human gland they appear large.

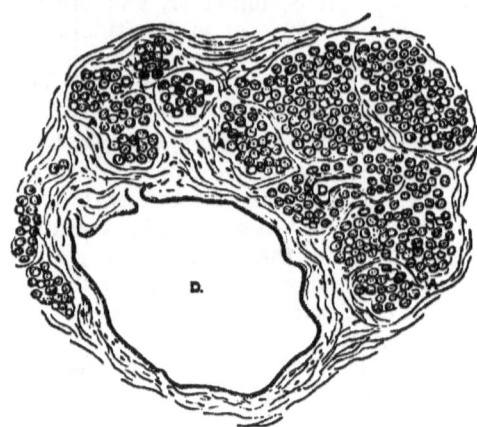

FIG. 365.—T.S. Mammary Gland. *D.* Duct; *A.* Group of acini with much connective tissue between, × 20

(i.) Harden small parts of the **mammary gland** in absolute alcohol. Select, when possible, the gland of a recently pregnant woman (or animal). Stain the sections in hæmatoxylin and mount in balsam, *i.e.*, to get a general view of the gland structure. Stain in bulk, embed and cut in paraffin.

(ii.) Harden very small pieces of a fresh gland, *e.g.*, from a

FIG. 366.—T.S. Secreting Mammary Gland, Mouse. Fol's solution and safranin, × 300.

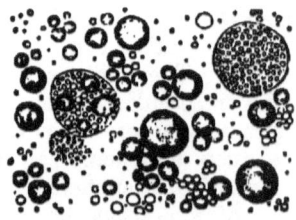

FIG. 367.—Colostrum.

pregnant cat or rabbit, in Flemming's mixture and stain the sections (very thin) in safranin.

1. V.S. Mammary Gland (*Hæmatoxylin*) (**L** and **H**) (fig. 365).

(*a.*) Observe the groups of acini, separated by a relatively large amount of somewhat loose connective tissue, and sections of the ducts (D). The sections should be made so as to include the nipple, when the larger ducts with their dilations will be seen. The ducts are large between the lobules, and within the latter the course of the finer ducts can readily be traced.

(*b*.) The globular acini, with a basement membrane lined by a single layer of somewhat flattened or cubical epithelium. In the inter-alveolar tissue many leucocytes and granular cells.

2. **Active Mammary Gland** (*Safranin*) (H).

(*a*.) Study specially the acini. Observe the large and tall columnar cells lining the acini, and in some of the cells clear refractive granules of fat. The lumen is wide, and is usually partially filled with the debris of the secretion-milk. Osmic acid is a good agent for showing the presence of fatty granules (fig. 366).

3. **Colostrum** (H), *i.e.*, the first milk secreted after delivery. If this can be obtained, examine it, and note, in addition to the ordinary milk-globules (Lesson I. 3), large coarsely granular nucleated refractile cells—**colostrum corpuscles**. The granules are sometimes pigmented, and are fatty (fig. 367).

UMBILICAL CORD AND PLACENTA.

4. **T.S. Umbilical Cord.**—Harden this in Müller's fluid or alcohol. Make T.S. by freezing, and stain them with hæmatoxylin or picro-carmine. Methyl-violet is also a good stain.

(*a*.) Note on the outside of the circular mass of tissue a thin layer of flattened cells derived from the amnion.

(*b*.) The cord itself, composed of **Wharton's jelly**, enclosing usually two umbilical arteries and a single vein with very thick muscular coats. They are completely surrounded by Wharton's jelly, which, however, in a cord at full time is very largely composed of fibrous tissue. Still numerous branched connective tissue corpuscles exist in the meshes, and there are also present numerous lymphoid-looking cells (Lesson XII. 10).

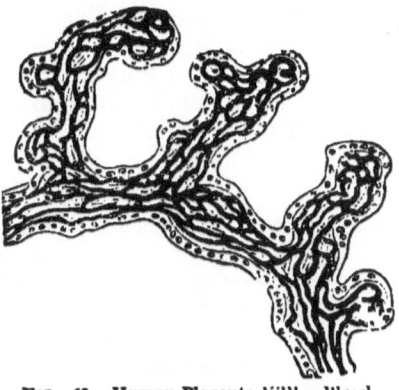

FIG. 368.—Human Placenta Villi. Blood-vessels black.

5. **Fresh Placenta** (H).—Tease a fragment of a placenta in normal saline. Note the *villi*, each long, tapering, and branched. In the interior capillary loops which occupy the greater part of the

villus, so that only a small amount of connective tissue intervenes between the vessels. Each villus is covered on its surface by a layer of epithelium, which, however, is thin at one part and thick at another. Especially at the ends of the villi are large granular masses of protoplasm containing many nuclei, but one cannot make out a separation of these masses into cells. They often contain vacuoles. The arangement of the blood-vessels may be followed from the distribution of the blood-corpuscles (fig. 368).

Small portions of a placenta are also to be hardened in Müller's fluid and stained in bulk in borax-carmine. Individual villi may be isolated in dilute alcohol or osmic acid.

6. Injected Placenta.—Examine a vertical section of a placenta with the fœtal blood-vessels injected, say blue, and the maternal vessels red. Observe how the one set interlocks with the other, yet both systems are closed and do not communicate with each other.

LESSON XXXVIII

TO MAKE PREPARATIONS RAPIDLY FROM FRESH TISSUES.

It is of the utmost importance that the student should be acquainted with the methods of making preparations from fresh tissues placed in his hands. The following is an outline of the work that each one can readily do for himself if supplied with a pithed frog, or other suitable material.

A. From a Frog.

1. Corneal Corpuscles.—With a sharp pair of scissors cut out the cornea. Divide it into two parts.
    (*a*.) Treat one by the lemon-juice method (p. 79).
    (*b*.) Treat the other part by placing it direct into .5 per cent. $AuCl_3$ (half an hour); wash in distilled water; place in a saturated solution of tartaric acid at 50° C. until the gold becomes reduced (p. 79).
    (*c*.) A cornea may be placed fresh in dilute methylene-blue (1 : 300 normal saline). Mount in picrate of ammonia glycerine (p. 192).

2. Corneal Lymph-Spaces.—Remove the eyelids, expose the surface of the other cornea, scrape off the epithelium, and rub it with solid silver nitrate. Cut out the cornea and expose it to light in water (p. 77).

3. **Tendons.**—These are best made from the tarsal tendons, which can readily be snipped off in considerable lengths.

(a.) *Fibrils.*—Tease a piece in baryta-water and mount in glycerine.

(b.) *Tendon Cells.*—(i.) Add dilute acetic acid to bring into view the rows of cells, then wash with water, and after all the acid is removed stain with logwood or picro-carmine. (ii.) Also tease a piece in normal saline containing a trace of methyl-violet.

(c.) Silver one of the tendons to show the endothelium covering its surface (p. 166).

(d.) Place a fresh tendon in ammoniacal carmine (10–15 mins.), wash and place in very dilute Delafield's logwood (10–15 mins.). Wash, tease; and mount in balsam. The tendon cells are red, their nuclei blue, and the tendon fibres rosy.

4. **Aponeurosis.**—The best is the femoral.

(a.) Remove the membrane and stretch it on a slide by the "semi-desiccation" method (p. 159), and after it is fixed to the slide apply a drop of acid methyl-green, or normal saline with methyl-violet. The nuclei are thereby stained, and the crests and ridges of the cells are made visible.

(b.) It may be fixed rapidly on a slide with absolute alcohol and stained with logwood. Or osmic acid may be used to fix it.

(c.) Show the effect of acetic acid.

5. **Areolar Tissue.**—(a.) Dissect out some from the intermuscular septa of the leg muscles. Stain with methyl-violet in normal saline. This stains the cells. Or use acetic-fuchsin (p. 92).

6. **Yellow Elastic Fibres.**—These are found in the septa between the lymph-sacs. Cut out a septum, fix it on a slide by "semi-desiccation," and then add acetic acid. Or make another preparation and stain it with a weak solution of methyl-violet-5B (p. 93).

7. **Pigment-Cells.**—(a.) These are found in the web of the frog's foot. Stretch the web between the toes, harden it in absolute alcohol for an hour or so, peel off the skin, and mount it in balsam (p. 173).

(b.) Or use the mesentery, or almost any blood-vessel; add dilute acetic acid and mount in glycerine.

8. **Hyaline Cartilage.**—(a.) Use either the episternum, scraping off the perichondrium, or make a section of the articular cartilage on the femur or tibia. Stain in hæmatoxylin. Or, before cutting, the cartilage may be hardened for an hour in absolute alcohol.

(*b.*) A silver nitrate preparation may also be made (p. 151).

9. Endothelium of Mesentery.—Place pieces of mesentery in $AgNO_3$ (.25 per cent.) for half an hour, wash in distilled water and expose to light in 50 per cent. alcohol.

Part may be afterwards stained in hæmatoxylin.

10. Endothelium of Great Lymph-Sac.—Open the abdomen from the front along the middle line, turn aside the intestines, and note the kidney. A thin membrane or septum stretches from this to the abdominal wall. With a fine pipette filled with silver nitrate solution perforate this membrane and allow silver nitrate to flow into the great lymph-sac. Expose the membrane to light, and then examine in glycerine to see endothelium and stomata (p. 239). One-half may be stained with hæmatoxylin to show the nuclei. Or expose the septum from behind as directed at p. 238.

11. Adipose Tissue.—Use the yellow-coloured fat bodies found in the abdominal cavity.

(*a.*) Tease a piece in glycerine.
(*b.*) Use osmic acid (p. 169).

See also other methods in Lesson XII.

12. Striped Muscle.—In this one must demonstrate—

(*a.*) Sarcolemma (p. 193).
(*b.*) Nuclei, *e.g.*, by acetic acid (p. 194).
(*c.*) Sarcous substance with its cross stripes. Harden for half an hour in alcohol and stain with hæmatoxylin or picro-carmine, or both. Mount in glycerine. Osmic acid also "fixes" the striation.
(*d.*) Fibres may be isolated by means of 33 per cent. caustic potash, but they must be examined in the same solution.

13. Cardiac Muscle.—Isolated cells are obtained by the 33 per cent. caustic potash method. The fresh tissue teased, stains well in picro-carmine.

14. Smooth Muscle.—Use

(*a.*) Frog's bladder (p. 190). In addition, spread out the bladder on a slide, expose it to the vapour of glacial acetic acid, wash away the epithelium, stain with violet-B, and mount in picrate-glycerine (*S. Mayer*).
(*b.*) Intestine. The muscular coat alone is to be used, after scraping away the mucous coat. Treat it as above.

15. Epithelium.—Scrape any epithelial surface, diffuse the scrapings in normal saline and examine fresh, and seal up with paraffin wax.

Squamous.—Use cornea.
Columnar.—Use intestine.
Ciliated.—Mucous membrane of palate.

Stain others with acid methyl-green or picro-carmine.

Make cover-glass preparations (p. 140).

Before staining pass the cover-glass three times through the flame of a Bunsen-burner.

16. Medullated Nerve-Fibres.—Expose the sciatic nerve.

(a.) Tease out a few fibres and show them fresh in normal saline. Seal up the preparation with paraffin wax.

(b.) Tease a piece in 1 per cent. osmic acid, cover with a watch-glass, and after half an hour mount in glycerine. This blackens the myeline. A part of this may be stained in picro-carmine (p. 206).

(c.) Tease a piece in .5 per cent. silver nitrate for Ranvier's crosses (p. 207).

(d.) Harden a piece of nerve in alcohol for a quarter of an hour. Tease, place in ether for 10 minutes, transfer to alcohol. Stain with logwood and mount in glycerine. This shows the axis-cylinder and nuclei of sheath.

(e.) Place a piece of fresh nerve in collodion to show axis-cylinder. This preparation only lasts for a short time (p. 211).

(f.) Show with silver nitrate the endothelial sheath on one of the small nerves to be found in the dorsal lymph-sac, stretching between the back muscles and the skin (p. 207).

17. Peripheral Nerve-Cells.—Use the rapid gold chloride (p. 79) or methylene-blue method (p. 222).

(a.) Cells along course of aorta. Cut out the abdominal part of the aorta.

(b.) Cells of sympathetic (p. 216) or

(c.) Cells of spinal ganglia, or other ganglia (p. 215).

(d.) Interauricular septum of the heart. The heart must be distended and kept fixed in this position (p. 233).

18. Retina.—Carefully dissect out the eyeball, remove retina.

(a.) Place a part in 1 per cent. osmic acid and then make teased preparations. One should show the pigmented epithelium, rods with outer segments blackened, pieces of the several layers, the glistening fatty globule rendered brownish.

(b.) Harden a piece in absolute alcohol and tease in very dilute eosin. Mount in glycerine.

(c.) Place small pieces in very dilute methylene-blue (p. 222) to show the nervous elements. Mount it in picro-glycerine (p. 222).

19. Blood-Vessels.—A new frog will be required. Proceed as directed at p. 230.

(a.) Inject AgNO$_3$ (.5 per cent.), and isolate arteries, veins, and capillaries; best from the intestine. Mount in balsam.

(b.) Show the effect of acetic acid on one of the larger vessels of the mesentery.

(c.) Methylene-blue injected into the vessels shows the lining endothelium and nerve-fibres.

20. Blood and Blood-Corpuscles.

(a.) Osmic acid and picro-carmine (p. 110).

(b.) Cover-glass preparations stained by eosin in glycerine and then in logwood. Or other dyes or double stain (p. 114).

(c.) Cover-glass preparations stained by methylene-blue for the nuclei. Pass the cover-glass three times through the flame of a Bunsen-burner before staining.

(d.) The colourless corpuscles may be stained in the methylene-blue specimens, but they may be specially stained by eosin-glycerine or indulin-glycerine, which stain certain granules in their protoplasm (p. 402).

21. Fibrin (p. 119).

22. Marrow.—Squeeze out same.

(a.) Examine fresh in NaCl (.6 per cent.) with methyl-violet.

(b.) Cover-glass preparations stained with eosin-glycerine and logwood (p. 188). The easiest way is to pass the cover-glass with its adherent film of marrow three times through the flame of a Bunsen-burner before staining.

23. Motor Nerves to Muscles.—Gold method or methylene-blue (p. 219).

B. From a Mammal.

If a mammal, *e.g.*, rat or guinea-pig, or part thereof, be given from which to prepare specimens, the methods are much the same as those described above.

24. Areolar Tissue.—(a.) To show its histological elements the best plan is to inject under the skin, by means of a hypodermic syringe, some fluid which will form an artificial œdema, *e.g.*, methyl-violet in normal saline, osmic acid (1 per cent.), silver nitrate 1 : 300. The first of these fluids does not alter the tissues. Excise a piece and examine it in the same fluid.

(b.) Use the "semi-desiccation" method with a small piece snipped off and spread out on a slide.

(c.) Cell-spaces by means of AgNO$_3$ (p. 162).

In addition to the methods described at p. 161 and p. 162, use acetic acid to show elastic fibres; magenta to stain the latter.

25. Tendons, *e.g.*, rat (p. 168), and methods at p. 397.

26. **Diaphragm.**—(*a.*) Show its endothelial covering by means of $AgNO_3$, and (*b*) its lymphatics (p. 237).

27. **Adipose Tissue.**—One must demonstrate the cells (Lesson XII.).

  (*a.*) Fresh, unaltered in normal saline, using either part of omentum or fat-cells from under skin. Seal up the preparation with paraffin wax.
  (*b.*) Action of osmic acid.
  (*c.*) Action of alcohol and ether to remove fats, and show empty envelopes.
  (*d.*) Harden fat-cells in alcohol and stain in logwood to show nuclei.
  (*e.*) Subcutaneous injection of silver nitrate (1 : 500) to show general characters of cells.

28. **Granular Cells** ("Mastzellen").—They occur in large numbers in the omentum of the rat. Place a small part of the omentum in a watch-glass containing aniline-water and 20–30 drops of a concentrated alcoholic solution of dahlia or gentian. Heat for short time as directed in Lesson X. 14. Wash in distilled water and then in acid-alcohol until nearly everything is decolorised except the granular cells. The tissue may also be stained with lithium-carmine. Mount in balsam. The nuclei of the cells are red, and only the granules in the protoplasm of the granular cells are blue.

The mucous membrane of a dog's tongue hardened in alcohol and treated in the same way shows numerous granular cells.

29. **Red Marrow of Bone.**—Methods (p. 188). Use the ribs and heads of long bones of guinea-pig.

The dry cover-glass method is excellent. Dry them in flame of spirit-lamp, or pass them three times through the flame of a Bunsen-burner. Stain with eosin-glycerine and then with methylene-blue or logwood. Many of the smaller cells will show eosinophile granules.

For studying the formation of the elements of the blood in red marrow or other situations, the following method of Foà is good.[1] The red marrow or blood is "fixed" in the following fluid after it cools:—100 cc. Müller's fluid is heated with 2 grams of mercuric chloride. Keep in a thermostat (2–3 hours) at 35° C. Harden in alcohol, cut sections, and stain (1–3 minutes) with the following:—

Hæmatoxylin solution (*Böhmer's*)	25 cc.
Safranin 1 per cent. watery-alcoholic solution	20 ,,
Water	100 ,,

Then stain in weak picric acid. Xylol-balsam.

[1] *Zeits. f. wiss. Mikrosk.*, ix., 1892, p. 227.

30. Blood Crystals (p. 120).—If it be a guinea-pig or rat use the defibrinated blood to obtain blood crystals.

(*a.*) Add water = hæmoglobin crystals.
(*b.*) Add a small quantity of ether = hæmoglobin crystals.
(*c.*) Add amyl nitrite = methæmoglobin crystals.

31. Blood—Ehrlich's Granules.—Perhaps it might be well to give here a short resumé of some of the results of Ehrlich and his pupils.[1] Cover-glass preparations of the blood of different animals are made, and they are either exposed to the air to dry (or they may be carefully heated for several hours at 120° C. or passed several times through the flame of a Bunsen-burner). On being dried rapidly in the air, there is no coagulation of the cell-proteids, and thus the cells retain their natural tendency to stain with dyes. As hæmoglobin is soluble in water it is better to use the dyes dissolved in glycerine.

Leucocytes.—The "granules" present in the protoplasm in the varieties of white blood-corpuscles vary in their reaction to staining reagents. Thus some are stained by what Ehrlich calls acidophile dyes, of which eosin is one. It is not enough that the granules are stained by one of these dyes. As a general rule, granules which are stained by all the following solutions belong to his α-granulation class and are "eosinophilous granules."

(1.) Eosin in glycerine.
(2.) Glycerine saturated with indulin.
(3.) Concentrated watery solution of orange.

The eosinophile cells (α granules) are always present in the leucocytes of frog's blood, marrow of frog (numerous)—very few in spleen—numerous also in the mesentery. In the rabbit they occur in the blood, marrow (very numerous), spleen (few).

Make a cover-glass preparation and dry it either at 120° C. (several hours) or rapidly in the flame of a Bunsen. Stain for an hour (or longer) with eosin-glycerine, wash in water, dry and mount in balsam. Or stain cover-glass preparations in glycerine (30 cc.) containing 2 grams each of aurantia, indulin, and eosin. Or a saturated alcoholic solution of bluish-eosin may be used.

If an eosin-indulin glycerine solution be used, the α-granulations are purplish-red and the nuclei well stained bluish-black by the indulin.

The **granular** cells ("Mastzellen"), which occur so abundantly in the connective tissue of the frog and some other animals, also occur in the blood of the frog, triton, and tortoise. In man, according to Ehrlich, they are found only pathologically. The granules

---

[1] *Farbenanalyt. Unters. z. Histol. u. Klinik des Blutes*, by P. Ehrlich, Berlin, 1891.

in these cells are stained by a fluid composed of 100 cc. water, 50 cc. absolute alcohol saturated with dahlia, 10-12.5 cc. glacial acetic acid. The leucocytes are stained blue, while the granules have what Ehrlich calls a metachromic red-violet tint. They correspond to Ehrlich's γ-granulations, and have been specially investigated by Westphal (*loc. cit.*, p. 17).

The δ-granulations occur especially in the mononuclear leucocytes of human blood. They are stained by basic dyes.

The ε-granulations, or neutrophile granules, occur in the polynucleated elements of human blood. They are stained only by neutral dyes, *e.g.*, acid-fuchsin and methyl-blue. Ehrlich classifies dyes as *acidophile*, *e.g.*, eosin, aurantia, and indulin; *neutrophile*, *e.g.*, acid-fuchsin or fuchsin-S, methyl-blue; *basophile*, *e.g.*, dahlia, gentian-violet, fuchsin.

Ehrlich also calls granules which attract acid dyes "oxyphile"— a term adopted by Wright and Bruce,[1] whose method is described below. According to the latter observers, the nucleus of the leucocyte is invariably basophile, while the granules of normal leucocytes are oxyphile.

Staining of Oxyphilous or Eosinophilous Granules.—Cover-glass preparations are fixed either by dry heat (Ehrlich's method) or by chemical reagents (osmic acid, $HgCl_2$).

Float the cover-glass on a 1 per cent. watery solution of eosin ($\frac{1}{4}$-1 min.). If it be desired to stain even more rapidly, add a trace of acetic acid to the fluid, when the preparation rapidly becomes over-stained. The surplus dye can be removed from all parts of the cells, except the oxyphile or eosinophilous granules, by dipping the cover-glass into a very dilute solution of sodic carbonate.

Basophilous Granules.—These are best stained with Loeffler's methylene-blue, which stains all basophilous elements, *e.g.*, nuclei and basophile granules. It may dissolve out oxyphilous granules. If the specimen has been already stained with eosin (the excess extracted by weak alkali), then only a second or so is required to stain with methylene-blue. Thus with care it is possible to stain the oxyphile and basophile elements of the leucocyte.

These observers deny the existence of so-called neutrophile granules. They believe them to be really oxyphilous in their behaviour to stains.

32. Stained Leucocytes.—Either one's own blood or the blood of an animal, or the leucocytes of lymph-glands, may be used. Use the dry cover-glass method, passing the cover-glass three times through the flame of a Bunsen-burner before staining. Excellent preparations of the nuclei stained blue are obtained by methylene-blue

---

[1] *Brit. Med. Jour.*, Feb. 1893.

alone, or first stain in eosin and then in methylene-blue or hæmatoxylin. For leucocytes the blood of the horse is specially valuable, as the white cells are so large (*Sherrington*).

33. Fibrin, Hæmin, Cartilage, Muscle, Nerve, and the other tissues are treated as recommended for frog's tissues, and the same is the case with organs.

34. Salivary Glands and Pancreas.

Tease a small part of the parotid, or other salivary gland, or the pancreas, in aqueous humour to see the fresh condition of these glands. In the guinea-pig's glands note the zymogen granules.

A permanent preparation may be made by exposing small pieces to the vapour of osmic acid and mounting in glycerine. In this way the zymogen granules are preserved (Lesson XXIII. p. 265).

N.B.—In this Lesson the methods stated expressly exclude complicated methods of hardening and section-cutting.

ADDENDA.

A. Altmann's Researches on "Granula" in Cells.

For those who wish to study Altmann's views on the constitution of the protoplasm and nuclei of cells, and the methods of displaying what he calls his "granula," we must refer to his monograph, entitled *Die Elementarorganismen und ihre Beziehung zu den Zellen*, Leipsig, 1890, which contains twenty-one beautiful plates.

B. Obreggia's Method for Paraffin Sections.

The method of Obreggia (*Neurologisches Centralblatt*, 1890) has been applied by Gulland (*Journal of Pathology*, 1893) to paraffin sections. The hardened tissues are embedded in paraffin, and ribbons of sections are cut with a rocking microtome or Minot's form. The sections are at once transferred to glass plates coated with the following solution :—

Syrupy solution of powdered candy-sugar made with boiling distilled water . . . . 30 cc.
Absolute alcohol . . . . 20 ,,
Transparent syrupy solution of pure dextrin made with distilled water . . . . 10 ,,

Pour this solution over the plates two or three days before they are used; run off the excess; allow the plates to dry slowly in a horizontal position and protected from dust.

Arrange the series of sections in rows on the plates. Place the plates in a paraffin-oven, which is kept at a temperature slightly above the melting-point of the paraffin employed, and leave it there for a few minutes, when the embedded tissues stick fast to the prepared surface.

Remove the surplus paraffin with xylol or naphtha, and then wash with methylated spirit or absolute alcohol.

The spirit is run off, and the plates are covered with the following celloidin solution:—

    Phytoxylin . . . . . . 6 grm.
    Absolute alcohol . . . . . 100 cc.
    Pure ether . . . . . . 100 ,,

The plates are placed horizontally. After the thin sheet of celloidin solidifies, run a knife along between the rows of sections, and allow further evaporation to take place.

When the sections are required plunge the plate into water; the ribbons float off as the sugar is dissolved. The ribbons may be stained with any reagents except those which dissolve or overstain celloidin. Stain with very dilute Ehrlich's acid-hæmatoxylin (p. 69), and then wash in dilute eosin. Dehydrate the sections, and clarify in a mixture of xylol 3 parts and carbolic acid crystals 1 part, and mount in balsam. We have tried this method, as recommended by Gulland, and find that it works very well. Moreover, we found that the sections, after being floated off, can be kept in 80 per cent. alcohol until they are required.

APPENDIX.

A.—SOME WORKS OF REFERENCE.

A.—Systematic Histology.

Schwann, Mikrosk. Untersuch., 1838 (translated by the Sydenham Society 1847).—**R. Virchow,** Die Cellular Pathologie (translated by **Chance**), 1860.—**Henle,** Handbuch der systematischen Anatomie des Menschen, 3rd ed., 1866-83.—**W. Krause,** Allgemeine und mikroskopische Anatomie, Hannover, 1876.—**F. Leydig,** Lehrbuch der Histologie des Menschen und der Thiere, Hamm, 1857; his Untersuchungen, 1883; and his Zellen. Gewebe, Bonn, 1885.—**L. Ranvier,** Traité technique d'histologie, Paris, 2nd ed., 1889.—**G.** Schwalbe, Lehrbuch der Neurologie, Erlangen, 1881; Lehrb. d. Anat. d. Sinnesorgane.—**S. Stricker,** Handbook of Histology (translated by the New Sydenham Society), 1871-73.—**L. Beale,** The Structure of the Elementary Tissues, London, 1881.—**A. Kölliker,** A Manual of Human Microscopic Anatomy, London, 1860; and his Icones Histolog., Leip., 1864; vol. i. of his Handbuch d. Gewebelehre, Leipzig, 1889.—**Rindfleisch,** A Manual of Pathological Anatomy (translated by **R. Baxter**), London, 1873.—**C. Toldt,** Lehrbuch der Gewebelehre, Stuttgart, 3rd ed., 1888.—**E. Klein** and **E. Noble-Smith,** Atlas of Histology, London, 1872.—**H. Frey,** Handbuch der Histologie und Histochemie des Menschen, Leipzig, 1876, Grundzüge, 1885.—**Cadiat,** Traité d'anat. gén., Paris, 1879.—**Brass,** Kurzes Lehrbuch d. Histologie, Leipzig, 1888.—**Heitzmann,** Mikrosk. Morphology, 1882.—**Purser,** Man. of Hist., Dublin.—**E. Klein,** Elements of Hist., London, 1883.—**W. Flemming,** Zellsubst. u. Zelltheilg, Leipzig, 1882.—**Bizzozero,** Hand. d. klin. Mikroskop., Erlang., 2nd ed., 1888.—**Carnoy, Gilson,** and **Denys,** Biol. Cellul., Louvain, 1884-88.—**Renaut,** Traité d'Histologie, Paris, 1885.—**Frommann,** Unters. ü. thier. u. pflanz. Zellen, Jena, 1884.—**Wiedersheim,** Lehrb. d. vergl. Anat., Jena, 1888.—**S. L. Schenk,** Grundriss der Histologie d. Menschen, Vienna, 1885.—**Orth,** Cursus d. norm. Histol., 4th ed., 1886.—**S. Mayer,** Histolog. Taschenbuch, Prag., 1887.—**Stöhr,** Lehrb. d. Histol., 5th ed., Jena, 1892.—**Lee** and **Henneguy,** Traité de méth. d'Anat., Paris, 1888.—**Schäfer,** Essentials of Histology, 3rd ed., 1892.—**Owsjannikow,** Text-Book of Histology (Russian), 1888.—**Fusari** and **Monti,** Compendio di Istologia

APPENDIX. 407

generale, Torino, 1891.—**Ellenberger**, Vergleich. Histol. der Hausthiere, vol. i., Berlin, 1887, vol. ii., 1892.—**Altmann**, Die Elementarorganismen u. ihre Beziehung z. d. Zellen, Leipzig, 1890 (with 21 plates). An abstract of the methods by which Altmann prepares his "granula" in cells will be found in Zeit. f. mik. Anat., vii. p. 199, 1890.—Quain's Anatomy, 10th ed., edited by **Schäfer and Thane**, 1892-93.—**Obersteiner**, Anleitung b. Studium d. Baues d. nerv. Centralorg., 2nd Germ. ed., 1892. English ed. translated by **Hill** from 1st Germ. ed.—**Edinger**, Zwölf Vorlesung. v. d. Bau d. Central Nerv. System, 2nd ed., and English trans. by **Vittum and Riggs**, 1890.

B.—**The Microscope, Microscopical Technique, and Manuals of Practical Histology.**

Some of the works already mentioned contain descriptions of microscopical methods, *e.g.*, those of Ranvier, Stöhr, Orth.

Dippel, Das Mikroskop und seine Anwendung.—**Beale**, How to Work with the Microscope, London, 1880.—H. **Frey**, Das Mikroskop., 8th ed., 1886.—**Carpenter's** The Microscope and its Revelations, edited by **Dallinger**, 7th ed., 1892.—J. **Hogg**, The Microscope, 12th ed., 1887.—**Naegeli and Schwendener**, The Microscope in Theory and Practice (translated from 2nd Germ. ed.), 1892.—**Gage**, The Microscope and Microscopical Methods, Philadelphia, 1892.—**Amstrom**, Anleit. z. Benutz. d. Polaris-Mikroskop., Leip., 1892.—**Rutherford**, Outlines of Histology, 1876.—**Schäfer**, Pract. Histol., London, 1877.—W. **Stirling**, Text-Book of Pract. Histol., London, 1881.—**Foster and Langley**, Pract. Phys., 5th ed., 1884.—**Friedländer and Martinotti**, Tecnica microscopica, Turin, 1885.—**Garbini**, Manuale per la tecnica mod. del Mikroscopio, 3rd ed., 1891.—**Fol**, Lehr. d. vergleich. mikros. Anatomie, Leip., Pt. i., 1885.—**Behrens**, Tabellen z. Gebrauch b. mik. Arbeiten, Braunschweig, 2nd ed., 1892.—**Fearnley**, Pract. Histol., 1887.—W. **Stirling**, Histological Memoranda, Aberdeen, 1880.—**Lee and Henneguy**, Traité de méth. de l'Anat., Paris, 1888.—**Friedländer and Eberth**, Mik. Technik., 4th ed., Berlin, 1888.—**Gierke**, Färberei z. mik. Zwecken, Braun., 1887.—**Renaut**, Traité d'Histol. pratique, Paris, 1889.—**Behrens, Kossel, and Schiefferdecker**, Das Mikroskop., Pt. i., Braunschweig, 1889, Pt. ii., 1891.—**Ramón y Cayal**, Manual de Histología normal, Valencia, 1889.—**Böhm and Oppel**, Taschenb. d. mik. Technik., München, 1890.—B. **Rawitz**, Leitfaden f. hist. Untersuch., Jena, 1889.—**Neelsen**, Grundriss d. Path.-Anat. Technik, Stuttgart, 1892.—**Squire**, Methods and Formulæ used in the Preparation of Animal and Vegetable Tissues for Micros. Exam., London, 1892.—**Strassburger**, Das botan. Practicum, and English trans.—**Zimmermann**, Die botan. Mikrotechnik, Tübingen, 1892.

C.—**Journals.**

Archiv f. mik. Anatomie, Bonn, formerly edited by **M. Schultze** and now by **Hertwig, La Valette, St George, and Waldeyer**.—Archiv für Anat. u.

Physiol., edited by **Du Bois-Reymond**.—**Virchow's** Archiv.—Quarterly Microscopical Journal, London.—Journal of the Royal Microscopical Soc., London.—Jour. of Anat. and Physiol., edited by **Humphry, Turner,** and **M'Kendrick**.—La Cellule, Louvain.—Journal de Micrographie, Paris, edited by **Pelletan**.—Zeits. f. wiss. Mikrosk., edited by **Behrens**.—Zoolog. Anzeiger, edited by **V. Carus**.—Internats. Monats. f. Anat. u. Phys., edited by **Krause**.—Journal of Physiology, edited by **Foster**.—Archiv ital. de Biologie, edited by **Mosso**.—Proceedings and Transactions of the Royal Society.—Comptes rendus de l'Acad. des Sciences.—Sitzb. d. k. Akad. d. Wissenschaft, Wien.—Archiv f. d. gesammte Physiologie, edited by **Pflüger**.—Journal of Morphology, edited by **Whitman**, in which will be found elaborate papers by Minot on the Uterus and Placenta, and on the Ear by H. Ayers.

B.—TABLES OF MAGNIFYING POWER OF OBJECTIVES AND OCULARS.

Objective.	Magnifying Power of Objective alone.	Eyepiece No. 1 (A) magnifies 5 times; combined with Objective magnifies	Eyepiece No. 2 (B) magnifies 7½ times; combined with Objective magnifies	Eyepiece No. 4 (D) magnifies 20 times; combined with Objective magnifies
1 inch	10	50	75	200
4/5 ,,	25	125	187	500
1/4 ,,	40	200	275	800
1/5 ,,	50	250	300	1000
1/6 ,,	60	300	450	1200
1/8 ,,	80	400	600	1600
1/10 ,,	100	500	750	2000
1/12 ,,	120	600	900	2400

This table is calculated for a 10-inch tube, and gives approximately the magnifying power; but if accuracy be required, each combination of lenses and objectives must be measured by the method already described at page 19. (After *Gibbes*.)

HARTNACK'S DRY LENSES.

Objective.	Ocular.			
	No. 1.	No. 2.	No. 3.	No. 4.
4	40	50	65	100
7	150	220	300	450
8	250	360	400	600
9	360	430	520	850

Magnifying Power of Zeiss's Objectives and Oculars.

Objective.	Length of Tube 155 mm. Ocular.			
	No. 1.	No. 2.	No. 3.	No. 4.
A	7	11	15	22
A*	...	4-12	7-17	10-24
A, AA	38	52	71	97
B, BB	70	95	130	175
C, CC	120	145	195	270
D, DD	175	230	320	435
E	270	355	490	670
F	405	540	745	1010

The lens A* is a particularly useful low-power lens, as by merely rotating a collar, a great variety of magnifying power is obtained without changing the lens.

Magnifying Power of Leitz's Objectives and Oculars.

Number of Objective.	Magnifying Power of Objective without Ocular.	Length of Tube 160 mm. Ocular.					
		$0=5.6$.	$I.=6.9$.	$II.=8.5$.	$III.=12.7$.	$IV.=16.3$.	$V.=19.1$.
1	3.2	16	21	25	39	50	60
2	5	26	34	40	63	81	95
3	9	47	62	72	114	146	171
4	11	58	75	88	139	179	210
5	26	137	179	208	330	423	496
6	34	180	234	272	431	554	649
7	47	250	325	380	600	770	900
8	60	318	414	480	762	978	1146

C.—LIST OF MAKERS OF MICROSCOPES, &c.

Microscopes.

American Makers and Agents.

Bausch & Lomb Optical Co., Rochester, N. Y.
J. W. Queen & Co., 1010 Chestnut Street, Philadelphia.
Williams, Brown & Earle, N. E. cor. Tenth and Chestnut Sts., Philadelphia.
Eimer & Amend, 205-211 Third Avenue, New York.
Joseph Zentmayer, 209 S. Eleventh Street, Philadelphia.

APPENDIX.

Foreign Makers.

Powell & Lealand, 170 Euston Road, N. W., London.
R. & J. Beck, 68 Cornhill, E. C., London.
Swift & Son, 81 Tottenham Court Road, London.
C. Baker, 244 High Holborn, London.
H. Crouch, 66 Barbican, London.
Carl Zeiss, Jena.
E. Hartnack, Waisenstrasse, 39, Potsdam.
W. & H. Seibert (successors to Gundlach), Wetzlar.
C. Reichert, Bennogasse, 26, Vienna.
Ernest Leitz, Wetzlar.
F. W. Shieck, Hallesche Strasse, 14, Berlin, S.W.
Nachet et Fils, Rue St. Severin, 17, Paris.
C. Verick, Rue des Ecoles, Paris.

MICROTOMES.

To be had from most of the above Firms, and also from—

Zimmermann, Albert Strasse, Leipzig (Maker of Minot's Microtome).
R. Jung, Heidelberg (Maker of Thoma's Microtome).
Schanze, Pathologisches Institut, Liebig Strasse, Leipzig.
W. Hume, Lothian Street, Edinburgh.
A. Fraser, Lothian Street, Edinburgh (Maker of Cathcart's Microtome).
J. Gardner, Teviot Place, Edinburgh (Maker of Rutherford's Microtome).
Cambridge Scientific Instrument Co.
Kanthack, Golden Square, off Regent Street, London.

CHEMICALS AND HISTOLOGICAL REAGENTS.

Hopkin & Williams, 16 Cross Street, Hatton Garden, London, E.C.
R. & J. Beck, London.
Baker, 244 High Holborn, London.
Southall Brothers & Barclay, Dalton Street, Birmingham.
Dr. Georg Grübler, Bayersche Strasse, 12, Leipzig.

D.—WEIGHTS AND MEASURES, EQUIVALENTS.

Centigram	= .154 English grains.
Decigram	= 1.543 ,,
Gram	= 15.432 ,,
Kilogram	= 2.2 lbs. (avoird.).
1 fluid ounce	= 28 cubic centimetres.
1 fluid drachm	= 3.9 ,, ,,
1 inch	= 2.539 centimetres.
1 foot	= 3.047 decimetres.
1000 micros (μ)	= 1 millimetre.
10 millimetres (mm.)	= 1 centimetre.
100 centimetres (cm. or ctm.)	= 1 metre (unit of length).
1μ = 0.000039 inch	= $\frac{1}{25600}$th inch (approximately).
1 cm.	= 0.3937 inch.
1 metre	= 39.3704 inches.

INDEX.

Abbe's condenser, 10.
Absolute alcohol, 28.
Absorption of fat, 283.
Acid-alcohol, 65.
Acid-fuchsin, 74, 154.
Acids, 30.
Adenoid reticulum, 236.
—— tissue, 172, 398.
Adipose tissue, 168, 398, 401.
Agminated follicles, 274.
Air-bubbles, 101.
Albo-carbo light, 23.
Albuminous glands, 258.
Alcohol, 27.
—— dilute, 25.
Alkalies, 93.
Alum carmine, 65.
Ammoniacal carmine, 63.
Ammonium bichromate, 25, 29.
—— chromate, 25, 29.
Amœboid movement, 111, 117.
Angle of aperture, 13.
Aniline-blue, 73.
—— blue-black, 76.
—— dyes, 72.
—— oil, 73.
Anodon's muscle, 201.
Aorta, 226.
Apochromatic lenses, 14.
Aponeurosis, 397.
Aqueous humour, 24.
Arachnoid, 328.
Areolar tissue, 159, 161, 397, 400.
Arrector pili, 322.
Arsenic acid, 37.
Arteriole, 229.
Artery, 223.
—— elastic fibres in, 233.
—— endothelium of, 228.
Articular cartilage, 150.
Atrophic fat-cells, 171.

Auerbach's plexus, 277.
Axis-cylinder, 206, 211.

Bacteria, 104.
Balsam, 82, 85, 87.
Baryta-water, 26.
Basement membrane, 311.
Bayerl's fluid, 36.
Benzo-azurin, 131.
Bile-ducts, 290.
—— auto-injection of, 291.
Bismarck brown, 75.
Bladder, 313.
—— cells of, 190, 313.
—— frog's, 134, 190.
—— crayfish's, 115.
Blood, colourless corpuscles of, 110, 117, 122, 402.
—— circulation of, 231.
—— division of, 113.
—— effects of reagents on, 112, 114.
—— feeding of, 113.
—— colourless corpuscles of, glycogen in, 113.
—— migration of, 113.
Blood-corpuscles, 400.
—— amphibians, 106.
—— bird, 110.
—— cover-glass preparations, 114.
—— crenation of, 119.
—— crystals from, 120, 402.
—— double staining, 114.
—— enumeration of, 121.
—— fish, 110.
—— frog, 106.
—— human, 115.
—— leukæmic, 121.
—— pseudo-membrane, 114.
—— tablets, 106, 123.
Blood, effect on, of acetic acid, 107, 112.

Blood aniline dyes, 121.
—— of boracic acid, 109.
—— of hydrochloric acid, 108.
—— of magenta, 109.
—— of osmic acid, 110.
—— of syrup, 108.
—— of tannic acid, 109.
—— of water, 108.
Blood-plates, 106, 123.
Blood-serum, 24.
Blood-vessels, 223.
—— injection of, 230.
—— development of, 228.
Bone, 174.
—— blood-vessels of, 179, 181.
—— cancellated, 180.
—— corpuscles, 177.
—— decalcified, 177, 181.
—— development of, 182.
—— marrow of, 186.
—— perforating fibres, 178.
—— polarised light, 181.
Borax-carmine, 64.
Boveri's fluid, 213.
Bowman's glands, 375.
Bronchus, 297.
Brownian movement, 100.
Brunner's glands, 278.

CABINET, 2.
Calcified cartilage, 183.
Caliciform cells, 273.
Camera lucida, 17.
—— Abbe's, 17,
—— Chevalier's, 18.
—— Malassez's, 18.
Camera, Zeiss's, 17.
Camera obscura shade, 23.
Canada balsam, 85.
Capillaries, 224, 231.
Capillary attraction method, 240.
Carbolic acid and xylol, 83.
Cardiac glands, 267.
—— muscle, 199.
Carmine, 63, 64.
—— acid-chloral, 282.
—— and Dahlia fluid, 67.
—— Frey's, 641.
Carter's injection, 89.
Cartilage, 146.
—— articular, 150.
—— cellular, 146.
—— costal, 148.
—— cuttlefish, 151.
—— encrusting, 184.
—— epiphysial, 184.

Cartilage, fibrous, 151.
—— hyaline, 147, 397.
—— parenchymatous, 146.
—— transition, 155.
Caustic potash, 25.
Cayal's methods, 222.
Cedar-wood oil, 83.
Cell, animal, 141.
Celloidin, 45.
Cell-spaces, 162, 168.
—— of cornea, 361.
Cellular cartilage, 146.
Cement of tooth, 251.
Central nervous system, staining of, 338.
—— freezing method, 339.
—— Golgi's methods, 344.
—— Pal's method, 343.
—— Vessale's method, 344.
—— Weigert's method, 338.
—— Weigert-Pal method, 340.
Central tendon, 129, 237.
Centrifuge, 94.
Cerebellum, 349.
—— blood-vessels of, 351.
—— Golgi's method, 355.
Cerebrum, 351.
—— blood-vessels of, 355.
—— Golgi's method, 357.
—— Weigert's method, 357.
Ceruminous glands, 371.
Chalice-cells, 139, 273.
Choroid, 362.
—— pigment-cells of, 363.
Chromic acid, 25, 28.
—— and nitric acid, 36.
Chromo-acetic acid, 31.
—— aceto-osmic acid, 32.
—— -formic acid, 31.
—— -osmic acid, 37.
Ciliary muscle, 363.
—— motion, 135.
—— —— effects of reagents on, 136.
—— processes, 363.
Ciliated epithelium, 135.
—— isolated cells, 138.
Circulation of blood, 231.
—— in frog's tongue, 234.
Circumvallate papillæ, 247.
Clarifying reagents, 82.
Clarke's column, 330.
Clasmatocytes, 162.
Cloves, oil of, 83.
Coarsely-granular cells, 162.
Cochineal, 67.
Cochlea, 371.

Cockroach, salivary glands of, 265.
Cohnheim's areas, 200.
Collateral fibres, 346.
Collodion, 211.
Colostrum, 395.
Columnar epithelium, 131.
Condenser, 10.
Cones, 368.
Connective tissues, 156.
Convoluted tubules, 305.
Copper acetate, 338.
Cornea, 358.
—— cell-spaces of, 361.
—— corpuscles, 359, 396.
—— fibrils of, 369.
—— lymph-spaces, 396.
—— nerves of, 359.
Corrosive sublimate, 33.
Cortex cerebri, 352.
Cortical arteries, 355.
Corti's organ, 373.
Costal cartilage, 148.
Cotton fibres, 102.
Cover-glasses, 1.
—— to clean, 98.
Cover-glass tester, 23.
Crab's muscle, 196.
Crayfish blood, 115.
Crenation, 119.
Crista acustica, 374.
Creosote, 83.
Crusta petrosa, 251.
Cutting sections in series, 53, 60.
Cuttlefish cartilage, 151.

Dahlia, 73.
Dammar lac, 86.
Decalcifying fluids, 36.
Decussation of pyramids, 347.
Demilunes, 258, 260, 265.
Dentine, 251.
Descemet's membrane, 359.
Diaphragm, 129, 237, 401.
Diaphragms, 7.
Digestion methods, 26.
Dilute alcohol, 25.
Dissecting case, 2.
—— microscopes, 22.
Dissociating fluids, 24.
Dogiel's method, 167.
Drawing materials, 3.
Dry cover-glass preparations, 114.
Duck's bill, 378.
Ductule, 257.
Dura mater, 328.

Ear, 371.
—— cartilage, 154.
Ebner's fluid, 37.
Ehrlich-Biondi fluid, 81, 140.
Ehrlich's hæmatoxylin, 69.
Eimer, organ of, 381.
Elastic fibres, 157, 397.
—— Herxheimer's method, 161.
—— Martinotti's reaction, 161.
Embedding, 40.
—— boxes, 45.
—— in celloidin, 45.
—— in gum, 41.
—— in paraffin, 41, 44.
—— interstitial, 41.
Enamel, 251.
End-bulbs, 378.
Endocardium, 223.
Endothelium, 129, 398.
—— of arteries, 228.
End-plates, 219.
Eosin, 72.
Eosin-hæmatoxylin, 70.
Eosinophilous cells, 122.
Epidermis of man, 317, 325.
—— of newt, 126.
Epididymis, 387.
Epiglottis, 153.
Epiphysis, 184.
Epithelial cells, fibrillation of, 327.
Epithelium, 124, 398.
—— ciliated, 135.
—— columnar, 131.
—— germinating, 239.
—— glandular, 133.
—— secretory, 133.
—— squamous, 124.
—— transitional, 133.
Erlicki's fluid, 29.
Eternod's rings, 77.
Eye, 358.
—— blood-vessels of, 369.
—— triton's, 370.
Eyelid, 369.

Fallopian tube, 390.
Farrant's solution, 85.
Fat-cell, 169.
—— absorption of, 283.
—— action of reagents on, 169.
—— atrophic, 171.
—— development of, 171.
Fenestrated membranes, 158, 160, 227.
Fibres of Tomes, 253.
Fibrin, 119, 123.

Fibro-cartilages, 151.
Filiform papillæ, 246.
Fixatives, 60.
Fixing fluids, 27.
Flemming's fluid, 32.
Fol's solution, 32.
Formatio reticularis, 348.
Free nerve-endings, 376.
Freezing fluid, 50.
Fresh tissues, examination of, 93, 396.
Frog's bladder, 134.
—— heart, nerve-cells of, 233.
—— tongue, 140.
Frommann's lines, 207, 211.
Fuchsin, 74.
Fundus glands, 268.
Fungiform papillæ, 247.

GAMBOGE, 100.
Ganglion, spinal, 214.
—— sympathetic, 216.
Gasserian ganglion, 215.
Gastric gland-cells, 268.
Gaule's method, 61.
Genital corpuscles, 378.
Gentian violet, 73.
Germinal epithelium, 389.
Germinating epithelium, 239.
Gland-ducts, 265.
Glandular epithelium, 133.
Glia-cells, 343.
Glisson's capsule, 285.
Glycerine, 85.
—— jelly, 85.
Glycogen, 293.
Goblet-cells, 138, 140, 273.
Gold chloride, 78.
Golgi's methods, 78, 220, 344.
—— slow method, 344.
—— sublimate method, 345.
—— rapid method, 345.
—— —— for retina, 369.
Goll's column, 330.
Graafian follicles, 389.
Gram's method, 105.
Grandry's corpuscles, 377.
Granular cells, 156, 293, 401.
Guanin cells, 173.
Gustatory cells, 250.

HÄM-ALUM, 71.
Hamatein, 71.
Hæmatoxylin, 68, 71.
—— acid, 69.
—— Böhmer's, 68.
—— Delafield's, 69.

Hæmatoxylin, Ehrlich's, 69.
—— eosin, 70.
—— glycerine, 70.
—— Hamilton's, 69.
—— Heidenhain's, 70.
—— Kleinenberg's, 69.
—— Kultschitzky's, 340.
—— nucleus-staining, 69.
—— Weigert's, 338.
Hæmin, 126.
Hæmoglobin, 120.
Hæmolymph, 115.
Hair, human, 323.
—— blood-pigment in, 327.
—— development of, 327.
—— elements of, 324.
—— rabbit's, 324.
Hair-follicles, 321.
—— coverings of, 321.
—— development of, 327.
—— double-staining of, 327.
Hairs, tactile, 327.
Half-drying method, 159.
Hamilton's hæmatoxylin, 69.
Hardening fluids, 27, 33.
Hard palate, 126.
Hartnack's dry lenses, 408.
Hassall's corpuscles, 242.
Hayem's fluid, 122.
Heart, 223.
—— valves, 225.
Heidenhain's method, 70, 259, 282.
Henle's tubule, 305.
Herbst's corpuscles, 379.
Hermann's fluid, 384.
Herxheimer's method, 161, 326.
Horny epidermis, 127.
Hot stage, 117.
Howship's lacunæ, 184.
Hyaline cartilage, 147.
Hypodermic syringe, 161.
Hypophysis cerebri, 357.

ILLUMINATION, artificial, 21.
—— direct and oblique, 7.
Immersion lenses, 12.
Impregnacion doble, 222, 346.
Incremental lines, 252.
Indigo-carmine, 67.
Indirect cell-division, 141.
Inflammation, 233.
Injection mass, 89.
—— methods of, 91.
Intercellular bridges, 127.
—— channels, 127.
Intercostal nerve, 207.

Interglobular spaces, 252.
Interlobular artery, 310.
Intervertebral disc, 151.
Intestinal glands, 283.
Intestine, large, 280.
—— small, 272.
Iodine green, 74.
Iodised serum, 25.
Iris, 363.
Irregular tubules, 305.
Irrigation, 107.

KARYOKINESIS, 141, 145.
Kidney, 303.
—— blood-vessels of, 306.
—— convoluted tubules, 308.
—— fresh, 311.
—— glomerulus, 307.
—— injected, 309.
—— irregular tubules, 308.
—— isolated tubules, 311.
—— medullary ray, 309.
Kleinenberg's hæmatoxylin, 69.
—— fluid, 30.
Klein's fluid, 29.
Kochs-Wolz lamp, 22.
Kronecker's fluid, 24.
Kühne's method, 211.
Kultschitzky's hæmatoxylin, 340.
—— method, 344.

LABELS, 4.
Lachrymal gland, 371.
Lacteal, 273, 282.
Landois' fluid, 26.
Large intestine, 280.
Leitz's lenses, 409.
Lens, crystalline, 364.
—— epithelium of, 369.
Lenses, apochromatic, 15.
—— dry, 11.
—— immersion, 11.
Leucocytes, 122, 402, 403.
Leukæmia, 121.
Lieberkühn's glands, 273, 280.
—— mitosis in, 283.
Ligamentum nuchæ, 157.
Linen fibres, 102.
Lithium carmine, 65.
Liver, 285.
—— bile capillaries of, 291, 292.
—— bile ducts, 290.
—— blood-vessels of, 285, 289.
—— cells, 133.
—— connective tissue of, 292.
—— frog's, 288.

Liver, glycogen in, 293.
—— granular cells of, 293.
—— human, 288.
—— iron in, 293.
—— methods for, 286.
—— pigment in, 293.
—— pig's, 286.
—— rabbit's, 288.
Löffler's blue, 93.
Logwood, 71.
Lugol's solution, 93.
Lungs, 294.
—— blood-vessels of, 294.
—— dried, 300.
—— elastic fibres in, 300, 302.
—— fœtal, 300.
—— fresh, 300.
—— frog's, 302.
—— newt's, 302.
Lymph, 112.
—— channels, 237.
—— gland, 234, 235, 245.
Lymphatics, 234.

MACROPHAGES, 285.
Macula lutea, 370.
Magenta, 74.
Magnifying power, 19.
—— powers of objectives, 408.
Malpighian capsule, 304.
—— pyramid, 303.
Mammary gland, 393.
—— active, 395.
Marchi's method, 212, 347.
Margarine crystals, 170.
Marrow, 186, 401.
Martinotti's methods, 161.
Mastzellen, 67, 156, 162, 401.
May's methods, 219.
Mayer's embedding bath, 37.
Measures, 410.
Medulla oblongata, 347.
—— olivary bodies, 348.
Medullary ray, 303.
Medullated nerve-fibres, 202, 399.
Meissner's corpuscles, 380.
—— plexus, 277.
Membrana tympani, 371.
Mercuric chloride, 33.
—— method for nervous system, 345.
Merkel's corpuscles, 377.
Methylated spirit, 28.
Methylene-blue, 73, 131, 192, 222, 370, 284.
Methyl-green, 74.
Methyl-mixture, 26.

Methyl-violet, 73.
Micrococci, 104.
Micrometer, 20.
Micro-organisms, 103.
Microphages, 285.
Microscope, 5.
—— choice of, 15.
—— dissecting, 23.
—— illumination of, 7.
—— lamp, 21.
—— magnifying power of, 19.
—— makers of, 409.
—— parts of, 15.
Microtomes, 49.
—— Cambridge, 53.
—— Cathcart's, 53.
—— Jung's, 56.
—— makers of, 410.
—— Malassez's, 56.
—— Minot's, 55.
—— Ranvier's, 59.
—— Roy's, 53.
—— Rutherford's, 50.
—— Swift's, 59.
—— Thoma's, 56.
—— Williams', 58.
Migratory cells, 157.
Milk, 98.
Mitosis, 141, 145.
Mounting block, 89.
—— fluids and methods, 85.
Mole's nose, 381.
Mucigen, 139, 260.
Muco-salivary glands, 257, 262, 264.
Mucous cells, 265.
—— glands, 256, 258.
—— tissue, 171, 174.
Müller's fluid, 29.
—— and spirit, 29.
Multipolar nerve-cells, 217, 343.
Muscle, 189, 398.
—— striped, 189, 398.
Myelin, 205.
—— drops, 205.
Myeloplaxes, 187.
Myocardium, 223.

NAIL, 326.
—— double-stained, 327.
Needles, 2.
Nerve-cells, 213, 399.
—— cover-glass preparations of, 218.
—— crayfish, 223.
—— in frog's heart, 221, 233.
—— multipolar, 217, 343.
—— pyriform, 221.

Nerve-cells, spinal cord, 217.
—— sympathetic, 217, 222.
—— transverse markings on, 207.
Nerve-endings, 376.
Nerve-plexuses, 276.
Nerve-fibres, 202.
—— axis-cylinder of, 205, 208, 211.
—— degeneration of, 213.
—— in osmic acid, 206, 211.
—— intercostal, 207.
—— living, 212.
—— medullated, 202, 399.
—— non-medullated, 203, 209.
—— size of, 212.
—— spinal cord, 213.
—— to muscle, 219.
—— transverse markings on, 207.
Nerve-plexus in intestine, 277.
Nerve-trunks, 203.
Neuroglia, 341, 343.
Neurokeratin, 210.
Newt's cartilage, 145.
Nitric acid, 31.
Nodes of Ranvier, 202.
Non-medullated nerve-fibres, 203.
Non-striped muscle, 189.
—— cement of, 191.
—— grooving on, 193.
—— plexus in, 192.
Normal fluids, 24.
—— saline, 24.
Nose, 374, 376.
Nuclear stains, 63.

OBJECTIVES, 9.
Obreggia's method, 404.
Oculars, 5, 10.
Odontoblasts, 253.
Œsophagus, 255.
—— and stomach, 271.
Oikoid, 109.
Olfactory bulb, 376.
—— cells, 375.
Omentum, 129.
Onion, 103.
Opaque injections, 289.
Optic nerve, 368.
—— entrance of, 370.
Origanum oil, 83.
Osmic acid, 25, 32.
Osmico-bichromate mixture, 345.
Ossification, 182.
Osteoblasts, 178.
Osteoclasts, 183.
Ovary, 388.
Ovum, 390.

PACINIAN corpuscle, 378.
Palate, hard, 126.
—— glands of, 251.
—— soft, 250.
Palm of hand, 317.
Pal's method, 343.
Pancreas, 262, 404.
—— cells of, 265.
—— nerves of, 266.
Papillæ foliatæ, 249.
—— of skin, 315.
—— of tongue, 246.
Paraffin, 41.
Penicillium, 103.
Penis, 313.
Perènyi's fluid, 31.
Perforating fibres, 178.
Pericardium, 223.
Pericœsophageal membrane, 163.
Periodontal membrane, 252.
Periosteum, 177.
Peyer's patch, 272, 274, 284.
Phagocytosis, 284.
Phenylene-brown, 75.
Phloroglucin, 37.
Photophore, 27.
Pia mater, 228.
Picric acid, 30, 36.
Picrin-glycerine, 192.
Picro-carmine, 66, 80.
Picro-glycerine mixture, 192.
Picro-lithium carmine, 66.
Picro-nitric acid, 31.
Picro-sulphuric acid, 30.
Pigment-cells, 173, 397.
—— choroid, 363.
Pipettes, 4.
Pituitary body, 357.
Placenta, 395.
—— injected, 396.
Plasma-cells, 156.
Polariscope, 200.
Potassic bichromate, 25, 29.
Potato starch, 100.
Preparation of tissues, 38, 396.
Prickle cells, 127, 318.
Pulp cavity, 253.
Purkinje's cells, 351.
—— fibres, 225.
Pyloric glands, 270.
Pyloro-duodenal region, 271.
Pyriform nerve-cells, 221.

RABL's fluid, 31.
Ranvier's crosses, 207, 211.
—— fluid, 25.

Ranvier's nodes, 206.
Razor, 3.
Rectified spirit, 28.
Red marrow, 187.
Respiratory organs, 295.
Retina, 365, 370, 399.
—— cones of, 367.
—— frog's, 368.
—— Golgi's methods for, 369.
—— macula lutea, 370.
Retro-lingual membrane, 201.
Rice-starch, 100.
Ripart and Petit's fluid, 24.
Rosanilin, 74.

SAFRANIN, 75.
Salivary corpuscles, 125.
—— glands, 256, 404.
—— of cockroach, 265.
Sarcolemma, 193.
Sarcostyles, 198.
Scalp, 323.
Schiefferdecker's fluid, 26.
Schizomycetes, 105.
Schreger's lines, 252.
Sebaceous glands, 323.
Section cutting, 48.
—— flatteners, 59.
—— in series, 60.
—— lifter, 3.
—— to place on slide, 86.
Semicircular canals, 374.
Semidesiccation method, 159.
Seminiferous tubules, 382.
Sensory nerve-terminations, 376.
Septum cisternæ, 238.
Serial sections, 60.
Serous fluids, 113.
—— glands, 257, 261.
—— nerves in, 265.
Serum, action of, 124.
—— and osmic acid, 78.
Sharpey's fibres, 178.
Silver lines, 129.
—— nitrate, 76.
Skin, 315.
—— blood-vessels of, 325.
—— cutis vera, 315.
—— elastic fibres in, 326.
—— epidermis of, 325.
—— finger, 317.
—— fœtal, 320.
—— injected, 325.
—— negro's, 320.
—— nerves of, 377.
—— sebaceous glands of, 322.

Skin, sweat-glands of, 319, 327.
—— touch-corpuscles of, 320.
—— to clean, 98.
Slides, 1.
Small intestine, 272.
—— absorption of fat, 283.
—— blood-vessels of, 275.
—— nerve-plexuses of, 276.
—— nerves of, 284.
Soft palate, 250.
Solitary follicles, 275, 280.
Spermatogenesis, 383, 385.
Spermatozoa, 386.
—— cover-glass preparations, 387.
—— frog's, 386.
—— human, 387.
—— newt's, 386.
Spiller's purple, 74.
Spinal cord, 328.
—— cells of, 217, 329.
—— collateral fibres, 346.
—— columns of, 330.
—— commissures of, 329.
—— cornua, 329.
—— degeneration of, 332, 347.
—— dry preparation of, 342.
—— fissures of, 328.
—— ganglia, 213.
—— grey matter of, 334.
—— human, 337.
—— longitudinal section, 338.
—— nerve-cells, 343.
—— nerve-fibres of, 341.
—— neuroglia of, 341.
—— staining of, 341.
—— substantia gelatinosa, 329.
—— tracts in, 542.
—— white matter of, 335.
Spleen, 242.
—— injected, 245.
Squames, 125.
Staining in bulk, 44.
—— general remarks, 81.
—— multiple, 80.
—— reagents, 63.
Starch, potato, 100.
—— rice, 100.
Stomach, 266.
—— blood-vessels of, 271.
—— cardiac end, 267.
—— double staining of, 271.
—— methods for, 267.
—— nerves of, 284.
—— pyloric end, 270.
Stomata, 239.
Stratified epithelium, 126.

Striped muscle, 193, 398.
—— Anodon's, 201.
—— cardiac, 199.
—— crab's, 196, 200.
—— discs, 195.
—— fibrillæ of, 195.
—— frozen, 200.
—— injected, 198.
—— isolated fibres, 194.
—— living, 200.
—— red, 199.
—— sarcolemma, 193.
—— tendon of, 196.
Sublingual gland, 259.
Submaxillary glands, 259, 262, 264.
Suprarenal capsule, 314.
—— nerves of, 315.
Sweat-glands, 319, 327.
Sympathetic nerve-fibres, 209.
—— ganglia, 216.
—— nerve-cells, 222.
—— Cayal's method, 222.
Syringes, 91.

TACTILE cells, 377, 381.
—— disc, 377.
—— hairs, 327.
Taste-buds, 249.
Teasing, 26.
Tendon, 163, 397.
—— nerves in, 221.
Terminal corpuscles, 377.
Testis, 387.
Thymus gland, 241.
Thyroid gland, 301.
Tizzoni's reaction, 293.
Tongue, 246.
—— circulation in, 234.
—— frog's, 140.
—— gland of, 251.
—— injected, 248.
—— nerves of, 251.
—— papillæ of, 246.
Tonsils, 239.
Tooth, 251.
—— development of, 253.
—— softened, 252.
Touch-corpuscle, 380.
Trachea, 294.
Transitional epithelium, 133.
Tubules isolated of kidney, 311.

UMBILICAL cord, 395.
Unna's method, 327.
Ureter, 312.
Uterus, 392.

VALENTIN's knife, 49.
Valves of heart, 225.
Vas deferens, 388.
Vegetable cells, 103.
Veins, 224, 230.
Vermiform appendix, 281, 284.
Vessale's method, 344.
Vesuvin, 76.
Villus of intestine, 272, 280, 282.
—— injected, 275.
Volkmann's canals, 176.
Von Ebner's fluid, 37.

WAGNER's touch-corpuscles, 380.
Warm stages, 117.
Weigert's hæmatoxylin, 338.
—— method for staining central nervous system, 338.

Weigert-Pal method, 340.
Weights, 410.
Welsbach light, 22.
Westphal's fluid, 67.
Wharton's jelly, 171.
White fibro-cartilage, 152.
—— zinc cement, 89.
Wool, 102, 324.
Works of reference, 406.

XYLOL, 83.
—— balsam, 86.

YEAST, 104.

ZEISS's lenses, 409.
Zinc cement, 89.
Zooid, 109.

Catalogue No. 8.   December, 1895.

CLASSIFIED SUBJECT CATALOGUE

OF

MEDICAL BOOKS

AND

Books on Medicine, Dentistry, Pharmacy, Chemistry, Hygiene, Etc., Etc.,

PUBLISHED BY

P. BLAKISTON, SON & CO.,

Medical Publishers and Booksellers,

1012 WALNUT STREET, PHILADELPHIA.

SPECIAL NOTE.—The prices given in this catalogue are absolutely net, no discount will be allowed retail purchasers under any consideration. This rule has been established in order that everyone will be treated alike, a general reduction in former prices having been made to meet previous retail discounts. Upon receipt of the advertised price any book will be forwarded by mail or express, all charges prepaid.

We keep a large stock of Miscellaneous Books, not on this catalogue, relating to Medicine and Allied Sciences, published in this country and abroad. Inquiries in regard to prices, date of edition, etc., will receive prompt attention.

Special Catalogues of Books on Pharmacy, Dentistry, Chemistry, Hygiene, and Nursing will be sent free upon application.

☞ SEE NEXT PAGE FOR SUBJECT INDEX.

SUBJECT INDEX.

☞ Any books not on this Catalogue we will furnish a price for upon application.

SUBJECT.	PAGE
Anatomy	3
Anesthetics	3
Autopsies (see Pathology)	16
Bandaging (see Surgery)	19
Biology (see Miscellaneous)	14
Brain	4
Chemistry	4
Children, Diseases of	6
Clinical Charts	6
Consumption (see Lungs)	12
Deformities	7
Dentistry	7
Diagnosis	17
Diagrams (see Anatomy, page 3, and Obstetrics, page 16).	
Dictionaries	8
Diet and Food (see Miscellaneous)	14
Dissectors	3
Domestic Medicine	10
Ear	8
Electricity	9
Emergencies (see Surgery)	19
Eye	9
Fevers	9
Gout	10
Gynecology	21
Headaches	10
Heart	10
Histology	16
Hospitals (see Hygiene)	11
Hygiene	11
Insanity	4
Journals	11
Kidneys	12
Lungs	12
Massage	12
Materia Medica	12
Medical Jurisprudence	13
Microscopy	13
Milk Analysis (see Chemistry)	4
Miscellaneous	14
Nervous Diseases	14
Nose	20
Nursing	15
Obstetrics	16
Ophthalmology	9
Osteology (see Anatomy)	3
Pathology	16
Pharmacy	16
Physical Diagnosis	17
Physical Training (see Miscellaneous)	14
Physiology	18
Poisons (see Toxicology)	13
Popular Medicine	10
Practice of Medicine	18
Prescription Books	18
Railroad Injuries (see Nervous Diseases)	14
Refraction (see Eye)	9
Rheumatism	10
Sanitary Science	11
Skin	19
Spectacles (see Eye)	9
Spine (see Nervous Diseases)	14
Students' Compends	22, 23
Surgery and Surgical Diseases	19
Syphilis	21
Technological Books	4
Temperature Charts	6
Therapeutics	12
Throat	20
Toxicology	13
U. S. Pharmacopœia	16
Urinary Organs	20
Urine	20
Venereal Diseases	21
Veterinary Medicine	21
Visiting Lists, Physicians'. *(Send for Special Circular.)*	
Water Analysis (see Chemistry)	11
Women, Diseases of	21

☞ *The prices as given in this Catalogue are net. Cloth binding, unless otherwise specified. No discount can be allowed under any circumstances. Any book will be sent, postpaid, upon receipt of advertised price.*

☞ *All books are bound in cloth, unless otherwise specified. All prices are net.*

ANATOMY.

MORRIS. Text-Book of Anatomy. 791 Illus., 214 of which are printed in colors. Clo., $6.00; Lea., $7.00; Half Russia, $8.00.

"Taken as a whole, we have no hesitation in according very high praise to this work. It will rank, we believe, with the leading Anatomies. The illustrations are handsome and the printing is good."— *Boston Medical and Surgical Journal.*

Handsome Circular of Morris, with sample pages and colored illustrations, will be sent free to any address.

CAMPBELL. Outlines for Dissection. Prepared for Use with "Morris's Anatomy" by the Demonstrator of Anatomy at the University of Michigan. *Just Ready.* $1.00

HEATH. Practical Anatomy. A Manual of Dissections. 8th Edition. 300 Illustrations. $4.25

HOLDEN. Anatomy. A Manual of the Dissections of the Human Body. 6th Edition. Carefully Revised by A. HEWSON, M.D., Demonstrator of Anatomy, Jefferson Medical College, Philadelphia. 311 Illustrations. Cloth, $2.50; Oil-Cloth, $2.50; Leather, $3.00

HOLDEN. Human Osteology. Comprising a Description of the Bones, with Colored Delineations of the Attachments of the Muscles. The General and Microscopical Structure of Bone and its Development. With Lithographic Plates and numerous Illus. 7th Ed. $5.25

HOLDEN. Landmarks. Medical and Surgical. 4th Ed. $1.00

MACALISTER. Human Anatomy. Systematic and Topographical, including the Embryology, Histology, and Morphology of Man. With Special Reference to the Requirements of Practical Surgery and Medicine. 816 Illustrations, 400 of which are original.
Cloth, $5.00; Leather, $6.00

MARSHALL. Physiological Diagrams. Life Size, Colored. Eleven Life-Size Diagrams (each seven feet by three feet seven inches). Designed for Demonstration before the Class.

In Sheets, Unmounted, $40.00; Backed with Muslin and Mounted on Rollers, $60.00; Ditto, Spring Rollers, in Handsome Walnut Wall Map Case (send for special circular), $100.00; Single Plates—Sheets, $5.00; Mounted, $7.50. Explanatory Key, .50. *Descriptive circular upon application.*

POTTER. Compend of Anatomy, Including Visceral Anatomy. 5th Edition. 16 Lithographed Plates and 117 other Illustrations.
.80; Interleaved, $1.25

WILSON. Human Anatomy. 11th Edition. 429 Illustrations, 26 Colored Plates, and a Glossary of Terms. $5.00

OBERSTEINER. Anatomy of the Central Nervous Organs. 198 Illustrations. $5.50

ANESTHETICS.

BUXTON. On Anesthetics. 2d Edition. Illustrated. $1.25

TURNBULL. Artificial Anesthesia. The Advantages and Accidents of; Its Employment in the Treatment of Disease; Modes of Administration; Considering their Relative Risks; Tests of Purity; Treatment of Asphyxia; Spasms of the Glottis; Syncope, etc. 3d Edition, Revised. 40 Illustrations. $3.00

BRAIN AND INSANITY.

BLACKBURN. A Manual of Autopsies. Designed for the Use of Hospitals for the Insane and other Public Institutions. Ten full-page Plates and other Illustrations. $1.25

GOWERS. Diagnosis of Diseases of the Brain. 2d Edition. Illustrated. $1.50

HORSLEY. The Brain and Spinal Cord. The Structure and Functions of. Numerous Illustrations. $2.50

HYSLOP. Mental Physiology. Especially in Relation to Mental Disorders. With Illustrations. *Just Ready.* $4.25

LEWIS (BEVAN). Mental Diseases. A Text-Book Having Special Reference to the Pathological Aspects of Insanity. 18 Lithographic Plates and other Illustrations.

MANN. Manual of Psychological Medicine and Allied Nervous Diseases. Their Diagnosis, Pathology, Prognosis, and Treatment, including their Medico-Legal Aspects; with chapter on Expert Testimony, and an Abstract of the Laws Relating to the Insane in all the States of the Union. Illustrations of Typical Faces of the Insane, Handwriting of the Insane, and Micro-photographic Sections of the Brain and Spinal Cord. $3.00

REGIS. Mental Medicine. Authorized Translation by H. M. Bannister, M.D. $2.00

STEARNS. Mental Diseases. Designed especially for Medical Students and General Practitioners. With a Digest of Laws of the various States Relating to Care of Insane. Illustrated.
Cloth, $2.75; Sheep, $3.25

TUKE. Dictionary of Psychological Medicine. Giving the Definition, Etymology, and Symptoms of the Terms used in Medical Psychology, with the Symptoms, Pathology, and Treatment of the Recognized Forms of Mental Disorders, together with the Law of Lunacy in Great Britain and Ireland. Two volumes. $10.00

WOOD, H. C. Brain and Overwork. .40

CHEMISTRY AND TECHNOLOGY.

Special Catalogue of Chemical Books sent free upon application.

ALLEN. Commercial Organic Analysis. A Treatise on the Modes of Assaying the Various Organic Chemicals and Products Employed in the Arts, Manufactures, Medicine, etc., with concise methods for the Detection of Impurities, Adulterations, etc. 2d Ed. Vol. I, Vol. II, Vol. III, Part I. *These volumes cannot be had.* Vol. III, Part II. The Amins. Pyridin and its Hydrozins and Derivatives. The Antipyretics, etc. Vegetable Alkaloids, Tea, Coffee, Cocoa, etc. $4.50
Vol. III, Part III. *In Press.*

ALLEN. Chemical Analysis of Albuminous and Diabetic Urine. Illustrated. *Just Ready.* $2.25

BARTLEY. Medical and Pharmaceutical Chemistry. A Text-Book for Medical, Dental, and Pharmaceutical Students. With Illustrations, Glossary, and Complete Index. 4th Edition, carefully Revised. *Just Ready.* Cloth, $2.75; Sheep, $3.25

BLOXAM. Chemistry, Inorganic and Organic. With Experiments. 8th Ed., Revised. 281 Engravings. Clo., $4.25; Lea., $5.25

CALDWELL. Elements of Qualitative and Quantitative Chemical Analysis. 3d Edition, Revised. $1.50

CAMERON. Oils and Varnishes. With Illustrations, Formulæ, Tables, etc. $2.25

CAMERON. Soap and Candles. 54 Illustrations. $2.00

CLOWES AND COLEMAN. Elementary Qualitative Analysis. Adapted for Use in the Laboratories of Schools and Colleges. Illustrated. $1.00

GARDNER. The Brewer, Distiller, and Wine Manufacturer. A Hand-Book for all Interested in the Manufacture and Trade of Alcohol and Its Compounds. Illustrated. $1.50

GARDNER. Bleaching, Dyeing, and Calico Printing. With Formulæ. Illustrated. $1.50

GROVES AND THORP. Chemical Technology. The Application of Chemistry to the Arts and Manufactures. 8 Volumes, with numerous Illustrations.
Vol. I. Fuel and Its Applications. 607 Illustrations and 4 Plates.
Cloth, $5.00; Half Morocco, $6.50
Vol. II. Lighting. Illustrated. Cloth, $4.00; Half Morocco, $5.50
Vol. III. Lighting—Continued. *In Press.*

HOLLAND. The Urine, the Gastric Contents, the Common Poisons, and the Milk. Memoranda, Chemical and Microscopical, for Laboratory Use. 5th Ed. Illustrated and interleaved, $1.00

LEFFMANN. Compend of Medical Chemistry, Inorganic and Organic. Including Urine Analysis. 4th Edition, Rewritten. .80; Interleaved, $1.25

LEFFMANN. Progressive Exercises in Practical Chemistry. Illustrated. 2d Edition. $1.00

LEFFMANN. Analysis of Milk and Milk Products. Arranged to Suit the Needs of Analytical Chemists, Dairymen, and Milk Inspectors. $1.25

LEFFMANN. Water Analysis. Illustrated. 3d Edition. $1.25

MÜTER. Practical and Analytical Chemistry. 4th Edition. Revised to meet the requirements of American Medical Colleges by CLAUDE C. HAMILTON, M.D. 51 Illustrations. $1.25

OVERMAN. Practical Mineralogy, Assaying, and Mining. With a Description of the Useful Minerals, etc. 11th Edition. $1.00

RAMSAY. A System of Inorganic Chemistry. Illus. $4.00

RICHTER. Inorganic Chemistry. 4th American, from 6th German Edition. Authorized translation by EDGAR F. SMITH, M.A., PH.D. 89 Illustrations and a Colored Plate. $1.75

RICHTER. Organic Chemistry. 2d American Edition. Trans. from the 6th German by EDGAR F. SMITH, M.A., PH.D. Illus. $4.50

SMITH. Electro-Chemical Analysis. 2d Edition, Revised. 28 Illustrations. $1.25

SMITH AND KELLER. Experiments. Arranged for Students in General Chemistry. 3d Edition. Illustrated. *Just Ready.* .60

STAMMER. Chemical Problems. With Explanations and Answers. .50

SUTTON. Volumetric Analysis. A Systematic Handbook for the Quantitative Estimation of Chemical Substances by Measure, Applied to Liquids, Solids, and Gases. 6th Edition, Revised. With Illustrations. $4.50

SYMONDS. Manual of Chemistry, for Medical Students. 2d Edition. $2.00

TRIMBLE. Practical and Analytical Chemistry. Being a Complete Course in Chemical Analysis. 4th Ed. Illus. $1.50

WATTS. Organic Chemistry. 2d Edition. By WM. A. TILDEN, D.SC., F.R.S. (Being the 13th Edition of Fowne's Organic Chemistry.) Illustrated. $2.00

WATTS. Inorganic Chemistry. Physical and Inorganic. (Being the 14th Edition of Fowne's Physical and Inorganic Chemistry.) With Colored Plate of Spectra and other Illustrations. $2.00

WOODY. Essentials of Chemistry and Urinalysis. 4th Edition. Illustrated. *In Press.*

⁎ *Special Catalogue of Books on Chemistry free upon application.*

CHILDREN.

GOODHART AND STARR. Diseases of Children. From the 3d English Edition. Rearranged and Edited, with Notes and Additions, by LOUIS STARR, M.D. *Out of Print.*

HALE. On the Management of Children in Health and Disease. .50

HATFIELD. Compend of Diseases of Children. With a Colored Plate. 2d Edition. *Just Ready.* .80; Interleaved, $1.25

MEIGS. Infant Feeding and Milk Analysis. The Examination of Human and Cow's Milk, Cream, Condensed Milk, etc., and Directions as to the Diet of Young Infants. .50

MONEY. Treatment of Diseases in Children. Including the Outlines of Diagnosis and the Chief Pathological Differences Between Children and Adults. 2d Edition. $2.50

MUSKETT. Prescribing and Treatment in the Diseases of Infants and Children. $1.25

POWER. Surgical Diseases of Children and their Treatment by Modern Methods. Illustrated. *Just Ready.* $2.50

STARR. The Digestive Organs in Childhood. The Diseases of the Digestive Organs in Infancy and Childhood. With Chapters on the Investigation of Disease and the Management of Children. 2d Edition, Enlarged. Illustrated by two Colored Plates and numerous Wood Engravings. $2.00

STARR. Hygiene of the Nursery. Including the General Regimen and Feeding of Infants and Children, and the Domestic Management of the Ordinary Emergencies of Early Life, Massage, etc. 4th Edition. 25 Illustrations. $1.00

CLINICAL CHARTS.

GRIFFITH. Graphic Clinical Chart. Printed in three colors. Sample copies free. Put up in loose packages of fifty, .50. Price to Hospitals, 500 copies, $4.00; 1000 copies, $7.50. With name of Hospital printed on, .50 extra.

TEMPERATURE CHARTS. For Recording Temperature, Respiration, Pulse, Day of Disease, Date, Age, Sex, Occupation, Name, etc. Put up in pads of fifty. Each, .50

DEFORMITIES.

REEVES. Bodily Deformities and Their Treatment. A Hand-Book of Practical Orthopedics. 228 Illustrations. $1.75

HEATH. Injuries and Diseases of the Jaws. 187 Illustrations. 4th Edition. Cloth, $4.50

DENTISTRY.

Special Catalogue of Dental Books sent free upon application.

BARRETT. Dental Surgery for General Practitioners and Students of Medicine and Dentistry. Extraction of Teeth, etc. 2d Edition. Illustrated. $1.00

BLODGETT. Dental Pathology. By ALBERT N. BLODGETT, M.D., late Professor of Pathology and Therapeutics, Boston Dental College. 33 Illustrations. $1.25

FLAGG. Plastics and Plastic Filling, as Pertaining to the Filling of Cavities in Teeth of all Grades of Structure. 4th Edition. $4.00

FILLEBROWN. A Text-Book of Operative Dentistry. Written by invitation of the National Association of Dental Faculties. Illustrated. $2.25

GORGAS. Dental Medicine. A Manual of Materia Medica and Therapeutics. 5th Edition, Revised. $4.00

HARRIS. Principles and Practice of Dentistry. Including Anatomy, Physiology, Pathology, Therapeutics, Dental Surgery, and Mechanism. 12th Edition. Revised by F. J. S. GORGAS, M.D., D.D.S. 1086 Illustrations. Cloth, $6.00; Leather, $7.00

HARRIS. Dictionary of Dentistry. Including Definitions of Such Words and Phrases of the Collateral Sciences as Pertain to the Art and Practice of Dentistry. 5th Edition. Revised and Enlarged by FERDINAND F. S. GORGAS, M D., D.D.S. Cloth, $4.50; Leather, $5.50

HEATH. Injuries and Diseases of the Jaws. 4th Edition. 187 Illustrations. $4.50

HEATH. Lectures on Certain Diseases of the Jaws. 64 Illustrations. Boards, .50

RICHARDSON. Mechanical Dentistry. 6th Edition. Thoroughly Revised by DR. GEO. W. WARREN. 600 Illustrations. Cloth, $4.00; Leather, $5.00

SEWELL. Dental Surgery. Including Special Anatomy and Surgery. 3d Edition, with 200 Illustrations. $2.00

TAFT. Operative Dentistry. A Practical Treatise. 4th Edition. 100 Illustrations. Cloth, $3.00; Leather, $4.00

TAFT. Index of Dental Periodical Literature. $2.00

TALBOT. Irregularities of the Teeth and Their Treatment. 2d Edition. 234 Illustrations. $3.00

TOMES. Dental Anatomy. Human and Comparative. 235 Illustrations. 4th Edition. $3.50

TOMES. Dental Surgery. 3d Edition. 292 Illustrations. $4.00

WARREN. Compend of Dental Pathology and Dental Medicine. With a Chapter on Emergencies. Illustrated. .80; Interleaved, $1.25

WARREN. Dental Prosthesis and Metallurgy. 129 Ills. $1.25

WHITE. The Mouth and Teeth. Illustrated. .40

⁎ *Special Catalogue of Dental Books free upon application.*

DICTIONARIES.

GOULD. The Illustrated Dictionary of Medicine, Biology, and Allied Sciences. Being an Exhaustive Lexicon of Medicine and those Sciences Collateral to it: Biology (Zoology and Botany), Chemistry, Dentistry, Parmacology, Microscopy, etc., with many useful Tables and numerous fine Illustrations. 1633 pages.
Sheep or Half Dark Green Leather, $10.00; Thumb Index, $11.00
Half Russia, Thumb Index, $12.00

GOULD. The Medical Student's Dictionary. Including all the Words and Phrases Generally Used in Medicine, with their Proper Pronunciation and Definition, Based on Recent Medical Literature. With Tables of the Bacilli, Micrococci, Leucomains, Ptomains, etc., of the Arteries, Muscles, Nerves, Ganglia, and Plexuses, etc.
Half Dark Leather, $2.75; Half Morocco, Thumb Index, $3.50

GOULD. The Pocket Pronouncing Medical Lexicon. (12,000 Medical Words Pronounced and Defined.) Containing all the Words, their Definition and Pronunciation, that the Medical, Dental, or Pharmaceutical Student Generally Comes in Contact With; also Elaborate Tables of the Arteries, Muscles, Nerves, Bacilli, etc., etc., a Dose List in both English and Metric System, etc., Arranged in a Most Convenient Form for Reference and Memorizing.
Full Limp Leather, Gilt Edges, $1.00; Thumb Index, $1.25

** Sample Pages and Illustrations and Descriptive Circulars of Gould's Dictionaries sent free upon application.

HARRIS. Dictionary of Dentistry. Including Definitions of Such Words and Phrases of the Collateral Sciences as Pertain to the Art and Practice of Dentistry. 5th Edition. Revised and Enlarged by FERDINAND J. S. GORGAS, M.D., D.D.S. Cloth, $4.50; Leather, $5.50

LONGLEY. Pocket Medical Dictionary. Giving the Definition and Pronunciation of Words and Terms in General Use in Medicine and Collateral Sciences, with an Appendix, containing Poisons and their Antidotes, Abbreviations used in Prescriptions, and a Metric Scale of Doses.   Cloth, .75; Tucks and Pocket, $1.00

CLEVELAND. Pocket Medical Dictionary. 33d Edition. Very small pocket size.   Cloth, .50; Tucks with Pocket, .75

MAXWELL. Terminologia Medica Polyglotta. By Dr. THEODORE MAXWELL, Assisted by Others. $3.00
The object of this work is to assist the medical men of any nationality in reading medical literature written in a language not their own. Each term is usually given in seven languages, viz.: English, French German, Italian, Spanish, Russian, and Latin.

TREVES AND LANG. German-English Medical Dictionary.
Half Russia, $3.25

EAR (see also Throat and Nose).

HOVELL. Diseases of the Ear and Naso-Pharynx. Including Anatomy and Physiology of the Organ, together with the Treatment of the Affections of the Nose and Pharynx which Conduce to Aural Disease. 122 Illustrations. $5.00

BURNETT. Hearing and How to Keep It. Illustrated. .40

DALBY. Diseases and Injuries of the Ear. 4th Edition. 28 Wood Engravings and 7 Colored Plates. $2.50

HALL. Compend of Diseases of Ear and Nose. Illustrated.
.80; Interleaved, $1.25

PRITCHARD. Diseases of the Ear. 3d Edition. Many Illustrations and Formulæ.   *In Press.*

ELECTRICITY.

BIGELOW. Plain Talks on Medical Electricity and Batteries. With a Therapeutic Index and a Glossary. 43 Illustrations. 2d Edition. $1.00

MASON. Electricity; Its Medical and Surgical Uses. Numerous Illustrations. .75

STEAVENSON AND JONES. Medical Electricity. 2d Edition. 103 Illustrations. *Preparing.*

EYE.

A Special Circular of Books on the Eye sent free upon application.

ARLT. Diseases of the Eye. Clinical Studies on Diseases of the Eye, Including the Conjunctiva, Cornea and Sclerotic, Iris and Ciliary Body. Authorized Translation by LYMAN WARE, M.D. Illustrated. $1.25

FICK. Diseases of the Eye and Ophthalmoscopy. Translated by A. B. HALE, M. D. 157 Illustrations, many of which are in colors. *In Press.*

FOX AND GOULD. Compend on Diseases of the Eye and Refraction, Including Treatment and Surgery. 2d Edition. 71 Illustrations and 39 Formulæ. .80; Interleaved, $1.25

GOWERS. Medical Ophthalmoscopy. A Manual and Atlas with Colored Autotype and Lithographic Plates and Wood-cuts, Comprising Original Illustrations of the Changes of the Eye in Diseases of the Brain, Kidney, etc. 3d Edition. $4.00

HARLAN. Eyesight, and How to Care for It. Illus. .40

HARTRIDGE. Refraction. 96 Illustrations and Test Types. 7th Edition. $1.00

HARTRIDGE. On the Ophthalmoscope. 2d Edition. With Colored Plate and many Wood-cuts. $1.25

HANSELL AND BELL. Clinical Ophthalmology. Colored Plate of Normal Fundus and 120 Illustrations. $1.50

HIGGENS. Ophthalmic Practice. Illustrated. $1.50

MACNAMARA. On the Eye. 5th Edition. Numerous Colored Plates, Diagrams of Eye, Wood-cuts, and Test Types. $3.50

MEYER. Ophthalmology. A Manual of Diseases of the Eye. Translated from the 3d French Edition by A. FREEDLAND FERGUS, M.B. 270 Illustrations, 2 Colored Plates. Cloth, $3.50; Sheep, $4.50

MORTON. Refraction of the Eye. Its Diagnosis and the Correction of its Errors. With Chapter on Keratoscopy and Test Types. 5th Edition. $1.00

PHILLIPS. Spectacles and Eyeglasses. Their Prescription and Adjustment. 2d Edition. 49 Illustrations. *Just Ready.* $1.00

SWANZY. Diseases of the Eye and Their Treatment. 4th Edition. 164 Illustrations. 2 Colored and 1 Plain Plate, and a Zephyr Test Card. Cloth, $2.50; Sheep, $3.00

WALKER. Students' Aid in Ophthalmology. Colored Plate and 40 other Illustrations and Glossary. *Just Ready.* $1.50

FEVERS.

COLLIE. On Fevers. Their History, Etiology, Diagnosis, Prognosis, and Treatment. Colored Plates. $2.00

GOUT AND RHEUMATISM.

DUCKWORTH. A Treatise on Gout. With Chromo-lithographs and Engravings. Cloth, $6.00
GARROD. On Rheumatism. A Treatise on Rheumatism and Rheumatic Arthritis. Cloth, $5.00
HAIG. Causation of Disease by Uric Acid. A Contribution to the Pathology of High Arterial Tension, Headache, Epilepsy, Gout, Rheumatism, Diabetes, Bright's Disease, etc. *New Ed. In Press.*

HEADACHES.

DAY. On Headaches. The Nature, Causes, and Treatment of Headaches. 4th Edition. Illustrated. $1.00

HEALTH AND DOMESTIC MEDICINE (see also Hygiene and Nursing).

BUCKLEY. The Skin in Health and Disease. Illus. .40
BURNETT. Hearing and How to Keep It. Illustrated. .40
COHEN. The Throat and Voice. Illustrated. .40
DULLES. What to Do First in Accidents and Poisoning. 4th Edition. New Illustrations. $1.00
HARLAN. Eyesight and How to Care for It. Illustrated. .40
HARTSHORNE. Our Homes, their Situation, Construction, Drainage, etc. Illustrated. .40
OSGOOD. The Winter and its Dangers. .40
PACKARD. Sea Air and Bathing. .40
PARKES. The Elements of Health. *Just Ready.* $1.25
RICHARDSON. Long Life and How to Reach It. .40
WESTLAND. The Wife and Mother. A Hand-Book for Mothers. $1.50
WHITE. The Mouth and Teeth. Illustrated. .40
WILSON. The Summer and its Diseases. .40
WOOD. Brain Work and Overwork. .40
STARR. Hygiene of the Nursery. 4th Edition. $1.00
CANFIELD. Hygiene of the Sick-Room. $1.25

HEART.

SANSOM. Diseases of the Heart. The Diagnosis and Pathology of Diseases of the Heart and Thoracic Aorta. With Plates and other Illustrations. $6.00

HYGIENE AND WATER ANALYSIS.

Special Catalogue of Books on Hygiene sent free upon application.

CANFIELD. Hygiene of the Sick-Room. A Book for Nurses and Others. Being a Brief Consideration of Asepsis, Antisepsis, Disinfection, Bacteriology, Immunity, Heating and Ventilation, and Kindred Subjects. $1.25

COPLIN AND BEVAN. Practical Hygiene. A Complete American Text-Book. 138 Illustrations. $3.25

FOX. Water, Air, and Food. Sanitary Examinations of Water, Air, and Food. 100 Engravings. 2d Edition, Revised. $3.50

KENWOOD. Public Health Laboratory Work. 116 Illustrations and 3 Plates. $2.00

LEFFMANN. Examination of Water for Sanitary and Technical Purposes. 3d Edition. Illustrated. *Just Ready.* $1.25

LEFFMANN. Analysis of Milk and Milk Products. Illustrated. $1.25

LINCOLN. School and Industrial Hygiene. .40

MACDONALD. Microscopical Examinations of Water and Air. 25 Lithographic Plates, Reference Tables, etc. 2d Ed. $2.50

McNEILL. The Prevention of Epidemics and the Construction and Management of Isolation Hospitals. Numerous Plans and Illustrations. $3.50

PARKES. Practical Hygiene. 8th Edition. Edited by J. Lane Notter. 10 Lithographic Plates and over 100 other Illustrations. $4.50

PARKES. Hygiene and Public Health. By Louis C. Parkes, M.D. 4th Edition. Enlarged. Illustrated. $2.50

PARKES. Popular Hygiene. The Elements of Health. A Book for Lay Readers. Illustrated. *Just Ready.* $1.25

STARR. The Hygiene of the Nursery. Including the General Regimen and Feeding of Infants and Children, and the Domestic Management of the Ordinary Emergencies of Early Life, Massage, etc. 4th Edition. 25 Illustrations. $1.00

STEVENSON AND MURPHY. A Treatise on Hygiene. By Various Authors. In Three Octave Volumes. Illustrated.
Vol. I, $6.00; Vol. II, $6.00; Vol. III, $5.00

* Each Volume sold separately. Special Circular upon application.

WILSON. Hand-Book of Hygiene and Sanitary Science. With Illustrations. 7th Edition. $3.00

WEYL. Sanitary Relations of the Coal-Tar Colors. Authorized Translation by HENRY LEFFMANN, M.D., PH.D. $1.25

* *Special Catalogue of Books on Hygiene free upon application.*

JOURNALS, ETC.

OPHTHALMIC REVIEW. A Monthly Record of Ophthalmic Science. Publ. in London. Sample number .25; per annum $3.00

NEW SYDENHAM SOCIETY PUBLICATION. Three to six volumes each year. Circular upon application. Per annum $8.00

KIDNEY DISEASES.

RALFE. Diseases of the Kidney and Urinary Derangements. Illustrated. $2.00
THORNTON. The Surgery of the Kidney. 19 Illus. Clo., $1.50
TYSON. Bright's Disease and Diabetes. With Especial Reference to Pathology and Therapeutics. Including a Section on Retinitis in Bright's Disease. Illustrated. $2.50

LUNGS AND PLEURÆ.

HARRIS AND BEALE. Treatment of Pulmonary Consumption. *In Press.*
POWELL. Diseases of the Lungs and Pleuræ, including Consumption. Colored Plates and other Illus. 4th Ed. $4.00

MASSAGE.

KLEEN AND HARTWELL. Hand-Book of Massage. Authorized translation by MUSSEY HARTWELL, M.D., PH.D. With an Introduction by Dr. S. WEIR MITCHELL. Illustrated by a series of Photographs Made Especially by DR. KLEEN for the American Edition. $2.25
MURRELL. Massotherapeutics. Massage as a Mode of Treatment. 5th Edition. $1.25
OSTROM. Massage and the Original Swedish Movements. Their Application to Various Diseases of the Body. A Manual for Students, Nurses, and Physicians. Third Edition, Enlarged. 94 Wood Engravings, many of which are original. $1.00

MATERIA MEDICA AND THERAPEUTICS.

ALLEN, HARLAN, HARTE, VAN HARLINGEN. A Hand-Book of Local Therapeutics, Being a Practical Description of all those Agents Used in the Local Treatment of Diseases of the Eye, Ear, Nose and Throat, Mouth, Skin, Vagina, Rectum, etc., such as Ointments, Plasters, Powders, Lotions, Inhalations, Suppositories, Bougies, Tampons, and the Proper Methods of Preparing and Applying Them. $3.00
BIDDLE. Materia Medica and Therapeutics. Including Dose List, Dietary for the Sick, Table of Parasites, and Memoranda of New Remedies. 13th Edition, Thoroughly Revised in accordance with the new U. S. P. 64 Illustrations and a Clinical Index. Cloth, $4.00; Sheep, $5.00
BRACKEN. Outlines of Materia Medica and Pharmacology. By H. M. BRACKEN, Professor of Materia Medica and Therapeutics and of Clinical Medicine, University of Minnesota. $2.75
DAVIS. Materia Medica and Prescription Writing. $1.50
FIELD. Evacuant Medication. Cathartics and Emetics. $1.75
GORGAS. Dental Medicine. A Manual of Materia Medica and Therapeutics. 5th Edition, Revised. $4.00
MAYS. Therapeutic Forces; or, The Action of Medicine in the Light of Doctrine of Conservation of Force. $1.25
MAYS. Theine in the Treatment of Neuralgia. ½ bound, .50

NAPHEYS. Modern Therapeutics. 9th Revised Edition, Enlarged and Improved. In two handsome volumes. Edited by ALLEN J. SMITH, M.D., and J. AUBREY DAVIS, M.D.
Vol. I. General Medicine and Diseases of Children. $4.00
Vol. II. General Surgery, Obstetrics, and Diseases of Women. $4.00

POTTER. Hand-Book of Materia Medica, Pharmacy, and Therapeutics, including the Action of Medicines, Special Therapeutics, Pharmacology, etc., including over 600 Prescriptions and Formulæ. 5th Edition, Revised and Enlarged. With Thumb Index in each copy. Cloth, $4.00; Sheep, $5.00

POTTER. Compend of Materia Medica, Therapeutics, and Prescription Writing, with Special Reference to the Physiological Action of Drugs. 6th Revised and Improved Edition, based upon the U. S. P. 1890. .80; Interleaved, $1.25

SAYRE. Organic Materia Medica and Pharmacognosy. An Introduction to the Study of the Vegetable Kingdom and the Vegetable and Animal Drugs. Comprising the Botanical and Physical Characteristics, Source, Constituents, and Pharmacopeial Preparations. With chapters on Synthetic Organic Remedies, Insects Injurious to Drugs, and Pharmacal Botany. A Glossary and 543 Illustrations, many of which are original. $4.00

WARING. Practical Therapeutics. 4th Edition, Revised and Rearranged. Cloth, $2.00; Leather, $3.00

WHITE AND WILCOX. Materia Medica, Pharmacy, Pharmacology, and Therapeutics. 3d American Edition, Revised by REYNOLD W. WILCOX, M.A., M.D., LL.D. Clo., $2.75; Lea., $3.25

MEDICAL JURISPRUDENCE AND TOXICOLOGY.

REESE. Medical Jurisprudence and Toxicology. A Text-Book for Medical and Legal Practitioners and Students. 4th Edition. Revised by HENRY LEFFMANN, M.D. Clo., $3.00; Leather, $3.50

"To the student of medical jurisprudence and toxicology it is invaluable, as it is concise, clear, and thorough in every respect."—*The American Journal of the Medical Sciences.*

MANN. Forensic Medicine and Toxicology. Illus. $6.50

MURRELL. What to Do in Cases of Poisoning. 7th Edition, Enlarged. $1.00

TANNER. Memoranda of Poisons. Their Antidotes and Tests. 7th Edition. .75

MICROSCOPY.

BEALE. The Use of the Microscope in Practical Medicine. For Students and Practitioners, with Full Directions for Examining the Various Secretions, etc., by the Microscope. 4th Ed. 500 Illus. $6.50

BEALE. How to Work with the Microscope. A Complete Manual of Microscopical Manipulation, containing a Full Description of many New Processes of Investigation, with Directions for Examining Objects Under the Highest Powers, and for Taking Photographs of Microscopic Objects. 5th Edition. 400 Illustrations, many of them colored. $6.50

CARPENTER. The Microscope and Its Revelations. 7th Edition. 800 Illustrations and many Lithographs. $5.50

LEE. The Microtomist's Vade Mecum. A Hand-Book of Methods of Microscopical Anatomy. 881 Articles. 4th Edition, Enlarged. *In Press.*
MACDONALD. Microscopical Examinations of Water and Air. 25 Lithographic Plates, Reference Tables, etc. 2d Edition. $2.50
REEVES. Medical Microscopy, including Chapters on Bacteriology, Neoplasms, Urinary Examination, etc. Numerous Illustrations, some of which are printed in colors. $2.50
WETHERED. Medical Microscopy. A Guide to the Use of the Microscope in Practical Medicine. 100 Illustrations. $2.00

MISCELLANEOUS.

BLACK. Micro-Organisms. The Formation of Poisons. A Biological Study of the Germ Theory of Disease. .75
BURNETT. Foods and Dietaries. A Manual of Clinical Dietetics. 2d Edition. $1.50
DAVIS. Biology. Illustrated. $3.00
GOWERS. The Dynamics of Life. .75
HAIG. Causation of Disease by Uric Acid. A Contribution to the Pathology of High Arterial Tension, Headache, Epilepsy, Gout, Rheumatism, Diabetes, Bright's Disease, etc. *New Ed. In Press.*
HARE. Mediastinal Disease. Illustrated by six Plates. $2.00
HENRY. A Practical Treatise on Anemia. Half Cloth, .50
LEFFMANN. The Coal-Tar Colors. With Special Reference to their Injurious Qualities and the Restrictions of their Use. A Translation of THEODORE WEYL'S Monograph. $1.25
TREVES. Physical Education: Its Effects, Value, Methods, Etc. .75
LIZARS. The Use and Abuse of Tobacco. .40
PARRISH. Alcoholic Inebriety from a Medical Standpoint, with Cases. $1.00

NERVOUS DISEASES.

GOWERS. Manual of Diseases of the Nervous System. A Complete Text-Book. 2d Edition, Revised, Enlarged, and in many parts Rewritten. With many new Illustrations. Two volumes.
Vol. I. Diseases of the Nerves and Spinal Cord. $3.00
Vol. II. Diseases of the Brain and Cranial Nerves; General and Functional Disease. $4.00
GOWERS. Syphilis and the Nervous System. $1.00
GOWERS. Diagnosis of Diseases of the Brain. 2d Edition. Illustrated. $1.50
GOWERS. Clinical Lectures. A New Volume of Essays on the Diagnosis, Treatment, etc., of Diseases of the Nervous System. *Just Ready.* $2.00
FLOWER. Diagram of the Nerves of the Human Body. Exhibiting their Origin, Divisions, and Connections, with their Distribution to the Various Regions of the Cutaneous Surface and to all the Muscles. 3d Edition. Six large Folio Maps or Diagrams. $2.50

MEDICAL BOOKS. 15

HORSLEY. The Brain and Spinal Cord. The Structure and Functions of. Numerous Illustrations. $2.50

OBERSTEINER. The Anatomy of the Central Nervous Organs. A Guide to the Study of their Structure in Health and Disease. 198 Illustrations. $5.50

ORMEROD. Diseases of the Nervous System. 75 Wood Engravings. $1.00

OSLER. Cerebral Palsies of Children. A Clinical Study. $2.00

OSLER. Chorea and Choreiform Affections. $2.00

PAGE. Injuries of the Spine and Spinal Cord. In their Surgical and Medico-legal Aspects. 3d Edition. *Preparing.*

PAGE. Railroad Injuries. With Special Reference to Those of the Back and Nervous System. $2.25

THORBURN. Surgery of the Spinal Cord. Illustrated. $4.00

WATSON. Concussions. An Experimental Study of Lesions Arising from Severe Concussions. Paper cover, $1.00

WOOD. Brain Work and Overwork. .40

NURSING.

Special Catalogue of Books for Nurses sent free upon application.

CANFIELD. Hygiene of the Sick-Room. A Book for Nurses and Others. Being a Brief Consideration of Asepsis, Antisepsis, Disinfection, Bacteriology, Immunity, Heating and Ventilation, and Kindred Subjects for the Use of Nurses and Other Intelligent Women. $1.25

CULLINGWORTH. A Manual of Nursing, Medical and Surgical. 3d Edition 18 Illustrations. .75

CULLINGWORTH. A Manual for Monthly Nurses. 3d Ed. .40

DOMVILLE. Manual for Nurses and Others Engaged in Attending the Sick. 7th Edition. With Recipes for Sick-room Cookery, etc. .75

FULLERTON. Obstetric Nursing. 40 Ills. 4th Ed. $1.00

FULLERTON. Nursing in Abdominal Surgery and Diseases of Women. Comprising the Regular Course of Instruction at the Training-School of the Women's Hospital, Philadelphia. 2d Edition. 70 Illustrations. $1.50

HUMPHREY. A Manual for Nurses. Including General Anatomy and Physiology, Management of the Sick-Room, etc. 13th Edition. Illustrated. $1.00

SHAWE. Notes for Visiting Nurses, and all those Interested in the Working and Organization of District, Visiting, or Parochial Nurse Societies. With an Appendix Explaining the Organization and Working of Various Visiting and District Nurse Societies, by HELEN C. JENKS, of Philadelphia. $1.00

STARR. The Hygiene of the Nursery. Including the General Regimen and Feeding of Infants and Children, and the Domestic Management of the Ordinary Emergencies of Early Life, Massage, etc. 4th Edition. 25 Illustrations. $1.00

TEMPERATURE CHARTS. For Recording Temperature, Respiration, Pulse, Day of Disease, Date, Age, Sex, Occupation, Name, etc. Put up in pads of fifty. Each .50

VOSWINKEL. Surgical Nursing. 111 Illustrations. $1.00

✱ *Special Catalogue of Books on Nursing free upon application.*

OBSTETRICS.

BAR. Antiseptic Midwifery. The Principles of Antiseptic Methods Applied to Obstetric Practice. Authorized Translation by HENRY D. FRY, M.D., with an Appendix by the Author. $1.00

CAZEAUX AND TARNIER. Midwifery. With Appendix by MUNDÉ. The Theory and Practice of Obstetrics, including the Diseases of Pregnancy and Parturition, Obstetrical Operations, etc. 8th Edition. Illustrated by Chromo-Lithographs, Lithographs, and other full-page Plates, seven of which are beautifully colored, and numerous Wood Engravings. Cloth, $4.50; Full Leather, $5.50

DAVIS. A Manual of Obstetrics. Being a Complete Manual for Physicians and Students. 2d Edition. 16 Colored and other Plates and 134 other Illustrations. $2.00

LANDIS. Compend of Obstetrics. 5th Edition, Revised by WM. H. WELLS, Assistant Demonstrator of Clinical Obstetrics, Jefferson Medical College. With many Illustrations, 80; Interleaved, $1.25.

SCHULTZE. Obstetrical Diagrams. Being a series of 20 Colored Lithograph Charts, Imperial Map Size, of Pregnancy and Midwifery, with accompanying explanatory (German) text illustrated by Wood Cuts. 2d Revised Edition.
Price in Sheets, $26.00; Mounted on Rollers, Muslin Backs, $36.00

STRAHAN. Extra-Uterine Pregnancy. The Diagnosis and Treatment of Extra-Uterine Pregnancy. .75

WINCKEL. Text-Book of Obstetrics, Including the Pathology and Therapeutics of the Puerperal State. Authorized Translation by J. CLIFTON EDGAR, A.M., M.D. With nearly 200 Illustrations. Cloth, $5.00; Leather, $6.00

FULLERTON. Obstetric Nursing. 4th Ed. Illustrated. $1.00

SHIBATA. Obstetrical Pocket-Phantom with Movable Child and Pelvis. Letter Press and Illustrations. $1.00

PATHOLOGY AND HISTOLOGY.

BLACKBURN. Autopsies. A Manual of Autopsies Designed for the Use of Hospitals for the Insane and other Public Institutions. Ten full-page Plates and other Illustrations. $1.25

BLODGETT. Dental Pathology. By ALBERT N. BLODGETT, M.D., late Professor of Pathology and Therapeutics, Boston Dental College. 33 Illustrations. $1.25

GILLIAM. Pathology. A Hand-Book for Students. 47 Illus. .75

HALL. Compend of General Pathology and Morbid Anatomy. 91 very fine Illustrations. .80; Interleaved, $1.25

STIRLING. Outlines of Practical Histology. 368 Illustrations. 2d Edition, Revised and Enlarged. With new Illustrations. $2.00

VIRCHOW. Post-Mortem Examinations. A Description and Explanation of the Method of Performing Them in the Dead House of the Berlin Charity Hospital, with Special Reference to Medico-Legal Practice. 3d Edition, with Additions. .75

PHARMACY.

Special Catalogue of Books on Pharmacy sent free upon application.

COBLENTZ. Manual of Pharmacy. A New and Complete Text-Book by the Professor in the New York College of Pharmacy. 2d Edition, Revised and Enlarged. 437 Illustrations. $3.50

BEASLEY. Book of 3100 Prescriptions. Collected from the Practice of the Most Eminent Physicians and Surgeons—English, French, and American. A Compendious History of the Materia Medica, Lists of the Doses of all the Officinal and Established Preparations, an Index of Diseases and their Remedies. 7th Ed. $2.00

BEASLEY. Druggists' General Receipt Book. Comprising a Copious Veterinary Formulary, Recipes in Patent and Proprietary Medicines, Druggists' Nostrums, etc.; Perfumery and Cosmetics, Beverages, Dietetic Articles and Condiments, Trade Chemicals, Scientific Processes, and an Appendix of Useful Tables. 10th Edition, Revised. $2.00

BEASLEY. Pocket Formulary. A Synopsis of the British and Foreign Pharmacopœias. Comprising Standard and Approved Formulæ for the Preparations and Compounds Employed in Medical Practice. 11th Edition. $2.00

PROCTOR. Practical Pharmacy. Lectures on Practical Pharmacy. With Wood Engravings and 32 Lithographic Fac-simile Prescriptions. 3d Edition, Revised, and with Elaborate Tables of Chemical Solubilities, etc. $3.00

ROBINSON. Latin Grammar of Pharmacy and Medicine. 2d Edition. With elaborate Vocabularies. $1.75

SAYRE. Organic Materia Medica and Pharmacognosy. An Introduction to the Study of the Vegetable Kingdom and the Vegetable and Animal Drugs. Comprising the Botanical and Physical Characteristics, Source, Constituents, and Pharmacopeial Preparations. With Chapters on Synthetic Organic Remedies, Insects Injurious to Drugs, and Pharmacal Botany. A Glossary and 543 Illustrations, many of which are original. $4.00

SCOVILLE. The Art of Compounding. A Text-Book for the Student and a Reference Book for the Pharmacist. $2.50

STEWART. Compend of Pharmacy. Based upon " Remington's Text-Book of Pharmacy." 5th Edition, Revised in Accordance with the U. S. Pharmacopœia, 1890. Complete Tables of Metric and English Weights and Measures. .80; Interleaved, $1.25

UNITED STATES PHARMACOPŒIA. 1890. 7th Decennial Revision. Cloth, $2.50 (postpaid, $2.77); Sheep, $3.00 (postpaid, $3.27); Interleaved, $4.00 (postpaid, $4.50); Printed on one side of page only, unbound, $3.50 (postpaid, $3.90).

Select Tables from the U. S. P. (1890). Being Nine of the Most Important and Useful Tables, Printed on Separate Sheets. Carefully put up in patent envelope. .25

WHITE AND WILCOX. Materia Medica, Pharmacy, Pharmacology, and Therapeutics. 2d American Edition. Revised by REYNOLD W. WILCOX, M.D., LL.D. Cloth, $2.75; Leather, $3.25

POTTER. Hand-Book of Materia Medica, Pharmacy, and Therapeutics. 600 Prescriptions and Formulæ. 5th Edition. Cloth, $4.00; Sheep, $5.00

** Special Catalogue of Books on Pharmacy free upon application.

PHYSICAL DIAGNOSIS.

TYSON. Hand-Book of Physical Diagnosis. For Students and Physicians. By the Professor of Clinical Medicine in the University of Pennsylvania. Illus. 2d Ed., Improved and Enlarged. $1.25

MEMMINGER. Diagnosis by the Urine. 23 Illus. $1.00

PHYSIOLOGY.

BRUBAKER. Compend of Physiology. 7th Edition, Revised and Illustrated. .80; Interleaved, $1.25

KIRKE. Physiology. (13th Authorized Edition. Dark Red Cloth.) A Hand-Book of Physiology. 13th London Edition, Revised and Enlarged. 516 Illustrations, some of which are printed in colors.
Cloth, $3.25; Leather, $4.00

LANDOIS. A Text-Book of Human Physiology, Including Histology and Microscopical Anatomy, with Special Reference to the Requirements of Practical Medicine. 5th American, translated from the last German Edition, with Additions by WM. STIRLING, M.D., D.SC. 845 Illus., many of which are printed in colors. *In Press.*

STARLING. Elements of Human Physiology. 100 Ills. $1.00

STIRLING. Outlines of Practical Physiology. Including Chemical and Experimental Physiology, with Special Reference to Practical Medicine. 3d Edition. 289 Illustrations. $2.00

TYSON. Cell Doctrine. Its History and Present State. 2d Edition. $1.50

YEO. Manual of Physiology. A Text-Book for Students of Medicine. By GERALD F. YEO, M.D., F.R.C.S. 6th Edition. 254 Illustrations and a Glossary. Cloth, $2.50; Leather, $3.00

PRACTICE.

BEALE. On Slight Ailments; their Nature and Treatment. 2d Edition, Enlarged and Illustrated. $1.25

CHARTERIS. Practice of Medicine. 6th Edition. Therapeutical Index and Illustrations. $2.00

FAGGE. The Practice of Medicine. Cloth, $7.00; Leather, $9.00

FOWLER. Dictionary of Practical Medicine. By various writers. An Encyclopædia of Medicine. Clo., $3.00; Half Mor. $4.00

HUGHES. Compend of the Practice of Medicine. 5th Edition, Revised and Enlarged.

    Part I. Continued, Eruptive, and Periodical Fevers, Diseases of the Stomach, Intestines, Peritoneum, Biliary Passages, Liver, Kidneys, etc., and General Diseases, etc.

    Part II. Diseases of the Respiratory System, Circulatory System, and Nervous System; Diseases of the Blood, etc.
Price of each part, .80; Interleaved, $1.25

    Physician's Edition. In one volume, including the above two parts, a Section on Skin Diseases, and an Index. 5th Revised, Enlarged Edition. 568 pp. Full Morocco, Gilt Edge, $2.25

ROBERTS. The Theory and Practice of Medicine. The Sections on Treatment are especially exhaustive. 9th Edition, with Illustrations. Cloth, $4.50; Leather, $5.50

TAYLOR. Practice of Medicine. Cloth, $2.00; Sheep, $2.50

PRESCRIPTION BOOKS.

BEASLEY. Book of 3100 Prescriptions. Collected from the Practice of the Most Eminent Physicians and Surgeons—English, French, and American. A Compendious History of the Materia Medica, Lists of the Doses of all Officinal and Established Preparations, and an Index of Diseases and their Remedies. 7th Ed. $2.00

BEASLEY. Druggists' General Receipt Book. Comprising a Copious Veterinary Formulary, Recipes in Patent and Proprietary Medicines, Druggists' Nostrums, etc.; Perfumery and Cosmetics, Beverages, Dietetic Articles and Condiments, Trade Chemicals, Scientific Processes, and an Appendix of Useful Tables. 10th Edition, Revised. $2.00

BEASLEY. Pocket Formulary. A Synopsis of the British and Foreign Pharmacopœias. Comprising Standard and Approved Formulæ for the Preparations and Compounds Employed in Medical Practice. 11th Edition. Cloth, $2.00

DAVIS. Essentials of Materia Medica and Prescription Writing. $1.50

PEREIRA. Prescription Book. Containing Lists of Terms, Phrases, Contractions, and Abbreviations Used in Prescriptions, Explanatory Notes, Grammatical Construction of Prescriptions, etc. 16th Edition. Cloth, .75; Tucks, $1.00

WYTHE. Dose and Symptom Book. The Physician's Pocket Dose and Symptom Book. Containing the Doses and Uses of all the Principal Articles of the Materia Medica and Officinal Preparations. 17th Ed. Cloth, .75; Leather, with Tucks and Pocket, $1.00

SKIN.

ANDERSON. A Treatise on Skin Diseases. With Special Reference to Diagnosis and Treatment, and Including an Analysis of 11,000 Consecutive Cases. Illus. Cloth, $3.00; Leather, $4.00

BULKLEY. The Skin in Health and Disease. Illustrated. .40

CROCKER. Diseases of the Skin. Their Description, Pathology, Diagnosis, and Treatment, with Special Reference to the Skin Eruptions of Children. 92 Illus. 2d Edition. Enlarged. $4.50

VAN HARLINGEN. On Skin Diseases. A Practical Manual of Diagnosis and Treatment, with special reference to Differential Diagnosis. 3d Edition, Revised and Enlarged. With Formulæ and 60 Illustrations, some of which are printed in colors. $2.75

SURGERY AND SURGICAL DISEASES.

CAIRD AND CATHCART. Surgical Hand-Book. 5th Edition, Revised. 188 Illustrations. Full Red Morocco, $2.50

DULLES. What to Do First in Accidents and Poisoning. 4th Edition. New Illustrations. $1.00

HACKER. Antiseptic Treatment of Wounds, Introduction to the, According to the Method in Use at Professor Billroth's Clinic, Vienna. With a Photo-engraving of Billroth in his Clinic. .50

HEATH. Minor Surgery and Bandaging. 10th Ed Revised and Enlarged. 158 Illustrations, 62 Formulæ, Diet List, etc $1.25

HEATH. Injuries and Diseases of the Jaws. 4th Edition. 187 Illustrations. $4.50

HEATH. Lectures on Certain Diseases of the Jaws. 64 Illustrations. Boards, .50

HORWITZ. Compend of Surgery and Bandaging, including Minor Surgery, Amputations, Fractures, Dislocations, Surgical Diseases, and the Latest Antiseptic Rules, etc., with Differential Diagnosis and Treatment. 5th Edition, very much Enlarged and Rearranged. 167 Illustrations, 98 Formulæ. Clo., .80; Interleaved, $1.25

JACOBSON. Operations of Surgery. Over 200 Illustrations.
Cloth, $3.00; Leather, $4.00
JACOBSON. Diseases of the Male Organs of Generation.
88 Illustrations. $6.00
MACREADY. A Treatise on Ruptures. 24 Full-page Lithographed Plates and Numerous Wood Engravings. Cloth, $6.00
MOULLIN. Text-Book of Surgery. With Special Reference to Treatment. 3d American Edition. Revised and edited by JOHN B. HAMILTON, M.D., LL.D., Professor of the Principles of Surgery and Clinical Surgery, Rush Medical College, Chicago. 623 Illustrations, over 200 of which are original, and many of which are printed in colors. *Just Ready.* Handsome Cloth, $6.00; Leather, $7.00

"The aim to make this valuable treatise practical by giving special attention to questions of treatment has been admirably carried out. Many a reader will consult the work with a feeling of satisfaction that his wants have been understood, and that they have been intelligently met."—*The American Journal of Medical Science.*

PORTER. Surgeon's Pocket-Book. 3d Ed. Lea. Cover, $2.00.
SMITH. Abdominal Surgery. Being a Systematic Description of all the Principal Operations. 80 Illus. 5th Edition. *In Press.*
VOSWINKEL. Surgical Nursing. 111 Illustrations. $1.00
WALSHAM. Manual of Practical Surgery. 5th Ed., Revised and Enlarged. With 380 Engravings. Clo., $2.75; Lea., $3.25
WATSON. On Amputations of the Extremities and Their Complications. 250 Illustrations. $5.50

THROAT AND NOSE (see also Ear).

COHEN. The Throat and Voice. Illustrated. .40
HALL. Diseases of the Nose and Throat. Two Colored Plates and 59 Illustrations. $2.50
HALL. Compend of Diseases of the Ear and Nose. Illustrated. .80; Interleaved, $1.25
HUTCHINSON. The Nose and Throat. Including the Nose, Naso-Pharynx, Pharynx, and Larynx. Illustrated by Lithograph Plates and 40 other Illustrations. 2d Edition. *In Press.*
MACKENZIE. The Pharmacopœia of the London Hospital for Diseases of the Throat. 5th Edition, Revised by Dr. F. G. HARVEY. $1.00
McBRIDE. Diseases of the Throat, Nose, and Ear. A Clinical Manual. With colored Illus. from original drawings. 2d Ed. $6.00
MURRELL. Chronic Bronchitis and its Treatment. (Authorized Edition.) A Clinical Study. $1.50
POTTER. Speech and its Defects. Considered Physiologically, Pathologically, and Remedially. $1.00
WOAKES. Post-Nasal Catarrh and Diseases of the Nose Causing Deafness. 26 Illustrations. $1.00

URINE AND URINARY ORGANS.

ACTON. The Functions and Disorders of the Reproductive Organs in Childhood, Youth, Adult Age, and Advanced Life, Considered in their Physiological, Social, and Moral Relations. 8th Edition. $1.75
ALLEN. Albuminous and Diabetic Urine. Illus. $2.25

MEDICAL BOOKS.

BEALE. One Hundred Urinary Deposits. On eight sheets, for the Hospital, Laboratory, or Surgery. Paper, $2.00
HOLLAND. The Urine, the Gastric Contents, the Common Poisons, and the Milk. Memoranda, Chemical and Microscopical, for Laboratory Use. Illustrated and Interleaved. 5th Ed. $1.00
LEGG. On the Urine. 7th Edition, Enlarged. Illus. $1.00
MEMMINGER. Diagnosis by the Urine. 23 Illus. $1.00
MOULLIN. Enlargement of the Prostate. Its Treatment and Radical Cure. Illustrated. $1.50
THOMPSON. Diseases of the Urinary Organs. 8th Ed. $3.00
THOMPSON. Calculous Diseases. The Preventive Treatment of, and the Use of Solvent Remedies. 3d Edition. .75
TYSON. Guide to Examination of the Urine. For the Use of Physicians and Students. With Colored Plate and Numerous Illustrations engraved on wood. 9th Edition, Revised. $1.25
VAN NUYS. Chemical Analysis of Healthy and Diseased Urine, Qualitative and Quantitative. 39 Illustrations. $1.00

VENEREAL DISEASES.

COOPER. Syphilis. 2d Edition, Enlarged and Illustrated with 20 full-page Plates. $5.00
GOWERS. Syphilis and the Nervous System. $1.00
HILL AND COOPER. Venereal Diseases. Being a Concise Description of Those Affections and Their Treatment. 4th Ed. .75
JACOBSON. Diseases of the Male Organs of Generation. 88 Illustrations. $6.00

VETERINARY.

ARMATAGE. The Veterinarian's Pocket Remembrancer. Being Concise Directions for the Treatment of Urgent or Rare Cases, Embracing Semeiology, Diagnosis, Prognosis, Surgery, Treatment, etc. 2d Edition. Boards, $1.00
BALLOU. Veterinary Anatomy and Physiology. 29 Graphic Illustrations. .80; Interleaved, $1.25
TUSON. Veterinary Pharmacopœia. Including the Outlines of Materia Medica and Therapeutics. 5th Edition. $2.25

WOMEN, DISEASES OF.

BYFORD (H. T.). Manual of Gynecology. With 234 Illustrations, many of which are from original drawings and photographs. *Just Ready.* $2.50
BYFORD (W. H.). Diseases of Women. 4th Edition. 306 Illustrations. Cloth, $2.00; Leather, $2.50
DÜHRSSEN. A Manual of Gynecological Practice. 105 Illustrations. *Just Ready.* $1.50
LEWERS. Diseases of Women. 146 Illus. 3d Edition. $2.00
WELLS. Compend of Gynecology. Illus. .80; Interleaved, $1.25
WINCKEL. Diseases of Women. Translated by special authority of Author, under the Supervision of, and with an Introduction by, THEOPHILUS PARVIN, M.D. 152 Engravings on Wood. 3d Edition, Revised. *In Preparation.*
FULLERTON. Nursing in Abdominal Surgery and Diseases of Women. 2d Edition. 70 Illustrations. $1.50

COMPENDS.

From The Southern Clinic.

"We know of no series of books issued by any house that so fully meets our approval as these ?Quiz-Compends?. They are well arranged, full, and concise, and are really the best line of text-books that could be found for either student or practitioner."

BLAKISTON'S ?QUIZ-COMPENDS?

The Best Series of Manuals for the Use of Students.

Price of each, Cloth, .80.   Interleaved, for taking Notes, $1.25.

☞ These Compends are based on the most popular text-books and the lectures of prominent professors, and are kept constantly revised, so that they may thoroughly represent the present state of the subjects upon which they treat.

☞ The authors have had large experience as Quiz-Masters and attaches of colleges, and are well acquainted with the wants of students.

☞ They are arranged in the most approved form, thorough and concise, containing over 600 fine illustrations, inserted wherever they could be used to advantage.

☞ Can be used by students of *any* college.

☞ They contain information nowhere else collected in such a condensed, practical shape. Illustrated Circular free.

No. 1. POTTER. HUMAN ANATOMY. Fifth Revised and Enlarged Edition. Including Visceral Anatomy. Can be used with either Morris's or Gray's Anatomy. 117 Illustrations and 16 Lithographic Plates of Nerves and Arteries, with Explanatory Tables, etc. By SAMUEL O. L. POTTER, M.D., Professor of the Practice of Medicine, Cooper Medical College, San Francisco; late A. A. Surgeon, U. S. Army.

No. 2. HUGHES. PRACTICE OF MEDICINE. Part I. Fifth Edition, Enlarged and Improved. By DANIEL E. HUGHES, M.D., Physician-in-Chief, Philadelphia Hospital, late Demonstrator of Clinical Medicine, Jefferson Medical College, Phila.

No. 3. HUGHES. PRACTICE OF MEDICINE. Part II. Fifth Edition, Revised and Improved. Same author as No. 2.

No. 4. BRUBAKER. PHYSIOLOGY. Seventh Edition, with new Illustrations and a table of Physiological Constants. Enlarged and Revised. By A. P. BRUBAKER, M.D., Professor of Physiology and General Pathology in the Pennsylvania College of Dental Surgery; Demonstrator of Physiology, Jefferson Medical College, Philadelphia.

No. 5. LANDIS. OBSTETRICS. Fifth Edition. By HENRY G. LANDIS, M.D. Revised and Edited by WM. H. WELLS, M.D., Assistant Demonstrator of Obstetrics, Jefferson Medical College, Philadelphia. Enlarged. 47 Illustrations.

No. 6. POTTER. MATERIA MEDICA, THERAPEUTICS, AND PRESCRIPTION WRITING. Sixth Revised Edition (U. S. P. 1890). By SAMUEL O. L. POTTER, M.D., Professor of Practice, Cooper Medical College, San Francisco; late A. A. Surgeon, U. S. Army.

? QUIZ-COMPENDS ?—Continued.

No. 7. WELLS. GYNECOLOGY. A New Book. By WM. H. WELLS, M.D., Assistant Demonstrator of Obstetrics, Jefferson College, Philadelphia. Illustrated.

No. 8. FOX AND GOULD. DISEASES OF THE EYE AND REFRACTION. Second Edition. Including Treatment and Surgery. By L. WEBSTER FOX, M.D., and GEORGE M. GOULD, M.D. With 39 Formulæ and 71 Illustrations.

No. 9. HORWITZ. SURGERY, Minor Surgery, and Bandaging. Fifth Edition, Enlarged and Improved. By ORVILLE HORWITZ, B. S., M.D., Clinical Professor of Genito-Urinary Surgery and Venereal Diseases in Jefferson Medical College; Surgeon to Philadelphia Hospital, etc. With 98 Formulæ and 71 Illustrations.

No. 10. LEFFMANN. MEDICAL CHEMISTRY. Fourth Edition. Including Urinalysis, Animal Chemistry, Chemistry of Milk, Blood, Tissues, the Secretions, etc. By HENRY LEFFMANN, M.D., Professor of Chemistry in Pennsylvania College of Dental Surgery and in the Woman's Medical College, Philadelphia.

No. 11. STEWART. PHARMACY. Fifth Edition. Based upon Prof. Remington's Text-Book of Pharmacy. By F. E. STEWART, M.D., PH.G., late Quiz-Master in Pharmacy and Chemistry, Philadelphia College of Pharmacy; Lecturer at Jefferson Medical College. Carefully revised in accordance with the new U. S. P.

No. 12. BALLOU. VETERINARY ANATOMY AND PHYSIOLOGY. Illustrated. By WM. R. BALLOU, M.D., Professor of Equine Anatomy at New York College of Veterinary Surgeons; Physician to Bellevue Dispensary, etc. 29 graphic Illustrations.

No. 13. WARREN. DENTAL PATHOLOGY AND DENTAL MEDICINE. Second Edition, Illustrated. Containing all the most noteworthy points of interest to the Dental Student and a Section on Emergencies. By GEO. W. WARREN, D.D.S., Chief of Clinical Staff, Pennsylvania College of Dental Surgery, Philadelphia.

No. 14. HATFIELD. DISEASES OF CHILDREN. Second Edition. Colored Plate. By MARCUS P. HATFIELD, Professor of Diseases of Children, Chicago Medical College.

No. 15. HALL. GENERAL PATHOLOGY AND MORBID ANATOMY. 91 Illustrations. By H. NEWBERRY HALL, PH.G., M.D., Professor of Pathology and Med. Chem., Chicago Post-Graduate Medical School; Mem. Surgical Staff, Illinois Charitable Eye and Ear Infirmary; Chief of Ear Clinic, Chicago Med. College.

No. 16. DISEASES OF NOSE AND EAR. Illustrated. Same Author as No. 15.

Price, each, Cloth, .80.   Interleaved, for taking Notes, $1.25.

Handsome Illustrated Circular sent free upon application.

In preparing, revising, and improving BLAKISTON'S ? QUIZ-COMPENDS ? the particular wants of the student have always been kept in mind.

Careful attention has been given to the construction of each sentence, and while the books will be found to contain an immense amount of knowledge in small space, they will likewise be found easy reading; there is no stilted repetition of words; the style is clear, lucid, and distinct. The arrangement of subjects is systematic and thorough; there is a reason for every word. They contain over 600 illustrations.

Moullin's Surgery.

Third Edition, Just Ready.

EDITED BY

JOHN B. HAMILTON, M.D.,

Professor of the Principles of Surgery and Clinical Surgery, Rush Medical College, Chicago, etc.

This is not only the latest, but the most uniform and complete one-volume Text-Book of Surgery. **The relative value of each subject has been carefully considered,** the constant aim of author and editor having been to make it **practical and useful.** It is systematically arranged and **pays special attention to treatment.**

Royal 8vo. 1250 Pages. 600 Illustrations.

Cloth, net, $6.00. Sheep, net, $7.00.

₊ Illustrated circular free upon application.

www.ingramcontent.com/pod-product-compliance
Lightning Source LLC
Chambersburg PA
CBHW032001300426
44117CB00008B/857